The Physical Geography of Africa

A.T. Grove

The
Physical Geography
of Africa

edited by
W. M. Adams
A. S. Goudie
and A. R. Orme

OXFORD UNIVERSITY PRESS
1996

Oxford University Press, Walton Street, Oxford OX2 6DP

Oxford New York
Athens Auckland Bangkok Bombay
Calcutta Cape Town Dar es Salaam Delhi
Florence Hong Kong Istanbul Karachi
Kuala Lumpur Madras Madrid Melbourne
Mexico City Nairobi Paris Singapore
Taipei Tokyo Toronto
and associated companies in
Berlin Ibadan

Oxford is a trade mark of Oxford University Press

Published in the United States
by Oxford University Press Inc., New York

British Library Cataloguing in Publication Data
Data available

Library of Congress Cataloging in Publication Data
Data available

ISBN 0–19–828875–1

10 9 8 7 6 5 4 3 2 1

Typeset by Graphicraft Typesetters Ltd., Hong Kong

Printed in Great Britain
on acid-free paper by
Butler & Tanner Ltd., Frome, Somerset

Dedicated to

A. T. Grove

Foreword

This is the first in a series of regional physical geography texts which will be produced by the Oxford University Press.

Their aim is to provide a relatively durable statement of physical conditions on the continents. Each volume will include a discussion of some of the main environmental factors (climate, tectonic history, soil types, etc., the history of environmental change), a discussion of some of the main types of environment (lakes, mountains, deserts, wetlands, etc.) with an emphasis on some of the main linkages between different components of the environment, and a consideration of some of the main environmental issues related to human use of the land (e.g. desertification, soil erosion, biodiversity loss, and conservation). We hope this will provide a coherent and informative basis for physical geography and that each volume will be a reference source for all those concerned with all aspects of the environments of the continents.

We dedicate this book to A. T. Grove as a tribute from his colleagues and students in recognition of the contribution he has made to our work and the understanding of the geography of Africa.

WMA
ASG
ARO

1995

Acknowledgements

The authors and publishers wish to thank the following who have kindly given permission for the use of copyright material.

Fig. 4.11: from J. Miller (1982), 'The Significance of Drought, Disease, and Famine in the Agriculturally Marginal Zones of West-Central Africa', *Journal of African History*, 23. By permission of Cambridge University Press.

Fig. 4.7: from P. Shaw (1984), 'A Historical Note on the Outflows of the Okovango Delta System', *Botswana Notes and Records*, 16. By permission of The Botswana Society.

Fig. 13.8: from I. Corbett (1993), 'The Modern and Ancient Pattern of Sandflow through the Southern Namib Deflation Basin, *International Association of Sedimentologists Special Publication*, 16: 45–60. By permission of Blackwell Science Ltd.

Fig. 13.14: from G. Kocurek, K. G. Havholm, M. Deynoux, and R. C. Blakey (1991), 'Amaigamated Accumulation resulting from Climatic and Eustatic Changes, Archar Erg. Mauritania', *Sedimentology*, 38: 751–72. By permission of Blackwell Science Ltd.

Fig. 13.16: from N. Lancaster (1981), 'Palaeoenvironmental Implications of Fixed Dune Systems in Southern Africa, *Palaeogeography, Palaeoclimatology, Palaeoecology*, 33: 327–46. By permission of Elsevier Science.

Fig. 13.18: from D. S. G. Thomas and P. A. Shaw (1991), *The Kalahari Environment* (Cambridge). By permission of Cambridge University Press.

Table 17.1: from M. J. Wells (1991), 'Introduced Plants of the Fynbos Biome of South Africa', in R. H. Groves and F. di Castri (eds.), *Biogeography of Mediterranean Invasions*, Cambridge. By permission of Cambridge University Press.

Figs. 17.7 and 17.8: from D. M. Richardson and B. W. van Wilgen (1992), 'Ecosystem, Community and Species Response to Fire in Mountain Fynbos: Conclusions from the Swarboskloof Experiment', in B. W. van Wilgen, *et al.* (eds.), *Fire in the South Africa Mountain Fynbos* (Berlin). By permission of Springer-Verlag GmbH & Co.

Although every effort has been made to trace and contact copyright holders, this has not always been successful. We apologize for any apparent negligence.

Contents

List of Figures

List of Plates

List of Contributors

Martin Adams is a Research Associate of the Overseas Development Institute in London, and an independent consultant. He has worked throughout Africa, particularly in the Sudan, Kenya, and in Southern Africa. His main interests are in land resources, land tenure, and agricultural development in semi-arid environments.

William Adams is a Lecturer in Geography at the University of Cambridge. His Ph.D. concerned the downstream impacts of dams in Nigeria, and was supervised by A. T. Grove. He has worked in Nigeria, Kenya, and Tanzania. His main interests are in conservation and sustainable development, particularly in the context of indigenous and modern use of water resources.

Harriet Allen is a Senior Lecturer in Geography at Homerton College, Cambridge. Her research interests and field-work are focused on the environmental history and biogeography of Mediterranean-type regions.

Olusegun Areola is Professor of Geography at the University of Ibadan. He has worked extensively on the soils and geomorphology of tropical Africa, particularly in Nigeria.

Andrew Goudie is Professor of Geography at the University of Oxford. He has worked extensively in Southern and Eastern Africa, and his main research interests are in the fields of geomorphology and environmental change.

Alan Grainger is Lecturer in Geography at the University of Leeds. His research interests include tropical land-use change, forest-resource appraisal, desertification, and global environmental change.

Francine Hughes is a Senior Research Associate and an affiliated Lecturer at the University of Cambridge. Her Ph.D. supervisor was A. T. Grove. Her main research interests concern the dynamics of floodplain ecosystems, and particularly the impacts of river control on riparian forests. She has worked on the Tana River in Kenya.

Mike Hulme is a Senior Research Associate in the Climatic Research Unit at the University of East Anglia, UK. He is a research climatologist specializing in global climate change, African climate and desertification, the validation of climate models, and the development of integrated models to evaluate climate-related policy. He has worked on regional projects in Northern, Eastern, and Southern Africa.

Nicholas Lancaster is a Research Professor at the Desert Research Institute of the University and Community College System of Nevada. He has studied desert geomorphology, especially dune processes and dynamics, in the Kalahari and Namib Deserts of Southern Africa, the western United States, and Tunisia.

Michael Meadows is Associate Professor in the Department of Environmental and Geographical Science at the University of Cape Town. He began his work in Africa in Malawi in 1976 as a student of Dick Grove, and has lived and worked in South Africa since 1983. He has research interests in the Quaternary palaeoecology and geomorphology of Central and Southern Africa.

Norman Myers is a Visiting Fellow at Green College Oxford and Visiting Professor at Universities from Utrecht to California. He has lived in Africa for twenty-five years.

Sharon Nicholson is a Professor in the Department of Meteorology, Florida State University, Tallahassee, Florida. She has studied many aspects of climate variability in Africa.

Celia Nyamweru is Associate Professor of Anthropology at St. Lawrence University. She taught for nineteen years at Kenyatta University, Nairobi, and has carried out research on Rift Valley lakes and on the eruptive activity of Ol Doinyo Lengai Volcano, Tanzania.

Antony Orme is Professor of Geography in the University of California, Los Angeles. He has worked extensively along the coasts of Southern and Eastern Africa, and his main research interests are in geomorphology, Quaternary studies, and environmental planning.

Michael Stocking is Professor of Natural Resource Development, School of Development Studies, University of East Anglia. He has carried out research on soil erosion in a wide range of African countries.

Michael Summerfield is Reader in Geomorphology at the University of Edinburgh. He has worked extensively in Southern Africa, and his main research interests are in large-scale tectonic geomorphology and long-term landscape development.

David Taylor is a Lecturer in the School of Geography and Earth Resources in the University of Hull. He has worked extensively in Eastern Africa, particularly in the fields of forest ecology and environmental change in montane environments.

Claudio Vita-Finzi is Professor of Neotectonics at University College London. His main research interests are the mechanisms of Holocene crustal deformation and the hydrological effects of solar oscillations.

Des Walling is Professor of Physical Geography at the University of Exeter. His main research interests are in the fields of hydrology, erosion, and sedimentation and he has worked on water-resource development and soil erosion assessment projects in several African countries.

Andrew Warren is Professor of Geography at University College London. He has worked in the Sudan, Libya, Algeria, Egypt, Niger, and Tunisia. His interests are in desert geomorphology and desertification.

A. T. Grove

Claudio Vita-Finzi

'The study of the *Earth* on which we inhabit', argued Robert Hooke, 'is so large that the efforts of Observers, Watchers, Compilers, and Namesetters however numerous would yield but a heap of Confusion.' What was required, 'in the collecting of Materials as well as in the use of them, was some End and Aim, some pre-design'd Module and Theory, some Purpose. I could wish', he remarked, 'that the Information of Experiments might be more respected than either the Novelty, the Surprizingness, the Pomp and Appearances of them.'

Three centuries later Moduling is indeed in the ascendant. But happily it has not suffocated exploration, or obscured the fact that, as Hooke himself says elsewhere, 'there are some Men who excel in their Observations and Deduction'. And, as the work of A. T. Grove resoundingly shows, these several activities are most fruitful if practised in unison. For he has carried out some outstanding feats of exploration, formulated and tested climatic models which are widely accepted, and retained a youthful capacity to be surprised by the Earth and our place on it, in a professional life where everything fits and nothing jars.

Alfred Thomas Grove, known universally and to many of his friends inexplicably as Dick, was born in Evesham, Worcestershire, on 8 April 1924, the son of a market gardener. He was educated at Prince Henry's Grammar School in Evesham. In 1941 at the age of 17 Grove entered St Catharine's College, Cambridge. A year later he was called up to the Royal Air Force and served in the RAFVR, attaining the rank of Flight Lieutenant. In 1945 he returned to Cambridge and in 1947 he graduated with a First Class Honours degree in Geography. During 1947–9 he was attached to the Nigerian Geological Survey to report on soil erosion. He returned as Demonstrator to the Department of Geography in Cambridge where, barring a spell of six months at the University of Legon in Ghana, a semester at UCLA, and countless field trips, he has remained. He was made lecturer in 1954 and is now a Senior Research Associate. He was Director of the Centre for African Studies during 1980–6. Elected a Fellow of Downing College in 1963 he has served successively as Tutor, Senior Tutor, and Vice Master. He has edited, chaired, and co-ordinated countless things unobtrusively and effectively.

Soon after taking up his Cambridge post Grove was introduced to his future wife Jean, then a research student, by Vaughan Lewis. Jean was planning to take an expedition to the Jotunheimen to look at cirque glaciers and Grove joined her on the first of many glaciological journeys on which, in the 1960s and 1970s, they took their growing family. The trips with Jean were fitted into an already crowded schedule of field-work which included work on geomorphology and climatic change in the Tibesti Mountains between Libya and Chad in 1957, and in northern Nigeria in 1960; on Quaternary landforms in East Africa in 1961, northern Nigeria in 1963, East and South-Central Africa in 1965, the Kalahari and South-West Africa in 1967, Botswana in 1972, and Malawi in 1977; and on Quaternary lake levels in Ethiopia in 1970, and 1974, Tanzania in 1978, Kenya and Sudan in 1980, and Patagonia in 1981. Following Grove's partial retirement his collaboration with Jean has gained momentum, notably on desertification and climatic change in a series of projects funded by the European Community.

An African Canvas

The range of topics on which Grove has published is as wide-ranging as his field-work and just as deceptively so. Running through his papers and books, whatever the level at which they are written, is a concern for the human response to a difficult and unreliable world. Africa led him into that painful pursuit and it has continued to supply much of the narrative.

The analysis may be sensitive but it remains detached. When *Africa South of the Sahara* came out in 1967 Grove found reasonable grounds for being optimistic about the future. By the time of the third edition in 1979 most African countries were facing grave financial difficulties. By 1989 action was needed 'to prevent the continent and its people receding to the margins of the modern world'. 'We seem to be witnessing', Grove wrote in *The Changing Geography of Africa*, 'processes that involve the disappearance for ever of much that is of value and interest in the geography of mankind.' Here is a zoologist talking without sentimentality about a threatened subspecies.

The newer book was intended for schools and colleges. It discarded regions and countries in favour of various fields of economic activity, and elements of the present crisis such as drought and aid. The need to communicate this concern may partly explain the care lavished by Grove on his textbooks in the first place. Extensive travel in Africa and wide and deep reading in its vast literature have given Grove an uniquely synoptic view of the physical geography of Africa. His introductory chapter on the geomorphological evolution of the Sahara and the Nile in the symposium by Williams and Faure (1980), like his more recent study of the African Rift System (1986), has the lightness of touch that comes from a secure grasp of a shifting literature.

The eloquence and unobtrusive didacticism of these general studies, often clothed in illustrations, are also deployed in Grove's more esoteric writings. Two memorable examples are the articles on the 'Geomorphology of Tibesti' and 'The Ancient Erg of Hausaland'. The latter, an excellent account of what palaeoclimatology stands to gain from regional analysis of shifts in wind direction, included a bold map of dunefields which, like the crescents that are said to scar the retinas of those who have gazed at a solar eclipse, has doubtless benevolently scarred many impressionable minds. The Tibesti paper, which RAFfishly made light of a tough journey ('arrived back in Tripoli, having suffered no serious mishaps'), included an aerial photograph of topographic alignments which has proved equally memorable.

The recurrence of themes is of course more an African dictate than a personal foible. Grove published a paper on farming systems and soil erosion in Nigeria in 1949, took up the subject in a series of papers in the 1950s, pursued it together with the vexed issue of desertification in his textbooks in the 1960s and 1970s, and is now confronting it in a Mediterranean context in the 1980s and 1990s.

Lakes and Dunes

The analysis of former climatic patterns will probably prove Grove's greatest contribution to earth science by its bearing on land degradation and human adaptive strategies. He recognized palaeowinds and palaeolakes as critical to the reconstruction of Pleistocene circulation patterns. Unlike oceanic indicators, they represent local conditions rather than a global average and, unlike isotopes or faunas, they reflect with reasonable faithfulness the very climatic factors we seek to recover. He was not alone, or the first, to do so but he set about recovering these data for a substantial chunk of the world with exemplary thoroughness.

The Tibesti area had provided abundant evidence of shifts in wind pattern and fossil landforms. Fossil lakes and dunes came together most productively in Grove's studies of the former Lake Chad, which begat a long series of collaborative writings with several of the authors of this volume in Africa, Australia, and southern Europe. What distinguishes this corpus from that of other groups? The accounts have something of Grove himself: dismissive of any physical difficulties encountered in their preparation, they include the essential information shorn of flummery yet embodying all the background information one might require. They were also solidly rooted in radiocarbon dates at a time when correlation by height or stratigraphic position was still prevalent in Quaternary studies. All three attributes have contributed to the durability of the work. In the words of H. H. Lamb (in Grove *et al.*, 1975: 199), 'Mr. Grove is exploiting [radiocarbon dating and other new techniques] in a programme that promises to clarify our fundamental understanding of how the world's climatic regime and global wind circulation work and link together.'

Chronology proved especially diagnostic in the lake work. For decades the fossil shorelines of East Africa had invited flamboyant correlation on the basis of preconceived notions or, at best, a handful of worked flakes at their periphery. The climatic record of low latitudes was thereby compromised. By dint of what Lamb called 'patient, long-continued, detailed work' Grove and his associates have replaced surmise by numbers. The early papers (e.g. Grove and Warren, 1968) cited a handful of dates, mostly obtained in overseas laboratories; the Ethiopian study (Grove *et al.*, 1975) relied on almost 100 ^{14}C ages for Ethiopia and East Africa. A review of Quaternary lake-level fluctuations ten years later cited 238 ^{14}C ages for lake-levels in intertropical Africa and, in a global survey of lake-levels since 30 000 BP containing 119 data points and 1265 dates, Africa contributed 64 data points with 563 dates (Street and Grove, 1976, 1979).

This is not simply a matter of quantity: the ages were carefully selected. More to the point, the quality of the African contribution could be vouched for because the bulk of it had been gathered by Grove and his associates. The findings invalidated such practices as the correlation of Middle Eastern lake sequences with those of North America and supported the correspondence between high lake-levels in Africa and interglacial or interstadial phases. They also scotched any lingering belief in a simple climatic interpretation of the lake record.

European Perspectives

The palaeoclimatic work, though dispassionate, has never been dissociated from its human implications. In a survey of semi-arid lands he presented for a symposium on resource development a few years ago Grove devoted a substantial section to climatic change before innocently citing a UN report which argued for an integrated approach to planning.

The soft sell is no longer needed. If anything some of the goods have to be recalled. The inverted commas around desertification in Grove's 1986 paper on Europe draw discreet attention to the risk of substituting melodrama for justifiable anxiety about loss of productivity in recent centuries. Grove's collaboration with his wife on medieval climate in Europe is being conducted with a sobriety and persistence not normally associated with the autumnal serenity of retirement.

The tenor of the enterprise is wholly contrary to Byron's languid 'retirement accords with the tone of my mind'. We should welcome Grove's interest in the European past also for entirely selfish reasons. Europe, unlike much of Africa, offers some prospect of yielding well documented palaeoclimatic sequences at high and low elevations and on a variety of terrains. The fine grinding of the Grove school may at last find grain worth the milling. Let us hope that the result will not lie unremarked in the grey binding of conference proceedings and interim reports. The relaxed and unassuming air of many of Grove's classic papers, and the limited circulation of the periodicals in which he has chosen to publish them, has undoubtedly muted their impact: the determination to communicate scientific ideas to a wide audience has been at the expense of the international dialogue with climatologists and geophysicists that is only now unfolding.

When in 1942 Dick Grove joined the Royal Air Force he was sent to Canada for pilot training; he spent much of his war service as a flying instructor and, in due course, as an instructor of instructors. *The Physical Geography of Africa* embodies work done by Grove, the research students he supervised, and *their* research students. And the reader will repeatedly recognize the hallmark of a classic Grove paper: an elegant, unruffled landing in an uncharted location despite treacherous (intellectual) cross-winds and an almost empty (funding) tank.

I thank Jean Grove, Brian Bird, and Jack Mabbutt for generous advice and revealing information.

References

Grove, A. T. (1949), 'Farming Systems and Soil Erosion on Sandy Soils in Southeastern Nigeria', *Bulletin Agricole du Congo Belge*, 40: 2050.

—— (1958), 'The Ancient Erg of Hausaland, and Similar Formations on the South Side of the Sahara', *Geographical Journal*, 124: 526–33.

—— (1960), 'Geomorphology of the Tibesti Region with Special Reference to Western Tibesti', *Geographical Journal*, 126: 18–31.

—— (1977), The Geography of Semi-arid Lands', *Philosophical Transactions of Royal Society of London*, B 278: 457–75.

—— (1978), *Africa* (3rd edn. of *Africa South of the Sahara*) (Oxford).

—— (1980), 'Geomorphic Evolution of the Sahara and the Nile', in M. A. J. Williams and H. Faure (eds.), *The Sahara and the Nile* (Rotterdam), 7–16.

—— (1986a), 'Geomorphology of the African Rift System', in L. E. Frostick *et al.* (eds.), *Sedimentation in the African Rifts* (London), 9–16.

—— (1986b), 'The Scale Factor in Relation to the Processes Involved in "Desertification" in Europe', in R. Fantechi and N. S. Margaris (eds.), *Desertification in Europe* (Dordrecht), 9–14.

—— and Pullan, R. A. (1963), Some Aspects of the Pleistocene Paleogeography of the Chad Basin, in F. Howell and F. Bourlière (eds.), *African Ecology and Human Evolution* (London), 230–46.

—— and Warren, A. W. (1968), Quaternary Landforms and Climate on the South Side of the Sahara', *Geographical Journal*, 134: 194–208.

—— Street, F. Alayne, and Goudie, A. S. (1975), 'Former Lake Levels and Climatic Change in the Rift Valley of Southern Ethiopia', *Geographical Journal*, 141: 177–202.

Grove, J. M., Grove, A. T., and Conterio, A. (1992), 'Little Ice Age Climate in the Eastern Mediterranean', in T. Mokami (ed.), *Proceedings of the International Symposium on Little Ice Age Climate* (Tokyo) 221–6.

Street, F. Alayne and Grove, A. T. (1976), 'Environmental and Climatic Implications of Late Quaternary Lake-level Fluctuations in Africa', *Nature*, 261: 385–90.

—— —— (1979), 'Global Maps of Lake-level Fluctuations since 30 000 yr BP', *Quaternary Research*, 12: 83–118.

1 Tectonics, Geology, and Long-Term Landscape Development

Michael A. Summerfield

Introduction

In this chapter I aim to provide a general background to some key aspects of the tectonic, geological, and landscape development of Africa, the framework which provides much of the setting for the physical geography of the continent. Limited to a necessarily brief review of an enormously diverse and intensively studied range of topics, my focus is on some of the concepts and data that have recently emerged and which illuminate our understanding of the tectonic and macro-scale morphological evolution of the African continent; relevant coverage of earlier material is to be found in two previous reviews (Summerfield, 1985a, b). Such a spotlight on new observations and ideas is particularly apposite because the continent of Africa has become something of a test-bed for tectonic and geomorphic models over the past decade or so with the development of ideas on continental breakup and the role of mantle plumes (White and McKenzie, 1989), and the beginnings of a better understanding of the denudational history of the continent through the application of novel analytical techniques, such as fission-track analysis (Brown *et al.*, 1990, 1994; Omar *et al.*, 1989), and the availability of digital topographic data and increasingly detailed information on offshore stratigraphy (Emery and Uchupi, 1984; Rust and Summerfield, 1990).

I begin by examining the macro-scale morphology of Africa and then briefly describe the structural development of the continent. I go on to assess the stratigraphic record and changing palaeogeography of Africa, in particular with respect to the information these can provide about its Mesozoic and Cenozoic topographic history, and then discuss the thermal and tectonic mechanisms which have been invoked to explain the evolution of

the present gross morphology. The chapter is concluded with an evaluation of the significance of these tectonic models and our increased knowledge of the denudational history of Africa for previous ideas about the post-Gondwana evolution of the African landscape.

First-order Morphology

The recent availability of digital elevation datasets means that it is now possible to assess with relative ease the statistical properties of the morphology of the African continent. Given that it lacks a significant area of Mesozoic or Cenozoic orogens with their associated zones of crustal thickening, the mean elevation of Africa is relatively high compared with the other major continental blocks. Its mean elevation with respect to sea-level is 651 m on the basis of one-degree digital elevation values (Cogley, 1985), but higher resolution 10-minute digital data produces a latitudinally weighted mean of 641 m. Of particular interest is the way elevation is distributed with respect to area in the African continent. This is clearly illustrated through a comparison of the hypsometry of Africa and South America which reveals the very low proportion of Africa at low relative altitudes (Figure 1.1). Indeed, the lower section of Africa's hypsometric curve has the steepest slope for any continent (Bond, 1979). A detailed assessment of the height–frequency distribution for the whole of the African continent reveals a concentration of elevation between 400 and 600 m, with a secondary peak in the distribution between 800 and 1000 m. The significance of such height–area distribution characteristics lies in the way they reflect at a fundamental level the tectonic and long-term landscape evolution of the continent.

Within the African continent itself the first-order

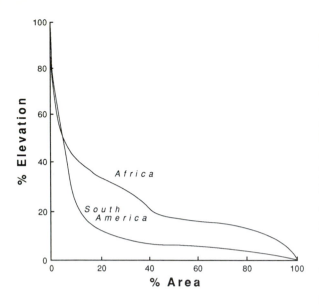

Fig. 1.1. Comparison of hypsometric curves constructed from one-minute digital elevation data for Africa and South America. Note that the base altitude is mean sea-level rather than the elevation of the continental margin as defined by Cogley (1985).

morphological distinction is between the broadly elevated south and east, and the lower-lying north and west (King, 1967) (Figure 1.2). Analysis of the hypsometric data shows that whereas about 80 per cent of Northern Africa (north of 5°N latitude) lies below the mean global elevation of the continents, around 70 per cent of Southern Africa (south of 5°N latitude) is above the global mean (Nyblade and Robinson, 1994). Large parts of Southern and Eastern Africa lie above 1000 m, forming a long wavelength ($\sim 10^7$ km^2), low amplitude (~ 500 m) positive topographic anomaly. Termed the Southern African Superswell, this feature extends into the adjacent oceanic areas of the African Plate suggesting that its source of buoyancy is located in the deep mantle (Nyblade and Robinson, 1994) (Figure 1.2). Superimposed on this very broad feature is a region of shorter wavelength–higher amplitude positive anomalies. These are concentrated in Eastern Africa where elevations exceed 2000 m along the margins of the rifted domal upwarps of the Cenozoic East African Rift System.

Structural Framework

More than any other continent, the post-late Palaeozoic tectonic evolution of Africa has been dominated by crustal extension and rifting (Lambiase, 1989) (Figure 1.3). The only significant compressional tectonics during this period has been confined to the Cenozoic Atlas orogeny in North Africa associated with the convergence of Africa and Eurasia, and the formation of the Cape Fold Belt (~ 280–230 Ma BP) along the southwestern perimeter of Gondwana. Final incorporation of Africa into the Gondwana component of the Pangaean supercontinent was accomplished during the Pan-African event (950–450 Ma BP) which resulted in the structural differentiation of the continent into cratons and mobile belts. Subsequent rifting has largely been focused within these pre-existing mobile belts and between the regions of cratonic terrane. For instance, the Tanganyika and Malawi Rift Zones follow Proterozoic mobile belts and bifurcate around the Tanganyika Craton. This concentration of rifting in zones of relative lithospheric weakness is a common feature of rift propagation globally (Dunbar and Sawyer, 1988).

Africa has experienced at least seven major rifting episodes since the early Permian (Lambiase, 1989) (Figure 1.3). The earliest of these preceded the breakup of Gondwana, the most significant involving Karoo rift and basin formation. Sedimentation in the Karoo Basin of Southern Africa was almost continuous from 280 to 100 Ma BP, and in its early stages basin subsidence was coeval with uplift of the Cape Fold Belt to the south. Four major phases of rifting were, however, clearly related to the breakup of Gondwana as the resulting rifts constitute the onshore extension of tectonic structures associated with the rupture of the supercontinent and the formation of the passive margins of Africa. Late Triassic–early Jurassic rifting was associated with the separation of north-west Africa and North America. The rift structures produced differ from those of Eastern and Southern Africa in that they were formed in a relatively short period of time and have not subsequently experienced extensional tectonics to any significant extent; compressional tectonics, notably that associated with the creation of the Atlas Mountains, has, however, affected a number of them. Initial opening of the Indian Ocean in the early to mid-Jurassic, including the separation of Madagascar from Africa about 160 Ma BP (Rabinowitz *et al.*, 1983), was associated with rifting events in Eastern Africa. Mid-Cretaceous structures were created immediately prior to, and during, the separation of South America from West Africa. Rifts and basins of this age occur over a broad area extending from the Benue Trough in the west, to Sudan in the east, and the Sirte Basin in Libya to the north. The continuity of rifts in the western part of this region has led to these structures being collectively termed the Central African Rift System (Fairhead, 1986). As with the East African Rift System and the Red Sea Rift, longitudinally continuous rifting in Central Africa has been accomplished by the growth

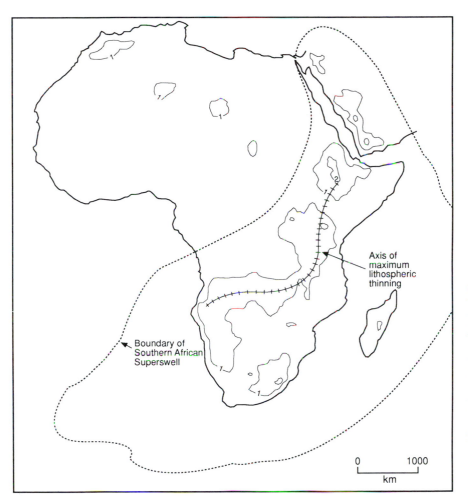

Axis of maximum lithospheric thinning

Boundary of Southern African Superswell

0   1000
km

Fig. 1.2. First-order morphology of Africa.

Contours are drawn at 1 km intervals and are generalized from one-minute digital elevation data (after Cogley, 1985). The delimitation of the Southern African Superswell is from Nyblade and Robinson (1994). The axis of maximum lithospheric thinning, which coincides with topography > 1 km elevation in Southern and Central Africa, is from Pavoni (1992). Note the relationship of this NW-trending region of high topography with the Walvis Ridge offshore.

and linkage of isolated zones of extension (Nelson *et al.*, 1992).

The Cenozoic structures of the East African Rift System (see Chapter 2) are directly related to the formation of the Red Sea and Gulf of Aden. Although it is now recognized that rifting occurred in East Africa during the the early Cenozoic (for instance, in the Turkana and Anza Basins (Reeves *et al.*, 1987)), the present episode of rifting began in the early Miocene (Frostick *et al.*, 1986) and seismic data indicate that the East African Rift System continues to propagate to the south-west towards the Kalahari Craton (Fairhead and Henderson, 1977). The rift structures form a linear feature, but individual basins are orientated at highly variable angles to the overall trend of the rift system. The relationship of late-Cenozoic rifts to earlier structures varies; in some locations, such as Lake Malawi, the late Cenozoic basin cuts across two Permo-Triassic basins, but in other cases late-Cenozoic extension has involved the reactivation of much earlier structures. For instance, the southern half of Lake Tanganyika overlies part of the Permo-Triassic Luama Rift.

The majority of African rifts do not have contemporaneous volcanic rocks associated with them, indicating that volcanism has not been integral to the rifting process (Lambiase, 1989). The major rift systems with associated volcanism are the eastern branch of the late-Cenozoic East African Rift System, the mid-Cretaceous Benue Trough, the late-Triassic–early-Jurassic rifts of northwest Africa and the Karoo System of Southern Africa. It is interesting to note that not all volcanic rifting events have led to the formation of ocean basins, nor has all sea-floor spreading around Africa been preceded by volcanic rift activity, as is illustrated, for example, by the non-volcanic separation of Madagascar and India from Africa in the early Jurassic.

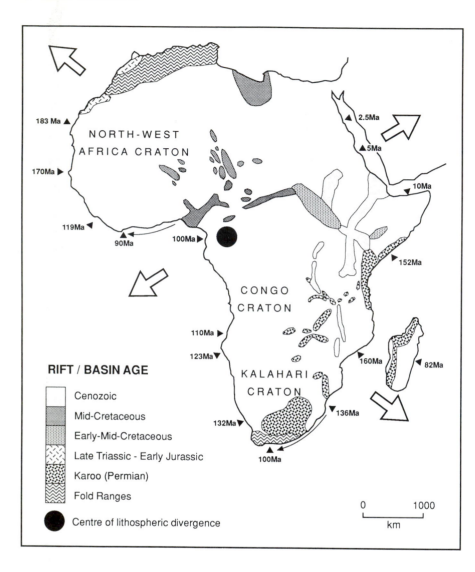

Fig. 1.3. Key structural features of Africa.

The demarcation of Phanerozoic rift systems is from Lambiase (1989). Ages refer to the probable time of initiation of basin generation, and the areas indicated represent general rifted areas rather than specific rift structures. Dates for the beginning of sea-floor spreading (drifting) are shown along the continental margin (from Uchupi and Emery (1991)); note that at a number of locations rifting began a significant time before the onset of sea-floor spreading. The large open arrows indicate the primary directions of large-scale lithospheric divergence, and the solid circle the centre of lithospheric divergence, according to Pavoni (1992).

If the present-day stress regime affecting the African continent is assessed from the orientation of fault systems and earthquake fault-plane solutions, a pattern of radial extension focused around a hypothetical West African 'spreading centre' is evident, according to Pavoni (1992) (Figure 1.3). He suggests that this stress pattern is related to shear traction being exerted on the base of the lithosphere by convecting mantle associated with a single megaplume, or several individual plumes. This interpretation appears to gain some support from the distribution of the positive topographic anomaly identified by Nyblade and Robinson (1994); certainly the most significant feature of the Bouguer gravity anomaly map of Africa is the prominent negative anomaly which extends from the termination of the Walvis Ridge on the southwest coast to the Ethiopian Highlands in the north-east and which probably reflects a major zone of lithospheric extension and thinning (Pavoni, 1992) (Figure 1.2).

Stratigraphy and Palaeogeography

Although substantial parts of the African continent have been subaerially exposed since the formation of the Gondwana supercontinent and have experienced prolonged denudation, significant accumulations of continental sediments have been deposited. Moreover, marine sedimentation has affected large areas of North Africa, as well as localities elsewhere along the coastal fringes of the continent (see Chapter 14). The continental sediments have presented considerable problems in terms of establishing a detailed chronology because they generally lack fossils and are highly oxidized. Nevertheless, problems of dating, correlation, and facies interpretation have been tackled through research undertaken as part of two IGCP programmes (nos. 127 and 210), the most significant conclusion of which being the identification of several marine horizons in strata previously considered to be entirely continental in origin (for instance, the Nubia Type Strata of north-east Africa) (Kogbe and Burollet, 1990).

Outside the folded strata associated with Hercynian and Alpine orogeny at the northern and southern extremities of the continent, the sedimentary cover of Africa consists largely of flat-lying to gently warped units lapping on to the crystalline basement. The Palaeozoic platform cover is concentrated in the north, with marine deposition extending to Guinea in the west and the Hoggar and Tibesti Plateaus in the south. In the south of the continent deposition forming the Karoo Supergroup extended from the Carboniferous to the early Jurassic and culminated in the vast outpourings of lava forming the Karoo flood-basalt province. Parts of the Karoo sequence extend as far north as the Zaire Basin and the East African Lakes Region, but Karoo sedimentation never extended as far as the Palaeozoic sequences of the north.

Mesozoic deposition was extensive and varied in North Africa, especially during the Jurassic. It is the Cretaceous cover that is particularly prominent, however, with initially continental facies being replaced by widespread marine deposition during the Cenomanian–Turonian (97–88.5 Ma BP) transgression before a return to predominantly continental sedimentation (Kogbe and Burollet, 1990). The Cenozoic has been largely characterized by the accumulation of extensive but shallow sequences of continental sediments within gentle downwarps, such as the Kalahari Basin of Southern Africa (Thomas and Shaw, 1990) and the Zaire and Chad Basins to the north.

The age and facies characteristics of sedimentary units over the African continent provide vital evidence of its palaeogeography, and, in particular, its surface uplift history since the breakup of Gondwana. Establishing the spatial pattern and timing of surface uplift is critical to the validation of the tectonic models of rifting and crustal deformation that have been applied to Africa, and the identification of the most recent episode of marine sedimentation is crucial in this regard as shoreline and shallow marine deposits provide the only unequivocal palaeodatum to which vertical movements of the land surface can be related. This approach has been used by

Sahagian (1988) who based his reconstruction of epeirogenic movements in Africa on shoreline deposits associated with the mid-Cretaceous global sea-level highstand (Figure 1.4). It can be applied successfully to North Africa since the Cenomanian (97–90.4 Ma BP) palaeogeography here was dominated by the Trans-Sahara Seaway which extended from the Benue Trough in the south to the Gulf of Sirte in the north. At this time a broad carbonate platform covered much of north-east Africa, and by the Turonian (90.4–88.5 Ma BP) a marine connection had been established between the South Atlantic and the Tethys Sea separating Africa from the southern flank of Eurasia. Elsewhere there is evidence of probable marine fossil fish of Cenomanian age from the Kwango Formation in the Zaire Basin (Cahen and Lepersonne, 1954), while evidence that a sea covering west-central Africa in the late Jurassic was connected to the Tethys Sea across what is now East Africa (Shackleton, 1978) suggests that this area was probably still low-lying, if not a shallow marine environment, until at least the mid-Cretaceous.

The paucity of Mesozoic and Cenozoic marine deposits in southern Africa renders the reconstruction of the surface uplift chronology for this area highly conjectural. Sedimentation from the late Carboniferous to the early Jurassic is represented by the Karoo Supergroup, only the lowest formation of which has indisputable marine deposits. These are the glaciomarine members of the Permo-Carboniferous Dwyka Formation which outcrop in the western part of the Karoo Basin (~290 Ma BP) (Visser, 1987, 1989). It has been argued that low elevations persisted in this region until at least the late Permian on the basis of deposition of the 'low-energy' sedimentary sequences composed of muds and sands that form the Beaufort Group (Dingle et al., 1983; Sahagian, 1988). However, since local slope gradients are not necessarily correlated with elevation (Summerfield, 1991a), such 'low-energy' sedimentation is not in itself evidence of low elevation, as is illustrated by Cenozoic deposition in the very low-gradient Kalahari Basin to the north which stands at an elevation of nearly 1000 m. On the basis of evidence from marine sediments, therefore, uplift of Southern Africa to its present high mean elevation can only be constrained to being postearly Permian. The only exceptions to this are the localized areas to the south of the Karoo Basin in the Cape Fold Belt where marine deposits of the Jurassic Uitenhage Group outcrop, and along the southern and eastern coast where marine sediments of Cenozoic age are found up to elevations of around 400 m (Dingle et al., 1983).

The most prominent region of significant postCenomanian crustal uplift represented in Figure 1.4 is

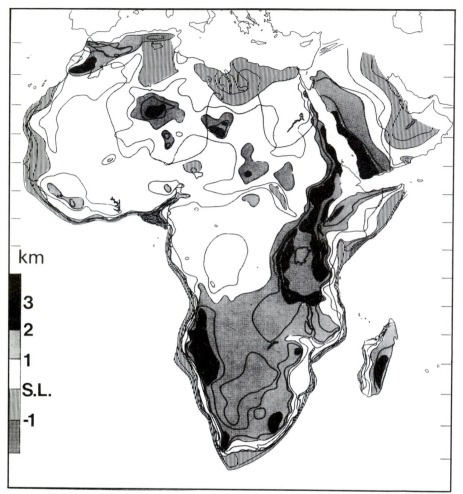

Fig. 1.4. African Cenomanian surface (~90 Ma BP) corrected for sediment loading by post-Cenomanian sediments, according to Sahagian (1988).

The Cenomanian shoreline is shown for reference. A correction for the decrease in global sea-level since the Cenomanian (~250 m based on estimates by Haq *et al.*, 1987; McDonough and Cross, 1991; and Sahagian, 1987), yields estimates of internally driven post-Cenomanian tectonic deformation. See text for discussion of limitations of these estimates.

km

3
2
1
S.L.
-1

that of the Red Sea region and the broad, elongated upwarp associated with the East African Rift System. The presence of Tertiary sediments in these areas, together with volcanic and structural evidence, actually enables the age of the initiation of uplift to be constrained to the mid-Cenozoic. Significant, although localized, post-Cenomanian surface uplift is also evident in North Africa in the development of the volcanic Hoggar and Tibesti Massifs. The extensive area of uplift indicated by Sahagian (1988) in Southern Africa cannot, however, be accepted as necessarily post-Cenomanian in age for the reasons outlined above. None the less, attempts to identify the surface uplift history of Southern Africa are vital if the diverse range of tectonic models proposed to account for the development of African topography are to be critically evaluated.

Tectonic and Thermal Mechanisms

The present landscape of Africa is very much a product of the breakup of Gondwana and the tectonic processes associated with this event. The other major influence since the creation of Africa as a distinct continental block has been the continuing episodes of rifting and volcanism that have occurred. Such tectonic and thermal mechanisms have to account for certain specific characteristics of present-day Africa. One is the series of broad upwarps rising to altitudes of several hundred metres or more above the surrounding terrain that occur over much of the continent. These are particularly evident as the long-recognized marginal upwarps which run parallel to the coastline in several areas (Birot *et al.*, 1982; Bremer, 1985; Huser, 1989; Obst and Kayser,

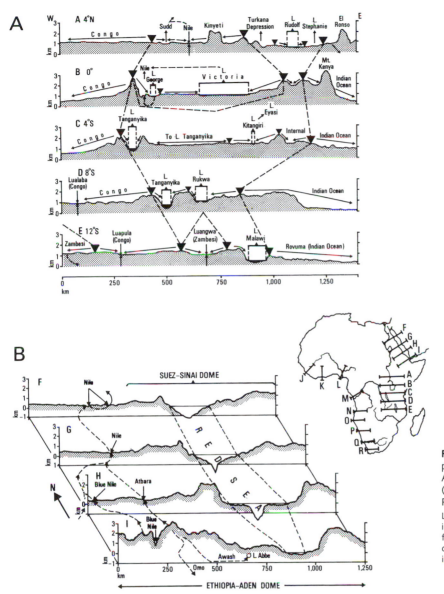

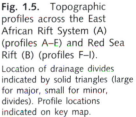

Fig. 1.5. Topographic profiles across the East African Rift System (A) (profiles A–E) and Red Sea Rift (B) (profiles F–I).

Location of drainage divides indicated by solid triangles (large for major, small for minor, divides). Profile locations indicated on key map.

1949; Sponemann and Brunotte, 1989) and which are flanked on their seaward side by a sharp topographic discontinuity in the form of a major escarpment, or series of escarpments (Figure 1.6). Since these features so strongly coincide with the new continental margins created by the breakup of Gondwana it would seem highly probable that they are genetically related to this rifting event. The most prominent upwarps, however, are those associated with the much younger East African Rift System, which, itself, appears to represent the site of incipient continental rifting (Figure 1.5A). A further

important component of African morphology are the areas of volcanism; some of these, such as the Karoo lavas of south-east Africa and the Etendeka lavas of Namibia, relate to the breakup of Gondwana in the Mesozoic, whereas others, such as the Tertiary Ethiopian flood basalts, are associated with the much more recent separation of Arabia from Africa and the opening of the Red Sea. By contrast, other centres of Cenozoic volcanic activity, including the Neogene volcanism of the Tibesti and Hoggar Massifs in North Africa, lack any association with continental rifting.

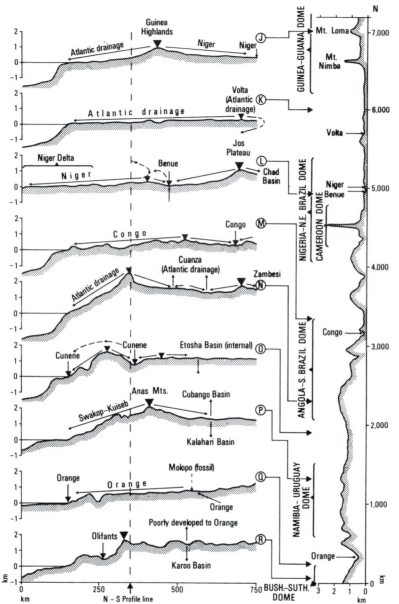

Fig. 1.6. Transverse topographic profiles along the Atlantic margin of Africa (profiles J–R) and longitudinal profile located approximately 200 km inland from coastline (350 km from profile L).

Alternate transverse profiles represent domal crest and inter-dome through sites located on longitudinal profile. Symbols for drainage divides same as for Figure 1.5. See key map in Figure 1.5. for transverse profile locations.

Hot Spots and Mantle Plumes

The Tibesti and Hoggar volcanic uplifts are only two of the numerous hot spots distributed across the African Plate (Figure 1.7). These anomalous centres of volcanic activity are thought to be associated with mantle plumes originating at depth within the sub-lithospheric mantle. The associated domal upwarps of up to 1000 km in diameter and 1 km or more in amplitude are generated by both mantle flow and thermal isostasy. It has been suggested that the relatively high concentration of hot spots on the African Plate is due to a very low rate of relative motion with respect to the underlying sub-lithospheric mantle from about 25 ± 5 Ma BP (Burke and Wilson, 1972; Bond, 1979). It has further been argued that the concentration of African hot spots on thinner, non-cratonic lithosphere implies that lithospheric thickness also plays a role (Gass *et al.*, 1978; Pollack *et al.*,

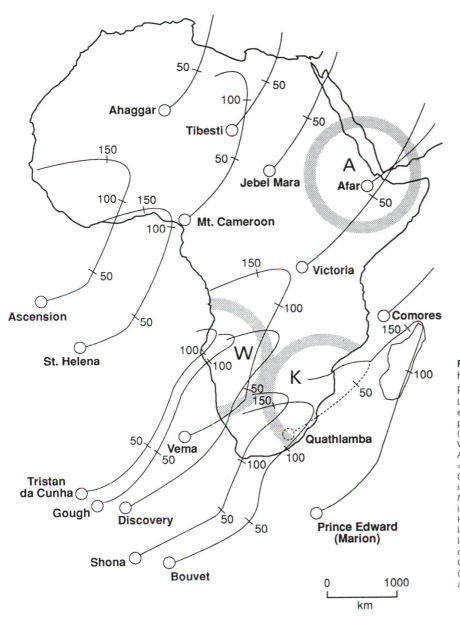

Fig. 1.7. Africa in relation to hypothesized major mantle plumes and hot-spot tracks.

Large circles represent assumed extent of mushroom head of plumes of abnormally hot mantle (radius 1000 km) according to White and McKenzie (1989): A = Afar plume; K = Karoo plume; W = Walvis plume. Hot-spot tracks (solid lines labelled at 50 Ma age increments) are primarily based on Morgan (1983) with additional information from Duncan (1981), Hartnady and le Roex (1985), and le Roex (1986). Present hot-spot locations are indicated by small open circles. The hypothesized Quathlamba hot spot and track (Hartnady, 1985) are indicated by a pecked circle and line.

1981), but statistical analysis questions this assertion (Summerfield, 1983, 1985*a*).

Although an association between the present location of African hot spots and specific topographic swells is clear, the wider role played by thermal uplifts in generating Africa's anomalous hypsometry is less certain. Although there is a concentration of hot-spot tracks in Southern Africa which is coincident with a high mean elevation (Figure 1.7), there is also high terrain in localities such as the Lesotho Highlands and the Namibia/ Angola Highlands which do not exhibit recent volcanism, but which do coincide with flood-basalt provinces dating to around the time of Gondwana breakup. Since thermal uplifts decay with a time constant of about 60 Ma it seems probable that such regions of high topography experienced crustal thickening through magmatic underplating at the time of continental rifting. Although there is a continuing debate as to the relative timing of mantle plume activity and rifting during continental rupture (Campbell and Griffiths, 1990; Cox, 1993; Hill, 1991; White and McKenzie, 1989), the notion of crustal thickening associated with rift-related magmatism provides a viable explanation of sustained high topography in the African continent which can only be reduced by denudation.

Rifting and Passive Margin Development

Over the past decade an enormous variety of rifting models have been proposed to account for the structural, stratigraphic, and morphological properties of new continental margins formed by continental breakup (Allen and Allen, 1990; Keen and Beaumont, 1990; Summerfield, 1991b). However, this diversity can be simplified into two end-member models. In active rifting unusually hot asthenosphere impinges on the base of the lithosphere causing it to be thinned and extended. The presence of hot asthenosphere close to the surface promotes volcanism and domal uplift, while extension gives rise to rifting which may eventually develop to the point where new ocean floor is formed as sea-floor spreading begins. In passive rifting the initiating mechanism is localized extension associated with the global pattern of plate movements which produces rifting as a result of tensional stresses in the lithosphere. In this model hot asthenosphere rises passively in response to thinning of the overlying lithosphere and this can give rise to volcanism and thermally induced surface uplift along the rift flanks. These two models predict quite distinct sequences for rifting, volcanism, and uplift. The anticipated order for passive rifting is rifting (with prevailing subsidence) followed by volcanism, and then uplift, whereas that for active rifting is volcanism–uplift–rifting (although uplift may precede volcanism). Many modifications have been made to the basic passive-rifting model originated by McKenzie (1978), these being mainly aimed at accounting for the significant surface uplift observed along the flanks of many (but not all) newly formed continental margins which the basic model fails to predict without making unrealistic assumptions as to the relative thicknesses of the crust and sub-crustal lithosphere (Gilchrist and Summerfield, 1994).

In Africa the domal upwarps and volcanism associated with the East African Rift System (Figure 1.6 (A)) appear to provide some support for the active-rifting mechanism, but as the detailed evaluation of this rift system by Rosendahl (1987) makes clear, there are considerable uncertainties as to the timing of uplift, rifting, and volcanism even in this relatively well-studied region. The high-standing shoulders of the Red Sea Rift with elevations exceeding 2 km and a structural relief of 4–5

km or more (Figure 1.6 (B)) also seem to suggest active rifting given that this structure dates only to the mid-Cenozoic. However, the presence of marine sediments which date to just before rifting along the flanks of parts of the Red Sea confirm that, in these localities at least, rift formation was not preceded by significant doming but, in fact, occurred more or less at the same time as rift-flank uplift (Garfunkel, 1988).

The lack of marine sediments which can be used as an unequivocal palaeodatum immediately prior to rifting makes it extremely difficult, and in many cases impossible, to decide from observational evidence whether the earlier rifting events which produced the breakup of Gondwana were associated with active or passive rifting. None the less, two important morphological features that characterize the African continent cannot be accounted for by either the active-, or symmetrical passive-rifting models. One is the persistence of marginal upwarps for more than 100 Ma after continental breakup (Figure 1.5). Since all the margins of Africa, apart from the Red Sea coast, were formed more than 100 Ma BP (Figure 1.3) we would now expect them to be low-lying because the thermal uplift associated with both rifting models described above would be expected to decay over about 60 Ma. The other feature of relevance is the variation in elevation along the African margins (Figure 1.5) and the fact that high margins in Africa appear to coincide with low-lying terrain on the conjugate margins of North and South America on the other side of the Atlantic Ocean (Etheridge et al., 1989). Such topographic asymmetry cannot be accounted for by symmetrical rifting, but is predicted by asymmetric rifting models in which the lithosphere extends by simple shear rather than pure shear and separates along a dipping detachment fault which penetrates to the base of the lithosphere (Etheridge et al., 1989; Lister et al., 1986). This produces an elevated upper-plate margin with a relatively narrow adjacent continental shelf in which the crust is thickened by magmatic underplating, and an opposing low-lying and highly faulted lower-plate margin with a broad continental shelf. The western margin of Southern Africa provides an example of such topographic asymmetry, with a high-elevation (upper-plate?) margin extending as far north as the major transform/transfer fault represented by the Rio Grande–Walvis fracture zone around latitude 30°S (Etheridge et al., 1989).

In the asymmetric rifting model high elevations in the upper plate are promoted by magmatic underplating as the base of the lithosphere is exposed to hot underlying asthenosphere. The significance of underplating, as opposed to thermally generated surface uplift is that

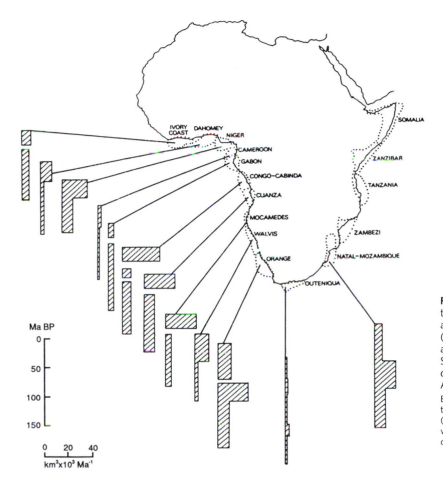

Fig. 1.8. Variations through time in volume of sediments accumulating in basins (delimited by dotted lines) around the margin of sub-Saharan Africa (equivalent data for North and East Africa not available).

Break points in histograms relate to principal seismic horizons. (Compiled by D. J. Rust from various sources and largely based on offshore seismic stratigraphy.)

the former is 'permanent' in the sense that it does not thermally decay and elevation can only be reduced by denudation. Both this model, and the mantle plume model of White and McKenzie (1989), provide a possible means for high topography being sustained along the rifted continental margins of Africa for well in excess of 100 Ma from the time of Gondwana breakup.

Long-term Landscape Development

Two main themes can be observed in traditional interpretations of long-term landscape evolution in Africa. One emphasizes the role of climate and climatic change and is exemplified in particular by a large body of work by French and German researchers (Büdel, 1982; Tricart, 1972). The other is most prominently represented by the approach of F. Dixey and L. C. King and focuses on the generation of cycles of erosion in response to tectonically driven changes in base level (see Partridge and

Maud (1987) for a recent review). Although climate, and especially late Cenozoic climatic change, has clearly been of great significance in the development of the African landscape, at the largest scale it has been the interplay between tectonics and surface geomorphic processes that has been more influential, and it is this relationship that I will focus on here.

A general criticism of the continent-wide, or even global, denudational schemes presented by workers such as King (1967) has been the lack of adequate dating control available, and the paucity of information on actual depths of denudation. Based primarily on an analysis of the African landscape, King's approach relied extensively on the correlation of land surfaces, often over great distances, without the benefit of the kinds of data that are now providing us with an increasingly detailed picture of the pattern of denudation across Africa since the breakup of Gondwana. One important source of information has come from estimates of rates of off-

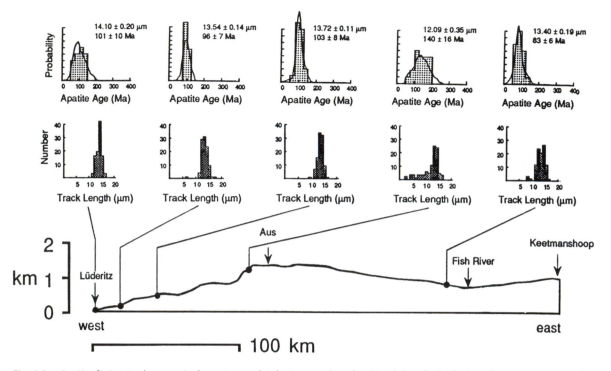

Fig. 1.9. Apatite fission-track ages, single-grain age distributions, and confined track-length distributions for outcrop samples from a transect from Luderitz to Keetmanshoop (southern Namibia).

Note that low variability in track-length distribution is confined to low elevations (see text for discussion). (*Source:* Brown *et al.*, 1990; and Brown, 1992.)

shore sedimentation. For instance, Robert (1987) has analysed the records of a number of Deep Sea Drilling Project (DSDP) sites in the South Atlantic to reconstruct changes in rates of sediment supply from the African hinterland. Although point data provided by boreholes is valuable, they have limited value in providing estimates of total volumes of sediment deposited. Where, however, borehole data are combined with seismic stratigraphy a three-dimensional view of offshore sedimentary bodies can be created and quantitative estimates made of spatial and temporal variations in sediment supply. As illustrated in Figure 1.8, these indicate an increase in the rate of sediment supply along much of the west coast of Africa during the late Cenozoic, although a rather different pattern is evident in Southern Africa with a peak in the mid- to late Cretaceous for both the Orange and Natal–Mozambique Basins. A more detailed analysis for the Orange Basin shows an apparent peak in sediment supply during the early Cenozoic (Rust and Summerfield, 1990).

A major shortcoming of such sediment volume data is that they do not indicate the source of the sediment. Thus, if used alone, it is not possible to distinguish

changes in overall rates of erosion in the source drainage basins from changes in source basin area. In some cases it is possible to identify a new pulse of sediment supply which is clearly related to a change in drainage organization in the African hinterland, such as the establishment of the present position of the outlet of the Zaire Basin in the Miocene (Emery and Uchupi, 1984), but in general other types of data are required to establish changing patterns of denudation across the African continent.

One such type of data is that from apatite fission-track analysis (Brown *et al.*, 1994). The number of fission tracks in basement and sedimentary apatites, together with the relative frequency of tracks of different lengths, can yield a cooling history for surface-outcropping rocks originally at depth within the crust. Where they have not been affected by significant thermal events, such a cooling history can be used to estimate a denudational history. Initial applications of this technique in Africa show it to have significant potential in establishing quantitatively well-constrained denudational histories (Brown, 1992; Brown *et al.*, 1990; Haack, 1976; Omar *et al.*, 1989). In southern Africa, for instance, fission-track

Fig. 1.10. Drainage systems of the African mainland.
Note the extensive area of internal drainage, especially in Western and Southern Africa, and the predominantly centripetal drainage pattern of the Zaire Basin.

analysis reveals several kilometres of crustal stripping along the continental margin broadly coinciding with the time of rifting, with more limited, but still substantial denudation in the interior (Brown, 1992; Brown *et al.*, 1990). In contrast to the interpretations of King, such data demonstrate that much, if not all, of the land surface of Southern Africa has experienced significant post-Gondwana denudation, a finding that renders the survival of Gondwanan land surfaces highly improbable.

The recognition that existing models of continental rifting and breakup cannot account for the long-term persistence of marginal upwarps along continental mar-

gins has led to a reassessment of the role of denudation in generating topography through the isostatic response to crustal unloading. Although earlier suggested by King (1955) and Pugh (1955) in the context of the generation of erosion surfaces in Africa, their initial formulation of the dynamics of such isostatic response was based on a misunderstanding of the way in which the lithosphere responds flexurally to changes in load (Gilchrist and Summerfield, 1991). Rather than being highly episodic, Gilchrist and Summerfield (1990) have demonstrated that where a marked topographic discontinuity is present along a passive continental margin in the form of a major escarpment separating an

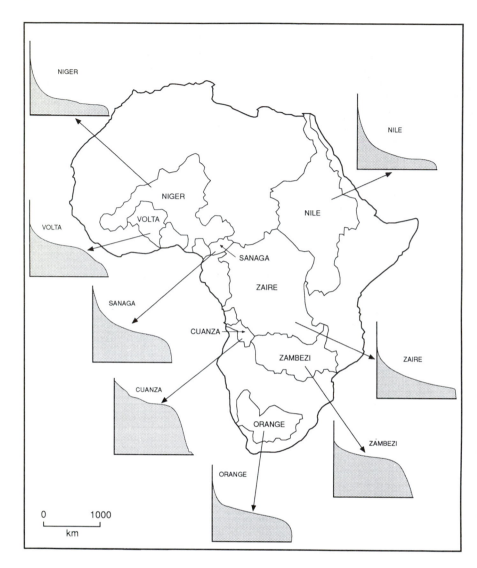

Fig. 1.11. Hypsometry of a range of drainage basins in Africa.

Hypsometric curves have been produced from National Geophysical Data Center, Boulder, Colo., 10-minute digital elevation data retrieved for the digitized basin areas indicated. For some basins, especially the Niger, Nile, and Orange, perimeters incorporate areas that do not currently contribute runoff to the basin outlet but which would have contributed runoff during more humid periods in the Quaternary. Note the small percentage area at low relative elevations indicating minimal coastal plain development, especially for the rifted margins of the west and east coasts.

elevated interior plateau from low-lying, coastal terrain, then a continuous flexural isostatic response will occur as the escarpment retreats, which is capable of generating a marginal upwarp with a magnitude of the order of several hundred metres. Subsequently other attempts have been made to model the flexural response to denudational unloading in Southern Africa (Stuwe, 1991; Ten Brink and Stern, 1992), but unfortunately these have used King's denudation chronology rather than applying the denudational record available from fission-track analysis and offshore sedimentary data.

One aspect of the interaction between tectonics and the surface denudational system that has until recently received insufficient attention is the relationship between regional tectonics and the evolution of African drainage. The organization of drainage across the African continent is anomalous in the sense that given the time elapsed since it became a discrete continental block a system more integrated with external drainage might be expected. As it is the present-day drainage pattern retains large areas of internal drainage and other basins, such as the Zaire, with a distinctly centripetal pattern (Figure 1.10). Clearly aridity has played an important role here, but the hypsometry of many African drainage basins also suggests that tectonics has had a profound influence (Figure 1.11). Although there is a large

Plate 1.1. Many of the lower reaches of African rivers display major knickpoints, as is the case with the Victoria Falls on the Zambezi at the border between Zimbabwe and Zambia.

literature on the evolution of specific basins, and in particular the identification of capture events (Thomas and Shaw (1988), for instance, on the Zambezi Basin), there is still much uncertainty given the ambiguity and paucity of the evidence. This is exemplified in the recent debate concerning the status of the mid-Cenozoic trans-African drainage system identified by McCauley *et al.* (1982). Although they regard it as a south-westward flowing system with headwaters originally in the highlands flanking the Red Sea and an outlet via the Benue Trough at the site of the modern Niger Delta, Burke and Wells (1989) have argued that this now buried drainage system may have been part of the Nile Basin.

At a continental scale the steep gradients of the lower reaches of many African rivers and the presence of major knickpoints such as the Victoria Falls on the Zambezi (Plate 1.1), the rapids on the Zaire, the Augrabies Falls on the Orange, and the Ruacana Falls on the Kunene, imply a continuing adjustment of the fluvial system to crustal deformation and the role of lithology in maintaining discontinuities in river long profiles. Although one explanation for the anomalous hypsometry of the whole African continent, as well as individual basins, is recent surface uplift of much of the continent as suggested, for instance, by Bond (1979), another view emphasizes the role of continuing flexural isostatic uplift along the margins of the continent as a result of denudation being largely focused in the coastal zone. This would be expected to have maintained a dual drainage system for a period after the breakup of Gondwana with aggressively eroding, high-gradient coastal streams progressively capturing the interior drainage (Summerfield, 1985). This process may account for the late-stage capture and accompanying anomalous hypsometry and drainage pattern of the Zaire Basin (Figure 1.11), rather than the shifting of drainage divides through crustal deformation in the continental interior as suggested by Faure and Lang (1991). Much further research is required on this aspect of the relationships between tectonics and landscape development on the African continent before we have a comprehensive picture of how that landscape has evolved since Africa became a discrete continent.

References

Allen, P. A. and Allen, J. R. (1990), *Basin Analysis: Principles and Applications* (Oxford).

Birot, P., Petit, M., and Lageat, Y. (1982), 'Les bourrelets intertropicaux', *Bulletin de l'Association Géographie Française*, 489: 261–9.

Bond, G. C. (1979), 'Evidence for some Uplifts of Large Magnitude in Continental Platforms', *Tectonophysics*, 61: 285–305.

Bremer, H. (1985), 'Randschwellen: A Link between Plate Tectonics and Climatic Geomorphology', *Zeitschrift für Geomorphologie Supplementband*, 54: 11–21.

Brown, R. W. (1992), 'A Fission-track Thermochronology Study of the Tectonic and Geomorphic Development of the Subaerial Continental Margins of Southern Africa', unpublished Ph.D. thesis, La Trobe University, Melbourne.

—— Rust, D. J., Summerfield, M. A., Gleadow, A. J. W., and De Wit, M. C. J. (1990), 'An Early Cretaceous Phase of Accelerated Erosion on the South-Western Margin of Africa: Evidence from Apatite Fission-track Analysis and the Offshore Sedimentary Record', *Nuclear Tracks and Radiation Measurements* 17: 339–50.

—— Summerfield, M. A., and Gleadow, A. J. W. (1994), 'Apatite Fission-track Analysis: Its Potential for the Estimation of Denudation Rates and Implications for Models of Long-term Landscape Development', in M. J. Kirkby (ed.), *Process Models and Theoretical Geomorphology* (Chichester), 23–53.

—— and Wells, G. L. (1989), 'Trans-African Drainage System of the Sahara: Was it the Nile?', *Geology*, 17: 743–7.

Büdel, J. (1982), *Climatic Geomorphology* (Princeton).

Burke, K. and Wilson, J. T. (1972), 'Is the African Plate Stationary?', *Nature*, 239: 387–90.

Cahen, L. and Lepersonne, J. (1954), 'Etat actuel des connaissances relatives aux séries mésozoiques de l'interieur du Congo', *Bulletin de Société Belge Géologigue Palaeontologie et Hydrologie*, 63: 20–37.

Campbell, I. H. and Griffiths, R. W. (1990), 'Implications of Mantle Plume Structure for the Evolution of Flood Basalts', *Earth and Planetary Science Letters*, 99: 79–93.

Cogley, J. G. (1985), 'Hypsometry of the Continents', *Zeitschrift für Geomorphologie Supplementband*, 53: 1–48.

Cox, K. G. (1993), 'Continental Magmatic Underplating', in K. G. Cox, D. McKenzie, and R. S. White (eds.), 'Melting and Melt Movement within the Earth', *Philosophical Transactions of the Royal Society London*.

Dingle, R. V., Siesser, W. G., and Newton, A. R. (1983), *Mesozoic and Tertiary Geology of Southern Africa* (Rotterdam).

Dunbar, J. A. and Sawyer, D. S. (1988), 'Continental Rifting at Pre-existing Lithospheric Weakness', *Nature*, 333: 450–2.

Duncan, R. A. (1981), 'Hotspots in the Southern Oceans: An Absolute Frame of Reference for Motion of the Gondwana Continents', *Tectonophysics*, 74: 29–42.

Emery, K. O. and Uchupi, E. (1984), *The Geology of the Atlantic Ocean* (New York).

Etheridge, M. A., Symonds, P. A., and Lister, G. S. (1989), 'Application of the Detachment Model to Reconstruction of Conjugate Passive Margins', in A. J. Tankard and H. R. Balkwill (eds.), 'Extensional Tectonics and Stratigraphy of the North Atlantic Margins', *American Association of Petroleum Geologists Memoir*, 46: 23–40.

Fairhead, J. D. (1986), 'Geophysical Controls on Sedimentation within the African Rift Systems', in L. E. Frostick, R. W. Renaut, I. Reid, and J. J. Tiercelin (eds.), 'Sedimentation in the African Rifts', *Geological Society Special Publication*, 25: 19–27.

—— and Reeves, C. V. (1977), 'Teleseismic Delay Times, Bouguer Anomalies and Inferred Thickness of the African Lithosphere', *Earth and Planetary Science Letters*, 36: 63–76.

Faure, H. and Lang, J. (1991), 'Dynamics of Continental and Paralic Sedimentation in Africa: Quaternary Models', *Journal of African Earth Sciences*, 12: 1–7.

Frostick, L. E., Renaut, R. W., Reid, I., and Tiercelin, J. J. (1986) (eds.), *Sedimentation in the African Rifts* (London).

Garfunkel, Z. (1988), 'Relation between Continental Rifting and Uplifting: Evidence from the Suez Rift and Northern Red Sea', *Tectonophysics*, 150: 33–49.

Gass, I. G., Chapman, D. S., Pollack, H. N., and Thorpe, R. S. (1978), 'Geological and Geophysical Parameters of Mid-plate Volcanism', *Philosophical Transactions of the Royal Society of London*, 288A: 581–96.

Gilchrist, A. R. and Summerfield, M. A. (1990), 'Differential Denudation and Flexural Isostasy in Formation of Rifted-margin Upwarps', *Nature*, 346: 739–42.

—— —— (1991), 'Denudation, Isostasy and Landscape Evolution', *Earth Surface Processes and Landforms*, 16: 555–62.

—— —— (1994), 'Tectonic Models of Passive Margin Evolution and their Implications for Theories of Long-term Landscape Development', in M. J. Kirkby (ed.), *Process Models and Theoretical Geomorphology* (Chichester), 55–84.

Haack, U. (1976), 'Rekonstruktion der Abkuhlungsgeschichte des Damara-Orogens in Sudwest-Afrika mit Hilfe von Spaltspuren-Altern', *Geologisches Rundschau*, 65: 967–1002.

Haq, B. U., Hardenbol, J., and Vail, P. R. (1987), 'Chronology of Fluctuating Sea-levels since the Triassic', *Science*, 235: 1156–67.

Hartnady, C. J. H. (1985), 'Uplift, Faulting, Seismicity, Thermal Spring and Possible Incipient Volcanic Activity in the Lesotho–Natal Region, S-E Africa', *Tectonics*, 4: 371–7.

—— and le Roex, A. P. (1985), 'Southern Ocean Hot Spot Tracks and the Cenozoic Absolute Motion of the African, Antarctic, and South American Plates', *Earth and Planetary Science Letters*, 75: 245–57.

Hill, R. I. (1991), 'Starting Plumes and Continental Breakup', *Earth and Planetary Science Letters*, 104: 398–416.

Huser, K. (1989), 'Die Sudwestafrikanische Randstufe', *Zeitschrift für Geomorphologie Supplementband*, 74: 95–110.

Keen, C. E. and Beaumont, C. (1990), 'Geodynamics of Rifted Continental Margins', in M. J. Keen and G. L. Williams (eds.), *Geology of the Continental Margin of Eastern Canada*, Geology of Canada, no. 2 (Ottawa), 391–472.

King, L. C. (1955), 'Pediplanation and Isostasy: An Example from South Africa', *Quarterly Journal of the Geological Society*, 111: 353–9.

—— (1967), *The Morphology of the Earth*, 2nd edn. (Edinburgh).

Kogbe, C. A. and Burollet, P. F. (1990), 'A Review of Continental Sediments in Africa', *Journal of African Earth Sciences*, 10: 1–25.

Lambiase, J. J. (1989), 'The Framework of African Rifting during the Phanerozoic', *Journal of African Earth Sciences*, 8: 183–90.

le Roex, A. P. (1986), 'Geochemical Correlation between Southern African Kimberlites and South Atlantic Hotspots', *Nature*, 324: 243–5.

Lister, G. S., Etheridge, M. A., and Symonds, P. A. (1986), 'Detachment Faulting and the Evolution of Passive Continental Margins', *Geology*, 14: 246–50.

McCauley, J. F., Schaber, G. G., Breed, C. S., Grolier, M. J., Haynes, C. V., Issawi, B., Elachi, C., and Blom, R. (1982), 'Subsurface Valleys and Geoarcheology of the Eastern Sahara Revealed by Shuttle Radar', *Science*, 218: 1004–20.

McDonough, K. J. and Cross, T. A. (1991), 'Late Cretaceous Sea-level from a Paleoshore', *Journal of Geophysical Research*, 96: 6591–607.

McKenzie, D. (1978), 'Some Remarks on the Development of Sedimentary Basins', *Earth and Planetary Science Letters*, 40: 25–32.

Nelson, R. A., Patton, T. L. and Morley, C. K. (1992), 'Rift-segment Interaction and its Relation to Hydrocarbon Exploration in Continental Rift Systems', *American Association of Petroleum Geologists*, 76: 1153–69.

Nyblade, A. A. and Robinson, S. W. (1994), 'The African Superswell', *Geophysical Research Letters*, 21: 765–8.

Obst, E. and Kayser, K. (1949), *Die Grosse Randstufe auf der Ostseite Sudafrikas und ihr Vorland* (Hamburg).

Omar, G. I., Steckler, M. S., Buck, W. R., and Kohn, B. P. (1989), 'Fission-track Analysis of Basement Apatites at the Western Margin of the Gulf of Suez Rift, Egypt: Evidence for Synchroneity of Uplift and Subsidence', *Earth and Planetary Science Letters*, 94: 316–28.

Partridge, T. C. and Maud, R. R. (1987), 'Geomorphic Evolution of Southern Africa Since the Mesozoic', *South African Journal of Geology*, 90: 179–208.

Pavoni, N. (1992), 'Rifting of Africa and Pattern of Mantle Convection beneath the African Plate', *Tectonophysics*, 215: 35–53.

Pollack, H. N., Gass, I. G., Thorpe, R. S., and Chapman, D. S. (1981), 'On the Vulnerability of Lithospheric Plates to Mid-plate Volcanism: Reply to Comments by P. R. Vogt', *Journal of Geophysical Research*, 86: 961–6.

Pugh, J. C. (1955), 'Isostatic Readjustment in the Theory of Pediplanation', *Quarterly Journal of the Geological Society*, 111: 361–9.

Rabinowitz, P. D., Coffin, M. F., and Falvey, D. (1983), 'The Separation of Madagascar and Africa', *Science*, 220: 67–9.

Reeves, C. V., Karanja, F. M. and MacLeod, I. N. (1987), 'Geophysical Evidence for a Failed Jurassic Rift and Triple Junction in Kenya', *Earth and Planetary Science Letters*, 81: 299–311.

Robert, C. (1987), 'Clay Mineral Associations and Structural Evolution of the South Atlantic: Jurassic to Eocene', *Palaeogeography, Palaeoclimatology, Palaeoecology,* 58: 87–108.

Rosendahl, B. R. (1987), 'Architecture of Continental Rifts with Special Reference to East Africa', *Annual Review of Earth and Planetary Sciences,* 15: 445–503.

Rust, D. J. and Summerfield, M. A. (1990), 'Isopach and Borehole Data as Indicators of Rifted Margin Evolution in Southwestern Africa, *Marine and Petroleum Geology,* 7: 277–87.

Sahagian, D. L. (1987), 'Epeirogeny and Eustatic Sea-level Changes as Inferred from Cretaceous Shoreline Deposits', *Journal of Geophysical Research,* 85: 3711–39.

—— (1988), 'Epeirogenic Motions of Africa as Inferred from Cretaceous Shoreline Deposits', *Tectonics,* 7: 125–38.

Shackleton, R. M. (1978), 'Structural Development of the East African Rift System', in W. W. Bishop (ed.), *Geological Background to Fossil Man* (Edinburgh), 19–28.

Sponemann, J. and Brunotte, E. (1989), 'Zur Reliefgeschichte de sudwestafrikanischen Randschwelle zwischen Huab und Kuiseb', *Zeitshcrift fur Geomorpholgie Supplementband,* 74: 111–25.

Stüwe, K. (1991), 'Flexural Constraints on the Denudation of Assymetric [*sic*] Mountain Belts', *Journal of Geophysical Research,* 96: 10 401–8.

Summerfield, M. A. (1981), 'Comments on "On the Vulnerability of Lithospheric Plates to Mid-plate Volcanism": Reply to Comments by P. R. Vogt, by H. N. Pollack, I. G. Gass, R. S. Thorpe, and D. S. Chapman', *Journal of Geophysical Research,* 88: 1248–50.

—— (1985*a*), 'Tectonic Background to Long-term Landform Development in Tropical Africa', in I. Douglas and T. Spencer (eds.), *Environmental Change and Tropical Geomorphology* (London), 281–94.

—— (1985*b*), 'Plate Tectonics and Landscape Development on the African Continent', in M. Morisawa and J. T. Hack (eds.), *Tectonic Geomorphology* (Boston), 27–51.

—— (1991*a*), 'Subaerial Denudation of Passive Margins: Regional Elevation Versus Local Relief Models', *Earth and Planetary Science Letters,* 102: 460–9.

—— (1991*b*), *Global Geomorphology* (London and New York).

Ten Brink, U. and Stern, T. (1992), 'Rift Flank Uplifts and Hinterland Basins: Comparison of the Transantarctic Mountains with the Great Escarpment of Southern Africa', *Journal of Geophysical Research,* 97: 569–85.

Thomas, D. S. G. and Shaw, P. A. (1988), 'Late Cenozoic Drainage Evolution in the Zambezi Basin: Geomorphological Evidence from the Kalahari Rim', *Journal of African Earth Sciences,* 7: 611–18.

—— —— (1990), 'The Deposition and Development of the Kalahari Group Sediments, Central Southern Africa', *Journal of African Earth Sciences,* 10: 187–97.

Tricart, J. (1972), *The Landforms of the Humid Tropics, Forests and Savannas* (London).

Uchupi, E. and Emery, K. O. (1991), 'Pangaean Divergent Margins: Historical Perspective', *Marine Geology,* 102: 1–28.

Visser, J. N. J. (1987), 'The Palaeogeography of Part of Southwestern Gondwana during the Permo-Carboniferous Glaciation', *Palaeogeography, Palaeoclimatology, Palaeoecology,* 61: 205–19.

—— (1989), 'The Permo-Carboniferous Dwyka Formation of Southern Africa: Deposition by a Predominantly Subpolar Marine Ice Sheet', *Palaeogeography, Palaeoclimatology, Palaeoecology,* 70: 377–91.

White, R. S. and McKenzie, D. (1989), 'Magmatism at Rift Zones: The Generation of Volcanic Continental Margins and Flood Basalts', *Journal of Geophysical Research,* 94: 7685–729.

2 The African Rift System

Celia K. Nyamweru

Introduction

The volume of literature related to the East African Rift System is immense. Mohr (1991: 17–21) lists 113 early sources published during the years 1834 to 1950, while in 1987 Rosendahl estimated that a full review of African rifting might demand a citation list of fifty printed pages (1987: 445). One might therefore be forgiven for assuming that any new contribution on the African rift valleys will be so technical that it will be comprehensible only to an audience of specialists, or that everything of value on this topic has already been said. Although it is true that many of the earlier models of the African Rift System have been revised, new questions have arisen, and the debate on its structure, evolution, and significance remains as lively as ever.

This chapter provides an introduction to what is arguably Africa's most interesting continental-scale landform. After a brief description of the extent and topography of the African Rift System, an historical account of the most important contributions to our knowledge of this feature is presented, with emphasis on the kinds of data collected and the analytical techniques employed. Regional studies serve to illustrate some of the more interesting modern research initiatives. The chapter ends with a summary of current opinions on rift processes and structures. The list of works cited provides interested readers with an entrance to the more specialized literature. Let them be warned: the rift may be spreading slowly, but the volume of paper published about it is spreading rapidly indeed.

Overview of the East African Rift System

The Cenozoic rift system of Eastern Africa extends from the Afar Depression in the north to beyond Lake Malawi in the south, a distance of about 5600 km (Figure 2.1). Close to the Equator it is made up of eastern and western rifts to either side of the Lake Victoria Basin (Figure 2.3). Different terminologies have been used in the description of this feature, for example East or Eastern African Rift System, Gregory or Kenya Rift, Western Rift or Western Branch. Despite attempts to standardize the terminology (Rosendahl, 1987), current usage shows considerable variation (Rosendahl *et al.*, 1992: 236). Rosendahl's terminology of eastern and western branches is broadly similar to the eastern and western rifts of many other authors. Rosendahl also defines rift zones: rift sectors circa 500–700 km long whose longitudinal boundaries generally coincide with offsets or changes in rift trend (1987: 451). The regional description that follows is organized according to rift zones, though with some divergence from the zones defined by Rosendahl.

The northernmost zone of Cenozoic rifting in Eastern Africa is the *Afar Depression* (Figures 2.1 and 2.2). It is a broadly triangular feature that extends north–south for over 600 km, from the Gulf of Zula on the Red Sea to the Main Ethiopian Rift at approximately 9°N. It extends about 400 km from east to west, inland from Djibouti on the Gulf of Aden. The Afar Depression is

I acknowledge the generous help of Barry Dawson, Jane Eaton, Cindy Ebinger, Nils Ekfelt, Bill Elberty, Jim Heirtzler, Paul Mohr, Martin Smith, Manfred Strecker, and Cathy Shrady in the preparation of this paper.

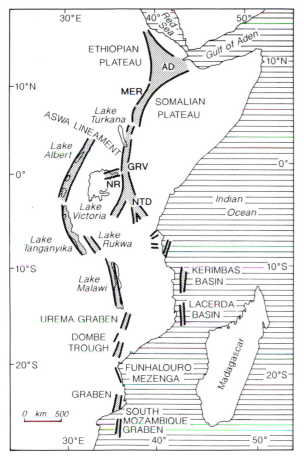

Fig. 2.1. Overview of the rift system in Eastern Africa.
Shading shows the extent of the rift system and some associated features. AD = Afar Depression; MER = Main Ethiopian Rift; GRV = Gregory Rift Valley; NR = Nyanza Rift; NTD = North Tanzania Divergence.

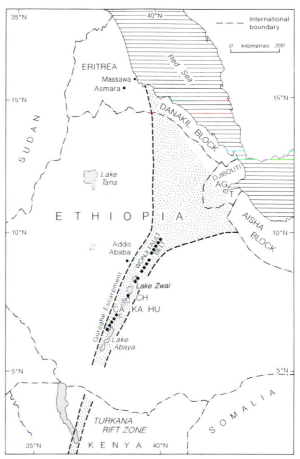

Fig. 2.2. The rift system in Ethiopia, Eritrea, and Djibouti.
The approximate alignment of the rift system is shown. T = Gulf of Tadjura; AG = Asal–Ghoubbet Rift; CH = Chike; KA = Kaka; HU = Hunkuolo; CA = Corbetti–Awasa.

bounded by major fault scarps rising 3000 m to the crest of the Ethiopian plateau in the west and 2000 m to the plateau in the south. To the north the Afar Depression is bounded by the Danakil Block, an uplifted ridge of ancient metamorphic rock that separates it from the Red Sea, while the Aisha Block separates it from the Gulf of Aden to the east. The floor of the Afar Depression slopes from about 1000 m above sea-level in the south to 120 m below sea-level in the north (Baker *et al.*, 1972: 5). The landscape is highly varied and includes deeply eroded marginal terrain, freshly formed small fault blocks, sedimentary basins (some containing saline lakes), and a variety of volcanic landforms ranging in age from the Miocene to the present day. Baker *et al.* (1972: 19) estimated the volume of Tertiary and Quaternary volcanic rocks of Ethiopia at about 350 000

km^3, including those of the Afar Depression, the Main Ethiopian Rift (below), and the surrounding plateaux. Seismic and volcanic activity continue today in parts of the Afar Depression, for example in the Asal–Ghoubbet Rift (see Regional Study 1).

South of 9°N the Afar Depression passes into the *Main Ethiopian Rift* (Figures 2.1 and 2.2). This graben structure, 60–70 km wide, trends south-west in the north and south-south-west in the south. In places the rift is defined by spectacular fault scarps, for example the Guraghe Escarpment, 1000 m high, that borders it to the west between 8 and 8.5°N. Elsewhere the rift margins are less well defined; for example east of Addis Ababa they take the form of a broad, gentle downwarp. The rift floor reaches its highest elevation of 1700 m to the north of Lake Zway at about 8.25°N.

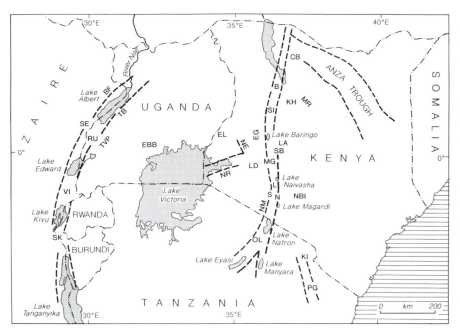

Fig. 2.3. The rift system in Kenya, Uganda, and northern Tanzania.

The approximate alignment of the rift system is shown. BF = Bunia Fault; TB = Toro–Bunyoro Fault; SE = Semliki Basin; RU = Ruwenzori; TVP = Toro–Ankole Volcanic Province; VI = Virunga Volcanic Province (includes Nyamuragira and Nyiragongo); SK = South Kivu Volcanic Province; CB = Chalbi Basin; B = The Barrier; SI = Silali; KH = Karisia Hills; MR = Mathews Range; EBB = Entebbe; EL = Mount Elgon; NE = Nandi Escarpment; NR = Nyanza Rift; EG = Elgeyo Escarpment (the Kerio Valley and the Tugen Hills lie between the Elgeyo Escarpment and Lake Baringo); SB = Subukia; LA = Laikipia Escarpment; MG = Menengai; LD = Londiani; L = Longonot; S = Suswa; N = Ngong Hills; NBI = Nairobi; OL = Ol Doinyo Lengai; KI = Kilimanjaro; PG = Pangani Graben.

The Main Ethiopian Rift is bounded to east and west by plateaux built of basalts up to 4000 m thick, the products of extensive fissure eruptions that began during the Eocene. Central volcanoes of different ages stand on the rift floor and shoulders. Among them are Pliocene trachyte volcanoes on the plateau east of the rift, including Kaka, Chilalo, and Hunkuolo (Woldegabriel *et al.*, 1990: 441). Quaternary caldera volcanoes on the rift floor include Awasa, O'a, Gademotta, and Corbetti that show a wide range of rock composition from basalts to rhyolites and obsidians.

Sedimentary basins on the rift floor, separated by volcanic and tectonic landforms, contain mostly shallow alkaline lakes. The rift floor is broken by closely spaced longitudinal faults, for example the Wonji Fault Belt, a zone 2–5 km wide of closely spaced recent faults and fissures that extends southwards from the Afar Depression along much of the Main Ethiopian Rift.

Though topographically not as spectacular as other sectors, the junction between the Ethiopian and Kenyan Rifts is one of the most interesting and controversial sectors of the rift. The main zone of rifting crosses a broad topographic low between the Ethiopian and Kenyan Highlands. The surface level of Lake Turkana is about 375 m above sea-level, while the Chalbi Basin to the east lies at about 370 m. Across this depression the *Turkana Rift Zone* (Figure 2.2) consists of a broad and complex set of troughs, trending approximately north–

south, that recent research has shown are of different ages, the locus of rifting in this area having shifted eastward with time (Cerling and Powers, 1977; Woldegabriel and Aronson, 1987).

The *Gregory Rift Zone* (Figure 2.3) runs from approximately 2.5°N to 2.5°S, including the whole of the Kenya Rift (south of Lake Turkana) as well as the Lake Natron Basin in northern Tanzania, a total length of about 1000 km. West of the Gregory Rift the *Nyanza Rift Zone* trends first east–west and then north-east–south-west for a total distance of about 130 km before disappearing beneath the waters of Lake Victoria (Pickford, 1982). The Gregory Rift trends approximately north–south, with local variations to both east and west of the dominant trend. It is approximately 60–70 km wide, including complexly faulted marginal platforms. Between the marginal structures the width of the inner graben floor varies from 17 to 35 km (Baker *et al.*, 1972: 26–7). West of Lake Baringo, at about 0.5 °N, the single graben is replaced by a double structure: the Baringo Graben is separated by the fault block of the Tugen Hills from the Kerio Valley farther west. One of the most impressive rift scarps, the 1500 m Elgeyo Escarpment, forms the western wall of the Kerio Valley. A comparable fault scarp, the Nguruman Escarpment, forms the western wall of the rift west of Lake Magadi. In many places the rift floor is cut by swarms of closely spaced normal faults (Plate 2.1), with throws of less than 150 m. Within the Gregory Rift are

Plate 2.1. Small faults arranged *en echelon* in the floor of the Eastern Rift Valley, south of Lake Baringo. (C. K. Nyamweru photograph.)

Plate 2.2. View to the south across saline Lake Magadi. Small fault scarps border the lake to east and west; in the background rise the volcanoes Shombole and Ol Doinyo Sambu. The Nguruman Escarpment is visible on the extreme right of the photograph. (Kenya Information Services photograph.)

several sedimentary basins containing shallow lakes ranging from hypersaline Lake Magadi (Plate 2.2) to freshwater Lake Naivasha and Lake Baringo. The salinity of the lakes is related to the elevation (and thus to the climate) of their basins; the rift floor reaches its highest level of about 1890 m in the Lake Naivasha Basin and falls to about 580 m in the Lake Magadi Basin. The loss of substantial amounts of water by subsurface seepage also explains why Lake Naivasha (salinity 0.245 p.p. thousand) and Lake Baringo (salinity 0.7 p.p. thousand) (S. Njuguna, personal communication) are significantly more dilute than Lake Nakuru (salinity 45

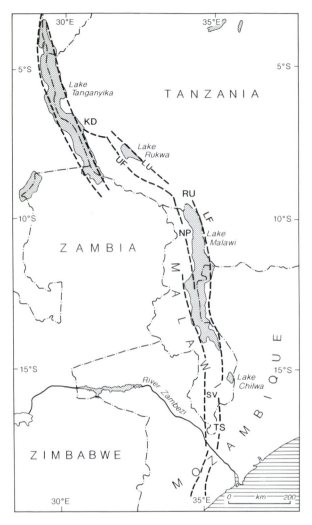

Fig. 2.4.   The rift system in Tanzania and Malawi.

The approximate alignment of the rift system is shown. LU = Lupa Scarp; UF = Ufipa Scarp; KD = Karema Depression; RU = Rungwe Volcanic Province; LF = Livingstone Fault; NP = Nyika Plateau; SV = Shire Valley; TS = Thyolo Scarp.

central volcanoes. Along the rift floor are a number of central volcanoes, including the dominantly trachytic Quaternary calderas Menengai, Longonot, Suswa, Silali, and The Barrier. Most recent eruptive activity probably took place on The Barrier in the 1880s and 1890s, but geothermal activity is widely distributed (Clarke *et al.*, 1990; Dunkley *et al.*, 1993). Miocene to Pliocene centres on the rift margins have been deeply eroded and downfaulted by later tectonism; examples are the Ngong Hills and Mount Londiani. Several large central volcanoes are located 50 km or more from the edge of the present-day rift, notably Mount Elgon, Mount Kenya, and Kilimanjaro.

A change in tectonic style and magma type takes place at about 2°S (Dawson, 1992). The rift loses its well-defined graben expression and the bounding faults are replaced by splay faults that separate tilted crustal blocks. This zone is described by Baker *et al.* as the *North Tanzania Divergence* (1972: 25). Farther south the faults die out in the plateau of the Tanzanian Craton, and this has been considered as the southward limit of the rift to the east of Lake Victoria. But fault structures continue to the south-east, through the Pangani Graben, and recent work has suggested a link between the faults of north-eastern Tanzania and the Kerimbas and Lacerda Basins of the Indian Ocean, implying that the continental rifting is continuous with the oceanic spreading ridges that separated Madagascar from the rest of Africa (Mougenot *et al.*, 1986, 1989).

To the west of Lake Victoria, the East African Rift System makes a broad curve along the western border of the Archaean Craton (Figures 2.3 and 2.4). In the north it trends south-west, becoming north–south at the Equator before taking up a south-easterly trend at about 6°S. Recent mapping of fault systems along this rift has suggested that it may be divided into as many as thirty-two basins, each of the order of 100 km long (Ebinger, 1989: 899). Magmatism along the Western Rift is restricted to four minor volcanic provinces: Toro-Ankole, Virunga, and South Kivu, which lie to the north of Lake Tanganyika, and the Rungwe volcanic province at the northern end of Lake Malawi.

The most northerly rift zone is the *Albert Rift Zone*, which extends from north of Lake Albert (Mobutu) to the South Kivu volcanic province (Rosendahl, 1987: 452). In the north, broadly north-east-trending faults define the West Nile Basin and control the course of the River Nile as it flows out of the northern end of Lake Albert. These structures intersect at about 4°N with a north-west-trending Precambrian structure, the Aswa Lineament. The Lake Albert Basin is bounded to the north-west by the Bunia Fault, an active 1300 m fault

p.p. thousand) and Lake Elmenteita (salinity 43 p.p. thousand) (Beadle, 1974: 262, 265) which lose water only by surface evaporation.

The formation of the Gregory Rift was accompanied by the production of large volumes of volcanic rock. Estimated volumes are not as large as in Ethiopia, but figures of over 200 000 km^3 have been suggested (Baker, 1986: 48). A wide range of rock compositions are represented with basalts less dominant than in Ethiopia. The Miocene plateau phonolites of Kenya have no equivalents in the Ethiopian volcanic succession. Volcanic landforms include lava and ignimbrite plateaux and

Plate 2.3. The Toro-Bunyoro fault scarp to the east of Lake Albert in the Western Rift Valley. (W. W. Bishop photograph.)

scarp, and to the south-east by a smaller scarp, the Toro-Bunyoro Fault (Figure 2.3) (Plate 2.3). South of Lake Albert the Western Rift contains the Semliki Basin to the west and the upthrust Ruwenzori Block to the east. The Ruwenzori Block is composed of Precambrian crystalline rock, and its eroded glacial peaks reach a maximum elevation of 5119 m. This section of the rift is highly seismic; minor tremors are frequent and two strong earthquakes have been recorded in the last thirty years. Estimates of M for the Toro earthquake of March 1966 range from 6.7 to 7.0 (Loupekine, 1971: 57), and in February 1994 an earthquake of M 6.2 was recorded (Smithsonian Institution, 1994: 11). Southwards the rift passes into the Lake Edward Basin, and then into the Virunga volcanic province. The Virunga Range is composed of basaltic shield volcanoes ranging in age from Miocene to Recent; two of them (Nyamuragira and Nyiragongo) have been the sites of frequent activity this century.

South of the Virunga volcanic province, the *Lake Tanganyika Rift Zone* contains at least ten separate basins. Rosendahl describes it as the 'archetypical continental rift' (1987: 452) with its major fault throws and striking half-graben structure (see Regional Study 5). Throws of major faults along the rift margins are up to 6000 m, and metamorphic basement lies below sea-level beneath many of the rift basins (Ebinger, 1989: 887). Major faulting has taken place in the total absence of magmatism, raising fundamental questions about the relationship of tectonism and volcanism in rift forma-

tion (see concluding section). The southern part of the Lake Tanganyika Rift Zone is paralleled, 50–100 km to the north-east, by the *Rukwa Rift Zone*. The Rukwa Basin is about 350 km long and 50 km wide and is bounded to the west by the Ufipa Scarp, a steep scarp over 1000 m high, and to the east by the lower, gentler Lupa Scarp (Mbede, 1993). The connection between the Rukwa and Tanganyika Rift Zones appears to follow oblique slip faults through the Karema Depression. To the south-east, the Rukwa Rift Zone passes into the Rungwe volcanic province at the northern end of Lake Malawi (Figure 2.4). Volcanicity here probably began during the late Miocene (Ebinger et al., 1989: 15 801), and the most recent eruptions may have taken place less than 200 years ago (Harkin, 1960).

The *Lake Malawi Rift Zone* includes both the lake basin and the Shire Valley to the south of the lake. It covers a distance of about 800 km and ranges in width from 40 to 90 km. The Malawi Rift Zone shows half-graben structures similar in size to those of Lake Tanganyika, with up to nine basins along its length (Rosendahl et al., 1992; 245). Basin subsidence, flank uplift, and sediment accumulation are greatest in the north and decrease southward. One of the largest fault scarps is the Livingstone Fault on the north-east side of the lake, with a throw of over 4500 m. On the west side of the lake the land rises in a series of step faults to the Nyika Plateau, whose summit is at an altitude of 2606 m. South of the lake, the eastern wall of the Shire Valley is formed by the Thyolo scarp, a north-west-trending fault with a throw of up to 1000 m (Crossley and Crow, 1980: 81).

South of Lake Malawi, fault structures continue for a distance of about 600 km through the Urema Graben and the Dombe Trough (Crossley and Crow, 1980: 77). Fault troughs still farther south have been identified, namely the Funhalouro Mazenga Graben and the Cretaceous Southern Mozambique Graben (Mougenot *et al.*, 1989: 403), but the relationship of these features to the Cenozoic faults of the East African Rift System is uncertain.

Research on the East African Rift System

Rift research methods may be broadly divided into geological and geophysical, with many subdivisions of each category. Many of the modern team projects are multidisciplinary in their approach. The earliest earth scientists to study the rift were geologists, beginning with reconnaissance surveys during the last decades of the nineteenth century. Among English-speaking scholars at least, J. W. Gregory, who in 1893–4 made an epic journey through the central Kenya Rift Valley and to Mount Kenya, is often considered to be the founder of African rift studies. However Gregory was not the first; Suess' great work *Die Brueche des oestlichen Afrika* (1891) predates Gregory's *The Great Rift Valley* (1896) by five years. Although he never set eyes on the features he described, Suess correctly interpreted the African Rift System as the result of crustal tension. During the late nineteenth and early twentieth centuries a number of other geologists carried out field-mapping and collected rock specimens in Eastern Africa, including many German scholars in German East Africa (now Tanzania) and Italians in Ethiopia (Mohr, 1991: 13–14).

Systematic geological mapping of much of eastern Africa was carried out between the 1920s and 1960s, mostly by geologists employed in colonial service, and the results were initially published in regional geological reports of the different territories. Several individuals made significant contributions to rift studies. Among them were B. H. Baker, L. A. J. Williams, and R. M. Shackleton in Kenya; E. J. Wayland, R. B. McConnell, and J. W. Pallister in Uganda; E. O. Teale, T. C. James, and J. B. Dawson in Tanganyika; F. Dixey and M. S. Garson in present-day Malawi; and G. Merla, G. Dainelli, and later P. Mohr in Ethiopia. Debate on tensional versus compressional origins for the rift system was active during the 1920s to 1940s, with Wayland (1923), Bullard (1936), and Willis (1936) supporting the compressional hypothesis that lost ground as detailed

mapping showed normal faulting to be widespread throughout the rift system. Work of this era also focused on the recognition of altitudinally consistent erosion surfaces and on tracing the disruption of former peneplains by tectonic and erosional processes (Shackleton, 1950; Dixey, 1956; Saggerson and Baker, 1965). One early line of research that is still current links Archaean and Precambrian structures and lineaments with Mesozoic and Cenozoic rift faulting (Dixey, 1956; McConnell, 1967, 1972). In general, however, papers written before the mid-1960s lack radiometric dates and the interpretations of rift origins lie outside the plate tectonic theory, rendering their conclusions of largely historic interest. A summary of the state of knowledge in 1965 is provided by UMC-UNESCO (1965), which includes a short article by Heezen comparing the East African rifts with the mid-oceanic ridges. A useful foundation to modern rift studies is the overview by Baker, Mohr, and Williams (1972). Research into the East African Rift System as the site of early hominid evolution added to our understanding of rift basins and the sediments within them; many useful papers were included in the volumes edited by Howell and Bourliere (1963); Bishop and Clark (1967); Bishop and Miller (1972); and Bishop (1978).

Field-mapping has continued to yield important information about rift lithology and structures. Mapping of much of the northern Gregory Rift Zone was carried out in the 1960s and 1970s under the auspices of the East African Geological Research Unit directed by B. C. King and W. W. Bishop. In particular, this work contributed significantly to our knowledge of the Baringo sector of the Gregory Rift and of the large caldera volcanoes in the rift floor (Martyn, 1967; King, 1978; Chapman *et al.*, 1978; Weaver, 1976/7; Carney, 1972; McClenaghan, 1971; Tallon, 1976; Truckle, 1976; Golden, 1978). A major mapping project, the Samburu-Marsabit Geological Mapping and Mineral Exploration Project, was carried out jointly by the British Geological Survey and the Kenya Department of Mines and Geology between 1980 and 1986. The project mapped an area of over 100 000 km^2 on a scale of 1 : 250 000, and published nine geological reports and maps. Members of the project team have also published a number of scientific papers (Key *et al.*, 1987, 1989; Hackman *et al.*, 1990). Between 1984 and 1992 the British Geological Survey, jointly with the Kenya Ministry of Energy, carried out a programme of regional geothermal-resource assessment along the Gregory Rift (Clarke *et al.*, 1990; Dunkley *et al.*, 1993; Smith and Mosley, 1993; Smith, 1994). Field-mapping has continued in Ethiopia, for example the Italian–French collaboration between Consiglio Nationale delle Ricerche (CNR) and Centre National de la Recherche

Scientifique (CNRS) that led to the compilation of a 1 : 500 000 geological map of the eastern Afar Depression (1971, 1975), and the work of Woldegabriel *et al.* (1990). In the Western Rift recent field-mapping has been completed by Ebinger *et al.*, 1989, and by Mbede (1993). Modern geologists have access to remotely sensed data, in particular images from the Landsat Thematic Mapper, Système Probatoire de l'Observation de la Terre (SPOT), and the Multispectral Scanner (MSS), which may provide information about structures not visible on the ground; for example, the Marigat Lineament south of Lake Baringo (Hackman *et al.*, 1990: 204).

Geological fieldwork in rift zones also involves detailed studies of faulted terrain, including geodetic measurement, well suited to areas where rapid spreading or uplift is occurring. In parts of the Afar Depression and the Main Ethiopian Rift, trilateration nets were established in the late 1960s and early 1970s and resurveyed in 1979 and again in 1984–5 and 1992. High-precision levelling lines were also surveyed (Ruegg and Kasser, 1987; Stein *et al.*, 1991; Regional Study 1). Elsewhere in the rift, rates of movement appear too slow for such methods to be widely applicable. Fault movements are known to have occurred in the Gregory Rift along the base of the Laikipia Escarpment, associated with the Subukia earthquake of January 1928. Field evidence of these movements has been analysed by Ambraseys (1991) and by Doser and Yarwood (1991). Stress-field analysis is another approach to regional tectonics that is yielding valuable results (Regional Study 3).

Petrology has also made great contributions to rift studies. For example, a concise introduction to rift valley magmatism is given by Dawson in the volume edited by Duff (1993: 682–5), and Wilson focuses attention on the East African Rift System in her discussion of continental rift-zone magmatism (1989: 325–74). She summarizes the analytical techniques that have been applied to rift volcanics, among them major-element geochemistry, trace-element geochemistry, analysis of rare-earth elements (REE) and of radiogenic and stable isotopes. Although more data are becoming available, the fundamental questions of magma generation in the continental rift zone – the nature of the magma source and the contribution of lithospheric melt versus asthenospheric sources – remain highly controversial. Comparisons of rift volcanics with rocks from other regions are made with the aid of geochemical and isotopic studies. Hart *et al.* (1989), from analyses of over 150 samples of Oligocene to Pleistocene lavas from the Main Ethiopian Rift and the western Afar Depression, have suggested that three isotopically distinct mantle sources may have contributed to the volcanism of this region at different times since the Oligocene. During the Oligocene and Miocene, there was little difference between the magma sources beneath the Main Ethiopian Rift and those of the Afar Depression, but from about 4–5 Ma BP onwards a difference shows, as continental breakup has proceeded farther in the Afar. No crustal or lithospheric source material may now remain below the Afar Depression, although lithospheric material continues to contribute significantly to the volcanic rocks erupted in the Main Ethiopian Rift (Hart *et al.*, 1989: 7744). Despite these results, the debate over oceanic versus continental crust on the floor of the Afar Depression remains lively (Mohr, 1989, 1992). Increasing attention is being paid to the metamorphic rocks of the rift margins, and the application of apatite fission-track analysis to the rocks of eastern Kenya has yielded provocative information about landscape evolution (Regional Study 4).

Geophysics has contributed greatly to understanding of the rift system. One of the earliest techniques applied was gravity measurement. Kohlschuetter (cited by Bullard, 1936 and Mohr, 1991: 16) was probably the first to carry out gravity measurements in East Africa, during 1899–1900; his recognition of a narrow belt of negative gravity anomalies along the Eastern Rift was supported by the later measurements of Bullard, and explained by the latter in terms of a compressional origin for the rift (Bullard, 1936: 513–17). Today many thousands of gravity measurements have been made across Eastern Africa, and detailed Bouguer anomaly maps and profiles are available that show a long-wavelength negative Bouguer gravity anomaly across the rift zone. This may be the result of hot, low-density material within the upper mantle. The situation is more complex, however, than is implied by the broad picture of topography and gravity values. Both positive- and negative-gravity anomalies characterize particular rift sectors, depending on whether basins are filled with basalt or with sedimentary rock. Recent work by Nyblade and Pollack presents a gravity model for western Kenya and north-eastern Tanzania that explains some of the anomaly in terms of the suture zone between the Tanzanian Craton and the rocks of the Mozambique Belt (1992: 265; Regional Study 2).

Seismology has also been widely applied to the African Rift System. Most earthquake data for this region during the early twentieth century are largely descriptive, though a seismoscope was set up in Massawa (Figure 2.2) as early as 1886 (Gouin, 1979: 44). Original records from this instrument, as from the seismograph that operated briefly in Asmara in 1913, are lost (ibid.: 241). A seismograph was operating at Entebbe (Uganda) between 1925 and 1932, and intermittently up to 1964 (Loupekine, 1965: 85). Loupekine (1971) compiled a record of earthquakes in Kenya between 1892 and 1969, and Gouin (1979) compiled one for Ethiopia that goes back several centuries. Permanent seismographic stations were established in Addis Ababa in 1958, in Nairobi in 1963, and in Djibouti in 1973. The observation and analysis of natural earthquakes (both teleseismic and microseismic) continues. Kebede and Kulhanek studied the distribution of seismic-wave energy released along the East African Rift System between 1963–89, and concluded that there was a seismicity gap between the Eastern Rift in Kenya and the southernmost rifts of Ethiopia (1991: 270). Maguire *et al.* (1988) provide a concise summary of earlier work on the seismicity of Kenya. Young *et al.* operated a fifteen station short-period seismic array for three months near Lake Bogoria in the Kenya Rift and recorded 572 microearthquakes, 81 per cent of them of M < 1.0. Most of these tremors appeared to be associated with large old faults on the rift shoulder and to originate at depths of less than 12 km (1991: 671–2). Molnar *et al.* (1970) have studied microearthquakes in Ethiopia, while other work on the seismicity of the Ethiopian Rift and the Afar Depression has been done by Abdallah *et al.* (1979).

Over the last twenty years, explosion seismology has become an important source of data on subsurface conditions, though it cannot fully replace the study of natural earthquakes, in particular as a source of information about upper mantle and deep crustal structures. Data obtained by both refraction and reflection seismology permit the creation of depth-velocity models which may be interpreted in terms of rock density, rigidity, temperature, and chemical composition. Seismic reflection studies provided knowledge on the Western Rift that contributed to the development of the half-graben model of rift structure (Regional Study 5).

Seismic refraction studies were first carried out along the Kenya Rift in 1971 and since then much work has been done there, as in the Afar Depression (Berckhemer *et al.*, 1975; Ruegg, 1975). The Kenya Rift International Seismic Project (KRISP) carried out large-scale seismic investigations, beginning in 1985 (Khan *et al.*, 1989; Morgan, 1991; Green *et al.*, 1991; KRISP Working Party, 1991; Achauer *et al.*, 1992; Dindi, 1993). KRISP is a multidisciplinary project investigating the nature and thickness of the crust and the lithosphere/asthenosphere interface beneath the Kenya Rift Zone. KRISP's methods include seismic-refraction profiles along and across the rift, and a technique known as seismic tomography which locates seismic-wave velocity perturbations, where P and S waves travel faster or slower than the regional

average at a particular depth. The results of the seismic refraction experiments showed that at 0.5°N the crust is about 40 km thick beneath the western flank of the rift, thinning to 30 km beneath the graben in the region of Lake Baringo, and thickening to 35 km beneath the eastern flank. Northwards along the rift from Lake Baringo to Lake Turkana the crust decreases in thickness to 20 km (KRISP, 1991). The seismic tomography experiments provided information about teleseismic wave velocities to a depth of 145 km. The results show that at depths of 10–35 km in the lower crust, earthquake waves travel significantly faster beneath the rift axis than they do beneath its flanks (Green *et al.*, 1991; Achauer *et al.*, 1992). This has been explained as the result of dense mantle material intruded along a narrow zone into the lower crust. In the upper mantle (depths of 35–105 km) there are relatively *low* velocities beneath the rift compared with its flanks, which has been explained as the result of 3–6 per cent partial melt related to a thermally driven upwarp of the asthenosphere (Green *et al.*, 1991: 203). The low velocities can also be explained by chemically altered, anomalously low-density mantle beneath the rift. Dawson and Smith (1992) describe xenoliths of such mantle from north Tanzanian volcanics.

Other geophysical techniques applied to African rifts include heat-flow measurements and aeromagnetic surveys. Heat-flow measurements in Eastern and Southern Africa indicate that in general, heat flow on older Archaean cratons is significantly less than on Proterozoic mobile belts (Nyblade *et al.*, 1990). Aeromagnetic surveys (frequently carried out in association with mineral prospecting) can yield information about sediment thickness and depth to magnetized basement rock, and such surveys helped to establish the existence of the buried Anza Trough in north-east Kenya (Reeves *et al.*, 1986/7, Regional Study 4). Geomagnetic research has been applied to the study of the spreading ridges along the Red Sea and the Gulf of Aden (Drake and Girdler, 1964). Palaeomagnetic study of rocks of the Afar Depression indicates that the Danakil Block has rotated approximately 10° counterclockwise since the Miocene (Tapponnier *et al.*, 1990; Acton *et al.*, 1991).

Regional Study 1: The Asal–Ghoubbet Rift: Active Sea-Floor Spreading?

The Afar Depression is a triple junction formed by the convergence of three zones of active crustal extension: the Red Sea, the Gulf of Aden, and the East African Rift.

The western part of the Gulf of Aden, which opened during the last 10 Ma, is now propagating westwards into eastern Afar at a rate of about 30 mm/yr (Acton *et al.*, 1991: 504). The surface expression of this propagation is the Asal–Ghoubbet Rift, a north-west to south-east-trending depression that extends into Djibouti for a distance of about 15 km from the head of the Gulf of Tadjura (Figure 2.2). Topographically this rift is similar to the axial trough of a mid-ocean spreading ridge:

The rift valley is 11.5 km wide, with an inner rift (also called the rift-in-rift) of 7.5 km width. The rift is 300–800 m deep and has walls lined by steep normal faults with throws of up to 150 m. A 1- to 2-km wide neovolcanic zone lies along the rift axis, displaying several eruptive centers, young basalt flows, rift-axis fissures, and in many places a central volcanic chain or horst. (Stein *et al.*, 1991: 21 792)

In 1978 the Asal–Ghoubbet Rift was the site of a swarm of shallow-focus earthquakes and the eruption of olivine tholeiite lava, described by Abdallah *et al.* (1979) and Allard *et al.* (1979) as events typical of those occurring on undersea mid-ocean rifts. Geodetic measurements have determined the deformation of the crust during and after this event. Up to 1.9 m of extension across the rift were measured between 1973 and 1978, and vertical movements over the same period consisted of about 0.7 m of inner-rift subsidence and 0.2 m of flank uplift (Stein *et al.*, 1991: 21, 272).

Seismic and gravity research in different parts of the Afar Depression support the interpretation of the region as one of crustal extension and thinning. Deep seismic soundings give crustal thickness on the Ethiopian Plateau of the order of 38 km, decreasing from 26 to 16 km from south to north across the Afar Depression (Berckhemer *et al.*, 1975: 89). Crustal thicknesses of 6 to 10 km have been estimated for the Gulf of Tadjura (Ruegg, 1975: 120). Based on gravity and topography measurements, Hayward and Ebinger (1995) estimate the thickness of the mechanical lithosphere as ranging from 58 km below the Ethiopian Plateau to 5–7 km below the tectonically active zones within the Afar Depression, implying drastic heating of the lithosphere beneath this depression.

Despite our detailed knowledge of some aspects of structure and process in the Afar, many questions remain. There is general broad agreement that the Red Sea (and so probably also the Gulf of Aden and the Afar Depression) formed in two stages, but there is still disagreement on the relative magnitude and timings of these stages, for instance Mohr lists fourteen different models of the amount and duration of extension of the southern Red Sea Basin (1989: 3). There are different estimates of the

rate at which spreading is taking place at the present time. Controversy also surrounds the nature of the crust under the Afar Depression; one model is concisely summarized as 'new igneous, not thinned continental' (Mohr, 1989: 1). In contrast, Makris and Ginsburg interpret seismic refraction profiles for the Afar in terms of underlying continental crust (1987). Part of the Afar controversy stems from the lack of data from western Afar compared with the intensive research in Djibouti. Since the recent independence of Eritrea is leading to greater security in the region, research in western Afar may contribute to a more balanced picture of this fascinating region.

Regional Study 2: The Gregory Rift Zone; The Influence of Crustal Anisotropies on Cenozoic Rifts

The volcanic and tectonic history of the Gregory Rift has been studied in detail for nearly a century and its broad chronology is well established (Baker *et al.*, 1972; King, 1978). Updoming over central Kenya largely preceded volcanism and rifting (Smith and Mosley, 1993: 597). The earliest volcanism in central Kenya is identified as the eruption of flood basalts between 23 and 15 Ma BP. Major faulting occurred between 12 and 7 Ma BP, and again between about 4 and 2.6 Ma BP (Bosworth *et al.* (1992: 11 854). The Gregory Rift is still seismically active, and volcanic activity has occurred during the last one to two hundred years (von Hohnel, 1894: 219–21; Dunkley *et al.*, 1993).

Our knowledge of the pattern of volcanism and tectonism along the Gregory Rift Valley provides a solid foundation for more specialized research. One such investigation concerns the influence of pre-existing rock structures and lines of weakness (broadly defined as 'crustal anisotropies') on the architecture of the modern rift valley. This is not a new idea; the influence of Archaean and Palaeozoic structures and rock fabric on Cenozoic rifting was stressed by Dixey (1956: 54) and is central to McConnell's interpretation of the African Rift System (1972). It has long been recognized that the broad alignment of the eastern and western branches of the rift follows the trend of Proterozoic mobile belts around the margin of the Archaean Tanzanian Craton (Versfelt and Rosendahl, 1989). Recent studies in the Gregory Rift Zone have focused on the relationship of the mobile belt/craton boundary to rift faulting and magmatism (Smith and Mosley, 1993). The north-east margin of the Tanzanian Craton in this region trends NNW–SSE and at the surface is marked approximately by the Nandi

Escarpment. It is interpreted as dipping eastwards across a zone about 100 km wide, under an 'extensive cover of overthrust and gravitationally collapsed nappes and imbricated thrust slices of Mozambique Belt lithologies' (Mosley, 1993; Smith and Mosley, 1993: 592). Eastward across the craton margin, a marked thinning of the crust (KRISP Working Party, 1991), an increase in heat flow (Nyblade et al., 1990) and a paired gravity anomaly (Nyblade and Pollack, 1992) typical for continental collision zones support this model and mark the transition to the mobile belt.

The Gregory Rift may be divided into three distinct sectors whose characteristics can be related to the nature of the underlying basement (Smith and Mosley, 1993). North of the Equator the rift lies over the relatively thinned crust of the mobile belt. Here the rift is 100 km wide, with linear zones of intense faulting in its floor, and volcanism is typically tholeiitic and transitional, characterized by mildly alkaline basalts and trachytes. In contrast, south of Lake Magadi the Gregory Rift is considered to be underlain by cratonic crust and is characterized by 'a 200-km wide, low-relief zone of normal faults and tilted fault blocks which define small asymmetric graben'. There is a lack of the rift-floor faulting seen farther north (Smith and Mosley, 1993: 598). Volcanism in this sector includes strongly under-saturated nephelinites and carbonatites. The central sector of the Gregory Rift, which overlies the craton margin, contains a well-defined trough 50–70 km wide, and shows both types of volcanism described above. Each of the three sectors of the Gregory Rift Valley has its own orientation (NNW in the north, NW in the centre, and a N-to-NNE trend in the south), and the changes in orientation are marked by zones of 'grid faulting' in the inner trough that directly overlie major NNW-trending shear zones in the basement.

Smith and Mosley further suggest that the influence of basement anisotropy on the rift has varied through time. Reconstructing the rift basins in existence during the upper Miocene, they suggest that rifting propagated southwards and that rift basins were smaller and developed later as the rift stresses encountered thicker and more rigid lithosphere (ibid.: 600). The distribution of early Miocene eruptive centres was in part controlled by the pattern of NW- and N-trending shear zones and thrusts in the Precambrian crystalline basement. With time, however, the controlling influence of individual basement structures becomes difficult to detect, and Smith and Mosley suggest that 'by Pliocene times the effects of a thicker cover and the warming and softening of the crust by magma intrusion combine to reduce the influence of basement structures on rifting' (ibid.: 601).

Regional Study 3: Eastern Africa: Changing Regional Stress Fields

Another recent development in our understanding of the Rift System involves detailed analysis of fault kinematics to provide information on the regional stress and palaeo-stress fields. Strecker et al. (1990: 300–1) have determined extension directions for faults cutting a number of isotopically dated rocks in the Gregory Rift Valley, and found that the direction of extension was broadly E–W during the Upper Pliocene and NW–SE for the youngest displacements, less than about 0.7 Ma BP. Bosworth et al. extended this work to cover a wider region (1992), based on the alignment of Quaternary eruptive centres on the eastern rift shoulder and on the orientation of borehole breakouts in petroleum exploration wells. A broadly similar picture appeared, according to which the regional extension direction across the Gregory Rift Valley was E–W from the Miocene (c.23 Ma BP) to the middle Pleistocene (c.0.6 Ma BP). Middle Pleistocene and younger faults, and current seismic and borehole breakout data, indicate a present-day NW–SE orientation for the extension direction. The regional extensional stress field over central Kenya appears to have rotated 45° in a clockwise direction over the last 0.6 Ma BP. Strecker et al. have explained this as a delayed response to stresses generated at the plate boundaries in the Red Sea and the Gulf of Aden (1990: 302). Evidence from boreholes suggests modern stress-field directions of approximately N–S in the Sudan and NE–SW in eastern Ethiopia; Bosworth et al. combine this with the Kenyan data to define an approximately arcuate regional pattern, centred on the Afar triple junction (1992: 11 861–2). From this they deduce 'a model of the East African Rift System in which the southern Ethiopian and Kenyan rifts either are being aborted or are entering a phase of oblique extension' (ibid.).

Regional Study 4: Eastern Kenya: The Contribution of Aeromagnetic Surveys and Apatite Fission-Track Analysis

Until recently it was assumed that for most of Mesozoic and Cenozoic time, the topography of much of Eastern Africa bordering the rift system remained relatively stable, undergoing broad uplift. Along the continental margins slow downwarping was accompanied by continental and shallow marine sedimentation. Evidence from several sources now provides a more dynamic picture of events in eastern Kenya. Aeromagnetic, seismic reflection,

Plate 2.4. The metamorphic rocks of the Mathews Range rise above the plateau surface in eastern Kenya. (Kenya Information Services photograph.)

well, and gravity data have shown the existence of a buried trough, the Anza Trough, that runs for about 600 km under the present topography of north-eastern Kenya (Figure 2.3). This trough is probably of Jurassic age, with an estimated sediment thickness of at least 10 km. Reeves *et al.* interpreted it as the failed arm of a triple junction that formed when the east coast of Africa was taking shape, following the opening of the Indian Ocean (1986/7: 310). After its formation, the Anza Trough underwent several episodes of extension; though it is now completely buried by younger sedimentary and volcanic rock, its existence is a powerful indication that the terrain east of the Gregory Rift Valley had its own dramatic tectonic history.

Information on the evolution of landforms in eastern Kenya has also been provided by the application of apatite fission-track analysis, a technique that allows the thermal history of a rock to be determined from measurements of the number and length of fission tracks in apatite crystals. The thermal history of intrusive igneous and metamorphic rocks can then be interpreted in terms of uplift and denudation of the land surface. This technique has shown that the metamorphic rocks of the Karisia Hills and the Mathews Range (Plate 2.4) underwent two episodes of rapid denudation, during early Cretaceous (100–120 Ma BP) and late Cretaceous to early Tertiary (62–65 Ma BP) times (Foster and Gleadow, 1992: 165). The reason for these short episodes of rapid denudation is unclear, but a number of factors have been suggested, including the uplift resulting from the separation of Madagascar, the opening of the Anza Rift, and possibly the initiation of the opening of the Atlantic Ocean (ibid.: 166). Foster and Gleadow interpret the summits of the Karisia Hills and the Mathews Range as the tops of westward-tilted blocks, downfaulted on the east to the Anza Rift. Though the interpretation of Foster and Gleadow's results is still speculative, their work illustrates the more dynamic picture of off-rift events that is now emerging.

Regional Study 5: The Western Rift: Seismic Reflection Studies and the Half-Graben Model

For much of this century less was known about the Western Rift than about the Eastern Rift, one reason being that much of the Western Rift in the Tanganyika and Malawi Basins is covered by many hundred metres of water. The application of marine seismology provided a way to study such basins, and academic interest was complemented by the interest of oil companies in possible petroleum source rocks. In the 1980s academic and commercial interests combined to carry out project CEGAL and later project PROBE, programmes of seismic reflection surveys in the Tanganyika, Malawi, and

Turkana basins. PROBE's multi-channel seismic profiles allowed the identification of crustal structures to a depth of about 6 km. These data supported the application of a model of asymmetric rift basin development to the Tanganyika and Malawi Basins (Rosendahl *et al.*, 1986; Rosendahl, 1987; Ebinger, 1989; Rosendahl *et al.*, 1992; and other sources). According to this model, rifts are made up of a sequence of asymmetrical basins, Rosendahl's 'fundamental units' (Rosendahl, 1987: 462). Each basin has the form of a half-graben, with one side formed by a main border fault, facing a more complex structure composed of monoclines, fissures, and steps (Rosendahl *et al.*, 1986: 33). The direction of the half-graben may alternate along the rift, and adjacent half-graben are joined by complex 'accommodation zones'.

While the asymmetry of many rift valleys in cross-section had been recognized and discussed by earlier writers, no overall model to explain this asymmetry had been put forward until the half-graben model provided a basis for studies of basin structure. This model has now been applied to rifts all over the world. Its most successful application has been in the big non-volcanic rift basins such as the Tanganyika and Malawi Basins. Even there, however, there are still disagreements on the precise application of the model; Rosendahl describes how, depending on the exact criteria applied, the Tanganyika Basin can be divided into 'about' twelve half-graben units, the Malawi Basin into 'about' nine such units (1992: 245).

The half-graben model is less successful in rifts where there are large volumes of volcanic rock that may obscure fault structures. The problem, however, is not just that the structures are inaccessible, but that the model itself does not explain the spatial distribution of magmatism within rifts. An early attempt at applying the half-graben model to the Turkana Basin interpreted magmatism as occurring particularly along accommodation zones, but re-interpretation of the data suggest a more complex picture (Rosendahl, 1992: 251–2).

Current opinions on the East African Rift System

The East African Rift System is recognized as a feature formed by normal faulting, the result of crustal tension. It is believed to be underlain by a region of upwelling asthenosphere (a mantle plume or hotspot) that has caused updoming and weakening of the lithosphere. This has brought about uplift of the pre-existing land surface, faulting along or close to the crest of the uplift, and in some regions the eruption of large volumes of volcanic rock. The sequence of the above events is not well established for all sectors of the rift, but in Ethiopia and Kenya it has been demonstrated that rifting and subsidence preceded the major phases of domal uplift (Baker *et al.*, 1972). Volcanism preceded faulting by several million years in some sectors of the rift, though elsewhere faulting has taken place in the absence of volcanism (see below).

Continental Rift Zones have been explained in terms of active and passive-rifting models (Keen, 1985; Reading, 1986; Summerfield (this volume)) and it is generally agreed that the East African Rift System represents the active model in which the driving force is an upwelling mantle plume. Due to the impingement of upwelling hot mantle material the lithosphere and the crust are weakened and thinned, which finally results in the surficial expression with the generation of normal faults. Despite the broad agreement on this, conflicting models of sub-surface structures and rift mechanisms have been put forward such as those of Mohr (1987) and Bosworth (1989).

The initiation of the East African Rift System probably occurred during early Cenozoic time, and the northern zones of rifting are demonstrably younger than those farther to the south. The evolution of the Rift System has occurred by means of discrete episodes of volcanism and tectonism, that vary considerably in date and nature from one part of the rift to another. However, the overall tectonic evolution suggests that rifting processes migrated from the Gulf of Aden to Tanzania. The formation of many Cenozoic rift bounding faults and rift zones in general has been guided by anisotropies in basement system rocks that were favourably oriented with respect to the prevailing tensional stress field. Furthermore, reactivated Mesozoic fault lines during Cenozoic rifting attest to the importance of crustal weak zones in the evolution of rift basins.

The East African Rift System can be broadly divided into high-volcanicity zones in which large volumes of magma have been produced, of which the Afar Depression, the Main Ethiopian Rift, and the Gregory Rift are the most important, and low-volcanicity zones in which magma production has been little or non-existent, notably the Albert, Tanganyika, and Malawi Rift Zones. It is debatable as to whether the style of faulting differs significantly between these two types of rift, and whether there are major differences in subsurface structures between them. It has been suggested that rifts may evolve from an initial half-graben to a fully developed full-graben structure, and also that the high-volcanicity zones may represent a more advanced stage of rift evolution than the low-volcanicity zones.

Plate 2.5. The active summit crater of Ol Doinyo Lengai volcano, Tanzania, in June 1993. (Martin Kuper photograph.)

The volume of volcanic rock produced along the East African Rift System is significantly greater than along other recently active continental rifts. These rocks vary greatly in chemical composition over both space and time. Though basalts are important, large volumes of phonolites, trachytes, and even rhyolites have also been produced (Wilson, 1989: 331). The coeval presence of basalts and rhyolites emphasizes the importance of crustal contamination and magma-chamber fractionation of magmas that are derived from the mantle. The basaltic rocks erupted along the East African Rift System are predominantly alkalic in composition. But transitional basalts dominate in Ethiopia, and the most recent basalts of the Afar Depression show evidence of an increasingly oceanic character (Hart *et al.*, 1989). The East African Rift System is also notable for the presence of carbonatite lavas, such as those of the active volcano Ol Doinyo Lengai in northern Tanzania (Plate 2.5), which have been described as 'the ultimate in alkali and volatile concentration' (Dawson, in Duff (ed.), 1993: 684). The relative importance of fractional crystallization and crustal contamination to the generation of these widely differing magmas remains uncertain.

Present day volcanism and tectonism vary in intensity between different sectors of the East African Rift System, but the wide distribution of active volcanoes and earthquakes supports its identification as an active, slowly spreading zone of continental extension. The petrologic observations indicating a gradient in rifting stages from south (young) to north (mature) are matched by a dramatic reduction in crustal thickness in the same direction as revealed by seismic refraction studies. This underscores the dynamics of the continental rifting process that may finally climax in continental separation and the generation of oceanic crust. However the future of the East African Rift System—oceanic-spreading system or continued slow continental extension—remains under debate.

References

Abdallah, A., Courtillot, V., Kasser, M., Le Dain, A.-Y., Lepine, J-C., Robineau, B., Ruegg, J.-C., Tapponnier, P., and Tarantola, A. (1979), 'Relevance of Afar Seismicity and Volcanism to the Mechanics of Accreting Plate Boundaries', *Nature*, 282: 17–23.

Achauer, U., Maguire, P. K. H., Mechie, J., Green, W. V., and the KRISP Working Group (1992), 'Some Remarks on the Structure and Geodynamics of the Kenya Rift', *Tectonophysics*, 213: 257–68.

Acton, G. D., Stein, S., and Engeln, J. F. (1991), 'Block Rotation and Continental Extension in Afar: A Comparison to Oceanic Microplate Systems', *Tectonics*, 3: 501–26.

Allard, P., Tazieff, H., and Dajlevic, D. (1979), 'Observations of Seafloor Spreading in Afar during the November 1978 Fissure Eruption', *Nature*, 279: 30–3.

Ambraseys, N. N. (1991), 'Earthquake Hazard in the Kenya Rift: The Subukia Earthquake 1928', *Geophysics Journal International*, 105: 253–69.

Baker, B. H. (1986), 'Tectonics and Volcanism of the Southern Kenya Rift Valley and its Influence on Rift Sedimentation', in L. E. Frostick *et al.* (eds.), *Sedimentation in the African Rifts* (London).

—— Mohr, P. A., and Williams, L. A. J. (1972), *Geology of the Eastern Rift System of Africa* (Boulder, Colo.).

Beadle, L. C. (1974), *The Inland Waters of Tropical Africa* (London).

Berckhemer, H., Baier, B., Bartelsen, H., Behle, A., Burkhardt, H., Gebrande, H., Makris, J., Mensel, H., Miller, H., and Vees, R. (1975), 'Deep Seismic Soundings in the Afar Region and on the Highland of Ethiopia', in A. Pilger and A. Rosler (eds.), *Afar Depression of Ethiopia* (Stuttgart).

Bishop, W. W. (1978) (ed.), *Geological Background to Fossil Man* (Edinburgh).

—— and Clark, J. D. (1967) (eds.), *Background to Evolution in Africa* (Chicago).

—— and Miller, J. A. (1972) (eds.), *Calibration of Hominid Evolution* (Edinburgh).

Bosworth, W. (1985), 'Geometry of Propagating Continental Rifts', *Nature*, 316: 625–7.

—— (1987), 'Off-axis Volcanism in the Gregory Rift, East Africa: Implications for Models of Continental Rifting', *Geology*, 15: 397–400.

—— (1989), 'Basin and Range Style Tectonics in East Africa', *Journal of African Earth Sciences*, 8: 191–201.

—— Strecker, M. R., and Blisniuk, P. M. (1992), 'Integration of East African Paleostress and Present-day Stress data: Implications for Continental Stress Field Dynamics', *Journal of Geophysical Research*, 97; 11 851–65.

Bullard, E. (1936), 'Gravity Measurements in East Africa', *Philosophical Transactions of the Royal Society of London*, 235: 445–531.

Carney, J. N. (1972), 'The Geology of the Area East of Lake Baringo, Kenya Rift Valley', Ph.D. thesis (London).

Cerling, T. E. and Powers, D. W. (1977), 'Paleorifting between the Gregory and Ethiopian Rifts', *Geology*, 5: 441–4.

Chapman, G. S., Lippard, S., and Martyn. J. E. (1978), 'The Stratigraphy and Structure of the Kamasia Range, Kenya Rift Valley', *Journal of the Geological Society of London*, 135: 265–81.

Clarke, M. C. G., Woodhall, D. G., Allen, D., and Darling, G. (1990), *Geological, Volcanological and Hydrogeological Controls on the Occurrence of Geothermal Activity in the Area Surrounding Lake Naivasha, Kenya* (Nairobi).

Consiglio Nazionale delle Ricerche and Centre National de la Recherche Scientifique (1971), *Geological Maps of Afar, 1, Northern Afar* (La Celle St. Cloud).

—— (1975), *Geological Maps of Afar, 2, Central and Southern Afar* (La Celle St. Cloud).

Crossley, R. and Crow, M. J. (1980), 'The Malawi Rift', *Atti Convegni Lincei*, 47: 77–87.

Dawson, J. B. (1992), 'Neogene Tectonics and Volcanicity in the North Tanzania Sector of the Gregory Rift Valley: Contrasts with the Kenya Sector', *Tectonophysics*, 204: 81–92.

—— (1993), Plateaus, Rift Valleys and Continental Basins in P. M. D. Duff (ed.), *Holmes' Physical Geography* (London), 664–97.

—— and Smith, J. V. (1988), 'Metasomatised and Veined Upper-mantle Xenoliths from Pello Hill, Tanzania: Evidence for Anomalously Light Mantle Beneath the Tanzanian Sector of the East African Rift Valley', *Contributions to Mineralogy and Petrology*, 100: 510–27.

Dindi, E. W. (1993), 'A 2D P-wave Refraction Model for the Crustal Structure Between Lake Turkana and Chandler's Falls Based on KRISP90 Data', *Proceedings of the Fifth Conference on the Geology of Kenya*, 13–16.

Dixey, F. (1956), 'The East African Rift System', *Overseas Geology and Mineral Resources*, supp. bull. 1.

Doser, D. I. and Yarwood, D. R. (1991), 'Strike-slip Faulting in Continental Rifts: Examples from Sabukia (*sic*), East Africa (1928), and Other Regions', *Tectonophysics*, 197: 213–24.

Duff, P. McL. D. (1993) (ed.), *Holmes' Principles of Physical Geology* (London).

Dunkley, P. N., Smith, M., Allen, D. J., and Darling, W. G. (1993) *The Geothermal Activity and Geology of the Northern Sector of the Kenya Rift Valley* (Keyworth, Nottingham).

Ebinger, C. J. (1989), 'Tectonic Development of the Western Branch of the East African Rift System', *Bulletin of the Geological Society of America*, 101: 885–903.

—— Deino, A., Drake, B., and Tesha, A. L. (1989), 'Chronology of Volcanism and Rift Basin Propagation: Rungwe Volcanic Province, East Africa', *Journal of Geophysical Research*, 94: 15 785–803.

Fairhead, J. D. (1976), 'The Structure of the Lithosphere beneath the Eastern Rift, East Africa, deduced from Gravity Studies', *Tectonophysics*, 30: 269–98.

Foster, D. A. and Gleadow, A. J. W. (1992), 'The Morphotectonic Evolution of Rift-margin Mountains in Central Kenya: Constraints from Apatite Fission-track Thermochronology', *Earth and Planetary Science Letters*, 113: 157–71.

Golden, M. (1978), 'The Geology of the Area East of Silale, Rift Valley Province, Kenya', Ph.D. thesis (London).

Gouin, P. (1979), *Earthquake History of Ethiopia and the Horn of Africa* (Ottawa).

Green, W. V., Achauer, U., and Meyer, R. P. (1991), 'A Three-dimensional Seismic Image of the Crust and Upper Mantle beneath the Kenya Rift', *Nature*, 354: 199–203.

Gregory, J. W. (1896), *The Great Rift Valley* (London, 1968).

Hackman, B. D., Charsley, T. J., Key, R. M., and Wilkinson, A. F. (1990), 'The Development of the East African Rift System in North-central Kenya', *Tectonophysics*, 184: 189–211.

Harkin, D. A. (1960), 'The Rungwe Volcanics at the Northern End of Lake Nyasa', *Geological Survey of Tanganyika Memoir*, 2: 1–172.

Hart, W. K., Woldegabriel, G., Walter, R. C., and Mertzman, S. A. (1989), 'Basaltic Volcanism in Ethiopia. Constraints on Continental Rifting and Mantle Interactions', *Journal of Geophysical Research*, 94: 7731–48.

Hayward, N. J. and Ebinger, C. J. (1995), 'Variations in the Along-Axis Segmentation of the Afar Rift System', *Tectonics* (in review).

Heezen, B. C. (1965), 'The World Rift System', in *East African Rift System*, UMC-UNESCO, 116–17.

Hetzel, R. and Strecker, M. R. (1994), 'Late Mozambique Belt Structures in Western Kenya and Their Influence on the Evolution of the Cenozoic Kenya Rift', *Journal of Structural Geology*, 16: 189–201.

Howell, F. C. and Bourliere, F. (1963) (eds.), *African Ecology and Human Evolution* (Chicago).

Kebede, F. and Kulhanek, O. (1991), 'Recent Seismicity of the East African Rift System and its Implications', *Physics of the Earth and Planetary Interiors*, 68: 259–73.

Keen, C. E. (1985), 'The Dynamics of Rifting: Deformation of the Lithosphere by Active and Passive Driving Forces', *Geophysical Journal of the Royal Astronomical Society*, 80: 95–120.

Keller, G. R., Khan, M. A., Morgan, P., Wendlandt, R. F., Baldridge, W. C., Olsen, K. H., Prodehl, C., and Braile, L. W. (1991), 'A Comparative Study of the Rio Grande and Kenya Rifts', *Tectonophysics*, 197: 355–71.

Key, R. M., Rop, B. P., and Rundle, C. C. (1987), 'The Development of the Late Cenozoic Alkali Basaltic Marsabit Shield Volcano, Northern Kenya', *Journal of African Earth Sciences*, 6: 475–91.

—— Charsley, T. J., Hackman, B. D., Wilkinson, A. F., and Rundle, C. C. (1989), 'Superimposed Upper Proterozoic Collision-controlled Orogenies in the Mozambique Orogenic Belt of Kenya', *Precambrian Research*, 44: 197–225.

Khan, M. A., Maguire, P. K. H., Henry, W., Higham. M., Prodehl, C., Mechie, J., Keller, G. R., and Patel, J. (1989), 'A Crustal Seismic Refraction Line along the Axis of the S. Kenya Rift', *Journal of African Earth Sciences*, 8: 455–60.

King, B. C. (1978), 'Structural Evolution of the Gregory Rift Valley', in Bishop (ed.), *Geological Background to Fossil Man*.

KRISP Working Party (1991), 'Large-scale Variation in Lithospheric Structure along and across the Kenya Rift', *Nature*, 354: 223–7.

Loupekine, I. S. (1965), 'Seismology in East Africa', in *East African Rift System*, UMC-UNESCO, 85.

—— (1971), *A Catalogue of Felt Earthquakes in Kenya, 1892–1969* (Nairobi).

McClenaghan, M. P. (1971), 'The Geology of the Ribkwo Area, Baringo District, Kenya', Ph.D. thesis (London).

McConnell, R. B. (1967), 'The East African Rift System', *Nature*, 215: 578–81.

—— (1972), 'Geological Development of the Rift System of Eastern Africa', *Bulletin of the Geological Society of America*, 83: 2549–72.

Maguire, P. K. H., Shah, E. R., Pointing, A. J., Cooke, P. A. V., Khan, M. A., and Swain, C. J. (1988), 'The Seismicity of Kenya', *Journal of African Earth Sciences*, 7: 915–23.

Makris, J. and Ginzburg, A. (1987), 'The Afar Depression: Transition between Continental Rifting and Sea-floor Spreading', *Tectonophysics*, 141: 199–214.

Martyn, J. (1967), 'Pleistocene Deposits and New Fossil Localities in Kenya', *Nature*, 215: 476–80.

Mbede, E. I. (1993), 'Tectonic Development of the Rukwa Rift Basin in South-west Tanzania', Ph.D. thesis (Berlin).

Mohr, P. (1987), 'Structural Style of Continental Rifting in Ethiopia: Reverse Decollements', *Eos*, 68: 721–30.

—— (1989), 'Nature of the Crust under Afar: New Igneous, Not Thinned Continental', *Tectonophysics*, 167: 1–11.

—— (1991), 'The Discovery of African Rift Geology: A Summary', in A. B. Kampunzu and R. T. Lubala (eds.), *Magmatism in Extensional Structural Settings: The Phanerozoic African Plate* (Berlin).

—— (1992), 'Nature of the Crust beneath Magmatically Active Continental Rifts', *Tectonophysics*, 213: 269–84.

Molnar, P., Fitch, T. J., and Asfaw, L. M. 1970, 'A Microearthquake Study in the Ethiopian Rift', *Earthquake Notes*, 41: 37–44.

Morgan, P. (1991), 'A Deep Look at African Rifting', *Nature*, 354: 188.

Mosley, P. N. (1993), 'Geological Evolution of the Late Proterozoic 'Mozambique Belt' of Kenya', *Tectonophysics*, 221: 223–50.

Mougenot, D., Hernandez, J., and Virlogeux, P. (1989), 'Structure et volcanisme d'un rift sous-marin: le fosse des Kerimbas (marge nord-mozambique)', *Bulletin de Société Géologique de la France*, 8: 401–10.

—— Recq, M., Virlogeux, P., and Lepvrier, C. (1986), 'Seaward Extension of the East African Rift', *Nature*, 321: 599–603.

Nyblade, A. A. and Pollack, H. N. (1992), 'A Gravity Model for the Lithosphere in Western Kenya and North-eastern Tanzania', *Tectonophysics*, 212: 257–67.

—— Pollack, H. N., Jones, D. J., Podmore, F., and Mushayandebvu, M. (1990), 'Terrestrial Heat Flow in East and Southern Africa', *Journal of Geophysical Research*, 95: 17 371–84.

Pickford, M. (1982), 'The Tectonics, Volcanics and Sediments of the Nyanza Rift Valley, Kenya', *Zeitschrift für Geomorphologie*, 42: 1–33.

Reading, H. G. (1986), 'African Rift tectonics and Sedimentation, an Introduction', in Frostick *et al.* (eds.), *Sedimentation in the African Rifts*.

Reeves, C. V., Karanja, F. M., and MacLeod, I. N. (1986/7), 'Geophysical Evidence for a Failed Jurassic Rift and Triple Junction in Kenya', *Earth and Planetary Science Letters*, 81: 299–311.

Rosendahl, B. R. (1987), 'Architecture of Continental Rifts with Special Reference to East Africa', *Annual Review of Earth and Planetary Science*, 15: 445–503.

—— Kilembe, E., and Kaczmarick, K. (1992), 'Comparison of the Tanganyika, Malawi, Rukwa and Turkana Rift Zones from Analyses of Seismic Reflection Data', *Tectonophysics*, 213: 235–56.

—— Reynolds, D. J., Lorber, P. M., Burgess, C. F., McGill, J., Scott, D., Lambiase, J. J. and Derksen, S. J. (1986), 'Structural Expressions of Rifting: Lessons from Lake Tanganyika, Africa', in Frostick *et al.* (eds.), *Sedimentation in the African Rifts*.

Ruegg, J.-C. (1975), 'Main Results about the Crustal and Upper Mantle Structure of the Djibouti Region (T.F.A.I.)', in Pilger and Rosler (eds.), *Afar Depression of Ethiopia* (Stuttgart), 120–34.

Ruegg, J.-C. and Kasser, M. (1987), 'Deformation across the Asal–Ghoubbet Rift, Djibouti: Uplift and Crustal Extension 1979–1986', *Geophysical Research Letters*, 14: 745–8.

Saggerson, E. P. and Baker, B. H. (1965), 'Post-Jurassic Erosion Surfaces in Eastern Kenya and their Deformation in Relation to Rift Structure', *Journal of the Geological Society of London*, 121: 51–72.

Shackleton, R. M. (1950), 'A Contribution to the Geology of the Kavirondo Rift Valley', *Journal of the Geological Society of London*, 106: 345–92.

—— 'Precambrian Collision Tectonics in Africa', in M. P. Coward and A. C. Ries (eds.), *Collision Tectonics* (London).

Smith, M. (1994), 'Stratigraphic and Structural Constraints on Mechanisms of Active Rifting in the Gregory Rift, Kenya', *Tectonophysics*, 236: 3–22.

—— and Mosley, P., 'Crustal Heterogeneity and Basement Influence on the Development of the Kenya Rift, East Africa', *Tectonics*, 12: 591–606.

Smithsonian Institution (1994), *Bulletin of the Global Volcanism Network*, 19/2 (Washington, DC).

Stein, R. S., Briole, P., Ruegg, J.-C., Tapponnier, P., and Gasse, F. (1991), 'Contemporary, Holocene and Quaternary Deformation of the Asal Rift, Djibouti: Implications for the Mechanics of Slow Spreading Ridges', *Journal of Geophysical Research*, 96: 21 789–806.

Strecker, M. R., Blisniuk, P. M., and Eisbacher, G. H. (1990), 'Rotation of Extension Direction in the Central Kenya Rift', *Geology*, 18: 299–302.

Suess, E. (1891), 'Die Brueche des oestlichen Afrika', *Denkschriften der Akademic der Wissenschaften (Math. Nat.)*, 58: 555–84.

Tallon, P. (1976), 'The Stratigraphy, Palaeoenvironments and Geomorphology of the Pleistocene Kapthurin Formation, Kenya', Ph.D. thesis (London).

Tapponnier, P., Armijo, R., Manighetti, I, and Courtillot, V. (1990), 'Bookshelf Faulting and Horizontal Block Rotations between Overlapping Rifts in Southern Afar', *Geophysical Research Letters*, 17: 1–4.

Truckle, P. (1977), 'The Geology of the Area to the South of Lokori, South Turkana, Kenya', Ph.D. thesis (London).

Upper Mantle Committee-UNESCO (1965), *East African Rift System* (Nairobi).

Versfelt, J. and Rosendahl, B. R. (1989), 'Relationships between Pre-rift Structure and Rift Architecture in Lakes Tanganyika and Malawi, East Africa', *Nature*: 354–7.

von Hohnel, L. (1894), *Discovery of Lakes Rudolf and Stefanie* (London).

Wayland, E. J. (1923), 'Continental Drift and the Stressing of Africa', *Nature*, 112: 938–9.

Weaver, S. D. (1976/7), 'The Quaternary Caldera Volcano Emuru-angogolak, Kenya Rift, and the Petrology of a Bimodal Ferrobasalt-pantelleritic Trachyte Association', *Bulletin of Volcanology*, 40: 209–30.

Willis, B. (1936), *East African Plateaus and Rift Valleys* (Washington, DC).

Wilson, M. (1989), *Igneous Petrogenesis: A Global Tectonic Approach* (London).

Woldegabriel, G. and Aronson, J. L. (1987), 'Chew Bahir Rift: A "Failed" Rift in Southern Ethiopia', *Geology*, 15: 430–3.

—— and Walter, R. C. (1990), 'Geology, Geochronology and Rift Basin Development in the Central Sector of the Main Ethiopia Rift', *Bulletin of the Geological Society of America*, 102: 439–58.

Young, P. A. V., Maguire, P. K. H., Laffoley, N. d'A., and Evans, J. R. (1991), 'Implications of the Distribution of Seismicity Near Lake Bogoria in the Kenya Rift', *Geophysical Journal International*, 105: 665–74.

3 Climate: Past and Present

Andrew S. Goudie

Introduction

What makes the climate of Africa distinctive in comparison with other areas of the Earth's surface?

The first characteristic is undoubtedly that Africa is so symmetrically located with respect to latitude. Its northern (Cape Blanc 37°N) and southern (Cape Agulhas 35°S) extremities extend almost equidistant from the Equator. Broadly similar series of climates can be traced northward and southward from the hot, moist equatorial belt. The Sudan has its counterpart in Zimbabwe, the Sahara in the arid tracts of the Namib and Kalahari, and the Mediterranean coast in the south-west of South Africa.

A second characteristic arising from this fact is that Africa is the most tropical of all the continents. Only the extremes reach far enough poleward to be directly influenced by the mid-latitude westerlies and their associated disturbances. Indeed, the Sahara and the belt of country to the south is hotter than any other part of the world of comparable size. As Kendrew (1961: 39) put it, 'Africa is the only continent in which the 50 °[F] isotherm (sea-level) never appears. The greater part of the continent has more than 9 months with means above 70 °[F].'

A third characteristic arising from Africa's latitudinal position is that it is strongly influenced by the subtropical anticyclonic belts of both hemispheres, so that it possesses extensive areas of dry climate both to the south and north of the Equator.

Fourthly, because so much of Africa consists of elevated plateaux, altitude plays a highly significant role in reducing temperatures over extensive areas. Most of Africa to the south and east of the Congo Basin lies at altitudes between 1000 and 3000 m.

Fifthly, because of its tectonic history, there are no great cordillera to act as major climatic divides. This contrasts with the situation in the Americas and Eurasia. Thus in Southern Africa the continuity of the subtropical high-pressure ridge is greater than in South America, with the result that there is not the same exchange of air between high and low latitudes. A high zonal index is a feature of Africa. Indeed, the absence of extensive cordillera means that the climatic pattern of Africa in many respects resembles that of the 'ideal' or 'hypothetical' continent of many textbooks.

Sixthly, notwithstanding the above point, Africa possesses hemispherical asymmetry in climate. This stems from the fact that Africa north of the Equator is not only much broader than Africa to the south of the Equator, but it also lacks a true ocean boundary to the north and north-east, being bordered instead by the great Eurasian landmass.

Finally, by way of introduction, Africa possesses some noteworthy deviations from the standard world pattern of climatic distribution (Trewartha, 1981: 106):

1. anomalously low rainfalls occur over much of eastern tropical-equatorial Africa (especially in Somalia and Kenya) where, on the basis of a windward position and of uplands rising abruptly, one might have anticipated much wetter conditions;
2. the dry coastal belt in Ghana and Togo on the Gulf of Guinea in the central portion of a coast which otherwise has a tropical wet climate;
3. a zone along the Guinea coast with a high-sun secondary rainfall minimum; and

I am grateful to Richard Washington for some comments on an early draft, and to Peter Hayward and Ailsa Allen for drawing the figures.

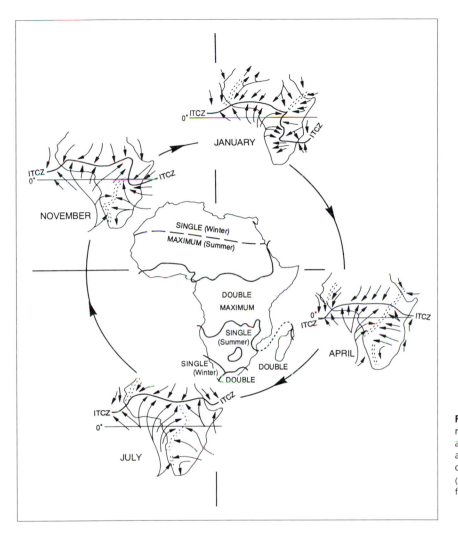

Fig. 3.1. Schematic representation of the mean annual cycle of precipitation and surface-wind systems over Africa.
(Modified from Rasmusson, 1988, fig. 1.2.)

4. the restricted latitudinal spread of wet-and-dry climates to the north of the Equator in comparison to the south because of the unusual southward penetration of the Saharan dry climate.

General Atmospheric Circulation

Although in the introduction to this chapter reference has been made to the uniqueness or distinctiveness of African climate, the continent is none the less hugely affected by the global patterns of the atmospheric heat engine. As in so much of the tropical world, the basic pattern of air movement is equatorwards, being from the north-east in the northern hemisphere and from the south-east in the southern one (Figure 3.1). Converging air meets in the equatorial zone, and here air rises, moving back towards the poles at high levels, only to descend in the vicinity of the two tropics. The air that descends on the poleward sides of these circulation systems, commonly called Hadley Cells, is dry and contributes to the development of the subtropical deserts. By contrast, the converging air of the low latitudes (often referred to as the Intertropical Convergence Zone (ITCZ)) produces rain. However, the positions of the Hadley Cells and of the ITCZ vary with the seasons. This is related to the tilt of the Earth's axis and the annual circuit of the Sun made by the Earth, which means that the input of solar energy reaches a maximum in the northern hemisphere in June and in the southern hemisphere in December. The zones of subsiding air, the high-pressure systems associated with them, and the equatorial westerlies in between, travel north and south with the seasons.

During the northern hemisphere summer the equatorial trough of low pressure migrates northward and

Plate 3.1. The Cape region of South Africa, which is in the zone of winter rains, shows the effects of relief enhancement of rainfall (photo: A. S. Goudie).

low pressure extends throughout most of Africa north of the Equator. At the same time an intense heat low forms over the Sahara and a pronounced high-pressure centre dominates over Southern Africa. A marked reversal takes place in the southern hemisphere summer, when a high-pressure cell extends over most of Northern Africa and the equatorial trough of low pressure moves into the lower latitudes and into Africa south of the Equator. The very northernmost part of the continent comes under the influence of eastward moving mid-latitude low-pressure systems and westerly winds. Thus in January the north-easterly dry trade wind called the Harmattan is dominant throughout most of North and West Africa while north-east trades originating over the Indian Ocean (the north-east monsoon) prevail over Eastern Africa.

However the extreme northern and southern extremities of Africa are seldom affected by the poleward movements of equatorial air. They are areas where winter rainfall predominates, when cool, moist air and associated frontal depressions bring precipitation from higher latitudes.

Another crucial general control on precipitation

patterns is the presence of cold ocean waters offshore. Atmospheric stability is increased by the low temperatures of the upwelling waters of the Canaries and Benguela Currents, contributing to the aridity experienced along the coasts of Morocco and Mauritania in north-west Africa, and of Angola, Namibia, and South Africa in south-west Africa respectively.

In recent years the influence that sea-surface temperatures exercise over climate variability on the continents has become increasingly apparent. Many general circulation model (GCM) experiments have demonstrated the effects of tropical sea-surface temperature anomalies, and connections between Sahel rainfall levels and Atlantic Ocean surface temperatures have been proposed. Cold water in the North Atlantic off West Africa may, for example, inhibit the northward movement of the ITCZ and so reduce the incidence of rainfall in the Sahel zone. Moreover, the relationship between wet-and-dry spells in the Sahel and sea-surface temperatures has proved to be a potentially powerful predictive tool for forecasting future seasonal rainfall levels in that area (Parker et al., 1988; Rowell et al., 1992).

An indication of the spatial and seasonal patterns of precipitation is given in Figure 3.1. One should note the winter rainfall regimes of the 'Mediterranean' climatic zones to north and south, the low total rainfalls of the two main desert zones, the summer-season rainfall of the areas equatorwards of the deserts, and the double-rainfall peak in those areas around the Equator affected by the annual march of the ITCZ.

However, there are some complications with respect to the quantities of precipitation received during the course of the year that need explanation. The most obvious of these complications is perhaps created by relief. In particular, exceptionally high annual totals are received by the mountains and escarpments of West Africa, with parts of Cameroon Mountain, for instance, receiving over 10 000 mm of precipitation per annum.

Rainfall enhancement is also evident over the Ethiopian, East African, and Rift Valley highlands, the Atlas Mountains, the Tibesti and Hoggar Massifs of the central Sahara, the Guinean Highlands and Jos Plateau of West Africa, and around the great fringing escarpment of Southern Africa (Plate 3.1). However both the degree to which enhancement occurs and the altitudinal zonation of precipitation vary greatly according to such factors as the interaction of the topography with the regional wind and moisture fields and the more local mountain-valley and slope-wind systems that the topography gives rise to.

The disturbances which bring precipitation to Africa are closely linked in their type to the major wind systems. Thus on the poleward extremes in North and South

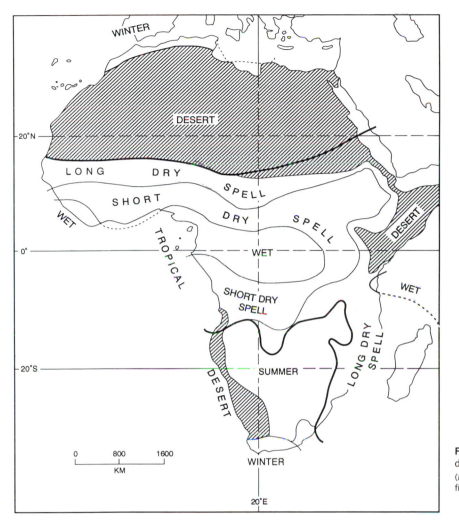

Fig. 3.2. The main climatic divisions of Africa. (Modified from Griffiths, 1987, fig. 10.)

Africa the rainfall is produced by low-pressure cells travelling in the westerlies. Along the sub-Saharan belt easterly waves superimposed in the African easterly jet produce precipitation and move westward. They are sometimes known as squall lines or cloud clusters and they originate as far east as the Sudan. The disturbances that produce rain in the equatorial regions are imperfectly understood, though in general intense falls are coupled with regional-scale convergence in the wind field. Localized rainfall can occur as a result of convection associated with intense surface heating.

Climate Regionalization

Following Griffiths (1987) it is possible to divide Africa into eight major climatic zones or regions, related essentially to precipitation and temperature characteristics (Figure 3.2).

The *tropical wet climate* type, which occurs as a large area straddling the Equator in West and Central Africa, and a small area on the east coast at about 5°S, is characterized by some rain in all months, and uniformly high temperatures throughout the year.

Surrounding this climatic type is the *tropical wet climate with short dry spells*, which has three to five dry months and a larger temperature range than in the wet climate. Precipitation and temperatures, however, remain high.

This merges into the zone of *tropical climate with long dry spells*, which has low rainfall for at least six months. Rainfall amounts are lower and more variable than in the two previous zones, temperatures show a larger seasonal swing, and drought and famine are recurrent problems.

Tropical-desert climates are drier still, but are not widespread, being most prevalent in the anomalous rainfall-deficit area of East Africa.

The remaining tropical climate zone is the *tropical-highlands climate*, where altitude is a major control of both precipitation and temperature, and where at high altitudes snow, and even glaciers, can occur.

There are also three subtropical climate zones. *The subtropical desert climate*, which is characteristic of the Sahara and the Namib, is the most extensive climatic type in Africa. Precipitation levels can be very low (e.g. *c.*25 mm per annum in the Namib), while in the interior diurnal and seasonal temperature ranges are high. Coastal zones along the western side of Africa, though exceedingly arid, have lower temperatures, lower ranges of temperature, and high frequencies of fog. These are all related to the relatively cold ocean currents offshore.

The *subtropical summer-rain climate* type is mainly found in the interior plateaux and basins of Southern and Central Africa, including much of the Kalahari semi-desert. Precipitation is usually concentrated in about three summer months.

Finally, at the northern and southern extremities of Africa is the zone of *subtropical winter rain*, which is often described as having a Mediterranean type of climate.

Diagrams showing precipitation and temperature patterns through the year are shown in Figure 3.3.

Distribution of Rainfall Over Africa

The average annual precipitation over Africa is about 725 mm (UNESCO, 1978: 258). There is, however, a great range, with as little as 1–2 mm in the driest parts of the Libyan Desert to around 10 000 mm in the foothills of Mount Cameroon (Figure 3.4). The pattern is to a great extent a latitudinal one as brought out by Table 3.1 (derived from UNESCO, 1978: table 78).

Table 3.1 Latitudinal variation in rainfall

Latitude (degrees)	Average annual precipitation (mm)
40–30 N	220
30–20 N	37
20–10 N	550
10– 0 N	1380
0–10 S	1320
10–20 S	1070
20–30 S	570
30–40 S	509

The Rainfall Anomaly of East Africa

As Trewartha (1981: 134) has remarked,

Undoubtedly the most impressive climatic anomaly in all of Africa is the widespread deficiency of rainfall in tropical East Africa . . . Nowhere else in a similar latitudinal and geographical location does there exist such a widespread water deficit. The abnormality of this rainfall deficiency in an equatorial-tropical region is all the more remarkable when one notes that this is the eastern side of a continent, in the latitudes of the tropical easterlies, where an almost continuous north–south highland parallels the coast at no great distance inland.

Kendrew (1961: 91) suggested some general reasons for this anomaly: the main wind systems are parallel to the coast rather than onshore; the south-east trades lose much of their vapour in heavy condensation on the mountains of Madagascar before reaching East Africa; the north-east trades have little moisture after their land passage; and, finally, the rain-bearing equatorial trough passes the region rapidly, being hastened far to the north in the northern summer and far to the south in the southern summer.

The Guinea Coast Rainfall Anomaly

The second most remarkable rainfall anomaly is that which exists along the mid-section of the Guinea coastlands centring on central and eastern Ghana, Togo, and Benin. It consists of a relatively dry area sandwiched between more humid climates to east and west, with the greatest intensity of dryness along the coast. The trend of the coast and the presence of upwelling cool water offshore may contribute to this situation.

Rainfall Variability and Reliability

It is often asserted that in a tropical continent the variability of rainfall from year to year will be greater than that in temperate latitudes. This is, however, a concept that has been largely discredited for Africa by Gregory (1969). It is also generally maintained that rainfall variability increases as average rainfall decreases. Certainly, rainfall variability is an important climatic characteristic but it is rather more complex in its patterning than often maintained. For example, Gregory (1969: 63) points out in the context of Ghana that the variability of rainfall in the Accra Plains (mean annual rainfall of *c.*760 mm) is 25 per cent, as is that in the uplands to the west of the Volta Lake (mean annual rainfall 1525 mm). He also points out that in Ghana, Northern Nigeria, and Sierra Leone variability falls within the limits to be found in Britain.

The question of periodicities and trends in rainfall is discussed in depth in Chapter 4, and large amounts of data are summarized in Nicholson *et al.* (1988).

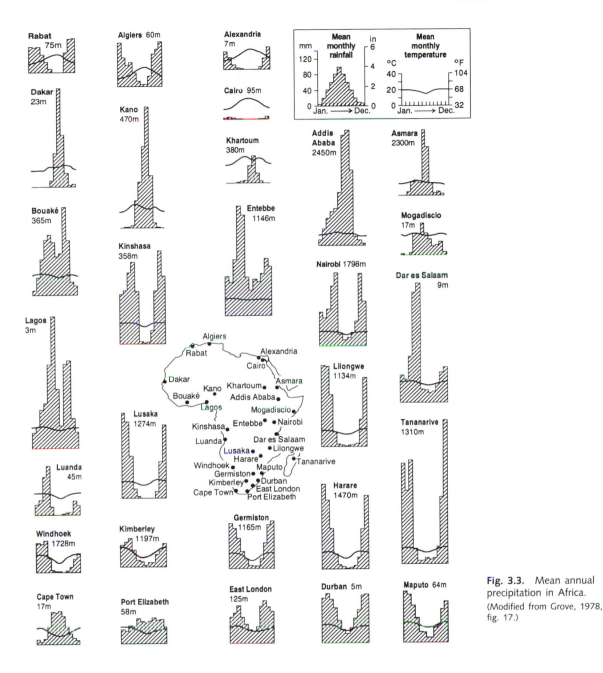

Fig. 3.3. Mean annual precipitation in Africa. (Modified from Grove, 1978, fig. 17.)

Temperature

Much of Africa shows only a small annual range of temperature (Figure 3.5). Indeed, about one-third of its area experiences an annual range that is below 6 °C. In the equatorial region there are some areas where the range is less than 3 °C. Much of Africa is also hot. Mean monthly maximum temperatures in excess of 32 °C are widespread, while about 30 per cent of the area is submitted to temperatures in excess of 38 °C. The highest values are found in the Sahara. Mean monthly minimum temperatures clearly show the effects of latitude, and only relatively small areas have values below 5 °C (the interior of Southern Africa and the northern Sahara and its margins).

The diurnal range in temperature values is often great

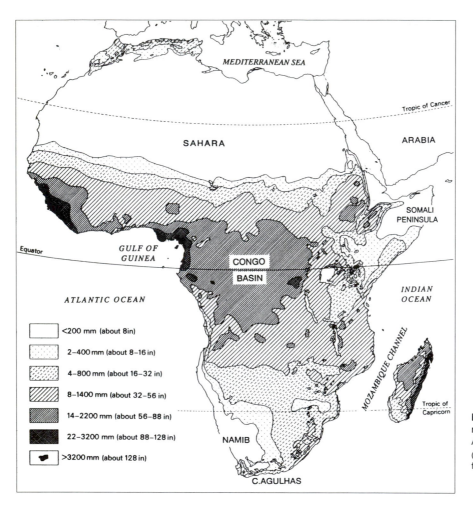

Fig. 3.4. Distribution of mean annual precipitation in Africa.
(Modified from Grove, 1978, fig. 17.)

and is often in excess of the annual range (Figure 3.5B). Over large areas the diurnal range exceeds 15 °C and reaches 19 °C in south-east Libya (Griffiths, 1972).

Evaporation

Because Africa has such a large share of its area in the tropics, its surface receives considerable inputs of solar heat and the radiation balance averages about 70 Kcal cm^{-2} yr^{-1} (UNESCO, 1978: 265). One consequence of this is that evaporation from the continent's surface is also considerable. A maximum of *potential evaporation* occurs in the Sahara, with values of over 2500 mm yr^{-1}, though for the continent as a whole the figure is about 1800 mm. Low values occur in the equatorial regions because of the humid climate, larger cloud cover, and lowered energy input. In the Gulf of Guinea region they may be little over 1000 mm. Likewise low values (around

1200 mm) are found at the higher latitude extremes. The overall latitudinal variation in potential evaporation is shown in Table 3.2 (derived from UNESCO, 1978: table 79).

When one considers *actual evaporation*, values show a great range because of the diversity of precipitation levels over the African continent—from a few mm in some parts of the Sahara to 1300 mm yr^{-1} in the middle portion of the Congo Basin. Table 3.3 gives estimates of the mean latitudinal amounts of evaporation (derived from UNESCO, 1978: table 80):

Climatic Hazards

Tropical Cyclones

Tropical cyclones (hurricanes) are rare over most of Africa and largely restricted in their distribution (Figure 3.6) to the eastern coast of Southern Africa, where they

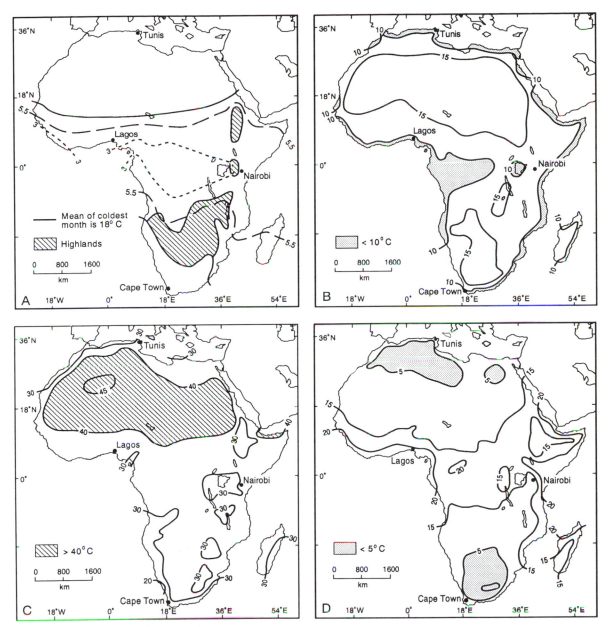

Fig. 3.5. Temperature conditions in Africa.
A. Annual temperature range (˚C). B. Mean annual temperature variation (˚C). C. Highest monthly maximum temperature (˚C). D. Lowest mean monthly minimum temperature (˚C). (Modified from Griffiths, 1972, figs. 12 and 13.)

form in summer (especially January and February) over the Indian Ocean. Usually between six and twelve storms are observed per annum, but few penetrate to 30˚S and also few penetrate inland (Preston-Whyte and Tyson, 1988). However, when such storms do cross the coast, copious rainfall and flooding occur. For example, in late January and early February 1984, Cyclone Domoina tracked over Mozambique, Swaziland, and South Africa, depositing as much as 900 mm of rainfall in a few days in some locations, and leading to severe flooding and

Table 3.2 Latitudinal variation in potential evaporation

Latitude (degrees)	Average annual potential evaporation (mm)
40–30 N	1700
30–20 N	2300
20–10 N	2100
10– 0 N	1400
0–10 S	1400
10–20 S	1600
20–30 S	1600
30–40 S	1400

Table 3.3 Latitudinal variation in actual evaporation

Latitude (degrees)	Average annual actual evaporation (mm)
40–30 N	200
30–20 N	30
20–10 N	400
10– 0 N	840
0–10 S	930
10–20 S	700
20–30 S	420
30–40 S	400

Note: The mean value for the whole continent is about 500 mm.

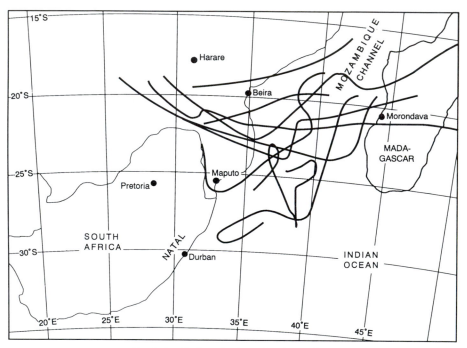

Fig. 3.6. Cyclone tracks in Southern Africa.
(Modified from Preston-Whyte and Tyson, 1988, fig. 11.4.)

geomorphological change (Plate 3.2), particularly in major river valleys like the Usutu (Goudie and Price Williams, 1984).

Hail

'In general', says Griffiths (1972: 28) 'hail is not a common phenomenon in Africa.' There are, however, exceptions to this: the area around Kericho, just south of the Equator in western Kenya, has as many as eighty hail storms per year 'a frequency that may not be exceeded anywhere in the world, save perhaps in a region of the Peruvian mountain chain' (ibid.).

Another major area for hail, not displayed in Figure 3.7 (derived from Griffiths), is the Highlands of Lesotho, where the point frequency of hail-days rises to over eight occurrences per year. On the other hand, areal frequencies of hail-days may be considerably higher with values of 100 days of hail per year being likely over the high plateau of Natal and Lesotho (Preston-Whyte and Tyson, 1988: 237).

Plate 3.2. The effects of Cyclone Domoina (1984) in Swaziland. (a) The breached railway line across the Lusutfu (Usutu) river; (b) floodplain sediment accretion in new point bars on the upper Ngwempisi River (photo: A. S. Goudie).

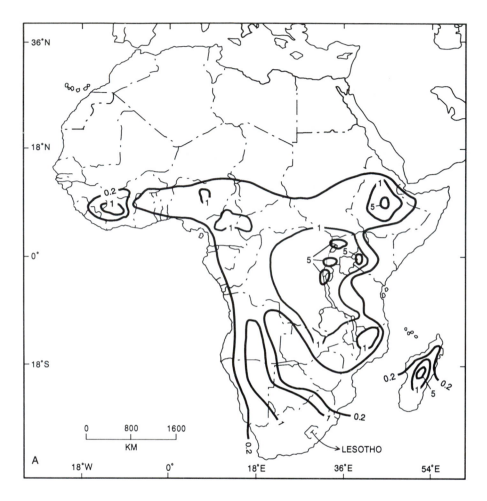

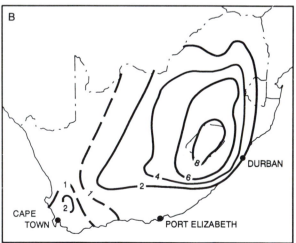

Fig. 3.7. A. Mean annual frequency of hail in Africa (10-year record).

(Modified from Griffiths, 1972, fig. 20.)

B. Frequency of hail days in Southern Africa.

(Modified from Schulze, in Preston-Whyte and Tyson, 1988.)

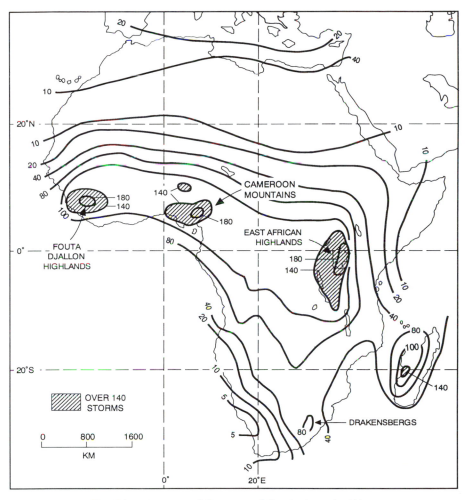

Fig. 3.8. Mean annual frequency of thunderstorms in Africa.
(Modified from Griffiths, 1972, fig. 21.)

Thunderstorms

Thunderstorms are highly significant features of the weather of Africa, with about 20 per cent of the continent experiencing more than 100 thunderstorm days each year and some areas having over 200. The distribution is shown in Figure 3.8. Note the concentration in the equatorial zone of East Africa and the high values associated with uplands in the Guinea Belt of West Africa. There is one anomalously high area of thunderstorm activity in the eastern interior of Southern Africa, where there are over eighty thunder days in the year.

Dust Storms

Wind-blown dust is an important phenomenon in the drier parts of Africa and the Sahara is the world's largest source of dust in the atmosphere, perhaps contributing as much as 50 per cent of the total (Middleton, 1986: 55). Some of the material from the Sahara is regularly transported away from the continent towards the Americas, Europe, and the Near East (Figure 3.9). There is evidence that as a result of a combination of low-rainfall qualities and increasing land-use pressures since the mid-1960s, the frequency of dust-storm events has increased in the Sahel and Sudan zones by four- to six-fold (Middleton, 1985). The main source areas include the Bodélé Depression alluvial plains in Niger and Chad, from which comes the Harmattan Dust Wind; an area that comprises southern Mauritania, northern Mali, and central-southern Algeria; southern Morocco, and western Algeria (Figure 3.10); the southern fringes of the

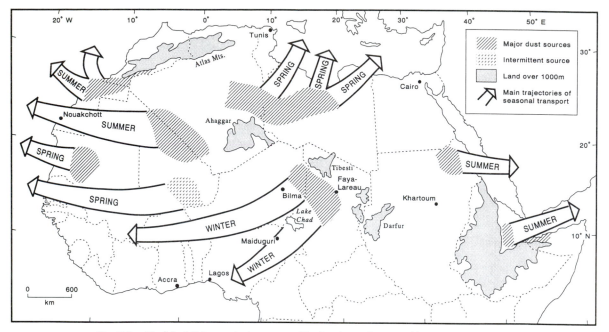

Fig. 3.9.   Model of Saharan dust sources with directions of seasonal long-range transport.
(After Middleton, 1986.)

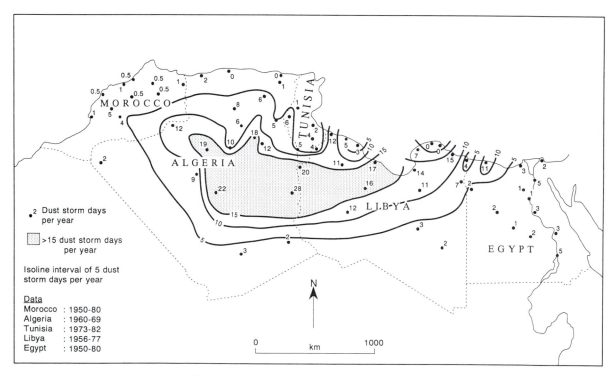

Fig. 3.10.   Distribution of dust storms in the northern Sahara.
(*Source*: Middleton, 1986.)

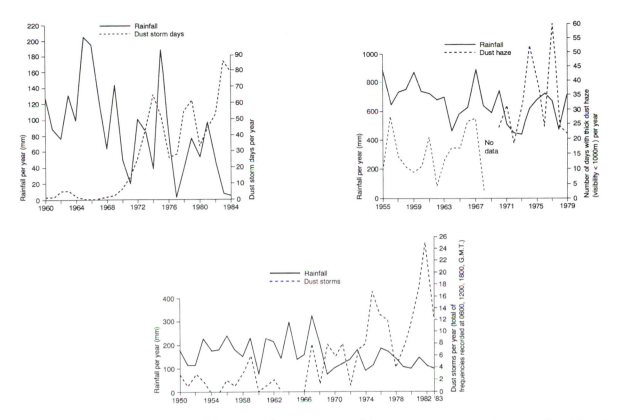

Fig. 3.11. A. Annual rainfall totals and dust storms frequencies at Nouakchott, Mauritania (1960–84). B. Annual rainfall totals and dust storm frequencies at Maiduguri, Nigeria (1955–79). C. Annual rainfall totals and dust storm frequencies at Khartoum, Sudan (1950–83).

Table 3.4 Average frequency of dust storms per annum

Maiduguri (Nigeria) (1955–79)	22.5
Nouakchott (Mauritania) (1960–84)	27.4
Sirte (Libya) (1956–77)	17.8
Khartoum (Sudan) (1968–78)	7.7
Tokar (Sudan) (1962–78)	47.2

Mediterranean Sea in Libya and Egypt; and northern Sudan. The Horn of Africa may also generate substantial amounts of dust, but Southern Africa seems a relatively insignificant source, though plumes of dust are generated from the Etosha and Magkadigadi Pans and from some surfaces in the Namib.

The data in Table 3.4 give the average frequency of dust storms per annum, defined as events when visibility was reduced to less than 1000 m.

Information on the weather systems that produce dust storms is given in Pye (1987: 101–6), while detailed studies of Saharan dust are given in Morales (1979).

Dust-storm activity shows considerable variability from year to year, but since the 1960s, there is an increasing incidence in dust storms, associated with runs of drought and increasing human disturbance of the ground surface (Figure 3.11).

Droughts

Droughts are perhaps the most serious climatic hazard of all and are discussed at great length for historical times and for the period of meteorological record in Chapters 4 and 5. Suffice it to say that the Sahel droughts since the mid-1960s and the drought of the early 1990s that afflicted Southern Africa have had profound implications for human welfare.

The Palaeoclimatology of Africa

Following the discussion of the present-day climatic situation in Africa, this chapter will now address the question of how the climatic environment has varied in the past, with particular reference to the Pleistocene and the early and mid-Holocene. The following chapter will then discuss the variable nature of African climates over recent decades and centuries. The issues that need to be considered are: the evidence for past climates; the effects of past climatic changes; the causes of such changes; and the dates of the major climatic events. Good local summaries of these issues are provided for the Sahara and its environs by Williams and Faure (1980), for East Africa by Hamilton (1982), and for Southern Africa by Vogel (1984) and Deacon and Lancaster (1988).

The first issue to discuss is the nature, consequences, and causes of the climatic changes of the late Cenozoic, and in particular to address the issue of declining temperatures over the last few tens of millions of years—the so-called 'Cenozoic climatic decline'.

The causes of this phenomenon are still not fully understood, but the trend seems to be associated with the breakup of the ancient supercontinent of Pangaea into the individual continents we know today. Round about 50 million years ago Antarctica separated from Australia and gradually shifted southwards into its present position centred over the South Pole. At the same time the continents of Eurasia and North America moved towards the North Pole. As more and more land became concentrated in high latitudes ice-caps could develop, surface reflectivity increased, and as a consequence climate probably cooled all over the world. Also, the Americas continued to move further west away from Africa, with accompanying expansion of the Atlantic Ocean and its cool east-side currents.

Late Cenozoic cooling may have promoted the expansion and development of some of the world's great deserts. Indeed in late Cenozoic times aridity became a prominent feature of the Saharan environment (Leroy and Dupont, 1994), probably because of the occurrence of several independent but roughly synchronous geological events (van Zinderen Bakker, 1984; Williams, 1985):

(i)   As the African plate moved northwards there was a migration of Northern Africa from wet equatorial latitudes (where the Sahara had been at the end of the Jurassic) into drier subtropical latitudes. Also, this caused a progressive narrowing and eventual closure of much of the Tethys, with accompanying reduction in maritime influences north and east of the African Plate.

(ii)   During the late Tertiary and Quaternary, uplift of the Tibetan Plateau had a dramatic effect on world climates, helping to create the easterly jet stream which now brings dry subsiding air to the Ethiopian and Somali deserts.

(iii)   The progressive buildup of polar ice-caps during the Cenozoic climatic decline created a steeper temperature gradient between the Equator and the Poles, and this in turn led to an increase in trade-wind velocities and their ability to mobilize sand into dunes.

(iv)   Cooling of the ocean surface may have reduced the amount of evaporation and convection in low latitudes, thus reducing the amount of tropical and subtropical precipitation.

Thus although the analysis of deep-sea cores in the Atlantic offshore from the Sahara indicates that some aeolian activity dates back to the early Cretaceous (Lever and McCave, 1983), it was probably around 2–3 Ma BP that a high level of aridity became established. From about 2.5 Ma BP the great tropical inland lakes of the Sahara began to dry out, and this is more or less contemporaneous with the time of onset of mid-latitude glaciation. Aeolian sands become evident in the Chad Basin at this time, and such palynological work as there is indicates substantial changes in vegetation characteristics (Servant-Vildary, 1973; Street-Perrott and Gasse, 1981).

If we consider the deserts of Southern Africa, we find that Seisser's investigation of offshore sediments (Seisser, 1978, 1980) has indicated that upwelling of cold waters intensified significantly from the late Miocene (7–10 Ma BP) and that the Benguela Current developed progressively thereafter. Pollen analysis of such sediments indicates that hyperaridity occurred throughout the Pliocene, and that the accumulation of the main Namib sand-sea started at that time. The deposits of the Kalahari Basin are more difficult to date and interpret in environmental terms and, as Thomas and Shaw (1991: 170) remark, the Kalahari sediments 'are much altered, are difficult to differentiate, contain few fossils and have low levels of preservation of organic matter. There is very little material suitable for radiometric dating.'

The Fossil Dunes of Africa

During the Pleistocene itself the dry areas of Africa showed repeated expansion and contractions of desert margins, as reflected in the extent of ergs and sand seas (see also Chapter 13).

In Southern Africa, the Kalahari sandveld is dominantly a fossil desert, now covered by a dense mixture of *acacia* and *mopane* forest, with grassland and shrubs. Relict dunes are widespread in Botswana, Angola, Zimbabwe, and Zambia (Thomas and Shaw, 1991) and may well extend as far north as the Congo rain forest zone (Figure 3.12). Several phases of dune activity have occurred in the late Pleistocene (Lancaster, 1989), though satisfactory dating has as yet proved elusive.

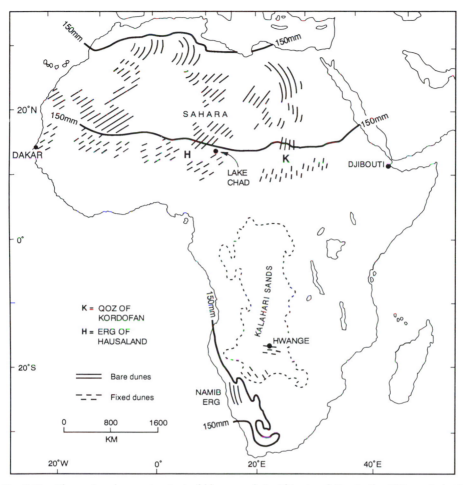

Fig. 3.12.   The past and present extent of blown sands in Africa in relation to the 150 mm isohyet. (Modified after Grove, 1978, fig. 6.)

North of the Equator, the fossil dune-fields extend south into the savanna and forest zone of West Africa, and have covered lateritic and other soils and invaded palaeolake basins. The so-called 'Ancient Erg of Hausaland' (Grove, 1958), extends into a zone where present rainfall is as high as 1000 mm per annum (Nichol, 1991). Many of the dunes in northern Nigeria are now cultivated, and in the vicinity of Lake Chad dunes have been flooded by rising lake waters. At one stage of the history of this area they blocked or altered the course of the River Niger. Indeed, in the middle Niger area there appear to be several ages of fossil dune, including old deeply-weathered linear dunes, and younger grey-brown and yellow dunes of lesser height. Details of age and pedogenesis are given in Völkel and Grunert (1990).

Farther east, in the Sudan, west of the White Nile, a series of fixed dunes known locally as *Qoz* cover most of the landscape up to the slopes of the Jebel Marra. The fixed dunes extend as far south as 10°N, and merge northwards, locally, with mobile dunes at about 16°N. They succeed in crossing the Nile, which thus probably dried up at the time of their formation. Again, as in West Africa there appear to have been at least two phases of dune activity. These two phases were interrupted by a relatively wet phase when extensive weathering and degradation took place. The first phase suggests a shift in the wind and rainfall belts of about 450 km southwards, and the second phase of dune building in Holocene times represents a shift of about 200 km (Grove and Warren, 1968).

Periods of aridity of the type indicated by these fossil dunes did, in marginal areas, have a marked effect on human activity, as witnessed by the clear hiatuses that exist in the archaeological record. As Wendorf *et al.* (1976: 113) remark, 'There are no traces anywhere in the Nubian Desert of any occupation, spring, or lacustrine deposits that are between the Aterian sites and the Terminal Palaeolithic in age. For this period of more than 30 000 years' duration the Western Desert of Egypt was apparently devoid of surface water and of any sign of life.'

However, while these ancient dunes indicate phases of past aridity, there is also evidence of formerly more humid conditions, as indicated by various lacustrine phenomena.

The Pluvial Lakes of Africa

During humid phases (lacustrals and pluvials) lake basins contained expanded bodies of water in many parts of Africa, including the driest parts of the Sahara. Large lakes occupied the salty *chotts* of North Africa, though their dates are the subject of controversy and may have been earlier than often thought (Fontes and Gasse, 1989). Lake deposits have also been found in some of the hyper-arid basins of Mali, northern Sudan, and the Western Desert in Egypt, though some of these may date back to the early Holocene (Petit-Maire, 1991; Pachur *et al.*, 1990; Brookes, 1989).

One of the largest and most spectacular of the pluvial lakes is that of Lake Chad. At more than one stage during the Pleistocene, Chad was considerably larger than it is at the present time. Chad at present stands at a height of 282 m above sea-level, but at some early stage a Chad river formed a 40 000 km² delta in association with a lake at 380–400 m. The lake then shrank during an arid phase of dune formation, but later again rose to 320–30 m, and formed a marked ridge or ridge complex, traceable over a distance of more than 1200 km. Between Maiduguri and Bama in north-east Nigeria this strandline is easily identifiable as a sand mound 12 m high (Grove and Warren, 1968).

To the east, the rift valleys of East Africa are occupied by numerous lakes around which occur late Pleistocene and early Holocene high strandlines. In Ethiopia, one of the biggest of the pluvial lakes, is 'pluvial Lake Galla'. This occurs to the south of Addis Ababa and is occupied by four shrunken remnants, Ziway, Langano, Abiyata, and Shala. However, when the lakes were larger, standing as much as 112 m above the present surface of Shala, the basin was occupied by one large sheet of water (Grove *et al.*, 1975). In Afar, Lake Abhé attained a surface area of 600 km² and a depth of more than 150 m (Gasse, in Rognon, 1976).

Farther south, the other lakes also display old strand-lines and lake sediments. Lake Awasa shows a series of terraces cut in volcanic debris at 10, 22, 33, and 40 m above the present lake surface, and lakes Margherita and Chamo have a 20–30 m terrace. Oyster (*Etheria*) shells have been found at 52 m above the present swamp level. Lake Chew Bahir is situated just to the north of the Kenya border, and increased discharge down the Sagan River appears to have led the lake, now either completely dry or seasonally flooded, to have reached a level of at least 20 m, creating fossil spits, and depositing incrustations of algal limestones on the old cliffs and islands (Grove *et al.*, 1975: 183) (Plate 3.3).

Even farther to the south similar evidence of high lake stands has been found in Kenya and Tanzania, and the chronology seems to have been similar to that in Ethiopia. A large body of water, for example, linked up Lakes Natron and Magadi (Roberts *et al.*, 1993). In Zambia, Lake Cheshi appears to have been considerably larger in the period 8000–4000 BP and rather lower between 15 000 and 13 000 BP (Stager, 1988), and Lake Albert (Uganda and Zaire) shows greater moisture from 9500–5000 BP (Ssemmanda and Vincens, 1993).

In Southern Africa relatively little research has been done on the pluvial lakes, and few dates are as yet available. However, some of the basins and lakes were much larger in the not too-distant past. In the northern part of the Kalahari sandveld tectonic adjustment and climatic change led to the formation of Palaeo-Makgadikgadi, encompassing the Okavango Delta, parts of the Chobe-Zambezi confluence, and the Ngami, Mababe, and Makgadikgadi basins of northern Botswana (Thomas and Shaw, 1991). Palaeo-Makgadikgadi may have covered an area of 120 000 km², making it second in size in Africa to Lake Chad at its Quaternary maximum. Despite considerable uncertainty about the dating of its highest stands, it is likely that its great volume was caused by the inflow of water from the Zambezi River. Another major lake basin in Southern Africa which was greatly expanded in the Late Pleistocene is the Etosha Pan of northern Namibia (Rust, 1984).

In recent decades, the isotopic dating of groundwater reserves has provided new information on those humid periods when groundwater was being recharged, and those when it was not. In the Sahara very little recharge occurred between 20 000 and 14 000 BP, indicating once again the existence of a period of very greatly reduced hydrological activity at the time of maximum glaciation (Sonntag *et al.*, 1980). Confirmatory dates come from the Sokoto basin of northern Nigeria (Geyh and Wirth, 1980), where there was a low rate of recharge from 20 000 to 10 000 BP.

The River Nile was transformed at this time. Alluviation occurred between 20 000 and 12 500 BP and sediments

Plate 3.3. Changing lake-levels in Lake Chew Bahir, southern Ethiopian Rift. (a) A fossil spit; (b) high-level calcareous stromatolites (photo: A. S. Goudie).

from this phase reveal that the river underwent metamorphosis to adopt an essentially braided form. It possessed a sandy floodplain occupied by a water flow that was much lower than that of the modern river, and perhaps only 10–20 per cent as large (Williams *et al.*, 1993: 112).

Post-glacial Times in the Sahara and Adjacent Regions

The dating of events during most of the African Quaternary is replete with difficulties, so that relatively little can be said with any certainty about events prior to the late Pleistocene. However, the reliability of information for the Holocene is more secure. (Street-Perrott and Perrott, 1993).

In the Sahara Desert it has for long been suspected that climatic conditions were wetter at some stage or stages in the Holocene than they are at present. This was deduced from facts such as the widespread distribution of rock paintings, and of human artefacts, in areas which are currently far removed from waterholes. Certain of the species represented in rock paintings, notably elephant, rhinoceros, hippopotamus, and giraffe

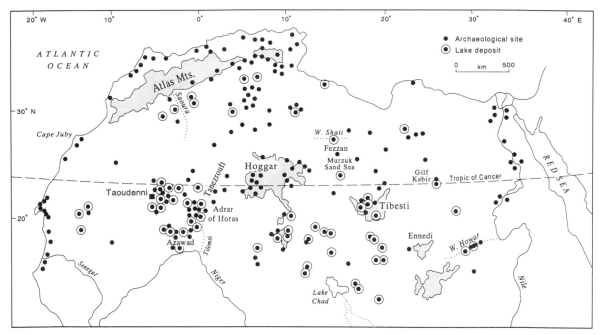

Fig. 3.13. The distribution of lake deposits and archaeological sites in Northern Africa at around 8000–9000 years BP. (Modified from Petit-Maire, 1991.)

were regarded as being representative of a moderately to strongly luxuriant savanna flora. Pollen analysis, though so far on a limited scale and subject to many doubts, has confirmed this essentially subjective archaeological evidence, and pollen of Aleppo Pine and other trees have been found in sediments of Holocene age in the Hoggar and other massifs. There are also now a large number of radio-carbon dates for lacustrine sediments in various parts of the desert which enables the sequence of events to be established with more certainty than hitherto. It seems likely on the basis of dates from Chad, Ténéré, Mali, the Nile valley, the Saoura valley, western Nubia, and the Hoggar that there were lacustral phases in the early Holocene and that during them vegetation was denser than at present.

At the Kharga Oasis and elsewhere in the Western Desert there are immense deposits of lime-rich spring tufas around or in which Neolithic tools have been found in great numbers. This indicates higher groundwater levels and a considerable population. The Neolithic was a time particularly favourable for human activities in the Sahara (Faure, 1966) (Figure 3.13).

A good example of early to mid-Holocene humidity in the dry heart of the eastern Sahara is provided by a study undertaken at Oyo (Ritchie and Haynes, 1987). Pollen spectra (Figure 3.14) dating from 8500 years BP

until around 6000 BP show that there were strong Sudanian elements in the vegetation, including pollen of tropical taxa. During this phase Oyo must have been a stratified lake surrounded by savanna vegetation similar to that now found 500 km to the south. After 6000 BP the lake became shallower and acacia thorn and then scrub grassland replaced the sub-humid savanna vegetation. At around 4500 years BP the lake appears to have dried out, aeolian activity returned, and vegetation disappeared except in wadis and oases. Roberts (1989) suggests that in effect the Sahara did not exist during most of the early Holocene. This is a point of view that is supported by the work of Petit-Maire (1989) in the western Sahara (p. 652): 'Biogeographical factors implicate total disappearance of the hyperarid belt at least for one or two milleniums before 7000 BP . . . The Sahel northern limit shifted about 1000 km to the north between 18 000 and 8000 BP and about 600 km to the south between 6000 BP and the present' (see Figure 3.13).

The Dating of African Lake Level Fluctuations

The dating of lake sediments by radiocarbon and other means enables lake history to be reconstructed for the

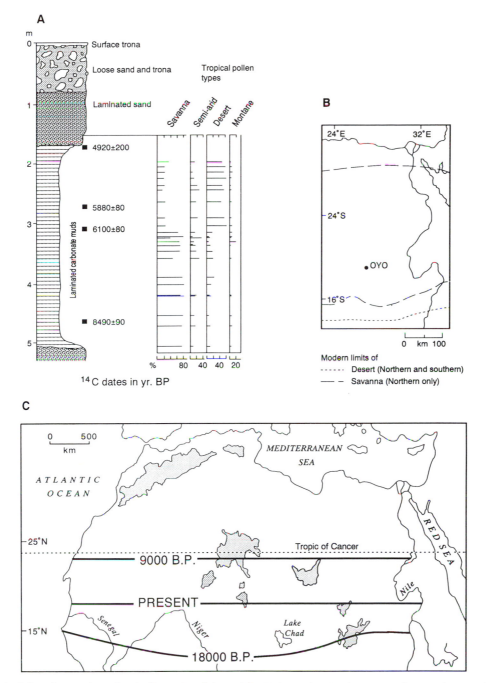

Fig. 3.14. A. Pollen diagram from Oyo in the eastern Sahara. (After Ritchie *et al.*, 1985, showing main features of stratigraphy and pollen sequence in the Holocene.) B. The location of Oyo. C. The changing position of the Sahara–Sahel limit. (*Source:* Petit-Maire, 1989, fig. 8.)

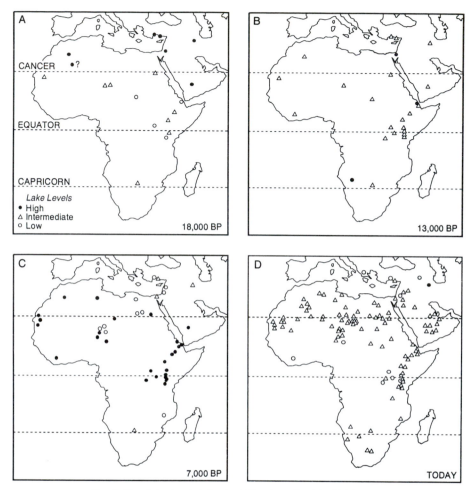

Fig. 3.15. Lake-level status in Africa since the late Pleistocene.
(*Source:* Street-Perrott *et al.*, 1985, figs. 8.5, 8.6, 8.8, and 8.11.)

last 30 000 years, and thereby gives a picture of humidity fluctuations over that time span. A synthesis of the available evidence was made by Street-Perrott *et al.* (1985), who believe in essence that the late Quaternary record in tropical Africa comprises two prolonged lacustral phases (from 27 000 to 21 000 BP and from 12 500 to 5000 BP) separated by an arid interval which spans the Last Glacial maximum (*c.*18 000 BP). In all they recognized five phases over the last 25 000 years:

(i)   25 000–17 000 BP

At around 18 000 BP water levels in most of northern intertropical Africa were low or intermediate and falling. Desiccation and deflation were widespread (Figure 3.15A).

(ii)   17 000–12 500 BP

This was a phase of prolonged and intense aridity, climaxing around 14 000–13 000 BP (Figure 15B). Lacustrine conditions may have occurred in some

northern upland environments (Tibesti and Jebel Marra?), perhaps as a result of orographic precipitation from westerly storms displaced southwards by the great northern ice sheets.

(iii)   12 500–10 000 BP

A transitional period. This coincides in part with an abrupt global cooling event at 11 000 to 10 000 BP called the Younger Dryas. Tropical lake-levels may have responded synchronously with the changes recorded in higher latitudes (Roberts *et al.*, 1993).

(iv)   10 000–5000 BP

A phase of Holocene warmth when moisture conditions showed a marked change. By 9000 BP a belt of high lake-levels extended from 4°S to 33°N (Figure 3.15C).

(v)   5000 BP–present

The belt of high lake levels from the previous phase disintegrated rapidly after 5000 BP and intermediate-level lakes became restricted to the comparatively narrow range (2°S–13°N) they occupy today (Figure 3.15D). The Saharan lakes, in particular, dried up.

There is still uncertainty with respect to the pattern in large areas and there need not have been a similar pattern in both hemispheres. For example, in the Kalahari the period from 16 000 to 13 000 BP was a humid episode, in contrast to the situation further north (Thomas and Shaw, 1991).

Glaciation in the African Mountains

Africa is more noted for its coral strands than for its icy mountains and under present conditions glaciers are of limited extent and area. However, in the Pleistocene some of the higher areas were glaciated, though the chronology is insecure and fragmentary, and there remains controversy in some areas as to whether or not the evidence for past glaciers is reliable.

The greatest development of glaciation was in the East African mountains (Mahaney, 1989), where an abundance of sound geomorphological and sedimentological evidence points to a much larger ice extent than now. The presently glaciated mountains, Kilimanjaro, Mt. Kenya, and Ruwenzori, have a total ice cover of around 10 km², but the same mountains, together with the currently unglaciated Aberdares and Mt. Elgon, had a maximum combined extent in the Pleistocene of 800 km² of glacier ice (Hastenrath, 1984). These mountains probably experienced multiple phases of glaciation, and Mahaney *et al.* (1991) have identified five Pleistocene and two Holocene glacial advance phases on Mt. Kenya.

Farther north there was extensive Pleistocene glaciation in the mountains of Ethiopia, with ice covering 10 km² in the Simen Mountains, possibly over 600 km² in the Bale Mountains, and 140 km² on Mount Badda (Messerli *et al.*, 1980).

It is also possible that there was limited glaciation in the Saharan mountains (e.g. Tibesti), but the extent of glaciation (if any) is Southern Africa is a matter of speculation. Whereas there is evidence for periglacial activity, most workers reject the notion of glaciation in the Drakensbergs, but there is a possibility of glaciation in the mountains of the western Cape (Gellert, 1991).

Quaternary Faunal and Flora Changes: Speculations on the Role of Climatic Change

The massive environmental changes suggested by fossil dunes, ancient lakes, and expanded glaciers have led to changes in floral and faunal distributions. The classic example of this is the distribution of the Nile crocodile (*Crocodylus niloticus*). It was formerly present in rivers from the eastern Cape Province to the Nile. Today it is found in pools in the Tibesti Massif in the heart of the Sahara, 1300 km from either Niger or Nile, and clearly isolated. There is no likelihood of natural migration there across the arid Saharan wastes, given present hydrological conditions, so that pluvial conditions presumably played a role. Fossil crocodile remains are widely reported from Holocene lake beds in currently hyper-arid areas (see, for example, Pachur *et al.*, 1990).

Another example illustrates the way in which the flora of the East African mountains has become isolated. The distinctive tree heath (*Erica arborea*) occurs in disjunct areas including the Ethiopian Highlands, the mountains of Ruwenzori, the Cameroon Mountains in West Africa, and the peaks of the Canary Islands off the west coast. Once again it seems likely that post-glacial changes in temperature and rainfall have led to this distribution. In general, because of its high mean elevation, the African continent would have been particularly severely affected by temperature depression in the glacials. The effect of a temperature depression of 5 °C would have been to bring down the main montane biomes from around 1500 m to 700 or 500 m (Moreau, 1963) (Figure 3.16). Instead of occupying a large number of islands as it does now, and as it must have done in interglacials, the montane type of biome would have occupied a continuous block from Ethiopia to the Cape, with an extension to the Cameroons. The strictly lowland biomes, comprising species today that do not enter areas above 1500 m, would, outside West Africa, have been confined to a coastal rim and to two isolated areas inland (the Sudan and the middle of the Congo Basin). Further massive changes in African biomes would have been occasioned by changes in humidity as well as of temperature. In West Africa, where the great sand ergs of the Sahara encroached as much as 500 km on the moister coastal regions, the southward movement of the vegetation belts cannot have failed to have had powerful effects on flora and fauna. Indeed, since the present West African rain forests reach inland by no more than 500 km, if the entire system of vegetation belts had shifted south as much as did the Saharan dunes, then the whole of the West African forests would have been eliminated against the coastline. The present richness, however, of the West African forests, and the existence of so many endemic species there, makes it virtually certain that this did not happen, but the effect of the southward advance must have been formidable.

An indication of the timing and degree of rain-forest disruption in West Africa is suggested by pollen analyses undertaken in Lake Bosumtwi in southern Ghana

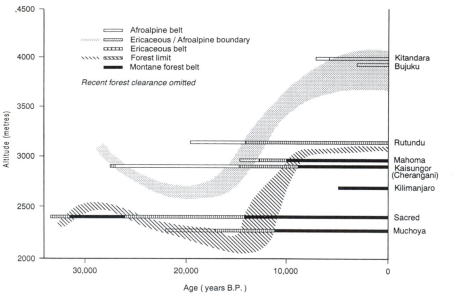

Fig. 3.16. Late Quaternary vegetational changes in selected sites from the East African mountains.
(Modified from Flenley, 1979.)

(Maley, 1989). These pollen show that at the time of the last glacial maximum, and especially between 20 000 and 15 000 BP, the lake had a very low level. Moreover, the principal pollen results show that before about 9000 BP forest was largely absent in the vicinity. Indeed, between the present and c.8500 BP arboreal pollen percentages oscillated from 75 to 85 per cent, while before 9000 BP they were generally below or close to 25 per cent. During the period from 19 000 to 15 000 BP arboreal pollen percentages reached minimum values of about 4 and 5 per cent. Trees had in effect been replaced at that time by herbaceous plants, *Gramineae* and *Cyperaceae*.

For about seventy years work has been undertaken to see how the environmental changes of East Africa might have influenced the distribution characteristics of the water fauna of the great rift valley lakes. During pluvial or lacustral phases the rivers and lake basins of East Africa would have been linked up to a greater extent than they are now. During dry interpluvial phases on the other hand the lakes would have become partly or wholly desiccated, and the linkage of water bodies would have been reduced. Such fluctuations would lead to alternations of contact and isolation for fauna, and complete lake desiccation would have led to the extermination of many species in any one basin. By a study of present and fossil fish types, and also of crocodiles, the consequences of these changes can be identified, and this helps to unfathom the anomalies of zoogeography (Beadle, 1981).

Lake Turkana, between Kenya and Ethiopia, is now not connected with the Nile, having no outlet. On the other hand, it has a fauna very similar to that in the Nile. The explanation seems to be that the lake level once stood much higher as evidenced by abandoned shorelines. When this was so the lake could have overflowed via a gorge, now dry, to the Sobat River, and hence to the Nile. This former connection would explain the faunal similarities. There are nevertheless, twelve endemic species of fish in Turkana, and the Nile perch has divided into two subspecies. The Nile perch also occurs in Lakes Chew Bahir, Abaya, and Chamo. These too were formerly connected to Turkana (Grove *et al.*, 1975: 183).

Desiccation of the lakes explains the differences between fossil and current faunas and also helps to explain the absence of some species from some of the lake basins. For example, Lake Edward in Uganda possesses no crocodiles, even though they are found in Lake Victoria, and in the Semliki River. It is difficult to explain the fact that the crocodiles have not managed to pass beyond the Semliki Gorge to Lake Edward, but it is probable that the falls of the Gorge, and the dense forest on either side, have acted as barriers. However, in fossil beds which border the shores of the Kazinga Channel, numerous teeth, scales, and bones of crocodiles have been found. Thus these reptiles once lived in large numbers in the lake, and so it appears likely that their present absence can be explained by the desiccation of the lake, and by the natural barriers which have prevented recolonization. Violent volcanic eruptions may also have led to their destruction.

Lake Victoria also largely dried up in the late Pleistocene (Kendall, 1969), and cores in other lake beds also indicate that many of the lakes became highly alkaline, if not actually dry, at some point. The only fish that could survive the very dry conditions of the interpluvials would be the lung- and mud-fish, for they can burrow into mud and live there for long periods. These two species are most widely distributed in the Nile and the lakes at the present time, for they still live both above and below both the Murchison and Semliki Falls. Unlike the lung- and mud-fish, many other fish were killed off by the aridity, and only partial recolonization has been possible because of the barriers presented by these two falls.

Elsewhere in Africa there are further zoogeographical anomalies of interest. For instance, there are some species of fish which are common to all the major basins of the Sudan belt: Senegal, Gambia, Volta, Niger, Chad, and Nile. The fish fauna is remarkably uniform over this enormous area, which is all the more surprising when one realizes that there are now more than 1600 km of desert separating the Nile from Lake Chad. This similarity may be explained by the presence of more rivers and lakes during the humid phases. These would provide the necessary linkages (Beadle, 1981: 201). Likewise, it was a surprise to zoologists when French expeditions during the early years of this century discovered that the Sahara itself supports a considerable, though a widely scattered, fresh-water vertebrate fauna. In permanent but isolated water-holes from Biskra to Tibesti there are the remains of former tropical African fish fauna species that are now very widespread in tropical Africa. They have become isolated as a result of a decrease in humidity (Beadle, 1981: 208).

Pleistocene Atmospheric Circulation

Having described the origins of deserts and of the late Cenozoic climate decline together with the nature of some of the environmental changes of the Pleistocene and Holocene it is worth considering the general atmospheric circulation pattern with which these last changes were associated. Thermal factors play a dominant role in determining its form, so that the thermal variations provoked by the growth and decay of the great Pleistocene ice sheets decisively influenced the patterns of the atmospheric circulation. For example, in theory an increased temperature gradient resulting from the presence of a greatly expanded northern hemisphere ice sheet, would result in stronger westerlies, an equatorial displacement of major circulation features, and an intensification and shrinking of the Hadley Cells and associated subtropical high-pressure zones. Temperature gradients would also influence the location of the transition between the Hadley Cell zone and the zone of extra-tropical Rossby Wave circulation. It might also influence the number and position of the Rossby Waves—the dominant control of mid-latitude weather.

At the glacial maximum there would also have been a decreased thermal contrast between the two hemispheres. At present the southern hemisphere, in comparison with the northern, is much cooler and its temperature gradient greater. This is because of the varying amounts and distribution of land and ocean in the two hemispheres. In the southern hemisphere the stronger temperature gradient produces a more intense circulation, and is probably largely responsible for the asymmetry that exists whereby the meteorological Equator (or Intertropical Convergence Zone—ITCZ), lies in the northern hemisphere. During a glacial phase intense continental glaciation in the northern hemisphere should have led to a displacement of this meteorological Equator to a position more coincident with the geographical Equator (i.e. southwards).

Furthermore surface temperature changes would affect precipitation levels through their influence on evaporation rates and on the stability of the atmospheric column.

What then was the likely situation round about 18 000 years ago? Nicholson and Flohn (1981) propose that the large changes in the northern hemisphere and the smaller changes in the southern hemisphere would have reduced the thermal contrast between the hemispheres. Moreover, the great continental ice sheets in the north would have increased the hemispheric temperature gradients causing circulation features to be displaced equatorwards, and intensifying the subtropical high-pressure zone. The result would have been a greater degree of westerly flow and mid-latitude cyclones over an area like North Africa, while increased oceanic upwelling resulting from the stonger subtropical high pressure (together with lower sea-surface temperatures) would both have reduced evaporation levels and the energy available for convection and storm formation. Futhermore, the decreased thermal contrast between the hemispheres would have displaced the ITCZ southwards, disrupting the annual march of the monsoon. However, it needs to be pointed out that the Nicholson and Flohn model is only one of several models that has been proposed to account for climatic conditions at the Last Glacial Maximum and that alternative explanations exist (see Deacon and Lancaster, 1988: 18–28).

Although this chapter has outlined the nature and

causes of present climatic conditions and those of the prehistoric era, historical and contemporary climatic changes and fluctuations are also of great significance and are the subject of the next chapter.

References

Beadle, L. C. (1981), *The Inland Waters of Tropical Africa: An Introduction to Tropical Limnology* (2nd edn.) (London).

Brookes, I. A. (1989), 'Early Holocene Basinal Sediments of the Dakleh Oasis Region, South-Central Egypt', *Quaternary Research*, 32: 139–52.

Deacon, J. and Lancaster, N. (1988), *Late Quaternary Palaeoenvironments of Southern Africa* (Oxford).

Flenley, J. R. (1979), *Equatorial Rain Forests: A Geological History* (London).

Fontes, J. Ch. and Gasse, F. (1989), 'On the Ages of Humid Holocene and Late Pleistocene Phases in North Africa', *Palaeogeography, Palaeoclimatology, Palaeoecology*, 70: 393–8.

Gellert, J. F. (1991), 'Pleistozän-Kaltzeitliche Vergletscherungen im Hochland von Tibet und im Sudafrikanishem Kapgebirge', *Eiszeitalter und Gegenwart*, 41: 141–5.

Geyh, M. A. and Wirth, K. (1980), '^{14}C Ages of Confined Groundwater from the Gwandu Aquifer, Sokoto Basin, Northern Nigeria', *Journal of Hydrology*, 48: 281–8.

Goudie, A. S. and Price-Williams, D. (1984), 'Cyclone Domoina, Swaziland, January 1984: Swaziland's Severest Natural Hazard of the Century', *Journal of Meteorology*, 9: 273–81.

Gregory, S. (1969), 'Rainfall Reliability', in M. F. Thomas and G. W. Whittington (eds.), *Environment and Land Use in Africa* (London), 57–82.

Griffiths, J. F. (1972) (ed.), *Climates of Africa* (Amsterdam).

—— (1987), 'Climate of Africa', in J. E. Oliver and R. W. Fairbridge (eds.), *The Encyclopedia of Climatology* (New York).

Grove, A. T. (1958), 'The Ancient Erg of Hausaland and Similar Formations on the South Side of the Sahara', *Geographical Journal*, 157: 13–24.

—— (1978), *Africa* (Oxford).

—— and Warren, A. (1968), 'Quaternary Landforms and Climate on the South Side of the Sahara', *Geographical Journal* 134: 194–208.

—— Street, F. A., and Goudie, A. S. (1975), 'Former Lake Levels and Climatic Change in the Rift Valley of Southern Ethiopia', *Geographical Journal*, 141: 177–94.

Hamilton, A. C. (1982), *Environmental History of East Africa: A Study of the Quaternary* (London).

Hasternrath, S. (1984), *The Glaciers of Equatorial East Africa* (Dordrecht, Neth.).

Kendall, R. L. (1969), 'An Ecological History of the Lake Victoria Basin', *Ecological Monographs*, 39: 121–76.

Kendrew, W. G. (1961), *The Climates of the Continents* (Oxford).

Lancaster, N. (1989), 'Late Quaternary Palaeoenvironments in the South-western Kalahari', *Palaeogeography, Palaeoclimatology, Palaeoecology*, 70: 367–76.

Leroy, S. and Dupont, L. (1994), 'Development of Vegetation and Continental Aridity in North-Western Africa during the Late Pliocene: The Pollen Record of ODP Site 658', *Palaeogeography, Palaeoclimatology, Palaeoecology*, 109: 295–316.

Lever, A. and McCave, I. N. (1983), 'Eolian Components in Cretaceous and Tertiary North Atlantic Sediments', *Journal of Sedimentary Petrology*, 53: 811–32.

Mahaney, W. C. (1989) (ed.), *Quaternary and Environmental Research on East African Mountains* (Rotterdam).

—— Harmsen, R., and Spence, J. R. (1991), 'Glacial and Interglacial Cycles and Development of the Afroalpine Ecosystem on East African Mountains: Glacial and Postglacial Geological Record and Paleoclimate of Mount Kenya', *Journal of African Earth Sciences*, 12: 505–12.

Maley, J. (1989), 'Late Quaternary Climatic Changes in the African Rain Forest: Forest Refugia and the Major Role of Sea-Surface Temperature Variations', in M. Leinen and M. Sarnthein (eds.), *Paleoclimatology and Palaeometeorology: Modern and Past Patterns of Global Atmospheric Transport* (Dordrecht, Neth.) 585–616.

Messerli, B., Winninger, M., and Rognon, P. (1980), 'The Saharan and East African Uplands during the Quaternary', in M. A. J. Williams and H. Faure (eds.), *The Sahara and the Nile* (Rotterdam), 87–118.

Middleton, N. J. (1985), 'Effect of Drought on Dust Production in the Sahel', *Nature*, 316: 431–4.

—— (1986), 'The Geography of Dust Storms', unpub. D.Phil. thesis, University of Oxford.

—— Goudie, A. S. and Wells, G. L. (1986), 'The Frequency and Source Areas of Dust Storms', in W. G. Nickling (ed.), *Aeolian Geomorphology* (Boston), 237–60.

Morales, C. (1979) (ed.), *Saharan Dust* (Chichester).

Moreau, R. E. (1962), 'Vicissitudes of the East African Biomes in the Late Pleistocene', *Proceedings Zoological Society of London*, 141: 395–421.

Nicholson, S. E. and Flohn, H. (1981), 'African Climate Changes in Late Pleistocene and Holocene and the General Atmospheric Circulation', *IAHS Publication*, 131: 295–301.

—— Jeeyoung, Kim, and Hoopingarner, J. (1988), *Atlas of African Rainfall and its Interannual Variability* (Tallahassee, Fla).

Pachur, H. J., Kröpelin, S. M., Hoelzmann, P., Goschin, M., and Altmam, N. (1990), 'Late Quaternary Fluvio-lacustrine Environments of Western Nubia', *Berliner Geowissenschaft Abhandlungen* A, 120.1: 203–60.

Parker, D. E., Folland, C. K., and Ward, M. N. (1988), 'Sea-surface Temperatures Anomaly Patterns and Prediction of Seasonal Rainfall in the Sahel Region of Africa', in S. Gregory (ed.), *Recent Climatic Change* (London), 166–78.

Petit-Maire, N. (1989), 'Interglacial Environments in Presently Hyperarid Sahara: Palaeoclimatic Implications', in M. Leinen and M. Sarnthein (eds.), *Palaeoclimatology and Palaeometeorology: Modern and Past Pattens of Global Atmospheric Transport* (Dordrecht, Neth.), 637–61.

—— (1991) (ed.), *Paléoenvironnements du Sahara: lacs holocènes à Taoudenni* (Mali) (Paris).

Preston-Whyte, R. A. and Tyson, P. D. (1988), *The Atmosphere and Weather of Southern Africa* (Cape Town).

Pye, K. (1987), *Aeolian Dust and Dust Deposits* (London).

Rasumsson, E. M. (1988), 'Global Climate Change and Variability: Effects on Drought and Desertification in Africa', in M. H. Glantz (ed.) *Drought and Hunger in Africa: Denying Famine a Future* (Cambridge).

Ritchie, J. C. and Haynes, C. V. (1987), 'Holocene Vegetation Zonation in the Eastern Sahara', *Nature*, 330: 645–7.

—— Eyles, C. H., and Haynes, C. V. (1985), 'Sediment and Pollen Evidence for an Early to Mid-Holocene Humid Period in the Eastern Sahara', *Nature*, 314: 352–5.

Roberts, N. (1989), *The Holocene* (Oxford).

—— Taibeb, M., Barker, P., Damnati, B., Icole, M., and Williamson, D. (1993), 'Timing of the Younger Dryas Event in East Africa from Lake-level Changes', *Nature*, 366: 146–8.

Rognon, P. (1976) (ed.), 'Oscillations climatiques au Sahara depuis 40 000 ans', *Revue de géographie physique et de géologie dynamique*, 18: 147–282.

Rowell, D. P., Folland, C. K., Maskell, K., Owen, J. A., and Ward, M. N. (1992), 'Modelling the Influence of Global Sea-surface Temperatures on the Variability and Predictability of Seasonal Sahel Rainfall', *Geophysical Research Letters*, 19: 905–8.

Seisser, W. G. (1978), 'Aridification of the Namib Desert: Evidence from Ocean Cores', in E. M. Van Zinderen Bakker (ed.), *Antarctic Glacial History and World Palaeoenvironments* (Rotterdam), 105–13.

—— (1980), 'Late Miocene Origin of the Benguela Upwelling System off Northern Namibia', *Science*, 208: 283–5.

Servant-Vildary, S. (1973), 'Le Plio-Quaternaire ancien du Tchad: Évolution des associations de diatomées stratigraphie, paléogéographie', *Cahiers ORSTOM Serie Géologique*, 5: 169–217.

Ssemmanda, I. and Vincens, A. (1993), 'Végétation et climat dans le Bassin du lac Albert (Ouganda, Zaïre) depuis 1300 ans BP: Apport de la palynologie', *Comptes rendus de l'academic des sciences*, 316, série II, No. 4: 561–7.

Street-Perrott, F. A. and Gasse, F. (1981), 'Recent Development in Research into the Quaternary Climatic History of the Sahara', in J. A. Allen (ed.), *The Sahara: Ecological Change and Early Economic History* (London), 7–28.

—— and Perrott R. A. (1993), 'Holocene Vegetation, Lake Levels and Climate of Africa', in H. E. Wright, J. E. Kutzbach, T. Webb, W. F. Ruddimer, F. A. Street-Perrott, and P. J. Bartlein (eds.), *Global Climates since the Last Glacial Maximum* (Minneapolis), 318–56.

—— Roberts, N. and Metcalfe, J. (1985), 'Geomorphic Implications of Late Quaternary Hydrological and Climatic Changes in the Northern Hemisphere Tropics', in I. Douglas and T. Spencer (eds.), *Environmental Change and Tropical Geomorphology* (London), 165–83.

Thomas, D. S. G. and Shaw, P. A. (1991), *The Kalahari Environment* (Cambridge).

Trewartha, G. T. (1981), *The Earth's Problem Climates* (Madison).

UNESCO (1978), *World Water Balance and Water Resources of the Earth* (Paris).

Van Zinderen Bakker, E. M. (1984), 'Elements for the Chronology of Late Cainozoic African Climate', in W. C. Mahaney (ed.) *Correlation of Quaternary Chronologies* (Norwich), 23–37.

Vogel, J. C. (1984) (ed.), *Late Cainozoic Palaeoclimates of the Southern Hemisphere* (Rotterdam).

Völkel, J. and Grunert, J. (1990), 'To the Problem of Dune Formation and Dune Weathering during the Late Pleistocene and Holocene in the Southern Sahara and the Sahel', *Zeitschrift für Geomorphologie*, 34: 1–17.

Wendorf, F., Schild, R., Said, R., Haynes, C. V. Gautier, A., and Kobusiewicz, P. (1976), 'The Prehistory of the Egyptian Sahara', *Science*, 193: 103–16.

Williams, M. A. J. (1985), 'Pleistocene Aridity in Tropical Africa, Australia and Asia', in I. Douglas and T. Spencer (eds.), *Environmental Change and Tropical Geomorphology* (London), 219–33.

—— and Faure, H. (1980) (eds.), *The Sahara and the Nile* (Rotterdam). Balkema.

—— Dunkerley, D. L., De Deckker, B., Kershaw, A. P., and Stokes, T. (1993), *Quaternary Environments* (London).

4 Environmental Change Within the Historical Period

Sharon E. Nicholson

Introduction

The African environment poses many hardships for the peoples who have inhabited the continent for millions of years. The ever-present heat and humidity in the low-latitudes, and the ever-present heat and aridity almost everywhere else, offer little human comfort. In between these extremes are the semi-arid regions, where there is little moderation of these conditions but where the environment fluctuates rapidly both within the course of the year and from year to year.

In such regions the course of human history is closely coupled with that of the environment; the same holds true for the continent's future. For these reasons, a valuable insight can be gained through knowledge of the region's environmental history and understanding of its link to human affairs. Would the recent droughts of the Sahel have been so disastrous if colonial development had not been geared to an environment more favourable than the present one, or had not a major rainfall decline occurred just at the time most countries gained

Much of the research for this article was carried on during the author's sabbatical at the University of Botswana. She has studied many aspects of climatic variability in Africa. She would like to acknowledge the support of the Fulbright Foundation, the National Science Foundation Grant No. ATM-9024340, and the National Institute of Development Research and Documentation. She would also like to thank numerous colleagues and friends in Botswana for their guidance and vigorous discussions, especially Paul Shaw, Peter Smith, Map Ives, M. B. M. Sekhwela, and Pauline Dube. The Meteorological Services in nearly all African countries generously provided data used in many of the analyses presented. Lastly, the author would like to thank Dick Grove for the enthusiasm for Africa he helped to impart and the encouragement he provided while she was a graduate student.

independence, or had the drought not corresponded with a period of political and economic struggles in the newly liberated nations? Knowledge of past environments can help us understand human history and can provide lessons to enable society to adapt better to environmental constraints. It can also help us to ascertain the degree to which recent environmental changes have been natural or induced by human activities, a controversy at the heart of such issues as desertification (see Chapter 19).

This chapter will summarize what is known about African climate during historical times. Since much of this knowledge is inferred from general environmental conditions, the material presented at the same time provides a general sketch of recent environmental history. The chapter will commence with a discussion of the methodology of historical climate reconstruction, particularly as applied to Africa. Then a few of the long-term indicators of climate and environment will be examined, followed by regional descriptions covering most of the last millenium until about 1900, when the instrumental record becomes quite detailed (see Chapter 5 in this volume). Finally the material will be synthesized as a general climatic chronology for the continent and will be evaluated in the context of global climatic history.

Since in Africa environmental changes are most directly related to rainfall, only this aspect of climate will be stressed. For a discussion of historical temperature fluctuations, the reader is referred to a recent summary by Tyson and Lindesay (1992). This chapter will also emphasize the last three centuries. For more detail on earlier historical fluctuations, one is referred to a summary of conditions in tropical Africa during the Little Ice Age by Grove (1992).

The Methodology of Historical Climate Reconstruction

The term 'historical' roughly refers to the period during which some form of written records are available. For the world at large, this is probably a few thousand years. For Africa, with the exception of Egypt, this is at most a few hundred to perhaps 1500 years.

One of the best methods of establishing a historical record of climate is written documentation, especially actual records of weather or related phenomena such as river flow. Lacking this, rigorous methodologies have been developed to utilize tree-rings to reconstruct past climate (Fritts, 1976). These have been tremendously successful in deriving knowledge of past climates in the United States and numerous other regions. In some instances high-resolution pollen sequences or geologic records in lakes provide information on time-scales ranging from years to decades. In Africa, however, this is rarely the case: most trees are not suitable for dendroclimatological studies; most peoples kept few written chronicles of their history; and there are formidable logistic problems in the field-research required for pollen and geological studies. Moreover, outside of Algeria and South Africa, climatic records rarely extend back more than a century. For these reasons, alternative methodologies must be developed which utilize the materials which are readily available for Africa.

Sources

The basis of the method of historical climate reconstruction utilized here is simple. Historical and geographical sources, such as archives, chronicles, travel journals, settlers' diaries, ships' logs, maps, and early geographical journals contain useful information on landscape and vegetation, lakes and rivers, famine, drought and floods, as well as climatic descriptions and, frequently, actual meteorological observations. Such information, as listed in Table 4.1, can be used as 'proxy' data for climate and weather observations. Geological, archaeological, and dendroclimatological studies provide additional information. Colonial weather observations and other weather observations lost from the modern climatic record complement these geological, historical, and geographical sources. Thus, the methodology is quite multidisciplinary.

Clearly, such observations are not subject to the scientific rigour associated with modern climatological and geographical methods and there are numerous difficulties in the use of such information to reconstruct climate. Local chronicles and historical descriptions record

Table 4.1. Types of data useful for historical climatic reconstructions, particularly with respect to Africa

I Landscape descriptions
1. Forests and vegetation: are they as today?
2. Conditions of lakes and rivers:
   (a) height of annual flood, month of maximum flow of the river
   (b) villages directly along lakeshores
   (c) size of the lake (e.g. as indicated on map)
   (d) navigability of rivers
   (e) desiccation of present-day lakes or appearance of lakes no longer existing
   (f) floods
   (g) seasonality of flow: condition in wet and dry seasons
3. Wells, oases, bogs in presently dry areas
4. Flow of wadis
5. Measured height of lake surfaces (frequently given in travel journals, but optimally some instrumental calibration or standard should accompany this)

II Drought, famine, and other agricultural information
1. References to famine or drought, preferably accompanied by the following:
   (a) where occurred and when occurred: as precisely as possible
   (b) who reported it; whether the information is second-hand
   (c) severity of the famine or drought; local or widespread?
   (d) cause of the famine
2. Agricultural prosperity:
   (a) condition of harvest
   (b) what produced this condition
   (c) months of harvests, in both bad years and good years
   (d) what crops grown
3. Wet cultivation in regions presently too arid

III Climate and meteorology
1. Measurements of temperature, rainfall, etc.
2. Weather diaries
3. Descriptions of climate and the rainy season: when do the rains occur, what winds prevail?
4. References to occurrence of rain, tornadoes, storms
5. Seasonality and frequency of tornadoes, storms
6. Snowfalls: is this clearly snow or may the reporter be mistakenly reporting frost, etc?
7. Freezing temperatures, frost, hail
8. Duration of snow cover on mountains (or absence)
9. References to dry or wet years, severe or mild winters, other unusual seasons

the fallible witness of memory or oral tradition; the observer selectively decides what information to include; and dates are often estimated. Weather reports in journals, diaries, or historical treatises on climate frequently contain second-hand information for which the location and date of origin may be obscured; they may also present theory as fact or observation. The climate of the whole West African coast might be judged from observations at one or two places, such as Dakar or Accra, during one or two possibly quite unrepresentative years. Old instrumental measurements lack a calibration for direct comparison with current observations.

Reports such as those listed in Table 4.1 are also subject to observer bias. It may be in a writer's interest to exaggerate the harshness or favourableness of a climate or environment. The information is relative and/or subjective: lacking an observing standard, the witness interprets it in terms of past experience. 'Forests' mentioned in accounts of the Sahel near Timbuktu may bear little resemblance to our concept of forests; the term may refer to sparse, isolated stands of trees which exist even today. Likewise, a European observer from a region of dense forests may consider any relatively dry landscape a desert. References to dry and wet years or severe winters are quite subjective as well as relative: they refer to departures from a past set of mean conditions. A 'dry' year in the seventeenth century may have been wet by present standards.

Similarly, to be useful, reports of fauna, vegetation, lakes, and rivers must be compared with present conditions. However, reports of changes of lake levels, river flow, or rainfall may provide information about short-term climatic anomalies even without a modern frame of reference. Likewise, the presence on a map of a lake in a now desiccated region, or knowledge of the precise location of a lakeshore, constitutes absolute evidence.

Reports of famine are quite widespread in historical literature on Africa, but it is another relative condition which is difficult to interpret. It essentially implies a bad harvest, a condition which can result from drought as well as from war, pestilence, bad agricultural management, or even floods. Even a famine for which drought is clearly indicated as the cause is not an unambiguous indicator of precipitation. The mean precipitation and its variability determine a region's sensitivity to famine and drought. Whether a marked reduction in rainfall produces a drought, with agricultural or societal impact, depends on the population to be fed, their habitual standard of living and their ability to accommodate food shortages, all of which may have been different centuries ago. A case in point is the 1992 drought in Southern Africa. While Botswana is the region's driest country and it experienced the greatest rainfall deficits, the impact of the drought was much less severe there than in surrounding countries such as Zambia or Zimbabwe.

The interpretation of regional integrators, such as lakes and rivers, poses a different set of problems. Consider Sahelian Lake Chad, for example. It is fed by the Logone and Chari Rivers, which emanate from much wetter regions to the south, but its level is also dependent on the highly variable Sahel rains falling directly on its surface. Likewise, the Okavango Delta and rivers of semiarid northern Botswana are greatly influenced by runoff from the Angolan highlands to the north, a comparatively humid region. The Nile flow in Egypt reflects both equatorial rainfall and rainfall which reaches it from the Ethiopian Highlands via the Blue Nile. When the rainfall determining the volume of water in a river or lake reflects a large region of highly inhomogeneous climate, the state of the river or lake is difficult to interpret in climatic terms. Furthermore, geomorphological processes (such as erosion or deposition of sediments) can evoke similar changes, as can anthropogenic factors.

There are special problems involved in reconstructing the historical climatology of Africa (Nicholson, 1979). Information is scarce, especially continuous records, so the researcher must often rely on types of information less commonly used in European reconstructions, and the indicators tend to be qualitative and not directly comparable with present data. The subjective nature of much of the available information is particularly problematic in Africa since here the observer is commonly foreign to the region. Finally, most of the data for Africa are landscape descriptions, which lead most directly to reconstruction of past environments. Their interpretation requires differentiating between various factors, including climate, geomorphic processes, and people, even though climate (especially rainfall) may be their prime determinant.

Methodologies

Overall, the process of utilizing the types of information in Table 4.1 to reconstruct climate is rendered difficult by the ambiguity of such indicators. This ambiguity centres on three questions: (1) is climate or some other factor responsible for the indicated environmental situation or cultural event; (2) is the indicated change of only local significance or widespread; and (3) if climate is clearly indicated as a factor in the change, what climatic element and region does it reflect? Clearly, cautious interpretation is required.

In circumventing these problems, two approaches to historical climate reconstruction can be considered. In the first, the task is to determine the absolute trend of variation—was the past wetter, drier, hotter, or cooler than today? This approach generally entails the compilation of long-term, nearly continuous series of one or more climate-related variables and use of occasional past measurements to 'calibrate' such time series. Examples of long-term records for Africa are given in the next section. In general, however, this approach is difficult to realize in Africa.

The second approach determines not the absolute climate of the past, but rather short-term climatic anomalies (such as a major drought) which occurred during historical times. With this approach, even fragmentary evidence without a direct comparison to the present becomes useful. Knowledge of such periods has numerous potential applications for understanding human behaviour or determining characteristics of climatic variability. This can, for example, shed light on typical time and space scales of drought or on teleconnections to climatic fluctuations in other regions. The unusual intensity and duration of Sahel drought, for example, is apparent in the historical record (Nicholson, 1989). This, and its occurrence in past centuries, strongly contradicts the once popular argument that the drought of the 1960s and 1970s in the Sahel was human-induced, because droughts of such magnitude and length had not previously occurred.

Ideally, these two approaches should be combined: the anomalies should be superimposed on the long-term trends inferred by other means. A combination of information can also help to verify the reconstructions. 'Convergence' of independent indicators (for example a local report of long drought and evidence in a lake core of contemporaneous desiccation) lends credence to both indicators. Credibility of the reconstructions can be judged in other ways, as well. If information derived from a number of different areas can be linked with known global or large-scale events, or if a phenomenon such as a drought is reported from many different locations, the interpretation of the proxy data is probably appropriate. Similarly, if the second approach produces non-random spatial anomaly patterns which are physically meaningful and comparable to anomalies in the modern record, the historical reconstruction is probably valid.

In this chapter, an attempt is thus made to integrate the available information into a long-term historical climate chronology for Africa. The greatest volume of information is available for West Africa and a reasonably detailed chronology can be derived. Such a synthesis has already been carried out (Nicholson, 1979, 1980*a*, 1981), so only a summary will be presented here. Much is also available for Southern Africa, although the most detailed refers only to relatively small portions of the Republic of South Africa. Material for this region will be emphasized here because a synthesis of recent works has not been carried out. Much less information has been evaluated for areas of North and East Africa. However, for these regions useful material abounds, especially Arabic geographical and historical literature, and could be exploited using the methodology described above.

Long-Term Chronologies

The long-term chronologies available for Africa include primarily lake and river levels and a few dendroclimatological studies. The latter are generally limited to the

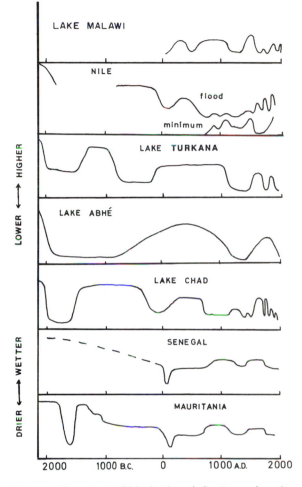

Fig. 4.1. Fluctuations of lake-levels and climate in Africa during the past 2000 years.
(*Source:* Nicholson, 1976, 1978, 1980*c*, based on numerous works described in this text.)

extra-tropical extremes of the continent. The only year-to-year record is that of the Nile flow; the record is nearly continuous since AD 622, but sporadic measurements were made in earlier centuries. A few miscellaneous reports supplement these. For example, there is evidence (Butzer, 1984) that the Gaap Escarpment in South Africa was probably relatively wet from *c.*1300 to 250 BP and evidence from Drotsky's cave (Brook *et al.*, 1990) suggests that a humid interval also commenced in the northern Kalahari *c.*1300 BP.

A few of the long-term chronologies are depicted in Figure 4.1. Only the fluctuations of Lake Chad have been well-dated radiometrically during historical times, but the dating of the fluctuations of the Nile is also accurate since it is derived from written records. The chronology for Senegal is based primarily on the work of Michel (1973) and that for Mauritania on the works of Munson (1971), Toupet (1973, 1977), Chamard (1973), Daveau (1965, 1969), Michel (1973), and Rognon (1974) (see Nicholson, 1976, for detail). These are also historically dated but highly generalized, while the recent dating for Lakes Abhé, Turkana, Bosumtwi, and Malawi is more speculative, although some radiometric dates are available (Gasse, 1977; Talbot *et al.*, 1984; Halfman and Johnson, 1988; Owen *et al.*, 1990).

Hydrologic Series

The Lake Chad curve (Figure 4.1), based on studies by Maley (e.g. 1976, 1981), is of first-order importance because it is accurately dated radiometrically and derived from high-resolution pollen and sediment analysis. The record clearly establishes that from about 1600 until some time in the eighteenth century its surface was about 4 m above the modern mean. During the twentieth century, the lake never stood more than 1 m above this mean and it is currently nearly desiccated. The levels of Lake Chad suggest increased precipitation in the Chad Basin (Niger, Chad, northern Cameroon, Central African Republic), including the semi-arid Sahel and Soudan zones south of the Sahara as well as the more humid and more southern Soudano-Guinean zone. Lake Chad stood at much higher levels for probably a few centuries prior to *c.* AD 700 or 800. Many of these same trends can be established for areas further west in the Sahel, such as areas of Mauritania and Senegal (Figure 4.1); particularly evident is the relatively humid period from about the eighth to thirteenth centuries (Nicholson, 1978, 1979; Grove, 1985), which is coincident with the 'global' warm period termed the Medieval Warm Epoch or Neo-Atlantic (Tyson and Lindesay, 1992). Historical descriptions of a watercourse between several sites currently in arid regions of Niger, oases there and in Chad (Akinjogbin, 1967), and lakes in those same regions (Servant, 1973) provide further evidence of this humid episode in the Sahel.

Extensive work has also been done on Lake Bosumtwi in Ghana, but little detail is provided for the historical period (Talbot and Delibrias, 1980; Talbot, 1983; Talbot *et al.*, 1984). The lake maintained relatively high stands *c.*1500 to 1000 BP, with this transgression reaching some 25 m above modern levels. Lake Bosumtwi regressed rapidly after about 1100 BP (i.e. about AD 850 ± 200 years), falling over 50 m within the span of at most 100 to 200 years. It reached and maintained a low level of −30 m during probably the last 500 years and is still recovering from this period of desiccation. The decline

*c.*1100 BP corresponds roughly to the onset of the humid period further north in the Sahel.

Further east, Butzer (1971) found evidence of a recent higher stage of Lake Turkana, on the border of Kenya and Ethiopia. He dated its recent evolution on the basis of estimated time of formation of geomorphological features, given two probable chronologies. The long-term chronology, which he indicates is more likely, would put a high stand from *c.*2200 to 800 BP (*c.*250 BC to AD 1150). The end of that transgression is roughly in agreement with recent work of Halfman and Johnson (1988), who inferred from carbonate concentrations in a core that a recent higher stage of Turkana persisted from around 1100 to 1500 BP. This is further support for the conclusion that drier conditions commenced in the low latitudes about the time that the onset of wetter conditions occurred in the Sahel, a climatic opposition which has been one of the most common patterns of the twentieth century (Nicholson, 1989, 1993).

The studies of Gasse (1975) show a final, recent transgression of Lake Abhé in the Afar which was radiometrically dated to 110 ± 50 BP. Although such recent dates are subject to considerable error, the accuracy is supported by the number of dates and by additional evidence, such as isotopic analysis and historical witness. The onset of the lake's transgression cannot be firmly established, but it likely occurred within the past few centuries and ended abruptly in the late nineteenth century. The preceding regression probably began toward 1000 BP. Gasse (1975) does not present evidence of the continuity of the recent high stand, but the lake probably followed a trend similar to that of Lakes Turkana, Stefanie, and others in Eastern Africa: lower levels throughout most of the nineteenth century, but a return to higher levels from about 1860 to 1895 (Nicholson, 1980a). The high stands of Lake Abhé late in the century are confirmed by oral tradition (Gasse, 1975).

A recent transgression of Lake Katwe, a self-contained salt lake lying in a crater at the foot of the Ruwenzori (Uganda), was dated radiometrically to 250 ± 50 BP (van Zinderen Bakker and Coetzee, 1972), with earlier transgressions dating to *c.*900 BP and 1200 BP. This roughly agrees with the Turkana and Abhé chronologies. Lake Kivu, nearby, was a closed basin from *c.*3000 to 1200 BP (Degens and Hecky, 1974; Hecky, 1978); thus its last transgression was at about the same time as Katwe's, toward AD 800.

These higher lake-levels of the sixteenth and seventeenth centuries occurred during a period of glacial advance in the Ruwenzori (de Heinzelin, 1962) (seventeenth and eighteenth centuries) and of lowered snowline in Ethiopia (Hövermann, 1962) (late sixteenth and

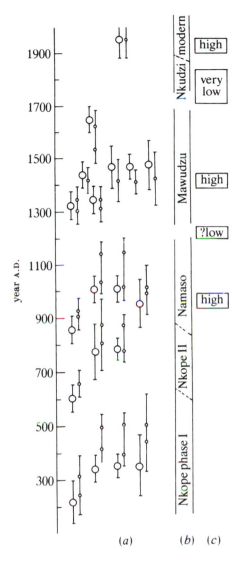

Fig. 4.2. Long-term fluctuations of Lake Malawi, based on radiocarbon-dated archaeological data.

(From Owen *et al.*, 1990). Large and small circles indicate uncorrected and corrected ¹⁴C dates respectively; right-hand axis indicates ceramic traditions. Also indicated on right are periods of high lake-level and inferred periods of low lake-level. (Modified from Owen *et al.*, 1990; Crossley *et al.*, 1984.)

early seventeenth centuries). A lowering of temperature alone, however, could not account for the increased volume of water in the lakes discussed above.

From AD 622, the levels of the Nile at the Roda gauge in Cairo were recorded annually, but with some gaps in the record (Toussoun, 1923; Riehl *et al.*, 1979). This includes both the flood levels, an indicator of precipitation in the Ethiopian Highlands, and the summer minimum

level, an indicator of equatorial rainfall. Historians have criticized these records on several grounds; these include, for example, falsification of records for revenue purposes, since taxes were levied according to flood height; silt accumulation in the gauge; and change of the length of the 'cubit', the measurement unit utilized. However, there is ample geological and archaeological evidence to suggest that the data presented by Toussoun (1923) are reasonably accurate. The flood height, for example, implies increased precipitation in the Ethiopian Highlands during the sixteenth through eighteenth centuries, and again in the late nineteenth century; Gasse (1975) reached a similar conclusion based on the fluctuations of Lake Abhé.

The minimum levels of the summer Nile, which are indicators of equatorial rainfall, show the opposite trend: very low in sixteenth through eighteenth centuries. This is not surprising because opposition between equatorial and subtropical latitudes is a commonly occurring rainfall anomaly pattern during the present century (Nicholson, 1986). However, both the summer minimum and the flood had high levels during the late nineteenth century.

Owen *et al.* (1990) and Crossley *et al.* (1984) have conducted extensive studies of Lakes Malawi and Chilwa; these provide a sketch of the lakes' long-term fluctuations. Their work includes numerous sediment cores and archaeological studies. The latter include extensive radiocarbon dates of settlements along the shores of Malawi. Gaps in the record are interpreted as periods of aridity with peoples moving with the receding lake-shore or to wetter highland regions (Figure 4.2).

The cores and other information for Lake Malawi indicate three high stands dated to *c.* AD 1980, 1390, and 950. The older two are radiometrically dated (Crossley *et al.*, 1984). Trees and charcoal dating to *c.*480 BP mark the middle high stand; synchronous evidence of good flow in the Shire River indicates that it is not due to blockage of lake overflow into the river (Crossley *et al.*, 1984). The gaps in the settlement record appear from the late fifteenth century to the mid-nineteenth (except for a single date *c.*1700) and in the twelfth and thirteenth centuries, i.e. about midway between high stands of the lake. Changes of ceramic traditions are noted at approximately the same times. Malawi previously reached high stands around the tenth century AD.

The Lake Chilwa record is not in complete agreement with that of Lake Malawi and their last major high stands do not coincide (Crossley *et al.*, 1984). That of Chilwa took place around 160 BP (i.e. *c.* AD 1750) Beach gravels representing a level of 631 m (9 m above modern levels) mark this high stand, but its date is roughly

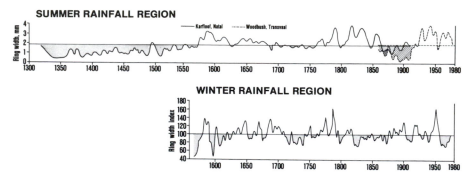

Fig. 4.3. South African tree-ring chronologies for the summer- and winter-rainfall regions. (*Source:* Tyson, 1986; based on Hall, 1976; Dyer, 1978; Dunwiddie and LaMarche, 1980.)

confirmed by archaeological evidence, including ceramic traditions, and by recent trade beads in the sediment layer just below the gravel. The date can also be approximated by the longevity of trees in the area, especially the baobabs which are killed off rapidly by floods. Many show an age of AD 1650 or earlier in areas above 631 m (the old shoreline indicated by the gravels), but none date back to prior to *c.* AD 1760 in lower areas presumably flooded by this transgression.

Thus, it appears that both Malawi and Chilwa were very low for an extended period between about AD 1750 and 1850. Chilwa attained extremely high levels *c.* AD 1650 and 1760, but Malawi reached its highest level of the last 1700 years some time around the fifteenth century during a high phase which probably ended by 1620. These changes paralleled those in many other parts of Africa. In Ethiopia, at least parts of equatorial Africa, and in the Chad Basin the climatic conditions of the sixteenth and seventeenth centuries were wetter than those of the present day. Long-term chronologies compiled for Senegal and Mauritania (Figure 4.1) and other evidence described in the next section support this conclusion for other Sahelian regions (see also Grove, 1985). A decline in rainfall began some time around 1700, but predominantly more humid conditions probably extended well into the late nineteenth century. In the Sahel, as in the regions of Lakes Malawi and Chilwa, a period of desiccation occurred in the late eighteenth and early nineteenth centuries, lasting until mid-century. Evidence for these trends is detailed in subsequent sections.

Dendrochronological Studies

Tree-ring chronologies from the southern and northern extremes of Africa add to this picture, but their interpretation is complex and the representativeness of some

is questionable because they are based on one or two trees (Tyson, 1986). For South Africa, a chronology for the summer rainfall region of the north-east is based on two specimens of *Podocarpus falcatus* whose records overlap for about half a century. It depicts fluctuations in Natal since about 1300 and in the Transvaal since about 1850 (Hall, 1976; Dyer, 1978). A second chronology from Dunwiddie and LaMarche (1980), based on forty-seven specimens of *Widdringtonia cedarbergensis*, represents the winter rainfall region of the south-west Cape since *c.*1550. A third series is available for the area of year-round rainfall in the south-east. Walter (1936) also carried out a study using several specimens of *Acacia giraffae* from Namibia, but the results are inconclusive.

Other studies relate to the Mediterranean climate north of the Sahara. Using a large number of trees from several locations, Stockton (1990) produces a tree-ring chronology for Morocco as a whole for the period 1750–1984, and a longer chronology from Col du Zad in central Morocco which begins around 1000. Munaut (1976, personal communication) presents a chronology commencing *c.*1700 which is based on numerous specimens of *Cedrus atlanticus* from Morocco. One for Algeria commencing *c.*1700 is derived from one specimen of *Pistacia atlantica* and is compared with the growth-rings in a *Pistacia atlantica* from the Negev desert (Fahn *et al.*, 1963). A study of one tree in Tunisia (*Quercus* sp.) has also been carried out (Ginestous, 1927), producing a record covering nearly 200 years. Waisel and Liphschitz (1968) analysed thirty-four specimens of *Juniperus phoenica* from the north-central Sinai, some of which had chronologies covering nearly 500 years.

The South African chronologies are depicted in Figure 4.3. Interpreting them conservatively, Tyson (1986) infers from them merely periods of more benign (i.e. above-normal growth) or harsher climatic environment (i.e. retarded tree growth), although rainfall plays a major role in the tree growth. The chronology for Natal indicates low growth relatively continually in the fourteenth through sixteenth centuries; a markedly apparent shift to high growth occurred c.1580. Chronologies for both summer and winter rainfall regions show above-normal growth for the seventeenth and early eighteenth centuries; peak growth centred on c.1630 in Natal and c.1670 in the south-west Cape. In the winter rainfall region of the south-west Cape, most favourable conditions were the seventeenth century, least favourable were the eighteenth and most of the nineteenth centuries, except for the period c.1760 to 1790, after which growth-rate drops rapidly. Growth was particularly poor c.1710–50 and in the 1820s and below normal growth persisted from then to the mid-1870s. Vigorous growth is again indicated for c.1874 to 1902. In contrast, highest growth occurred in Natal, in the summer-rainfall regime, in the late eighteenth century and first half of the nineteenth century (c.1769–1860), but conditions began to decline from c.1820 onward. The period c.1860 to 1910 or 1915 was one of generally unfavourable conditions for growth.

The tree-ring chronology of Stockton for Morocco as a whole for the period 1750–1984 (Stockton, 1990) shows several periods of drought but they seldom persist more than four years. Three of the longest were 1978–82 (six years), 1790–4 and 1816–20 (five years each). A lengthy drought also occurred in the 1970s and 1980s. The chronology shows little evidence of major decadal scale variations, although there are indications of relatively low growth in the mid-to-late nineteenth century and relatively high growth from about 1750 to 1790. Stockton's longer series from Col du Zad in central Morocco indicates five-year droughts c.1069–74 and 1626–31 and four-year droughts c.1404–07 and 1714–17.

Munaut's chronology (1976, personal communication) from the Rif and Middle Atlas regions of Morocco gives a more generalized picture. It indicates rapid tree growth before 1500, a marked drop of ring-width between 1500 and 1600, and slower growth after 1600. Further east a chronology from near Laghouat, Algeria, indicates very rapid growth from the mid-seventeenth to mid-eighteenth centuries (Fahn et al., 1963), a marked drop in growth-rate in the mid-eighteenth century, but increased growth late in the century until about 1800. This is roughly consistent with changes of precipitation suggested by other types of indicators (Nicholson, 1980a), but it is not clear to what extent the small ring-width in early years is related to the slower growth-rate of young trees.

Much further east, specimens of *Juniperus phoenica*

from the north-central Sinai have been used to estimate rainfall changes there from 1500 to 1900 (Waisel and Liphschitz, 1968). These also showed a drop in ring-width some time in the seventeenth century and suggested that rainfall was two to three times greater than today from about 1500 to the mid-seventeenth century.

West Africa

Famine and Drought and Other Information in Historical Chronicles

Detailed historical chronicles covering several centuries are available for several areas of West Africa (Figure 4.4). Those of Timbuktu and the Niger Bend (Cissoko, 1968) mention essentially every year from the early seventeenth century through the nineteenth century. During this period, the chronicles talk of the great prosperity of these empires, explaining that conditions were so good that famine simply could not break out. In contrast, a number of tremendous floods are mentioned during the later sixteenth and early to mid-seventeenth centuries. The Niger flood regularly reached Timbuktu, something which rarely occurred in either the nineteenth or twentieth century. Chronicles from Senegambia and southern Mauritania confirm such trends (Curtin, 1975; Nicholson, 1980a).

Further east, the Bornu chronicles of the Lake Chad region begin well before 1500. They provide detailed information on famines, droughts, and 'prosperous' periods which can be interpreted as periods with no major famines or similar hardships (Nicholson, 1980a). Such conditions lasted until the major famine of the 1680s. Afterward, conditions improved for a time until the mid-eighteenth century, then from about 1790 conditions worsened considerably and several famines occurred in Bornu in the early to mid-nineteenth century.

Geographical Descriptions

Geographical descriptions for earlier centuries abound in reports of European travellers and settlers, and in various historical sources. They, like the chronicles from West Africa, suggest that a more humid climate prevailed in the sixteenth and seventeenth centuries and that these conditions generally lasted throughout the eighteenth century, although a decline had begun toward 1700 and the eighteenth century was plagued by several severe famines.

In Chad, archaeological evidence indicates that the Tibesti Massif was populated from about 1500 to 1700 or later (Plote, 1974; Maley, 1973), indicating wetter conditions. Chapelle reports that elders living in Tibesti retained a memory of an earlier time when the massif

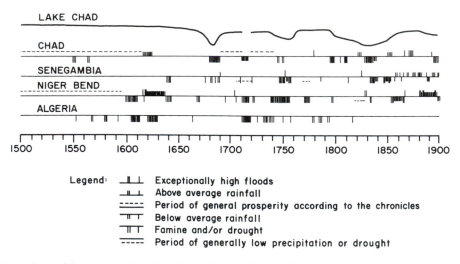

Fig. 4.4. Chronology of famine and drought in Chad, Senegambia, the Niger Bend, and northern Algeria, 1500 to 1900. (*Source:* Nicholson, 1978.)

was wooded; currently no region receives more than 100 mm annually. If the report is factual, it may relate to the time of the seventeenth century transgression of Lake Chad. Much evidence, including lake-levels, suggests that the wetter conditions persisted during at least part of the eighteenth century (Maley, 1976). In about the 1780s or 1790s it was reportedly still possible to travel from Lake Chad by boat as far north as Borkou, by way of the lakes of Djourab and Borkou (Plote, 1974); tremendous floods occurred in Agadez and near Murzuq in 1780 (Palmer, 1936). The described route would indicate that Chad's occasional outlet, the Bahr el-Ghazal valley, was completely flooded, an occurrence for which there is also sedimentary evidence (Shaw, 1975). Bones of elephants and hippopotami, dated at *c*.140 ± 90 BP, were found in these sediments and Browne (1799), who resided in nearby Darfur from 1793 to 1796, also mentions a forest near the Bahr el-Ghazal and talks of a small river which runs into it near Bornu.

Browne also describes the rainy season at Darfur near his residence at Kutum (western Sudan). The rains normally fell from mid-June to mid-September frequently and intensely, producing tremendous verdure in the countryside. Wood was plentiful; the city was surrounded by wadis and torrents filled with water. The region currently receives on the average about 300 mm of rain per year, insufficient to support woodlands, and in most years the rainy season is limited to July and August. Numerous radiocarbon dates from regions of Niger, the Sudan, and Ethiopia, which are now too dry for human occupation indicate at least temporary occupation around

waterholes around 150 to 350 BP (M. A. J. Williams, 1975, personal communication; Nicholson, 1980a).

Reports from the western Sahel further support the conclusion that the Sahelian environment was wetter than today during the sixteenth, seventeenth, and perhaps eighteenth centuries (e.g. Michel, 1969). Waterways and former mouths of the Senegal delta were navigated by the Portuguese in the sixteenth century; they were not blocked by sand until the end of the seventeenth century or later, about the time of the well-documented 1680s drought (Nicholson, 1980a). Chambonneau in 1677 describes thick mangrove stands and palm trees along the banks of the Senegal and in its valley from St. Louis to Cape Verde (Chamard, 1973). Adanson's (1759) description, and those of several others (Nicholson, 1980a), is in agreement: thick woods, thorny bushes near the river's mouth, with dense forests on the banks near Podor. His maps show lakes in northern Senegal and southern Mauritania which have since dried up and a forest in southern Mauritania up to *c*.18°N, where precipitation averaged about 200–50 mm during the twentieth century. Adanson's observations of the rainy season and Winterbottom's mid-eighteenth century observations of the Harmattan season there further support the hypothesis of more humid conditions (Winterbottom, 1969).

Reports of navigators and travellers also indicate that the climate of Senegal and Mauritania in the sixteenth century was wetter than today. Portuguese explorers of the time described a precipitation regime which was more abundant and earlier in the year than now (Daveau,

1969; Chamard, 1973), as well as numerous wadis, a regular rainy season, sedentary occupation, and rain-fed agriculture in regions of Mauritania which are now climatically arid and lack a perennial water source (Nicholson, 1980a). The details of the climate descriptions imply both an increase in transition season rainfall (the 'heug' rains) and a more northward position of the ITCZ.

A wetter Sahel during the seventeenth and eighteenth centuries is indirectly supported by climate and weather observations from the Guinea coast. Several descriptions of the rainy season and prevailing winds suggest an earlier northward advance of the ITCZ (hence longer Sahel rainy season) and drier conditions along the Guinea coast, especially in mid-summer (e.g. Hillier, 1697; Barbot, 1732; Bosman, 1705; see Nicholson, 1980a). Observations of Volta River hydrology also suggest a delayed retreat of the ITCZ, with flooding of the river occuring in November in the mid-eighteenth century, rather than during September–October as today (J. M. Grove and Johansen, 1968). Winterbottom (1969) actually measured rainfall at Freetown (Sierra Leone) during the three years 1793 to 1795, finding a mean of about 2000 mm, compared to a mean of 3400 mm during the twentieth century and 4800 mm during late nineteenth century. Such an opposition between the Sahel and Guinea coast is the most common mode of rainfall variability (Nicholson, 1980b).

Conditions During the Eighteenth and Nineteenth Century

In the eighteenth and nineteenth centuries detailed information on climate and weather, including actual measurements of rainfall, became much more abundant. These clearly indicate that the humid conditions declined in the Sahel toward 1800; that severe droughts occurred in the 1820s and 1830s; but that a return to humid conditions occurred and persisted during the last three decades of the nineteenth century. Around 1895 was another onset of another decline in rainfall, beginning a long downward trend which culminated in relatively dry conditions throughout most of the twentieth century (Grove, 1973, 1985).

The historical record also documents severe and extensive droughts in the 1680s, the 1710s and in the 1730s to 1750s throughout the Sahel (Nicholson, 1980a). Minor droughts also occurred in the 1770s. The droughts were reported from areas as widespread as Ouaddai (Chad), Darfur and Funj (Sudan), Agadez (Niger), Bornu (near Lake Chad), the Niger Bend and elsewhere in Mali,

Mauritania, Nigeria, Burkina Faso, Cape Verde, Benin, Ghana, Senegal, and Gambia. The famine of mid-century was so long and severe that reportedly half the population of the Sahel perished.

In the 1790s drought again occurred, signalling a decline in rainfall which eventually led in the 1820 and 1830s to one of the worst droughts ever experienced. In the 1790s sand-blocked streams which had earlier flowed near Murzuk and the level of Lake Chad began to fall continuously for most of the next three to four decades (Plote, 1974). Its fall is also recorded in the lake's sedimentary record (see Maley, 1976, 1981). Beginning in the 1790s the onset of rains in the Sahel occurred one or two months later than previously (Park, 1799; Jackson, 1820). In the nineteenth century the most severe drought probably occurred from c.1828 to 1839 and most of the first half of the century was relatively dry in the Sahel. The region near Lake Chad was probably hardest hit, but the chronicles of Tichitt and Oualata indicate that Oualata and Takrour (Mauritania), Araouane (Mali) and other areas of the western Sahel were also severely affected (Marty, 1927; Monteil, 1939).

In contrast, more than a decade of rainfall measurements at Christiansborg (near Accra, Ghana) indicate high rainfall from 1829 to 1842 (Nicholson, 1980a). Rainfall was particularly heavy in parts of Ghana in 1838 and 1839, with the summer dry season having been replaced by intense rainfall (J. M. Grove and Johansen, 1968), something which occurs occasionally today. But, in the early or mid-1840s drought occurred. Measurements at Saint Louis (northern Senegal) and harvests at Tichitt and Oualata also imply good but non-seasonal rainfall in the north-western Sahel during the period of drought (Nicholson, 1980a).

By the 1870s, measurements of rainfall and river flow and historical and geographical information became plentiful. All types of indicators suggest a temporary return to wetter conditions in the Sahel from about 1870 to 1895 (Nicholson, 1980a). These include reports of harvests from numerous regions of Mauritania and Mali; floods of the Niger and Senegal Rivers; measurements of the level of Lake Chad (Figure 4.5); detailed geographical descriptions of forests, wells, and wadis throughout the region; and rainfall measurements in several cities, especially in Senegal. Those at Freetown and elsewhere suggest rainfall during those decades was about 25 per cent above the twentieth century mean in much of West Africa. This humid period was evident throughout most of Africa, as described in the following sections. But it quickly gave way to markedly more arid conditions during most of the twentieth century.

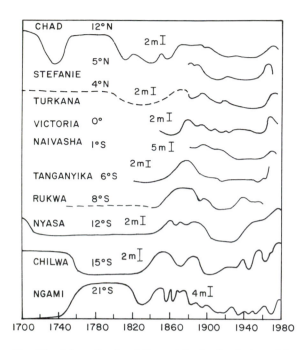

Fig. 4.5. Schematic of variations of the levels of African lakes in the nineteenth and twentieth centuries.

Both the trends and the magnitudes of the fluctuations are highly generalized. (Sketch based on Sieger (1887); Pike and Rimmington (1965); Nicholson (1976, 1978, 1980c); Vincent *et al.* (1979); Maley (1981); Crossley *et al.* (1984); Shaw (1985); Owen *et al.* (1990); Hastenrath (1988); and others.)

Southern Africa

Information useful in reconstructing historical climates becomes plentiful in Southern Africa in the nineteenth century, but information for earlier centuries is available for limited regions. This includes lake-level and other geological information, oral traditions, travellers' diaries and reports, and documentary sources, such as archives. The latter are so plentiful for some regions that essentially every year can be described with some degree of certainty. This is true for the Cape area of South Africa (Vogel, 1989) and parts of Angola, especially near Luanda (Miller, 1982).

In general, the trends for Southern Africa appear quite similar to those for West Africa. Major times of discontinuity appear to be relatively synchronous, as do major periods of anomalous conditions, such as the aridity early in the nineteenth century. This is not surprising, however, since the climatic teleconnections between the regions are strong and the modern climatic record shows that the major periods of wetter or drier conditions appear to be continental in scale, with similar trends often indicated for the Sahel and for the analogous semi-arid regions of Southern Africa (Nicholson, 1986).

Geographical Descriptions and Lake-levels

Numerous explorers frequented Southern Africa since the eighteenth century, most giving detailed geographical descriptions. Landscape descriptions and faunal evidence suggest that before about 1800 parts of the Karoo and dry north-western Cape, much of the Kalahari and other dry regions of South Africa were considerably less arid than subsequently, although the broad features of the wind and rainfall regimes were relatively constant during the last few centuries (Nicholson, 1981). One exception is the area just north of the Cape. It was within the winter rainfall province of the Cape before about 1800, but after that time the summer rainfall regime prevailing further north extended southwards to at least the Roggeveld. Prior to about 1850 snow and freezing temperatures were more common in South Africa than at present. Deep and lasting snow persisted throughout the winter on the Roggeveld and it was frequent and lasting on higher parts of central Namibia, although a rare occurrence there today.

Evidence is plentiful that a major decline in rainfall occurred after about 1800, with a progressive desiccation of lakes, rivers, and springs from about 1800 to 1830, and numerous, widespread droughts in the 1820s and 1830s (Nicholson, 1981; Tyson, 1986). This complements evidence based on landscape descriptions and fauna summarized by Schwarz (1920). In 1799, for example, Barrow (1804, 1801) described the Beer Vley, north of Willowmore, as a periodical stream running through a vast plain of rushy grasses, swamps, springs, and periodical rivers. He also describes a vegetation cover which could only be sustained by higher rainfall than today. In Schwarz's time, the vlei was completely dry, its surroundings dry, barren, and desolate. He mentions a report from le Vaillant, from about 1780, which corroborates Barrows's remarks: in the Great Fish River area near Craddock thrived a fauna reminiscent of East Africa; lions, buffalo, and numerous types of antelope roamed the plains, while hippopotami thrived in the rivers there. As with Beer's Vley, this region was typically barren Karoo in Schwarz's time, as well as today.

Schwarz (1920) claims that a major change of climate and environment took place in Southern Africa toward 1820. His claim of desiccation early in the nineteenth century is substantiated by the oral traditions and memories of both local tribesmen and immigrant settlers in the region. Nevertheless, signs of environmental conditions wetter than at present persisted until later. Up to about 1830 the Matlaring River ran at Kuruman and the Kuruman River flowed strongly in the desert to the west (Tyson, 1986). Moffatt (1842) and Livingstone (1858) both reported periods of heavy

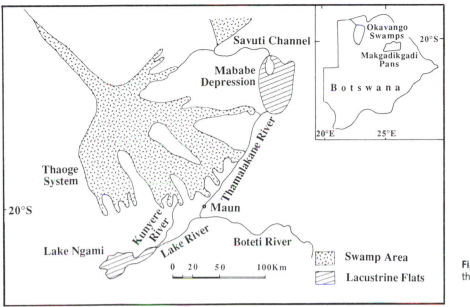

Fig. 4.6. Sketch map of the Okavango Delta system.

rainfall and expanses of surface water in now drier regions, but in is unclear if these conditions were typical.

Schwarz's contention of desiccation in Southern Africa, which is echoed by Wilson (1865), is strongly supported by extensive and detailed descriptions of Lake Ngami and the rest of the Okavango Delta system; these describe major changes in regional hydrology. The general sketch of this complex hydrologic system is presented in Figure 4.6. The system is fed primarily from the north by rainfall occurring in the more humid highland regions of Angola. During historical times it has had seven outlets (Shaw, 1984), three of which (the Thaoge, Kunyere, and Thamalakane Rivers) feed into Lake Ngami. The Magwegqana Spill connects it to the Linyati swamps and Zambezi River and, by way of the Savuti River, to the Mababe Depression. It also feeds the latter via the Mababe and Thamalakane Rivers. The system is further linked to the Makgadikgadi Pans to the south-east by way of the Boteti River. Lake Ngami, the Mababe Depression and these pans are filled only in years of particularly high inflow into the system. When the lake reaches critical levels of about 928 m above m.s.l., backflooding of the Lake and Kunyere Rivers serves as an outlet.

In the early nineteenth century, Lake Ngami had been so deep that it sustained waves which were powerful enough to throw hippopotami and fish to shore (Nicholson, 1981; 1980c). The Shua River had been navigable and the Makarikari Pans, today only occasionally flooded, never dried up. Around 1820 Ngami

dried up and many marginal areas, where woods previously lined river banks, turned into barren, arid land. After that time the pans received little flood water from the Boteti River. This desiccation of Ngami was probably only brief, however, as it was a substantial lake in the 1850s, as well as during most of the late eighteenth and early nineteenth centuries. A ring of substantial vegetation growth surrounding the lake basin atop a former beach ridge is a remnant of these wetter conditions.

Numerous reports of European explorers in the Lake Ngami region confirm its high stands of the mid-nineteenth century but at the same time suggest it had by then already commenced a long downward trend. Enough detail is given that the size and depth of lake can be estimated for quite a few years in the 1850s and 1860s (Shaw, 1985) and the general state of Ngami and nearby rivers can be estimated for most years since mid-century (Figure 4.7).

Livingstone in 1853 reported higher levels than did travellers who visited the region in subsequent years. Numerous reports of severe droughts in the region in the 1840s and 1850s (e.g. Hitchcock, 1979) support a climatic explanation for these observed fluctuations. Andersson (1857) refers to the presence of submerged tree stumps but also shows vegetated areas which were once the haunt of hippopotamus. Baines (1864) indicates on his map mud-flats that were previously inundated. Their discussion of inflows and outlets clearly indicate higher levels and more humid conditions than at present during much of the nineteenth century. The

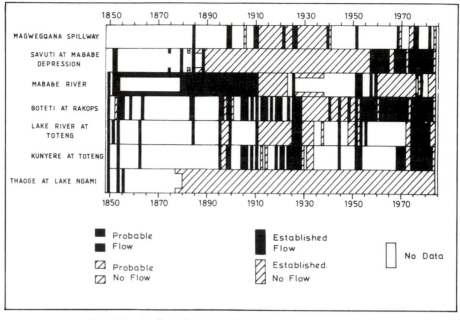

Fig. 4.7. Outflows from the Okavango System 1849–1983.
(*Source*: Shaw, 1984.)

Thaoge River, for example, was a major inflow to Ngami then and it was a small but navigable stream.

Lake Ngami in the early 1850s probably stood between 928 and 930 m above m.s.l. on a seasonal basis, but its level gradually declined over the next few decades, as did river flow and surface-water in the region (Shaw, 1984, 1985). Even then, however, the lake was probably in a state of decline. Its main tributary then, the Thaoge, had previously been a true swamp with many channels, but by the 1850s it had only a single distributary. Then the annual flood of the Thaoge arrived at the lake by June, slightly before the secondary input from the Kunyere and Lake Rivers. This created a swamp in a maze of channels to the north of the lake and backflooding of rivers. From the mid-1850s onward the flow of the Thaoge gradually diminished in volume and regularity, ceasing altogether by the 1880s. Lake Ngami shrank from a perennial lake of some 800 km² (the area bounded by a sand ridge at at *c*.929 m, see Grove, 1969) to an ephemeral body at most 250 km² between 1850 and 1880 (Thomas and Shaw, 1991), when it became then a closed-system lake fed entirely by the Kunyere and Lake River inlets.

To the east of Lake Ngami, the absence of water in the Mababe Depression at the time of the European explorers suggests, however, that the flow of the Savuti River (to the north of Ngami) was not significantly greater during *c*.1850 to 1900 than in recent years. Earlier in the nineteenth century, however, Mababe fed a substantial swamp and oral reports describe a river deep enough to hinder passage. Oral tradition suggests it was a veritable lake during the eighteenth century, in agreement with information on the former state of Ngami and the Thaoge (Shaw, 1984). Both the Savuti and Thaoge systems dried up at about the same time (Stigand, 1923). The Savuti flowed frequently until the late 1880s. The cessation of flow from Thaoge into Ngami can be more precisely fixed, being bracketed between 1877, when McKiernan sailed up the river, and 1881, when Ngami dried up (Shaw, 1985), but sporadic flow may have occurred as late as 1894 (Wilson, 1973).

The observations suggest that there was a marked reduction of outflow from the whole Okavango system during the course of the nineteenth century, but especially since mid-century. At that time, all seven outlets from the Okavango Delta were functioning at least at current or above-average stages (Shaw, 1984). Livingstone (1858) spoke of the desiccation of springs previously feeding pans further south. Large numbers of springs ceased to function, such as those at Khakhea, Tshabong, and probably the one at Kang. Similar changes occurred in the Molopo and Auob-Nossob valleys in south-western Kalahari and in the central Kalahari. By the end of the century, Lake Ngami, the lower Thaoge, Savuti, and Mababe Rivers had dried up and the wetland habitats of the lower Okavango, Boteti, and Savuti Rivers had shrunk dramatically (Thomas and Shaw, 1991). Shaw suggests that these changes were largely a result of human intervention or channel shifts in its upper regions, but the overwhelming evidence from other areas of Southern Africa (described here and in earlier works, e.g. Nicholson, 1981), as well as characteristics of the Okavango hydrology (Peter Smith, Botswana Department of Water Affairs, personal communication) strongly implicates climatic causes.

This decline in regional hydrology culminated in a series of severe droughts and economic depression in Botswana earlier in the twentieth century (Hitchcock, 1979). A general condition of relative desiccation characterized most of the twentieth century, although there were a few good years. The Savuti River, for example, reached the Mababe Depression in 1925. That year, plus 1910, 1927, and 1944, were good years throughout the Okavango system, in contrast to the very dry period prevailing in the 1930s (Shaw, 1984). In the 1950s, conditions improved but the levels of the mid-nineteenth century were not attained for any extensive periods. Flow resumed in the Boteti River as far downstream as Rakops and in the Savuti; water flowed again occasionally in the Mababe and Thamalakane. Lake Ngami contained enough water in the 1950s that residents of nearby Maun held speedboat races on the lake. A stand of relatively young vegetation surrounding the lake basin attests to the high levels of the 1950s. Since then a number of humid periods occurred, particularly c.1968/9 and in the late 1970s. These were responses to conditions of extremely high rainfall throughout the southern subcontinent (Nicholson et al., 1988), especially during 1974, 1976, and 1978. Following these humid periods, in c.1972 and 1980, Lake Ngami was filled. Drier conditions set in once again during the following decade and 1983 was one of the driest on record in the region. In recent years, the only flooding occurred in 1988/9, and at present Ngami is again an extensive, dry savanna landscape.

Further north, extensive evidence from the region of Lakes Malawi and Chilwa describes a roughly parallel progression of drier and more humid conditions (Figure 4.5). Historical fluctuations of some 14 m occurred in Lake Malawi, and of 12 m in Lake Chilwa (Crossley et al., 1984). Both lakes were low for an extended period some time between c.1750 and 1850. The time of this desiccation is established in numerous ways, including C14 and Pb210 dates, as well as historical reports. These lakes attained very high stands in the mid-nineteenth

century but then declined more or less continually to around 1900. Low levels were maintained throughout most of the first half of the twentieth century, although the lakes commenced rising again in the 1920s and by the 1940s or 1950s lake-levels were relatively high. These high stands, up to 475 m for Malawi and nearly 645 m for Chilwa, were maintained throughout most of the 1950s, 1960s, and 1970s, although interrupted by low phases during the extensive droughts of the 1960s and 1970s in much of Southern Africa.

The low stands (Figure 4.5) of both Lakes Malawi and Chilwa during the early nineteenth century are indicated in reports from European travellers but evidence for this desiccation derives also from sediment cores from both lakes (Owens et al., 1990). The reports for Lake Malawi are further substantiated by gigantic trees that were found standing in three feet of water in 1892 and 1893. The trees show that the lake surface was continuously below 469 m for a long period; the size and growth-rate of the trees suggests this would have been the late 1700s and early 1800s. Other evidence of low levels which can be ascribed to the early eighteenth century include lake core evidence of erosion, decline in diatom quantity and diversity, and beach sands. The lithology and microfossils in the cores indicate that, prior to the early 1800s, the lake-level had declined by 120–50 m in just over 250 years (i.e. post c.1500). The exposed littoral sands were reworked into aeolian dunefields along the windward shoreline during low stages. The local fauna of many island and rock outcrops in Lake Malawi also suggest that these areas were dry land within the last 200 to 300 years.

Numerous cores from Lake Chilwa contain a hardpan layer 1.2 m thick, which resulted from a prolonged low level in the early 1800s. A previous high stand was attained between about AD 1650 and 1760; beach gravels in a core show that the lake stood at 631 m, some 9 m above modern levels. Cores from the Nfera Lagoon, Lake Malombe, and Lake Chiwondo, which are hydrologically continuous with Malawi, also produce evidence of low phases spanning the late 1700s and early 1800s. Sediments from Lake Chiuta, about 100 km to the southeast of Chilwa, also indicate desiccation during an extensive period prior to the 1850s.

Both Lakes Malawi and Chilwa again attained high stands in the mid-1850s (Figure 4.5). Some sources (e.g. Pike and Rimmington, 1965) suggested these were significantly above the twentieth-century maximum achieved in 1878, but Lancaster (1979) demonstrates that levels could not have been quite so high. Both Lake Malawi and nearby Lake Malombe, which is a good indicator of Malawi's levels, were visited in the 1860s in separate expeditions by the explorers Livingstone and Kirk, who both took soundings of the lake to establish their levels (Crossley *et al.*, 1984; Owens *et al.*, 1990). Both exhibited extremely high stands in the early 1860s, although at a somewhat lower level in 1861.

Malawi's level declined by nearly 4 m over the next five decades (Pike, 1965; Drayton, 1979; Lancaster, 1979), maintaining a period of minimum during the 1900s and 1910s. In 1892/3 Swann observed the lake at 470 m, a low level attained since Livingstone's 1861 visit. Some decline occurred during the 1870s, especially in Lake Chilwa. Nevertheless, extensive historical evidence as well as continuous measurements of Lake Malawi beginning in 1896 demonstrate that lake-levels were relatively high from mid-century until just prior to 1900 (Crossley *et al.*, 1984; Pike 1965; Drayton, 1979).

Oral Tradition

The aridity of the early nineteenth century throughout Southern Africa is confirmed by oral traditions of many local peoples in regions presently occupied by the countries of South Africa, Namibia, Botswana, Malawi, and elsewhere. For example, both the Damaras and the Namaquas, and numerous white settlers claimed toward 1850 that the rains had been more abundant earlier in the century in present-day northern Namibia and Botswana (Nicholson, 1981). More southern areas of the Kalahari experienced a similar decline in rainfall. Around the 1780s and 1790s, the Matlaring and Kuruman were deep and strong rivers, but they continually dried up during the course of the next three or four decades (Campbell, 1822). Prior to the 1830s the Matlaring River ran at Kuruman, the rains fell much more abundantly, and the Kuruman River ran across what is now desert. By the early 1830s obvious signs of desiccation of these rivers and the surrounding countryside had appeared. In southern Namibia, the flow of the spring at Copper Berg continually decreased from *c*.1800 to *c*.1830.

Livingstone's and Andersson's mid-nineteenth century reports from Lake Ngami are also supplemented with oral tradition. As the Okavango Delta dried up late in the nineteenth century, numerous droughts reportedly occurred: 1862 and 1876 to 1877 at Shoshong (the Ngwato tribal capital), and in Kweneng (Jwaneng) in 1879 and 1896. The 1876/7 drought was so severe that both the missionary Hepburn and tribal rainmakers were called in to perform prayers and traditional ceremonies to end it. Hepburn was successful, causing the local chief to henceforth institute a day of prayer to

replace the rainmaking ceremony (Hitchcock, 1979). Mokwati, a descendant of a local chief, related tales of both higher levels and of a formerly dry lake bed (Thomas and Shaw, 1991).

An earlier desiccation of Ngami probably occurred from about 1670 to 1770 (Campbell and Child, 1971; Schwarz, 1920). One of the earliest reports came from a very old man who, around 1820, told the traveller Stigand that during his youth (*c*.1760) he heard old men of his tribe talking about a time when the lake was dry and instead a river with trees along its banks flowed through a grassy plain. The story was confirmed when the stumps of these trees emerged during a modern desiccation. The apparent age of the trees and other information suggest that Ngami remained dry for a century or longer, probably commencing some time in the mid-seventeenth century.

Abundant oral tradition complements the historical and geological information suggesting low phases of Lakes Malawi and Chilwa in the early 1800s. Johnston (1897) records a tradition from *c*.1825 that the northern end of Lake Malawi was so shallow between Deep Bay and Amelia Bay that a local chief could walk across areas that are now quite deep. The Ngonde king Mwangonde, whose reign is genealogically dated to about 1815 to 1835 (Owens *et al.*, 1990), walked across the northern end of the lake to Mwela to take a bride named Mapunda (Wilson, 1939). Further south, Chief Amapunda, walked across dry ground to Likoma Island by way of the north end of Malawi, an event dated to around the 1830s. Similarly, the locations of villages suggest the lake stood then at least 9 m below modern levels and references to tribal fighting around the 1840s also provide indications of low lake-levels.

In this same region, traditions of severe droughts and the drying up of rivers persisted from the late eighteenth century to the early 1800s. Rainmakers were called to combat severe drought sometime between *c*.1780 and 1840 (Owens *et al.*, 1990). Lake Rukwa totally dried up *c*.1770 and people emigrated from the lake shore to the highlands to the north-east, probably because of drier conditions between *c*.1740 and 1840. The North Rukuru River in northern Malawi dried up during the reign of Mwangonde (*c*.1815 to 1835). The tradition of the Ngoni clan of northern Malawi describes crossing the Zambezi near Zumbo on a sandy causeway during the dry season of 1835 (an historically established date).

The early nineteenth-century desiccation of Lake Chilwa is also confirmed by oral tradition. An account from the early 1900s (a time of extremely low lake-levels) indicates that a person could walk across the dried lake floor to Nchisi Island, occasionally breaking through the crust to ankle-deep mud and the hard surface below it. This lower surface would correspond to the hard-pan layer in the sediment core. Based on assumptions about the rate of deposition and the thickness of mud crust required to support a man, the sediment layer above the hard pan must have been 1 to 2 m thick and would have taken fifty to a hundred years to accumulate. This suggests the arid period marked by the hard pan ended sometime between 1800 and 1850 (Crossley *et al.*, 1984).

For parts of present-day South Africa, oral traditions (Ballard, 1986) reaffirm a chronology based on documentary sources and described below. There are frequent references to poor conditions in the Cape and Natal, frequent droughts and poor harvests early in the nineteenth century. Two major droughts occurred in the Cape in the 1800s and early 1820s. These affected also the Ciskei and Transkei and the summer rainfall regions of south-eastern Africa, especially Zululand and Natal. Evidence of the drought also comes from parts of Zambia, Zimbabwe, and the Tete region of Mozambique (Nicholson, 1980*a,c*). On the whole the period *c.*1800 to 1824 was likely one of reduced precipitation throughout much of south-eastern Africa. Famines occurred which were severe enough to create a breakdown of social, political, and economic institutions among the Nguni cattle-keeping cultivators, giving rise to Shaka's Zulu kingdom. An earlier drought *c.*1800–7 is explicit and vivid in Zulu oral tradition. Another period of relatively dry conditions probably occurred *c.*1700 to 1750. Then the Abakwa Qabe kingdom expanded from the drier, drought-stricken interior to the wetter, humid coastal lowlands, where the risk of drought is inherently lower.

Corroboration of the aridity of the early 1800s comes from oral traditions of Mozambique (Liesegang, 1978). In the eighteenth century a number of brief and localized droughts and famines had occurred: in 1717 in Changamire's territory, in 1744–5 in the Zumbo are, in 1758–9 in the Zambezi valley and neighbouring areas, and in Delagoa Bay in 1777. They became more frequent in the 1790s, with one reported in 1791 in Delagoa Bay and Inhambane and from *c.*1791–6 further north in the Bay. Famine occurred, probably as a result of drought, throughout most of the 1820s near Sena, and in some years in areas of south-eastern Zambia, Inhambane, and the Duma area of Zimbabwe. Reports of famine and drought continue in the mid-1850s in Sena, Tete, Quelimane, and the Limpopo Valley. Some areas were also affected from 1858 to 1863, and Lake Chiuta almost dried up at this time (Liesegang, 1978; Webster, 1979*a*). The next famines occurred in the 1870s and these were mentioned in most years of the

decade. Droughts were minor and localized in the 1880s, and limited to two years of the decade. But in 1896–7, most of the lowveld of southern Mozambique and much of the Transvaal suffered drought.

Documentary Sources

Weather chronologies for ten regions of Southern Africa have been derived by Nicholson (1981) making extensive use of documentary sources (Figure 4.8). The location of the regions is indicated in Figure 4.5 and a more thorough description of the sources and their content is given in Nicholson (1980*c*). In general, these chronologies are in good agreement with information in the geographical and historical sources described above, but more year-to-year information is available. The relatively poor conditions of the first half of the nineteenth century are apparent, as are the good conditions late in the century and the relatively poor conditions throughout much of the twentieth century. However, these are very broad generalizations; the length and timing of dry or wet episodes differs among the various regions and individual years show tremendous variability (Vogel, 1989), with floods well documented in some years. Also, a frequent opposition between the eastern and western Cape is apparent.

The most thorough study of historical climates in the region is that of Vogel (1989), using missionary, official, and archival records. Gathering sufficient sources to describe essentially every year between 1820 and 1900 (Figure 4.9), Vogel's work provides much more detail than that of Nicholson (1981) or other studies of the region (e.g. Hutchins, 1889; van der Merwe, 1937; Kokot, 1948), but it is somewhat localized, being limited to areas of the southern and eastern Cape. There is good agreement with the trends described in Figure 4.8, although it is not readily apparent to which of these ten regions Vogel's regions correspond. However, there are notable differences between Vogel's southern Cape chronology and Nicholson's western Cape region, with which it appears to overlap. The latter describes primarily the Cape Town area, and derived support from the Cape tree-ring chronology of Dunwiddie and LaMarche (1980), which indicates vigorous growth from *c.*1874 to 1902. It is interesting to note that Vogel demonstrates frequent opposition between the eastern and southern Cape, an opposition confirmed with the modern meteorological record. This might account for some of the discrepancies.

Particular anomalous periods in Vogel's Cape chronology are depicted in Figure 4.10. These indicate relatively dry conditions *c.*1825–9, *c.*1834–43, 1849–51, and

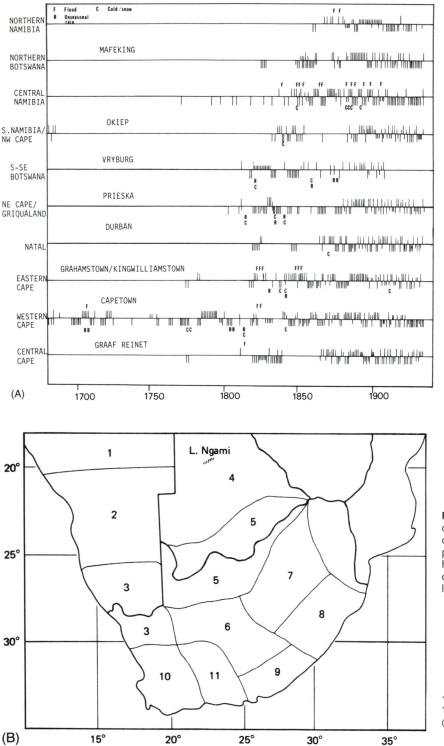

Fig. 4.8. Climate chronologies for ten regions of Southern Africa based primarily on reports of harvests, famine, and drought. See key map for location of regions.

1. Northern Namibia
2. Central Namibia
3. Southern Namibia–NW Cape
4. Northern Botswana
5. S–SE Botswana
6. NE Cape–Griqualand
7. Transvaal
8. Natal
9. Eastern Cape
10. Western Cape
11. Central Cape

(*Source*: Nicholson, 1981.)

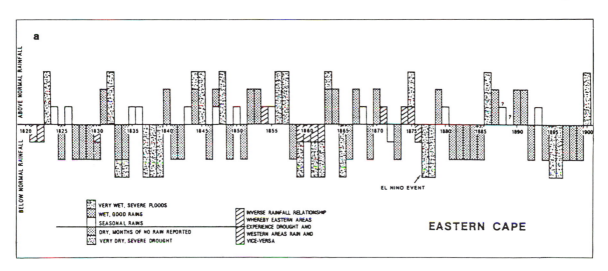

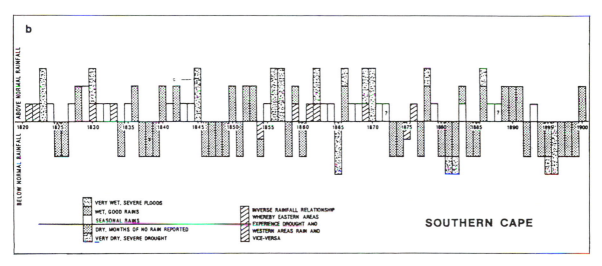

Fig. 4.9. Documentary-derived chronologies of precipitation variations for (a) eastern, and (b) southern Cape of South Africa, 1820–1900.
(*Source*: Vogel, 1989.)

much of the 1870s, early 1880s and in many regions *c.*1886–96. The wettest periods appear to be *c.*1830–3, 1844–8, and the 1850s. The 1860s were comparatively dry, but there were nearly as many regions with dry conditions as with relatively wet ones. Vogel points out that the tree chronology derived for the Cape Province by Dunwiddie and LaMarche (1980) is in general agreement with her results, especially for the drier years. Years of suppressed growth, with corroborative reports of dryness, include the early 1820s, the 1850s, early 1870s, mid-1880s, and mid-1890s. Years of increased growth, which cannot be unambiguously attributed to

increased rainfall, are frequent in the 1810s and just before, and again commencing around 1875.

Northern Africa and the Eastern Mediterranean

The primary proxy records of climate and weather in North Africa are information on harvests from Algeria and tree-ring chronologies described earlier. Drought became increasingly common in Algeria between *c.*1550 and 1630. Favourable conditions then prevailed until

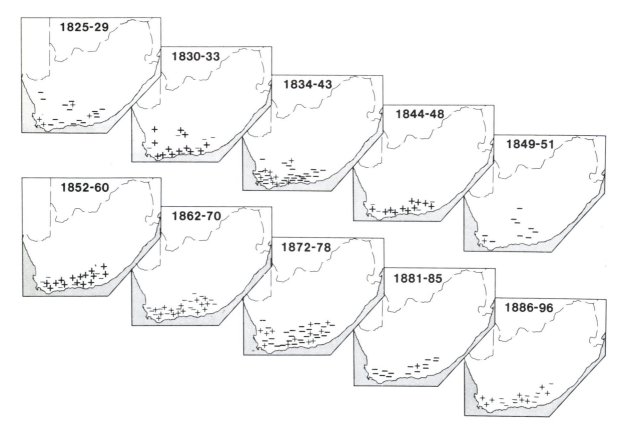

Fig. 4.10. Maps of periods of pronounced above-normal or below-normal rainfall, based on documentary evidence. Plus symbols denote generally wetter conditions; minus symbols denote generally drier conditions. (*Source:* Vogel, 1989.)

around 1710, after which time until at least 1800 numerous bad harvests and famines are reported (Nicholson, 1980*a*).

An interesting set of rainfall measurements is available from a nearby region which probably experiences rainfall fluctuations synchronously with much of North Africa. Measurements were made at Funchal, Madeira, during the seven years 1747 to 1753. The island, like North Africa, receives rainfall in winter when the subpolar low pressure systems are displaced to more southerly latitudes; it lies in the same storm tracks as North Africa. These show, at least during this 7-year period, a marked shift of the rainfall regime, with a December–January maximum instead of the current one in January to March. The mean for the period was 780 mm, compared with an annual mean today of 618 mm; that sum appeared to be normal for the mid-eighteenth century (Nicholson, 1976, 1980*a*).

The remaining information for North Africa is quite diverse and no well-dated long-term chronologies are available, except from the dendroclimatological studies described earlier. Nevertheless, there is extensive evidence of drier and wetter periods in numerous regions of North Africa during historical times. This includes river terraces, sedimentary deposits, bogs, and historical descriptions from Morocco, Algeria, Tunisia, Libya, and the Sinai (Brooks, 1932; Akinjogbin, 1967; Rognon, 1967; Vita-Finzi, 1969; Servant, 1973; Grove, 1973; Nicholson, 1976). This is complemented by a chronology from the Dead Sea, further east (Klein and Flohn, 1987). However, most reports are quite isolated and little detail is given, so that a firm synthesis is not possible.

Overall the evidence from North Africa and the Mediterranean suggests climatic fluctuations similar to those described for the Sahel. For example, much of the evidence suggests a relatively humid period within the

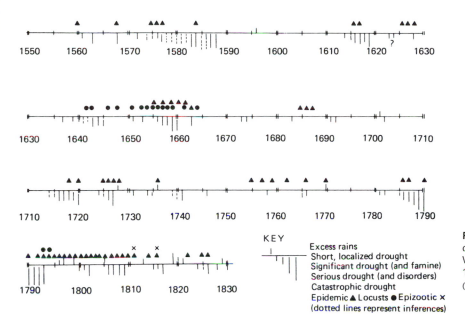

KEY

Excess rains
Short, localized drought
Significant drought (and famine)
Serious drought (and disorders)
Catastrophic drought
Epidemic ▲ Locusts ● Epizootic ✕
(dotted lines represent inferences)

Fig. 4.11. Time-line of drought and disease in West-Central Africa (Angola), 1560–1830.

(*Source*: Miller, 1982.)

time-frame of the ninth or tenth through thirteenth centuries and probably a major decline in rainfall in some regions toward 1500. Rainfall appears to have decreased in numerous regions toward 1700, with a possible return of wetter conditions in the late eighteenth century. Lower rainfall would have characterized the nineteenth century until about 1850 or 1870, with increased rainfall again from about 1870 to 1895 or 1900 (Nicholson, 1980a).

Eastern Africa and Other Equatorial Regions

Comparatively little historical information has been compiled for equatorial regions. For Eastern Africa the primary sources are lake-levels during the nineteenth century (Figure 4.5), Nile records for then and previous centuries, and meteorological observations commencing in the late nineteenth century. There is a virtual gap in coverage in central equatorial Africa, until weather records were made sporadically in the 1880s and 1890s. For western sectors, such as present-day Angola, there is extensive historical literature from the Portuguese colonial records and historical documents beginning in the sixteenth century. Also, the rainfall record at Luanda begins in 1858. The historical records have recently been compiled by Miller (1982), who has produced an extensive climate and weather chronology for the region. Similar effort could probably result in a similar but perhaps shorter chronology for eastern Africa. Webster (1979a,b) has commenced such a study but so far the dating method (correlation with Nile flow) is highly subjective.

Miller's detailed chronology is schematically summarized in Figure 4.11. References to famine and drought are much more common than to wet years and these show a surprising coincidence with periods of drought in West Africa, particularly in the Niger Bend region of the Sahel (Figure 4.4). Droughts were frequent in the 1570s and 1580s, the 1610s, and the 1640s and 1650s. They occurred only sporadically over the next fifty years, then again became frequent in the 1710s and, to a lesser extent, the 1720s. Droughts were again frequent, but not extreme, in the 1730s and 1740s. A long period of drought extended from about 1785 to 1794. Droughts were relatively common in the first few decades of the nineteenth century, especially in the 1800s and 1820s. Miller's chronology stops about 1830, but there are reports from other sources of a scarcity of rainfall between 1837 and 1841 (Dias, 1981). The rainfall record of Luanda suggests that relatively dry conditions again occurred, beginning in the late 1850s and continuing through the 1870s or 1880s, although a number of wet years occurred in the 1880s (Dias, 1981). Drought was severe in the 1870s and later in the 1910s.

The lake chronologies, from equatorial regions much further east, show a somewhat different picture (Figure 4.5). The few with available information suggest relatively low stages in the early to mid-nineteenth century. However, in contrast to the Angolan chronology, most indicate relatively high levels during the 1870s until towards the end of the nineteenth century. The decline in lake-levels commenced generally between 1880 and 1890; for example, Lake Rukwa by 1897 had shrunk to one-third of the size it had attained around 1880. This was coincident with a period of severe famine and drought throughout East Africa in the 1890s (Kjekshus, 1977). Lake-levels continued to fall until the early twentieth century, with relatively low stands being maintained until about the 1960s in most places. This is quite similar to the picture in Southern Africa.

Synthesis and Conclusions

A synthesis of the available historical records indicates that within the last few centuries numerous periods of wetter or drier conditions have occurred in Africa and that these have generally been continental in scale, although the precise timing (e.g. onset and end) are more regionally specific. These periods are illustrated in Figures 4.12 and 4.13, which are based on information presented in the foregoing references, together with material in this chapter.

One interesting period is the transition between the late eighteenth and early nineteenth centuries. This appears to have been a time of a major aridification in the subtropical latitudes of Africa in both hemispheres (Nicholson, 1981). Evidence for the transition is particularly abundant in the southern hemisphere (Figure 4.12), which experienced a trend towards increased dryness beginning probably in the 1780s or 1790s and culminating in severe and widespread droughts by the 1820s and 1830s. It is interesting to note that this is roughly synchronous with the end of the so-called 'Little Ice Age' in the global chronology of historical climates (Lamb, 1968; Wendland and Bryson, 1974; Grove, 1992).

Certainly the 1820s and 1830s were a period of increased aridity throughout most of the continent. During this time droughts affected most of the continent. However, they generally lasted only a few consecutive years in Southern Africa, but extreme drought probably persisted for nearly the entire two decades in Sahelian regions (Figure 4.13), a situation quite reminiscent of recent decades (Nicholson, 1993, 1994).

Relatively good conditions returned in mid-century and by the last decades of the nineteenth century (Figure 4.13), most of the regions experiencing the aridity of the 1820s and 1830s again experienced relatively humid conditions apparent both in environment and climate. However, the long decline at the onset of the nineteenth century was repeated at the beginning of the twentieth century. A continental-scale trend to increasing aridity again culminated in severe and widespread droughts, primarily in the 1910s. This period, however, was apparent not only in historical proxy data but also in the modern meteorological record (Nicholson, 1981).

The continental-scale anomalies apparent in Figure 4.13 are analogous to those most commonly occurring during the twentieth century, patterns established using high temporal and spatial resolution precipitation data (Nicholson, 1980b). These include two quite contrasting modes of behaviour: one in which the precipitation anomalies are of one sign over most of Africa, and one which comprises a strong opposition between the tropical and subtropical latitudes. The continental scale of the anomalies is strong evidence that the environmental changes which occurred, and which might individually have been interpreted as being non-climatic in origin, had underlying climatic causes. It is further significant that the major changes occurred at times of quasi-global climatic change. For example, the 1820s and 1830s were a period of anomalous global circulation and a last intensification of the 'Little Ice Age' conditions (Lamb, 1968).

Some of the modern analogues for these historical periods are depicted in Figure 4.14. The 1950s are quite reminiscent of the situation c.1870–95, with increased rainfall in the subtropics but anomalous low rainfall in the equatorial latitudes. An abrupt change occurred around 1960 and the pattern for the 1960s is quite the opposite of that for the 1950s. The rainfall decline of most of the continent, apparent in the 1820s and 1830s as well as in the 1910s, shows strong similarities to the 1980s.

Two major characteristics of African climate are illustrated by these periods. One is that more than one mode of anomalous behaviour exists, and the other is that both the spatial and temporal transitions between anomalously dry and wet conditions can be very abrupt. As a consequence, many regions with distinct climatic teleconnections may sometimes exhibit parallel fluctuations but sometimes be clearly in opposition to each other.

These characteristics, probably unknown to most researchers evaluating the palaeoclimate record, have posed problems for studies using proxy data which have relied on looking for simultaneous occurrence of environmental changes as a means of verifying climatic and environmental reconstructions. In view of them,

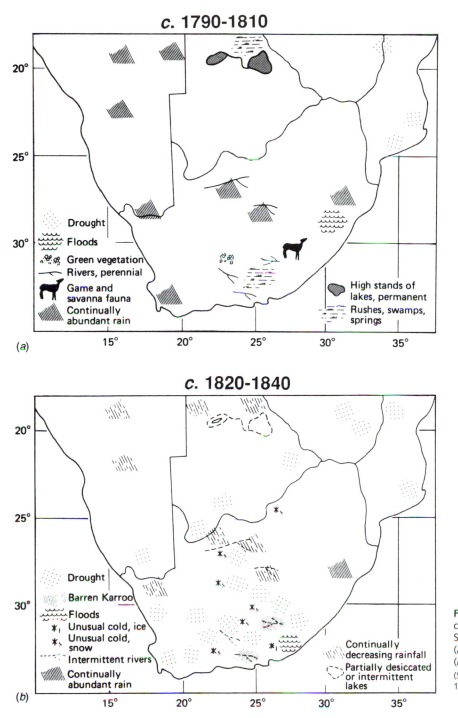

c. 1790-1810

(a)

c. 1820-1840

(b)

Fig. 4.12. Comparison of climate and environment in Southern Africa (a) c.1790–1810, and (b) c.1820–40.

(*Source:* material in Nicholson, 1980c, 1981, and in this chapter.)

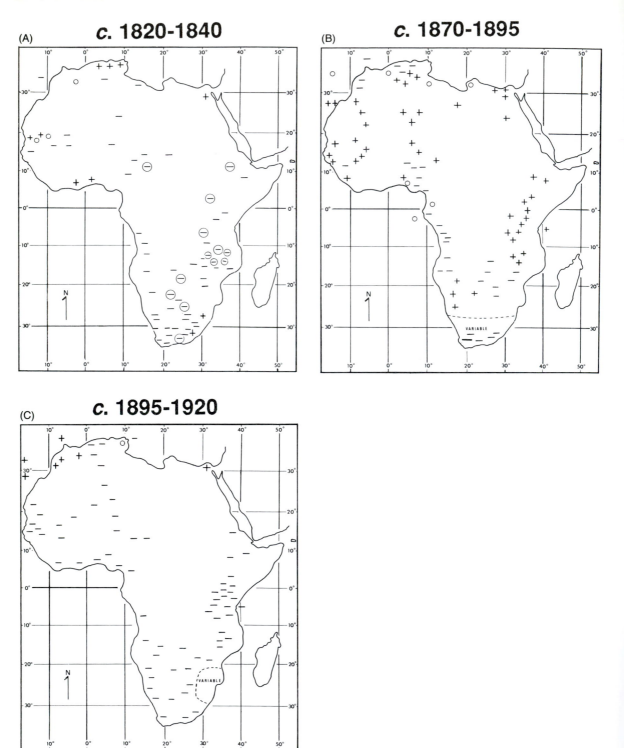

Fig. 4.13. African rainfall anomalies for three historical periods. Minus signs denote evidence of drier conditions; plus signs, above-average conditions; small circles, near-average conditions; circled symbols, regional integrators such as lakes and rivers.
(*Source:* material in Nicholson, 1978, 1980c, 1981, and in this chapter.)

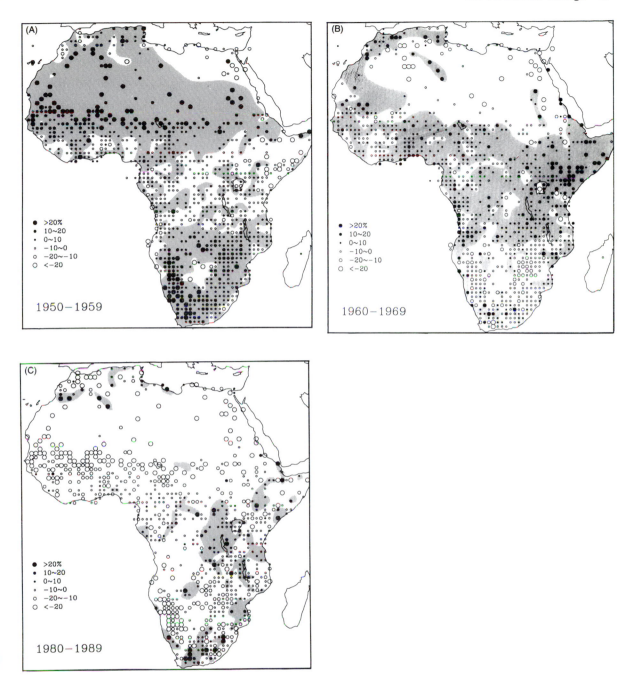

Fig. 4.14. Maps of rainfall anomalies for the African continent for the decades of the 1950s, 1960s, and 1980s, expressed as a percentage above or below the long-term mean for one-degree grid squares.

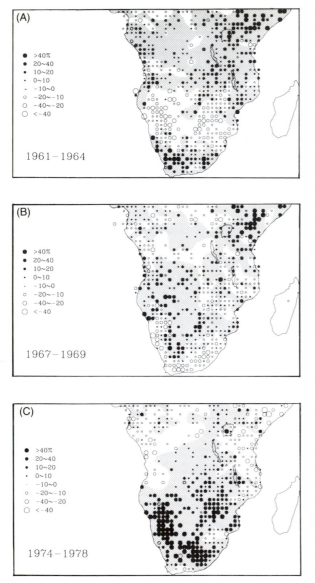

Fig. 4.15. Maps of rainfall anomalies in Southern Africa for the periods 1961–4, 1967–9 and 1974–8, expressed as a percentage above or below the long-term mean for one-degree grid squares.

however, seeming discrepancies, such as nearby lakes exhibiting opposing trends, can be resolved.

The modern record also provides some information about the magnitude of precipitation anomalies needed to account for the reconstructed environmental changes in Africa. None of the historical indicators provides a precise measurement of rainfall, but rough estimates

can be gleaned from several sources. Grove (1972) shows that the Nile discharge during the period c.1880–95 was about 35 per cent greater than for the period 1910–40. Rainfall at Freetown, Sierra Leone, declined by a comparable amount during the same period (Nicholson, 1981). Shaw (1985) calculates that the nineteenth century high stands of Lake Ngami required that it receive about 11 per cent of the Okavango's current total inflow, implying a rainfall increase of about this same order of magnitude. A hydrologic model indicates that the transgression of Lake Chilwa that ended around 1750 would have represented about a 30 to 50 per cent increase in rainfall, compared to the present century (Owen et al., 1990). Anomalies of this magnitude are not unusual today. In Southern Africa, rainfall for the periods 1967–9 and 1974–8, two periods of high lake-levels, was of the order of 20–40 per cent above normal; the rapid transgression of the East African lakes in the early 1960s was also associated with rainfall about 20–40 per cent above the mean (Figure 4.15). In the Sahel, rainfall during the 1950s was about 25–40 per cent above the long-term mean. Thus, the long-term changes of environment during the past millenium appear to represent primarily a more persistent occurrence of conditions commonly characterizing briefer periods of the twentieth century.

A final conclusion concerns the issue of long-term environmental change. So far, the major environmental changes which have occurred within the last century can be shown to have historical counterparts. This throws some doubt on the concept of human-induced desertification (see Chapter 19) over wide expanses of the continent (e.g. Eckholm, 1975) because evidence of this process is in part the large and extensive environmental changes which have affected Africa in recent decades. Other recent studies have also challenged this notion (e.g. Mainguet, 1991; Nelson, 1988; Tucker et al., 1991; and Helldén, 1991 take issue with much of the literature produced on desertification in the 1970s and early 1980s). The African environments are sensitive indicators of global climate and exhibit marked changes in response to its changes. The current period of aridity is not unlike that of the early nineteenth century and is not necessarily a sign of a continued or irreversible trend.

References

Adanson, M. (1759), *A Voyage to Senegal, the Isle of Gorée, and the River Gambia* (London).

Akinjogbin, I. A. (1967), *Dahomey and its Neighbours, 1708–1818* (Cambridge).

Andersson, C. J. (1857), *Lake Ngami, or Explorations and Discoveries During Four Years Wanderings in the Wilds of Southwest Africa* (New York).

Baines, T. (1864), *Explorations in Southwest Africa* (London).

Ballard, C. (1986), 'Drought and Economic Distress: South Africa in the 1800s', *Journal of Interdisciplinary History*, 17: 359–78.

Barbot, J. (1732), *A Description of the Coasts of North and South Guinea, and of Ethiopia Inferior, Vulgarly, Angola* (London).

Barrow, J. (1801), *Account of Travels into the Interior of Southern Africa*. i (London).

—— (1804), *Account of Travels into the Interior of Southern Africa*. ii (London).

Bosman, W. (1975), *A New and Accurate Description of the Coast of Guinea, Divided into the Gold, the Slave, and the Ivory Coasts* (London).

Brook, G., Burney, D., and Cowart, J. (1990), 'Desert Paleoenvironmental Data from Cave Speleothems with Examples from the Chijuahuan, Somali–Chalbi, and Kalahari Deserts', *Palaeogeography, Palaeoclimatology, Palaeoecology*, 76: 311–29.

Brooks, C. E. P. (1932), *Le Climat de Sahara et de l'Arabie* (Paris).

Browne, W. G. (1799), *Travels in Africa, Egypt, and Syria* (London).

Butzer, K. W. (1971), 'Recent History of an Ethiopian Delta', *University of Chicago Geographical Department Papers*, 136.

—— (1984), 'Late Quaternary Environments in South Africa', in J. C. Vogel (ed.), *Late Cainozoic Palaeoclimates of the Southern Hemisphere* (Rotterdam), 235–64.

Campbell, A. and Child, G. (1971), 'The Impact of Man on the Environment of Botswana', *Botswana Notes and Records*, 3: 91–109.

Campbell, J. (1822), *Travels in South Africa*, 2 vols. (London).

Chamard, P. (1973), 'Les paleoclimats du Sud-Ouest Saharien au Quaternaire recent, colloque de Nouakchott sur les problèmes de la desértification' (unpublished).

Cissoko, S. M. (1968), 'Famines et épidemies à Tombouctoo et dans la Boucle du Niger du XVIᵉ au XVIIIᵉ siècle', *Bulletin de l'Institut fondamental d'Afrique Noire*, 30: 806–21.

Crossley, R., Davison-Hierschmann, S., Owen, R. B., and Shaw, P. (1984), 'Lake Level Fluctuations during the Last 2000 Years in Malawi', in Vogel (ed.), *Late Cainozoic Palaeoclimates*, 305–16.

Curtin, P. D. (1975), *Economic Change in Pre-Colonial Africa: Supplementary Evidence* (Madison).

Daveau, S. (1965), 'Dunes ravinées, et dépôts du Quaternaire recent dans le Sahel mauritanien', *Revue de Géographie de l'Afrique occidentale-Dakar*, 1: 7–48.

—— (1969), 'La découverte du climat d'Afrique tropicale au cours des navigations portugaises', *Bulletin de l'Institut fondamental d'Afrique Noire*, 31: 953–88.

de Heinzelin, J. (1962), 'Carte des extensions glaciaires du Ruwenzori', *Biuletyn Peryglacialny*, 11: 133–9.

Degens, E. T. and Hecky, R. E. (1974), 'Palaeoclimatic Reconstruction of Late Pleistocene and Holocene Based on Biogenic Sediments from the Black Sea and a Tropical Lake', in *Les méthodes quantitatives d'étude des variations du climat au cours du Pleistocene* (Paris), 13–24.

Dias, J. (1981), 'Famine and Disease in the History of Angola', *Journal of African History*, 22: 349–78.

Drayton, R. (1979), *Study of the Causes of the Abnormally High Levels of Lake Malawi in 1979* (Blantyre, Malawi).

Dunwiddie, P. and LaMarche, V. (1980), 'A Climatically-responsive Tree-ring Record from *Widdrintonia cedarbergensis*', *Nature*, 286: 796–7.

Dyer, T. (1978), unpublished material.

Eckholm, E. P. (1975), 'Desertification: A World Problem', *Ambio*, 4: 137–45.

Fahn, A. and Wachs, N. (1963), 'Dendrochronological Studies in the Negev', *Israel Exploration Journal*, 13: 291–9.

Fritts, C. (1976), *Tree Rings and Climate* (New York).

Gasse, F. (1975), 'L'évolution des lacs de l'Afar Central (Ethiopie et T.F.A.I.) du Plio-Pleistocene à l'Actuel', Ph.D. thesis (Paris).

—— (1977), 'Evolution of Lake Abhé (Ethiopia and TFAI), from 70 000 BP', *Nature*, 265: 42–5.

Ginestous, G. (1927), 'Le chêne zeen d'ain Draham', *Bulletin de la direction générale de l'agriculture*, 3–12.

Grove, A. T. (1969), 'Landforms and Climatic Change in the Kalahari and Ngamiland', *Geographical Journal*, 135: 191–212.

—— (1972), 'Climatic Change in Africa in the Last 20 000 Years', in *Les problèmes de developement du Sahara septentrional, Ouarghla colloquium*, 2 vols., (Algiers).

—— (1973), 'Desertification in the African Environment', in D. Dalby and R. J. Harrison Church (eds.), *Drought in Africa, Report of the 1973 Symposium*, Centre for African Studies (London), 33–45.

—— (1985) (ed.), *The Niger and Its Neighbours* (Boston).

—— (1992), 'The Little Ice Age in Tropical Africa', in T. Mikami (ed.), *Proceedings of the International Symposium on the Little Ice Age Climate* (Tokyo), 46–51.

Grove, J. M. and Johansen, A. M. (1968), 'The Historical Geography of the Volta Delta, Ghana, during the Period of Danish Influence', *Bulletin de l'Institut fondamental d'Afrique Noire*, 30: 1374–421.

Halfman, J. and Johnson, T. (1988), 'High-resolution Record of Cyclic Change during the Past 4 ka Lake Turkana, Kenya', *Geology*, 16: 496–500.

Hall, M. (1976), 'Dendroclimatology, Rainfall and Human Adaptation in the Later Iron Age of Natal and Zululand', *Annals of the Natal Museum*, 22: 693–703.

Hastenrath, S. (1984), *The Glaciers of Equatorial East Africa* (Dordrecht).

Hecky, R. E. (1978), 'The Kivu-Tangangyika Basin: The Last 14,000 years', *Polskie Archiwum Hyrobiologii*, 25: 159–65.

Helldén, U. (1991), 'Desertification: Time for an Assessment?', *Ambio*, 20: 372–83.

Hillier, J. (1697), 'Two Letters from Cape Corse in Guinea', *Philosophical Transactions*, 687–707.

Hitchcock, R. K. (1979), 'The Traditional Response to Drought in Botswana', in M. T. Hinchey (ed.), *Proceedings of the Symposium on Drought in Botswana* (Gaborone), 91–7.

Hövermann, J. (1959), 'Über die Höhenlage der Schneegrenze in Äthiopien und ihre Schwankungen in historischer Zeit', *Nachrichten der Akademie der Wissenschaften in Göttingen*, 111–37.

Hutchins, D. E. (1889), *Cycles of Drought and Good Seasons in South Africa* (Wynberg, South Africa).

Jackson, J. G. (1820), *An Account of Timbuctoo and Housa* (London).

Johnston, H. H. (1897), *British Central Africa* (London).

Kjekshus, H. (1977), *Ecology Control and Economic Development in East African History: The Case of Tanganyika* (Los Angeles).

Klein, C. and Flohn, H. (1987), 'Contributions to the Knowledge of the Fluctuations of the Dead Sea Level', *Theoretical and Applied Climatology*, 38: 151–6.

Kokot, L. (1948), *An Investigation into the Evidence Bearing on the Recent Climatic Changes over Southern Africa* (Pretoria).

Lamb, H. H. (1968), *The Changing Climate* (London).

Lancaster, N. (1979), 'The Changes in the Lake Level', in A. J. McLachlan, C. Howard-Williams, and M. Kalk (eds.), *Monographae Biologicae, Lake Chilwa: Studies of Change in a Tropical Ecosystem* (The Hague), 27–34.

Liesegang, G. (1978), 'Famines and Smallpox in South-Eastern Africa 18th to 20th Centuries' (unpublished).

Livingstone, D. (1858), *Misssionary Travels and Researches in South Africa* (New York).

Mainguet, M. (1991), *Desertification, Natural background and Human Mismanagement* (Berlin).

Maley, J. (1973), 'Mécanisme des changements climatiques aux basses latitudes', *Palaeogeography, Palaeoclimatology, Palaeoecology*, 14: 193–227.

—— (1976), 'Les variations du lac Tchad depuis un millenaire: Consequences paléoclimatiques', *Palaeoecology of Africa*, 9: 44–7.

—— (1981), 'Etudes palynologiques dans le Bassin du Tchad et paléoclimatologie de l'Afrique nord tropicale de 30 000 ans à l'époque actuelle', *Travaux et documents de l'O.R.S.T.O.M.*, 129.

Marty, P. (1927), 'Chroniques de Oualatta et de Nema', *Revue des études islamiques*, 355–426, 531–75.

Michel, P. (1973), 'Les bassins des fleuves Sénégal et Gambie: Étude géomorphologique', *Mémoires de l'O.R.S.T.O.M.*, 63.

Miller, J. (1982), 'The Significance of Drought, Disease and Famine in the Agriculturally Marginal Zones of West-Central Africa', *Journal of African History*, 23: 17–61.

Moffatt, R. (1842), *Missionary Labours and Scenes in South Africa* (London).

Montiel, V. (1939), 'Chroniques du Tichitt', *Bulletin de l'Institut Fondamental d'Afrique Noire*, 1: 282–313.

Munson, P. J. (1971), 'The Tichitt Tradition: A Late Prehistoric Occupation of the South-western Sahara', Ph.D. thesis (Urbana-Champagne).

Nelson, R. (1988), *Dryland Management: The 'Desertification' Problem* (Washington, DC).

Nicholson, S. E. (1976), A Climatic Chronology for Africa: Synthesis of Geological, Historical, and Meteorological Information and Data, Ph.D. thesis (Madison).

—— (1978), 'Climatic Variations in the Sahel and Other African Regions during the Past Five Centuries', *Journal of Arid Environments*, 1: 3–24.

—— (1979), 'The Methodology of Historical Climate Reconstruction and its Application to Africa', *Journal of African History*, 20: 31–49.

—— (1980*a*), 'The Nature of Rainfall Fluctuations in Subtropical West Africa', *Monthly Weather Review*, 108: 473–87.

—— (1980*b*), 'Saharan Climates in Historic Times', in M. A. J. Williams and H. Faure (eds.), *The Sahara and the Nile* (Rotterdam), 173–200.

—— (1980*c*), 'Study of Environmental and Climatic Changes in Africa during the Past Five Centuries' (Worcester, Mass.), uupublished report to the National Science Foundation, 155 pp.

—— (1981), 'The Historical Climatology of Africa', in T. M. L. Wigley, M. J. Ingram, and G. Farmer (eds.), *Climate and History* (Cambridge), 249–70.

—— (1986), 'The Spatial Coherence of African Rainfall Anomalies: Interhemispheric Teleconnections', *Journal of Climate and Applied Meteorology*, 25: 1365–81.

—— (1989), 'African Drought: Characteristics, Causal Theories and Global Teleconnections', in A. Berger, R. E. Dickinson, and J. W. Kidson (eds.), *Understanding Climate Change* (Washington, DC), 79–100.

—— (1993), 'An Overview of African Rainfall Fluctuations of the Last Decade', *Journal of Climate*, 6: 1463–66.

—— (1995), 'Variability of African Rainfall on Interannual and Decadal Time Scales', *National Academy of Sciences Report on Climate Variations on 10–100 Year Time Scales*.

—— Kim, J., and Hoopingarner, J. (1988), 'Atlas of African Rainfall and its Interannual Variability' (Tallahassee, Fla.), unpublished report, 237 pp.

Owen, R., Crossley, R., Johnson, T., Tweddle, D., Kornfield, I., Davison, S., Eccles, D., and Engstrom, D. (1990), 'Major Low Levels of Lake Malawi and Their Implications for Speciation Rates in Cichlid Fishes', in *Proceedings of the Royal Society of London*, 519–53.

Palmer, H. R. (1936), *Bornu, Sahara, and Sudan* (London).

Park, M. (1799), *Travels in the Interior Districts of Africa (1795–97)* (London).

Pike, J. G. (1965), 'The Sunspot–Lake Level Relationship and the Control of Lake Nyasa', *Journal of Institute of Water Engineers*, 19: 221–6.

—— and Rimmington, G. T. (1965), *Malawi: A Geographical Study* (Oxford).

Plote, H. (1974), *L'Afrique Sahélienne se dessèche-t-elle?* (Orléans, France).

Riehl, H., El-Bakry, M., and Meitin, J. (1979), 'Nile River Discharge', *Monthly Weather Review*, 107: 1546–53.

Rognon, P. (1967), *Le massif de l'Atakor et ses bordures (Sahara Central), étude géomorphologique* (Paris).

—— (1974), 'Paleoclimatologie et sechèresse actuelle (1969–1974) au Sahel', *Revue de géographie physique et géologie dynamique*, 32: 55–69.

Schwarz, E. H. L. (1920), *The Kalahari or Thirstland Redemption* (Cape Town).

Servant, M. (1973), 'Séquences continentales et variations climatiques: Évolution du bassin du Tchad au Cénozoique supérieur', Ph.D. thesis (Paris).

Shaw, B. (1975), 'Economy and Society of the Maghreb during the Period of the Roman Empire', Ph.D. thesis (Cambridge).

Shaw, P. (1984), 'A Historical Note on the Outflows of the Okavango Delta System', *Botswana Notes and Records*, 16: 127–30.

—— (1985), 'The Desiccation of Lake Ngami: An Historical Perspective', *Geographical Journal*, 151: 318–26.

Sieger, R. (1887), 'Schwankungen der innerafrikanischen Seen', in *Jahresbericht, Verein der Geographie* (Vienna), 41–60.

Stigand, A. G. (1923), 'Ngamiland', *Geographical Journal*, 62: 401–19.

Stockton, C. (1990), 'Climatic Variability on the Scale of Decades to Centuries', *Climatic Change*, 16: 173–83.

Talbot, M. (1983), 'Lake Bosumtwi, Ghana', *Nyame Akuma*, 23: 162.

—— and Delibrias, G. (1980), 'A New Late Pleistocene–Holocene Water-Level Curve For Lake Bosumtwi, Ghana', *Earth and Planetary Science Letters*, 47: 336–44.

—— Livingstone, D., Palmer, P., Maley, J., Melack, J., Delibrias, G., and Gulliksen, S. (1984), 'Preliminary Results from Sediment Cores from Lake Bosumtwi, Ghana', *Palaeoecology of Africa and the Surrounding Islands*, 16: 173–92.

Thomas, D. and Shaw, P. (1991), *The Kalahari Environment* (Cambridge).

Toupet, C. (1973), 'L'évolution du climat de la Mauritanie du Moyen-Age jusqu'à nos jours', *Colloque de Nouakchott sur les problèmes de la desértification*, 20.

—— (1977), 'La sédentarisation des nomades en Mauritanie Centrale Sahelienne', Ph.D. thesis (Paris).

Toussoun, O. (1923), 'Mémoire sur l'histoire du Nil', *Mémoires de l'Institut d'Egypte*, 9: 63–213.

Tucker, C. J., Dregne, H. E., and Newcomb, W. W. (1991), 'Expansion and Contraction of the Sahara Desert from 1980 to 1990', *Science*, 253: 299–301.

Tyson, P. D. (1986), *Climatic Change and Variability in Southern Africa* (Oxford).

—— and Lindesay, J. A. (1992), 'The Climate of the Last 2000 Years in Southern Africa', *The Holocene*, 2: 271–8.

van der Merwe, P. J. (1937), *Die Noordwaarse Beweging van die Boere voor die Groot Trek* (Cape Town).

van Zinderen Bakker, E. M. and Coetzee, J. A. (1972), 'A Reappraisal of Late-Quaternary Climatic Evidence from Tropical Africa', *Palaeoecology of Africa*, 7: 151–81.

Vincent, C., Davies, T., and Beresford, A. (1979), 'Recent Changes in the Level of Lake Naivasha, Kenya, as an Indicator of Equatorial Westerlies over East Africa', *Climatic Change*, 2: 175–89.

Vita-Finzi, C. (1969). *The Mediterranean Valleys: Geological Changes in Historical Time* (Cambridge).

Vogel, C. (1989), 'A Documentary-Derived Climatic Chronology for South Africa', *Climatic Change*, 14: 291–308.

Waisel, Y. and Lipschitz, N. (1968), 'Dendrochronological Studies in Israel', *La-Yaaran*, 18: 1–22.

Walter, H. (1936), 'Die Periodizität von Trocken und Regenzeit in Deutsch-Suedwestafrika auf Grund von Jahresringmessungen an Bäumen', *Berichten der Deutschen Botanischen Gesellschaft*, 54: 608–11.

Webster, J. B. (1979*a*), 'Drought and Migration: The Lake Malawi Littoral as a Region of Refuge', in M. T. Hinchley (ed.), *Proceedings of the Symposium on Drought in Botswana*, 148–57.

—— (1979*b*), 'Noi! Noi! Famines as an Aid to Interlacustrine Chronology', in J. B. Webster (ed.), *Chronology, Migration and Drought in Interlacustrine Africa*, 1–37.

Wendland, W. M. and Bryson, R. A. (1974), 'Dating Climatic Episodes of the Holocene', *Quaternary Research*, 4: 9–24.

Williams, M. A. J. (1975), 'Late Pleistocene Tropical Aridity Synchronous in Both Hemispheres?', *Nature*, 253: 617–18.

Wilson, B. H. (1973), 'Some Natural and Man-made Changes in the Channels of the Okavango Delta', *Botswana Notes and Records*, 5: 132–53.

Wilson, G. (1939), *The Constitution of the Ngonde* (Lusaka, Zambia).

Wilson, J. (1865), 'Water Supply in the Basin of the River Orange or Gariep, South Africa', *Geographical Journal*, 35: 106–29.

Winterbottom, T. M. (1969), *An Account of the Native Africans in the Neighbourhood of Sierra Leone* (London).

5 Climate Change Within the Period of Meteorological Records

Mike Hulme

Introduction

Continuous meteorological records commenced in Africa during the first half of the nineteenth century, largely coinciding with the establishment of European settlement. It was only towards the end of the nineteenth century, however, that the observing network became sufficiently dense to enable large-scale analysis of climate and its variability from instrumental records. This chapter therefore discusses changes in African climate which have occurred over the last 100 years with primary attention being given to changes in rainfall, the climatic resource of greatest significance for the continent. The rainfall regimes of Africa are highly variable, both in time and space and, being strongly seasonal, differ markedly from those of higher latitudes. Whilst all of Africa is subject to droughts of varying lengths and intensities, it is only the Sahel region of Northern Africa which has witnessed sustained rainfall depletion in recent decades.

Rainfall is one of the most important natural resources for many of mainland Africa's forty-eight nations.

Achieving food security in these nations has been a continual struggle in recent years, hampered by civil war, political volatility, worsening terms of international trade, rapid population growth, and drought (de Waal, 1988). Of these hindrances drought is frequently given a pre-eminence which is not always deserved (see Warren, Chapter 19, this volume). Careful assessment of the magnitude and extent of African rainfall changes over recent decades is needed, together with an appreciation of those characteristics of the rainfall supply which are most essential for its effective utilization as a resource and those which are most sensitive to change. Rainfall variability is the most critical characteristic of African rainfall for resource applications. Failure to appreciate fully the greater and more complex temporal variability of African rainfall compared to rainfall in temperate latitudes has in the past led to misjudgements being made about the viability of hydrological and agricultural schemes (e.g. the South Chad Irrigation Project; Kolawole, 1987).

This chapter describes the characteristics of this rainfall variability using instrumental rainfall data from all forty-eight African nations. Some consideration is also given to changes in temperature characteristics, most notably changes in the diurnal cycle of temperature in recent years, and in mountain glacier movements. The chapter concludes by considering a number of explanations for climate change in Africa which have been proposed in recent years. These explanations include land cover changes within Africa, changes in ocean temperatures and circulation, and the enhancement of the planetary 'greenhouse effect' due to increasing emissions of 'greenhouse gases'.

The extensive work by Sharon Nicholson in the 1970s and early 1980s in compiling many of the original station time-series used in this chapter is fully acknowledged. Substantial additions and updates have been supported through contracts to the Climatic Research Unit from the US Department of Energy and the UK Department of the Environment. These extensive additions have been possible through the generous provision of data by National Meteorological Agencies and individuals throughout Africa too numerous to mention by name. Their efforts in maintaining observing networks and data archives in the face of numerous difficulties are to be applauded. This author, for one, is indebted to them.

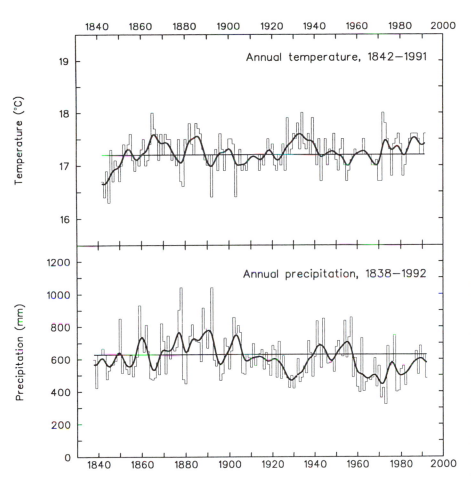

Fig. 5.1. Homogenized time-series of annual mean temperature (1842–1991), and annual precipitation (1838–1992) for Cape Town.

Individual (calendar) yearly values are smoothed with a filter which suppresses variations on time-scale of less than 10 years. The horizontal lines represent the full record mean. Data extracted from the CRU/DoE global data set described in Jones *et al.* (1985, subsequently updated).

The Origins of Instrumental Meteorological Records in Africa

The longest continuous instrumental meteorological records for Africa commence in the early nineteenth century and are representative of the coastal regions of the continent where Europeans first settled. The longest continuous temperature and precipitation time-series for a single location are for Cape Town and these are shown in Figure 5.1. Ensuring homogeneity in these records is as much a problem for African stations as for elsewhere in the world (Jones and Wigley, 1990). For example, the sporadic temperature records which exist for Cape Town before the Observatory was constructed in 1841 cannot be used since there is a systematic 2–3 °C warm bias in mean temperature, probably related to the absence of night-time measurements in these early years. Later, a step jump of about 0.5 °C occurred around 1910, perhaps reflecting a change in site, and a comparison of

the old Observatory site with the newer Airport site located outside the city boundary indicates a modest urban warming effect in recent decades. These latter two sources of inhomogeneity have been removed from the time-series shown in Figure 5.1. Some of the other earliest instrumental temperature and precipitation records for Africa are listed in Table 5.1.

By the 1890s about fifty climatological stations were well established in Africa, nearly all of them within 100 km of the coast (Figure 5.2). Subsequent network development reflected the success or otherwise of various European nations in establishing their administrations in the continent. By the 1920s most countries had at least a handful of reliable climatological stations in operation. The conventional ground observation network (Plate 5.1) for Africa reached its peak during the 1960s, since when the combined influence of the contraction of government services and the resurgence of civil war in numerous countries has greatly eroded the network

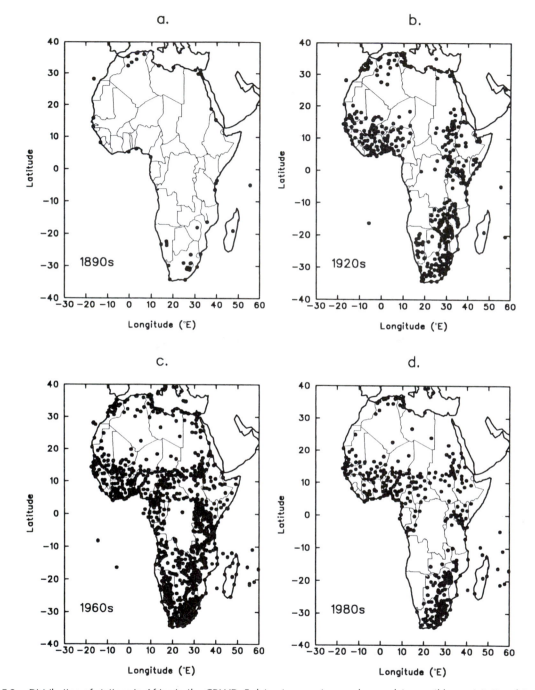

Fig. 5.2. Distribution of stations in Africa in the CRU/DoE dataset possessing nearly complete monthly precipitation data for four different decades.

These data are extracted from the CRU/DoE global dataset described in Jones *et al.* (1985, subsequently updated). They are therefore only illustrative of the observing networks existing in these decades—additional data and stations almost certainly exist for these decades, although they are not readily accessible.

Plate 5.1. A synoptic meteorological station at Dongola in northern Sudan, one of about twenty-five currently operating synoptic stations in Sudan. For many countries in Africa, it has proved hard to maintain such stations in the face of civil war or reductions in public expenditure, yet the observations they make are invaluable for a wide range of operational and research applications (photo: M. Hulme).

Table 5.1 Some of the longest instrumental temperature and precipitation records for Africa, with the dates of commencement of continuous record

Temperature record	Start date	Precipitation record	Start date
Cape Town, South Africa	1842	Cape Town, South Africa	1838
Lungi, Sierra Leone	1849	Algiers, Algeria	1838
Oran, Algeria	1852	Oran, Algeria	1841
Algiers, Algeria	1856	St. Louis, Senegal	1848
St. Louis, Senegal	1862	Lungi, Sierra Leone	1849
Cairo, Egypt	1870	Grahamstown, South Africa	1854

(Figure 5.2). This trend is worrying for the purposes of climate monitoring and climate change detection, since the maintenance of stations with long climate records is essential for meeting these two objectives. Although numerous climate variables are now monitored routinely from satellite platforms—and this is as true for Africa as for other continents—time-series of such remotely sensed variables are inevitably less than twenty years, and often less than ten years in duration. Such data are inadequate for assessments to be made of climate variability or change which occurs on time-scales of decades or longer.

Rainfall is the most important climate resource within Africa and the next two sections summarize its variability over, first, space and, second, time using data from the instrumental record. Numerous authors have undertaken this type of rainfall analysis for either the continent as a whole (e.g. Nicholson, 1986; Janowiak, 1988; Shinoda, 1990) or for specific regions and countries (e.g. Ogallo, 1989; Hulme, 1990; Janicot, 1990; Olaniran, 1991; Matarira and Jury, 1992; Fontaine and Bigot, 1993).

Spatial Patterns of Changing Rainfall

The most recent 30-year World Meteorological Organisation (WMO) normal period is now 1961–90, thus enabling two successive 30-year climatologies for Africa to be compared for the first time: 1931–60 and 1961–90. This comparison is important for two reasons. First, it provides a continent-wide illustration of the dependence of so-called climatological 'normals' upon the 30-year period which is chosen. This is a point which has frequently been made in recent years with respect to Africa (e.g. Todorov, 1985; Quinlan, 1986; Farmer, 1989), yet one which requires a more comprehensive spatial illustration. Second, such a comparison reveals the magnitude and pattern of rainfall change which has occurred over this 60-year period, thereby quantifying the variability of African rainfall on multi-decadal time-scales.

Changes in Mean Rainfall

The change between 1931–60 and 1961–90 in mean seasonal rainfall rates over Africa is shown in Figure 5.3 for boreal (June, July and August: JJA) and austral summers (December, January, and February: DJF). Rainfall change has been dominated by the reduction

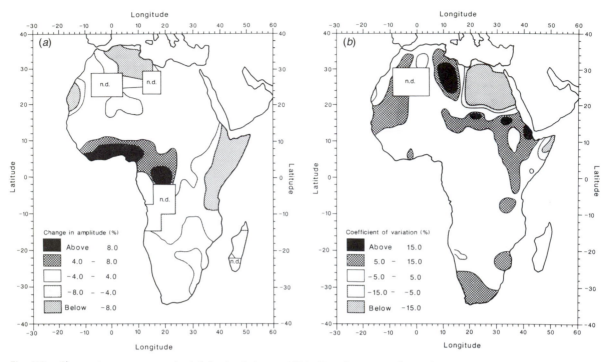

Fig. 5.3. Change in mean seasonal rainfall rates between 1931–60 and 1961–90 for (*a*) boreal summer (JJA); and (*b*) winter (DJF).
Only changes greater than ±0.2mm day⁻¹ are shown (n.d. = no data). (*Source*: Hulme, 1992.)

in JJA rainfall in the Sahel with widespread decreases of well over 0.4 mm day⁻¹ (locally over 30 per cent) between these two 30-year climatologies. Throughout the Sahel this magnitude of change represents more than 0.5 of the 1931–60 standard deviation and in parts of Mali a decrease of over 1.0 standard deviations. Such a large relative rainfall change between these two 30-year climatologies is unparalleled elsewhere in the world (Hulme *et al.*, 1992). In contrast, JJA rainfall rates have increased by over 0.4 mm day⁻¹ in the southern coastal regions of West Africa and parts of west Equatorial Africa. However, this represents an increase of only ~10 per cent in rainfall and generally is less than 0.5 of the 1931–60 standard deviation. Partly compensating the summer increase in these latter regions has been a decline in winter (DJF) rains of about 0.2 mm day⁻¹ (Figure 5.3), representing a decrease of ~15 per cent.

Austral summer (DJF) rainfall rates have declined in the southern hemisphere tropical margins with decreases of more than 0.4 mm day⁻¹ (~10 per cent) over parts of Botswana and Zimbabwe. Similar magnitude increases in DJF rainfall rates have been restricted to the interior of Tanzania and northern Madagascar. None of these changes in DJF rainfall are greater than 0.5 of the

1931–60 standard deviation and are therefore less significant than those changes affecting tropical North Africa. The greatest percentage *increases* in seasonal rainfall rates have occurred over equatorial East Africa with SON rainfall increasing by ~25 per cent (not shown), about 0.5 standard deviations. This latter change largely reflects the series of wet years which occurred in the early 1960s (Flohn, 1987). These maps of rainfall change should be further interpreted with respect to the regional time-series of rainfall anomalies shown later.

Because of the strong latitudinal alignment of rainfall gradients over much of Africa, in particular in the Sahel, these changes in mean rainfall can be converted into equivalent latitudinal shifts in isohyets. The maximum migration has occurred at about 12.5°N with southward shifts of JJA rainfall zones of just over 1° latitude (~120 km) between 1931–60 and 1961–90. Contrasting the two 5-year periods experiencing the greatest rainfall extremes this century (1950–4, wet and 1983–7, dry) yields a southerly latitudinal shift of just under 3° latitude (~330 km). This compares with shifts of up to 2.2° latitude (~240 km) in the position of the 200 mm annual isohyet in the Sahel during the decade 1980 to 1990 estimated by Tucker *et al.* (1991)

using Normalised Difference Vegetation Index (NDVI) data derived from satellites.

The spatial character of rainfall anomalies for individual years can also be assessed and Nicholson (e.g. 1978, 1986, 1994) has undertaken the most extensive analysis of such spatial characteristics for Africa. She identified six major rainfall anomaly types which characterize this spatial variability (Nicholson, 1986). They illustrate four major patterns, two with anomalies of the same sign (negative or positive) over most of the continent and two with anomalies in the low latitudes opposing those in the subtropical regions of both hemispheres. Figure 5.4 illustrates some of these anomaly types by showing the decile anomalies for four separate years extracted from the gridded dataset of Hulme (1994). The two driest years of the twentieth century over Africa—1913 and 1984—are characterized by uniform dryness over the whole continent with the exception of the extreme north and south-east of the continent (Nicholson's Anomaly Type 2). The severity of the 1984 drought in the Sahel is striking. 1950—a wet year except in equatorial latitudes—illustrates Nicholson's Anomaly Type 3, a mode of contrasting anomalies between low latitudes and the subtropics. Janicot (1992) also clearly identifies this dipole over West Africa. 1961 is the year of the 'East African rainfall anomaly' (Flohn, 1987) and displays contrasting rainfall anomalies between Equatorial/north-eastern Africa and West Africa.

Changes in Seasonality

Rainfall over most of Africa is strongly seasonal, reflecting the dominant role of the migrating Inter-Tropical Convergence Zone (ITCZ) in determining the rainfall seasons. Seasonality is predominantly unimodal (i.e. a single wet season), with a dual wet season restricted to parts of the immediate equatorial zone especially in East Africa. One measure of seasonality is the amplitude of the first harmonic which describes the annual cycle of rainfall. This amplitude can be standardized by expressing it as a percentage of the mean monthly rainfall for the year. Seasonality thus defined varies in Africa from close to zero in the central Sahara, to about 50 per cent in Equatorial Africa), to well over 100 per cent in the Sahel (Hulme, 1992).

Figure 5.5a presents the change in this index of seasonality between 1931–60 and 1961–90 expressed as the change in the percentage of the mean monthly rainfall for the respective period. Seasonality has increased most substantially (>8 per cent) in the southern coastal region of West Africa reflecting the increased JJA and decreased DJF rainfall noted above. Elsewhere in Africa, seasonality has generally decreased, most notably in the coastal regions of East Africa and parts of the North African littoral where decreases have been greater than 8 per cent. In the Sahel, rainfall seasonality has remained constant despite the decrease in the JJA rainfall rate. This reflects the continuing dominant role even during periods of lower rainfall of the northerly summer migration of the ITCZ in generating rainfall in this region.

Changes in Variability

The interannual variability of rainfall is an important indicator of the reliability of the rainfall resource in Africa. There are difficulties in defining an adequate measure which reflects changes in variability rather than changes in the mean rainfall. A measure of *relative* variability (the coefficient of variation—which is the standard deviation standardized by the mean) is therefore used to represent changes in reliability. Figure 5.5b presents this measure of annual rainfall variability change over the continent and shows that between 1931–60 and 1961–90 areas of increased variability outweighed areas of reduced variability. The latter are restricted to Egypt and eastern Libya and a small area of northern Somalia. The Sahel shows an increased rainfall variability, although with a contrast between the western (increases generally of less than 5 per cent) and eastern Sahel (increases over 5 per cent and locally over 15 per cent). Tunisia and western Libya and the extreme south of Africa have also seen increases in relative variability of over 5 per cent. For most of Equatorial Africa little change in relative annual variability has occurred.

Temporal Patterns of Changing Rainfall

The above comparison of 30-year climatologies needs to be placed within a longer historical context and this can be done by examining time-series of regional rainfall anomalies over the last 100 years. Numerous such studies have again been conducted, starting with those of Tanaka *et al.* (1975), Tyson *et al.* (1975) and Nicholson (1979), and more recently including Beltrando and Camberlin (1993) and Nicholson (1994). Rainfall Anomaly Indices (RAIs: Katz and Glantz, 1986; Bärring and Hulme, 1991) for three regions in Africa are shown here (Figure 5.6), these being constructed from individual station time-series. The three regions are the Sahel (defined as the area of tropical North Africa which received 100 mm to 600 mm mean annual rainfall for 1931–90), East Africa (Uganda, Kenya, and Tanzania), and south-western Africa (100 mm to 600 mm mean annual rainfall for 1931/2–1990/1). For the latter region the rainfall year was defined as July to June.

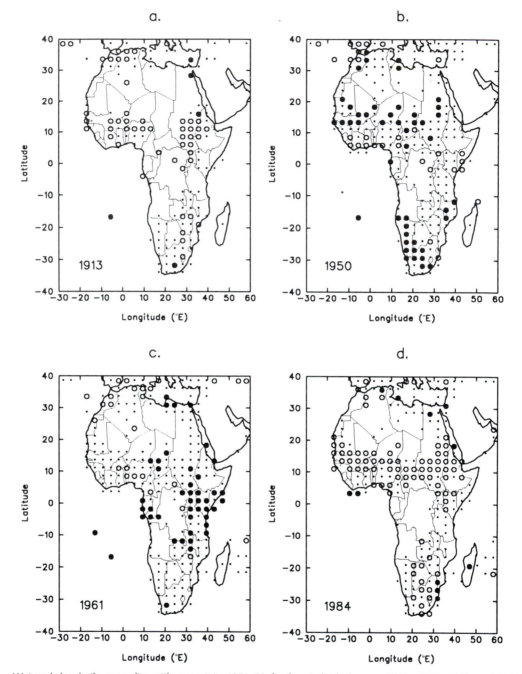

Fig. 5.4. Wet and dry decile anomalies with respect to 1951–80 for four individual years: 1913, 1950, 1961, and 1984. Anomalies are derived from the gridded global dataset at 2.5° latitude by 3.75° longitude resolution described in Hulme (1993). A decile anomaly indicates that the rainfall total for that year fell into the lowest (dry) or highest (wet) 10 per cent of years based on 30 years of data between 1951 and 1980.

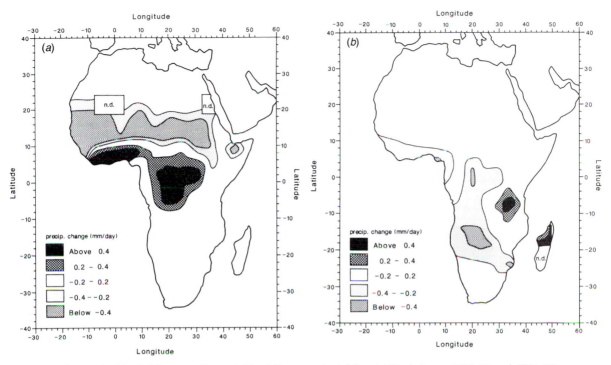

Fig. 5.5. Changes in (a) rainfall seasonality; and (b) relative annual rainfall variability between 1931–60 and 1961–90. Seasonality is measured as the amplitude of the first harmonic expressed as a percentage of the mean monthly rainfall for the respective time period; variability is measured as the coefficient of variation (n.d. = no data). (*Source*: Hulme, 1992.)

These three regions of Africa display quite different characteristics of rainfall variability over the twentieth century (Figure 5.6). The striking persistence of Sahelian rainfall anomalies, clearly revealed by its autocorrelation function, is in marked contrast to both the other two regions. This persistence has increased in recent decades (Folland *et al.*, 1991). Despite the near average year of 1988 in the Sahel, the subsequent years have again seen below average rainfall, continuing the drought sequence of the last twenty-five years. Equatorial East Africa shows little temporal organization in rainfall variability although, as remarked earlier, the extremely wet years of the early 1960s are noteworthy as is the cluster of wet years around 1905. The autocorrelation function of the East African series reveals little interannual persistence of rainfall in this region. The quasi-periodicity of Southern African rainfall over about an 18-year cycle, well documented by Tyson *et al.* (1975) and Tyson (1991), is only weakly revealed in the south-west African time-series shown here. Nevertheless, alternate wet and dry phases over recent decades are clearly evident from Figure 5.6. The coincidence in the early 1980s of a dry phase in south-western Africa, with severe rainfall deficits in the Sahel and two dry years in Equatorial East Africa

in 1983 and 1984, resulted in the near continent-wide drought between 1983 and 1987 noted previously, with related food security problems for the continent (Borton and Clay, 1988).

The contrasting longer-term trends in rainfall in these different regions is well illustrated in Figure 5.7 where the temporal evolution from 1901 to 1992 of the Rainfall Anomaly Indices for the Sahel and south-western Africa is plotted. These time series have been smoothed using a 30-year filter to highlight multi-decadal scale variability and are plotted against each other to emphasize their contrasting behaviour. The largely cyclical character of south-western African rainfall versus the prolonged desiccating trend in the Sahel is clearly evident over recent decades. Figure 5.7 also shows the equivalent rainfall analysis for the two most important headwater catchments of the River Nile—the Blue Nile Basin and the Lake Victoria catchment. The unique nature of the rainfall changes over tropical North Africa of the last two to three decades is again illustrated here, with rainfall over the Blue Nile Basin displaying a similar desiccating tendency to the Sahel. Rainfall over Lake Victoria since the end of the 1960s displays a much smaller drying trend, but the coincidence of these drying trends in

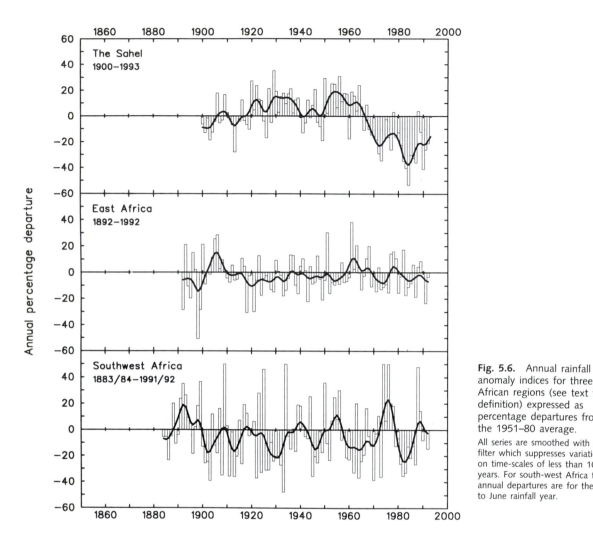

Fig. 5.6. Annual rainfall anomaly indices for three African regions (see text for definition) expressed as percentage departures from the 1951–80 average.

All series are smoothed with a filter which suppresses variations on time-scales of less than 10 years. For south-west Africa the annual departures are for the July to June rainfall year.

the two Nile catchments over recent years has resulted in reduced main Nile flows into Lake Nasser (Conway and Hulme, 1993).

One final analysis of interest which relates to the temporal variability of rainfall is shown in Table 5.2. During the 1950s and 1960s much of Africa regained its independence from the colonial European powers. Table 5.2 lists, for selected countries, the percentage change in mean annual rainfall from colonial to independent eras. For a belt of countries across the Sahel, this transition roughly coincided with a substantial (statistically significant) decline in their average rainfall resource. Elsewhere in Africa, however, no such sustained depletion in rainfall occurred and a number of countries have had modest (but not significant) increases in mean rainfall yield.

Changes in Other Climate Variables

Daily Rainfall

Fewer studies of African climate have considered temporal variations in the frequency and distribution of daily rainfall events of different magnitudes. Sivakumar (1992), for example, demonstrated how important are such secondary rainfall parameters for assessing the agricultural implications of climate variability in Africa. The comprehensive atlas of rainfall variability in Africa of Nicholson *et al.* (1988) contains data about mean daily rainfall for numerous locations, but no assessment of changes over time in rainfall intensities.

Two studies which *have* explicitly analysed time-series of daily rainfall magnitudes are those of Olaniran and Sumner (1989) and Hulme (1990), for Nigeria and

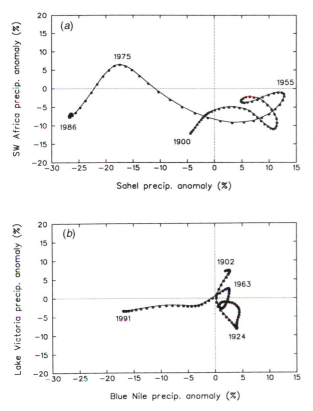

Fig. 5.7. Time evolution of the annual RAIs for (a) the Sahel versus south-western Africa (1901–92); and for (b) the Blue Nile Basin versus the Lake Victoria catchment (1902–91).

Time-series have been smoothed with a filter which suppresses variation on time-scales up to 30 years.

Table 5.2 Percentage change in mean annual rainfall from pre- to post-independence for selected African countries

Country	Rainfall change (%)	Periods compared
Niger	−21*	1922–59/1960–92
Sudan	−20*	1901–56/1957–92
Mali	−19*	1907–59/1960–92
Zimbabwe	− 7	1900–78/1979–90
Ghana	+ 1	1904–56/1957–89
Kenya	+ 3	1894–62/1963–91
Congo	+ 3	1933–59/1960–80
Botswana	+ 3	1923–65/1966–89
Tunisia	+ 8	1901–55/1956–89
Somalia	+16	1922–59/1960–86

Note: The dates of the two periods compared are shown. Only those changes marked by* are significant in a strict statistical sense.

rainfall events of small magnitude would tend to lead to the conclusion summarized above that rainfall intensities have increased. This is an area where substantially more work is needed, combining the careful scrutiny and digitizing of daily rainfall data with the statistical analysis of their changes over time.

Temperature

Since rainfall dominates many of Africa's resource management problems, it is not surprising that far less attention has been devoted to the analysis of temperature variations over the continent. Global-scale maps of mean temperature change, such as those presented in Hulme *et al.* (1992), can be used to identify changes in mean air temperature over 30-year periods. These indicate that between 1931–60 and 1961–90, much of Equatorial, West, and North Africa cooled slightly by up to 0.5 °C in mean annual temperature. Away from the Equator, however, seasonal temperature changes displayed different characteristics, with the Sahel and much of Southern Africa warming by up to 0.5 °C during boreal summer (JJA).

Regional time-series analyses of temperatures over Africa have also been scarce. The gridded global data set of Jones *et al.* (1985, updated) can be used to extract regional time series. Two such series, representing annual temperature for the Nile Basin in north-east Africa and for the SADC region of Southern Africa, are shown in Figure 5.8. In both cases warming has been observed over the last 100 years, in the case of the Nile Basin by about 0.5 °C between about 1910 and 1940 and in the case of SADC by a few tenths of a °C, again between 1910 and 1940 and also during the 1980s. The characteristics of this latter region parallel what has happened to global-mean temperature over the last century

Sudan respectively. Both studies found evidence to support an increase in mean rainfall intensity, even during the recent decades of reduced total rainfall volume. The storm which affected the Khartoum area of Sudan on 4/5 August 1988, and which yielded a record 200 mm of rain within 24 hours causing extensive flooding of the Sudanese capital (Hulme and Trilsbach, 1989), is a dramatic illustration of such a trend. A certain amount of caution, however, is needed when interpreting the results from analyses such as these which rely on daily rainfall data of dubious reliability. For example, Snidjers (1986) showed for Burkina Faso that observations of rainfall events of low magnitude (< 5 mm) are not homogenous over time. This is particularly so in recent years as the quality of climate observations has been placed under increasing pressure owing to the contraction of public-sector activities. The 'non-recording' of

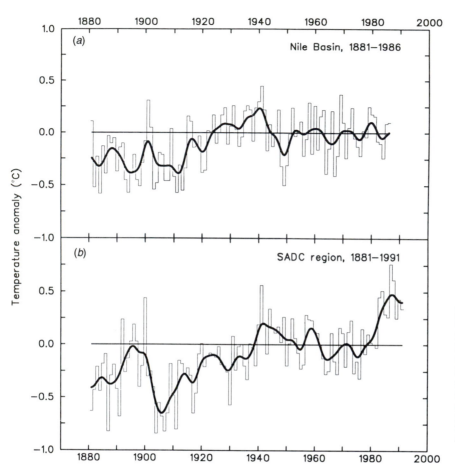

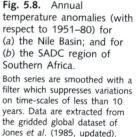

Fig. 5.8. Annual temperature anomalies (with respect to 1951–80) for (a) the Nile Basin; and for (b) the SADC region of Southern Africa.

Both series are smoothed with a filter which suppresses variations on time-scales of less than 10 years. Data are extracted from the gridded global dataset of Jones et al. (1985, updated).

(Jones and Briffa, 1992). This magnitude of warming (0.5 °C) may not appear to be very substantial, although in the case of the Nile Basin this results in a potential increase in evaporation losses from open-water surfaces such as the Equatorial Lakes and Lake Nasser of perhaps 2 to 3 per cent (Conway, 1993). An additional water loss of this magnitude from Lake Nasser (0.2 to 0.3 km³; nearly 10 per cent of Egypt's current budgeted water surplus) amounts to an important volume of water with respect to water management in Egypt (Abu Zeid and Hefny, 1992).

The above discussion refers to mean air temperature and this hides the sometimes contrasting trends which characterize minimum (night-time) and maximum (day-time) temperatures. Identifying changes in the diurnal cycle of temperature has, in recent years, become an important question since it has been suggested that warming due to an enhancement of the greenhouse effect resulting from human emissions of greenhouse gases might result in asymmetric warming between night and day (Gates et al., 1992). Although sufficiently long time-series of minimum and maximum temperatures are not always easily available for regions within Africa, a recent analysis by Jones and Lindesay (1993) of forty to fifty years of data from both Sudan and South Africa found some interesting results (Figure 5.9). Annually, the diurnal range of temperature has weakened in both countries—by about 0.3 °C in South Africa (1941 to 1991) and by about 0.7 °C in Sudan (1950 to 1987). While this annual pattern of diurnal temperature change is consistent with changes reported in other regions of the world (Karl et al., 1993), the seasonal changes are more complex. Sudan shows a strong decrease in the diurnal range in boreal winter (DJF) and lesser decreases in spring and autumn, but an increase in summer diurnality. South Africa's trend for reduced diurnality is also strongest in (austral) winter (Figure 5.9). The explanation for the trends in Sudan is probably

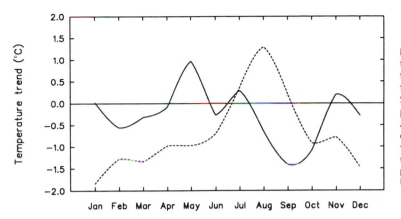

Fig. 5.9. Observed trends in the mean monthly diurnal temperature range derived from networks of stations in Sudan (1951 to 1987, pecked line) and South Africa (1941 to 1991, continuous line).

Trends calculated as mean linear trends over the respective time periods. (Data *source*: Jones and Lindesay, 1993.)

related to regional factors—increased dustiness in winter reducing daytime maxima and reduced rainfall in summer increasing daytime maxima because of reduced cloudiness. This shows the difficulty of relating results from such country-specific studies to global-scale forcing mechanisms.

The Glaciers of Equatorial East Africa

Movements in land glaciers represent an interesting indicator of climate change, although the complexities of the mass balance of mountain glaciers often make it difficult to relate glacier movements to changes in specific climatic parameters. Within Africa, the most substantial glaciers are found on the mountains of East Africa, particularly Mt Kenya and Mt Kilimanjaro. Work by Kruss (1983) and Hastenrath (1984) has clearly established the very substantial retreat of many of Mt Kenya's glaciers since the turn of the century. Indeed, Hastenrath (1984) reports that of the eighteen glaciers supported on Mt Kenya in 1899, only ten remained into the 1980s and even these had suffered substantial area and volume shrinkage.

While temperature increases at these high (> 4500 m) elevations may initially be thought to be responsible for this shrinkage, in fact it is the more complex changes associated with decreased cloudiness, and hence precipitation, which appear to be most important (Kruss and Hastenrath, 1987). Reduced cloudiness increases incident solar radiation (and also air temperature as albedos fall) which increases the energy available for ice wastage. It would appear therefore that over the first decades of this century decreased cloudiness at these altitudes has been largely responsible for glacier recession. Generalizing this conclusion to lower elevations and to other regions of Equatorial Africa is, of course, not legitimate.

Causes of Contemporary Climate Change in Africa

The supply of rainfall over Africa has changed substantially over the last 100 years, as it has over longer historical time-scales (see Chapter 4). This change has been most notable over tropical North Africa where rainfall during 1961–90 declined by up to 30 per cent compared to 1931–60. For the boreal summer months these changes in mean rainfall are statistically significant. For these Sahelian countries, annual rainfall yield during the independence era has declined by about 20 per cent compared to the colonial era earlier in the century. The tropical margins of Southern Africa have seen rainfall reduced by ~5 per cent over this period. Rainfall increases between these two periods have occurred in some areas of the continent, most notably in equatorial East Africa (+15 per cent) and in the southern coastal region of West Africa (+10 per cent). The variability of annual rainfall is high over much of the continent and this variability, when measured with respect to mean rainfall amount, has generally increased in Africa between 1931–60 and 1961–90. These variability changes, however, are not statistically significant when considering the continent as a whole. Rainfall seasonality is a fundamental characteristic of African rainfall and changes in seasonality also have occurred. One striking aspect of these changes is the contrast between the southern coastal region of West Africa (increased seasonality) and much of equatorial East Africa (decreased seasonality).

What explanations may be offered for the rainfall changes identified above? Possible causes may conveniently be summarized as falling into three broad areas: those related to land cover changes within the continent; those related to changes in the global ocean circulation and associated with patterns of sea-surface temperatures (SSTs); and those related to the changing composition of the global atmosphere.

Land-cover changes in Africa have occurred within two main biomes—tropical rain forests and acacia savanna. The destruction of tropical rain forest is primarily for logging, plantations, and first-generation cultivation, and Myers (1991) has estimated that during the late 1980s the process was occurring at an annual rate of between 0.5 and 1.0 per cent for tropical Africa. Theoretical and empirical evidence that contemporary tropical rain forest destruction significantly reduces regional-scale rainfall remains inconclusive for Africa, although clearer for the Amazon (Lean and Warrilow, 1989; Salati and Nobre, 1991). The progressive clearing of savanna-type acacia vegetation in the more sub-humid and semi-arid parts of the continent for fuelwood and increased intensity cultivation is a harder process to quantify. Callaghan *et al.* (1985) estimated the annual clearance rate of acacia woodland in central Sudan to be about 3.6 per cent, although this is likely to represent an upper limit. Reduction of dryland vegetation cover has led to the proposition of a land-surface–atmosphere feedback whereby reduced vegetation leads to decreases in soil moisture, increased sensible heat flux and hence reduced rainfall (reviewed in Nicholson, 1988). The Sahel is the prime candidate region for such a feedback mechanism to operate and the strong persistence of the recent rainfall decline provides circumstantial evidence in support of the idea. However, in such a strongly seasonal regime as the Sahel where one wet season is effectively decoupled from the next, it is hard to see how soil moisture can act as a year-to-year hydrological 'memory' and induce such persistence. Recent experiments by Rowell *et al.* (1992) suggest that the soil-moisture feedback is only a secondary factor in determining regional rainfall anomalies in the African Sahel.

An additional feedback mechanism involving vegetation change may operate through changes in surface albedo and roughness associated with vegetation succession. Modelling experiments of the Amazon Basin by Lean and Warrilow (1989) suggest that replacement of forest with grassland will reduce regional rainfall owing to reduced moisture convergence (through higher albedos) and decreased evaporation (through decreased surface roughness). The importance of the roughness mechanism for the Sahel is likely to be substantially less than for the Amazon, however, since only a small proportion of Sahelian rainfall is derived from local evaporation.

The influence of regional and global SSTs upon African rainfall has been most comprehensively assessed by Folland *et al.* (1991) for the Sahel, Nicholson and Entekhabi (1986) and Lindesay and Vogel (1990) for southern Africa, and Janowiak (1988) for the whole continent. The latter three studies found some association between regional rainfall anomalies and indices of the El Niño/Southern Oscillation (ENSO) phenomenon. This association was most notable for south-eastern Africa (negative anomalies shortly following ENSO events, as witnessed by the severe drought of 1991/2) and Equatorial East Africa (positive anomalies shortly following ENSO events). Such relationships account, however, for at most 25 per cent of the interannual variability in rainfall and are not sufficient to explain the quasi 18-year periodicity of south-west African rainfall shown in Figure 5.6. Work by Mason and Tyson (1992), however, suggests that this periodicity may be related to local SST anomalies around southern African coasts.

The relationship between global SST anomalies and Sahel rainfall anomalies is even more convincing. The mode of SST variation most closely associated with negative rainfall anomalies in the Sahel is one where the southern oceans (plus the Indian Ocean) are warm and the northern oceans are cool. The coherence of this relationship has led to the development of a seasonal forecasting capability for Sahel rainfall using April to June global SST anomalies (Folland *et al.*, 1991). The observed pattern of global SSTs over the last thirty years has predominantly expressed this mode of SST variation and modelling studies have supported this explanation for interdecadal rainfall fluctuations in the Sahel (Rowell *et al.*, 1992). Furthermore, the influence of South Atlantic SST anomalies on African rainfall suggests a strong dipole in rainfall anomalies between coastal West Africa and the interior Sahel (Druyan and Hastenrath, 1991; Janicot, 1992; Fontaine and Bigot, 1993), very much as shown here in the JJA rainfall change observed between 1931–60 and 1961–90 (Figure 5.3). An outstanding question therefore concerns the reason for the recent pattern of observed global SST anomalies. Street-Perrott and Perrott (1990), among others, have put forward the idea that changes in the strength of the thermohaline circulation of the Atlantic Ocean may account for the buildup of warmth in the southern oceans relative to the north. Evidence can be found for this mechanism operating on different time-scales including both millenia (Bond *et al.*, 1992) and decades (Read and Gould, 1992).

A third candidate explanation for contemporary changes in African rainfall concerns the changing composition of the global atmosphere. Empirical evidence for global-mean warming induced by rising greenhouse gas concentrations remains inconclusive, albeit suggestive (Wigley and Barnett, 1990). Global-mean precipitation is likely to increase with global warming at a rate of between 2 and 4 per cent per 1 °C of global-mean warming. This is a result of the intensification of the hydrological cycle, especially through increased evaporation over warmer ocean surfaces. The regional manifestation of such global precipitation change remains, however, highly uncertain. Rainfall decreases are anticipated in some regions for enhanced CO_2 conditions by all General Circulation Model (GCM) experiments, although these regions tend to vary from model to model.

Does the pattern of observed rainfall change in Africa from 1931–60 to 1961–90 bear any resemblance to rainfall change patterns generated by recent GCM greenhouse experiments? In fact, none of the recent GCM experiments which have modelled the effects of increased CO_2 concentrations have generated a pattern of rainfall change over Africa similar to that observed over the last sixty years (Hulme, 1992). This negative result is not surprising for two main reasons. First, the relatively poor simulation of present-day regional rainfall patterns by current GCMs (Hulme, 1991; Gates *et al.*, 1992) suggests that currently little confidence may be placed in simulations of future regional rainfall. If, indeed, global SSTs determine much of the interannual and inter-decadal variability of African rainfall, then a coupled ocean–atmosphere GCM which realistically simulates interannual variability in SSTs will be necessary to improve confidence. Second, the low signal-to-noise ratio anticipated for greenhouse-induced rainfall change (Wigley and Barnett, 1990) means that it will be very difficult to distinguish greenhouse-related rainfall anomalies from naturally induced variations in regional rainfall.

References

Abu Zeid, M. and Hefny, K. (1992), 'Water Resources Assessment and Management in Egypt during Conditions of Stress', *WMO Bulletin*, 41: 35–46.

Bärring, L. and Hulme, M. (1991), 'Filters and Approximate Confidence Intervals for Interpreting Rainfall Anomaly Indices', *Journal of Climate*, 4: 837–47.

Beltrando, G. and Camberlin, P. (1993), 'Interannual Variability of Rainfall in the Eastern Horn of Africa and Indicators of Atmospheric Circulation', *International Journal of Climatology*, 13, 533–46.

Bond, G., Heinrich, H., Broecker, W., Labeyrie, L., McManus, J., Andrews, J., Huon, S., Jantschik, R., Clasen, S., Simet, C., Tedesco, K., Klas, M., Bonani. G., and Ivy, S. (1992), 'Evidence for Massive Discharges of Icebergs into the North Atlantic Ocean During the Last Glacial Period', *Nature*, 360: 245–9.

Borton, J. and Clay, E. J. (1988), 'The African Food Crisis of 1982–86, in D. Rimmer (ed.), *Rural Transformation in Tropical Africa* (London), 140–67.

Callaghan, T. V., Bacon, P. J., Lindley, D. K., and el Moghraby, A. I. (1985), 'The Energy Crisis in the Sudan: Alternative Supplies of Biomass', *Biomass*, 8: 217–32.

Conway, D. (1993), 'A Grid-based Hydrologic Model of the Blue Nile and the Sensitivity of Nile Discharge to Climate Change', unpublished Ph.D. thesis, University of East Anglia, UK, 300 pp.

—— and Hulme, M. (1993), 'Recent Fluctuations in Precipitation and Runoff over the Nile Sub-Basins and Their Impact on Main Nile Discharge', *Climatic Change*, 25: 127–53.

de Waal, A. (1988), 'Famine Early Warning Systems and the Use of Socio-economic Data', *Disasters*, 12: 81–91.

Druyan, L. M. and Hastenrath, S. (1991), 'Modelling the Differential Impact of 1984 and 1950 Sea-surface Temperatures on Sahel Rainfall', *International Journal of Climatology*, 11: 367–80.

Farmer, G. (1989), 'Rainfall', in *The IUCN Sahel Studies, 1989* (Gland, Switz.), 1–25.

Flohn, H. (1987), 'East African Rains of 1961/62 and the Abrupt Change of the White Nile Discharge', in J. A. Coetzee (ed.), *Paleoecology of Africa and the Surrounding Islands* (Rotterdam), 3–18.

Folland, C. K., Owen, J. A., Ward, M. N., and Colman, A. W. (1991), 'Prediction of Seasonal Rainfall in the Sahel Region Using Empirical and Dynamical Methods', *Journal of Forecasting*, 10: 21–56.

Fontaine, B. and Bigot, S. (1993), 'West African Rainfall Deficits and Sea-surface Temperatures', *International Journal of Climatology*, 13: 271–86.

Gates, W. L., Mitchell, J. F. B., Boer, G. J., Cubasch, U., and Meleshko, V. P. (1992), 'Climate Modelling, Climate Prediction and Model Validation', in J. T. Houghton, B. A. Callander, and S. K. Varney (eds.), *Climate Change 1992: The Supplementary Report to the IPCC Scientific Assessment* (Cambridge), 97–134.

Hastenrath, S. (1984), *The Glaciers of Equatorial East Africa* (Dordrecht, Neth,.).

Hulme, M. (1990), 'The Changing Rainfall Resources of the Sudan', *Transactions of the Institute of British Geographers*, 15: 21–34.

—— (1991). 'An Intercomparison of Model and Observed Global Precipitation Climatologies', *Geophysical Research Letters*, 18: 1715–18.

—— (1992), 'Rainfall Changes in Africa: 1931–60 to 1961–90', *International Journal of Climatology*, 12: 685–99.

—— (1994), 'Validation of Large-Scale Precipitation Fields in General Circulation Models', in M. Desbois and F. Désalmand (eds.), *Global Precipitations and Climate Change* (Berlin), 387–405.

—— and Trilsbach, A. (1989), 'The August 1988 Storm over Khartoum: Its Climatology and Impact', *Weather*, 44: 82–90.

—— Marsh, R., and Jones, P. D. (1992), 'Global Changes in a Humidity Index Between 1931–60 and 1961–90', *Climate Research*, 2: 1–22.

Janicot, S. (1990), 'Variabilité des précipitations en Afrique de l'ouest et circulation quasi-stationnaire durent une phase de transition climatique', unpublished Ph.D. thesis, University of Paris.

—— (1992), 'Spatio-temporal Variability of West African Rainfall, Part I: Regionalisations and Typings', *Journal of Climate*, 5: 489–97.

Janowiak, J. E. (1988), 'An Investigation of Interannual Rainfall Variability in Africa', *Journal of Climate*, 1: 240–55.

Jones, P. D. and Briffa, K. (1992), 'Global Surface Air Temperature Variations during the Twentieth Century, Part I, Spatial, Temporal and Seasonal Details', *The Holocene*, 2: 165–79.

—— and Lindesay, J. (1993), 'Maximum and Minimum Temperature Trends of over South Africa and the Sudan', in *Proceedings of the Fourth International Conference on Southern Hemisphere Meteorology and Oceanography*, Hobart (American Meteorological Society, Boston), 359–60.

—— and Wigley, T. M. L. (1990), 'Global Warming Trends', *Scientific American*, Aug., 84–91.

—— Raper, S. C. B., Santer, B. D., Cherry, B. S. G., Goodess, C. M., Kelly, P. M., Wigley, T. M. L., Bradley, R. S., and Diaz, H. F. (1985), *A Grid Point Surface Air Temperature Data Set for the Northern*

Hemisphere, US Department of Energy, Report No. TR022 (Washington, DC).

Karl, T. R., Jones, P. D., Knight, R. W., Kukla, G., Plummer, N., Razuvayev, V., Gallo, K. P., Lindesay, J. and Peterson, T. C. (1993), 'A New Perspective on Recent Global Warming: Asymmetric Trends of Daily Maximum and Minimum Temperature', *Bulletin of the American Meteorological Society*, 74: 1007–23.

Katz, R. W. and Glantz, M. H. (1986), 'Anatomy of a Rainfall Index', *Monthly Weather Review*, 114: 764–71.

Kolawole, A. (1987), 'Environmental Change and the South Chad Irrigation Project (Nigeria)', *Journal of Arid Environments*, 13: 169–76.

Kruss, P. D. (1983), 'Climatic Change in East Africa: A Numerical Simulation From the 100 Years of Terminus Record at Lewis Glacier, Mount Kenya', *Zeitschrift Für Gletscherkunde und Glazialgeologie*, 19: 43–60.

—— and Hastenrath, S. (1987), 'The Role of Radiation Geometry in the Climate Response of Mount Kenya's Glaciers, Part I: Horizontal Reference Surfaces', *Journal of Climatology*, 7: 493–506.

Lean, J. and Warrilow, D. A. (1989), 'Simulation of the Regional Climatic Impact of Amazon Deforestation', *Nature*, 342: 411–13.

Lindesay, J. A. and Vogel, C. H. (1990), 'Historical Evidence for Southern Oscillation–Southern African Rainfall Relationships', *Internation Journal of Climatology*, 10: 679–90.

Mason, S. J. and Tyson, P. D. (1992), 'The Modulation of Sea-Surface Temperature and Rainfall Associations over Southern Africa with Solar Activity and the Quasi-biennial Oscillation', *Journal of Geophysical Research*, 97: 5847–56.

Matarira, C. H. and Jury, M. R. (1992), 'Contrasting Meteorological Structure of Intra-seasonal Wet and Dry Spells in Zimbabwe', *International Journal of Climatology*, 12: 165–76.

Myers, N. (1991), 'Tropical Forests: Present Status and Future Outlook', *Climatic Change*, 19: 3–32.

Nicholson, S. E. (1978), 'Climatic Variations in the Sahel and Other African Regions During the Past Five Centuries', *Journal of Arid Environments*, 1: 3–24.

—— (1979), 'Revised Rainfall Series for the West African Subtropics', *Monthly Weather Review*, 107: 620–3.

—— (1986), 'The Nature of Rainfall Variability in Africa South of the Equator', *Journal of Climatology*, 6: 515–30.

—— (1988), 'Land Surface–Atmosphere Interaction: Physical Processes and Surface Changes and Their Impact', *Progress in Physical Geography*, 12: 36–65.

—— (1994), 'Recent Rainfall Fluctuations in Africa and their Relationship to Past Conditions over the Continent', *The Holocene*, 4: 121–31.

—— and Entekhabi, D. (1986), 'The Quasi-periodic Behaviour of Rainfall variability in Africa and its Relationship to the Southern Oscillation', *Archiv für Meteorologie Geophysik und Bioklimatologie*, Ser. A, 34: 311–48.

—— Kim, J., and Hoopingarner, J. (1988), *Atlas of African Rainfall and its Interannual Variability* (Tallahassee, Fla.).

Ogallo, L. J. (1989), 'The Spatial and Temporal Patterns of East African Seasonal Rainfall Derived from Principal Components Analysis', *International Journal of Climatology*, 9: 145–68.

Olaniran, O. J. (1991), 'Rainfall Anomaly Patterns in Dry and Wet Years Over Nigeria', *International Journal of Climatology*, 11: 177–204.

—— and Sumner, G. M. (1989), 'Climatic Change in Nigeria: Variation in Rainfall Receipt per Rainday', *Weather*, 44: 242–8.

Quinlan, F. T. (1986), 'Comments on "Sahel: The Changing Rainfall Regime and the 'Normals' used for its Assessment"', *Journal of Climatology and Applied Meteorology Meteorology*, 25: 257.

Read, J. F. and Gould, W. J. (1992), 'Cooling and Freshening of the Subpolar North Atlantic Ocean since the 1960s', *Nature*, 360: 55–7.

Rowell, D. P., Folland, C. K., Maskell, K., Owen, J. A. and Ward, M. N. (1992), 'Causes and Predictability of Sahel Rainfall Variability', *Geophysics Research Letters*, 19: 905–8.

Salati, E. and Nobre, C. A. (1991), 'Possible Climatic Impacts of Tropical Deforestation', *Climatic Change*, 19: 177–96.

Shinoda, M. (1990), 'Annual Rainfall Variability and its Interhemispheric Coherence in the Semi-arid Region of Tropical Africa: Data Updated to 1987', *Journal of the Meteorological Society of Japan*, 67: 555–64.

Sivakumar, M. V. K. (1992), 'Empirical Analysis of Dry Spells for Agricultural Applications in West Africa', *Journal of Climate*, 5: 532–9.

Snijders, T. A. B. (1986), 'Interstation Correlations and Nonstationarity of Burkina Faso Rainfall', *Journal of Climatology and Applied Meteorology*, 25: 524–31.

Street-Perrott, F. A. and Perrott, R. A. (1990), 'Abrupt Climate Fluctuations in the Tropics: The Influence of Atlantic Ocean Circulation', *Nature*, 343: 607–12.

Tanaka, M., Weare, B. C., Navato, A. R., and Newell, R. E. (1975), 'Recent African Rainfall Patterns', *Nature*, 255: 201–3.

Todorov, A. V. (1985), 'Sahel: The Changing Rainfall Regime and the "Normals" used for its Assessment', *Journal of Climatology and Applied Meteorology*, 24, 97–107.

Tucker, C. J., Dregne, H. E. and Newcomb, W. W. (1991), 'Expansion and Contraction of the Sahara Desert from 1980 to 1990', *Science*, 253: 299–301.

Tyson, P. D. (1991), 'Climatic Change in Southern Africa: Past and Present Conditions and Possible Future Scenarios', *Climatic Change*, 18: 241–58.

—— Dyer, T. G. J. and Mametse, M. N. (1975), 'Secular Changes in South African Rainfall: 1880–1972', *Quarterly Journal of the Royal Meteorological Society*, 101: 817–33.

Wigley, T. M. L. and Barnett, T. P. (1990), 'Detection of the Greenhouse Effect', in J. T. Houghton, G. J. Jenkins, and J. J. Ephraums, *Climate Change: The IPCC Scientific Assessment* (Cambridge), 243–55.

6 Hydrology and Rivers

Des E. Walling

Introduction

The presence of European colonial powers in Africa during the early and middle years of the twentieth century, particularly the French in North and West Africa and the British in East and Southern Africa, meant that effective hydrometric networks were developed at a relatively early stage in several African countries. For example, Ball (1939) reports detailed measurements of suspended sediment and dissolved load transport for the River Nile in Egypt dating from the early part of the twentieth century and systematic gauging of the Nile flow started in 1912 under the the auspices of the Irrigation Department. Similarly, flow measurements commenced in the 1900s in South Africa and Senegal, and detailed recording of river flows was initiated on several rivers in Kenya, Morocco, and Tunisia in the 1920s. However, the coverage of the overall continent provided by these early measurement programmes was relatively sparse and in many cases their records have been disrupted by subsequent political and economic problems. As a result, any attempt to review in detail the hydrological characteristics of the African continent faces problems of limited data availability and several areas of uncertainty and even discrepancy exist. In its classic review of global water resources, the USSR Committee for the International Hydrological Decade (see UNESCO, 1978a) estimated that only 2030 river gauging-stations existed in Africa, whereas the equivalent numbers for Asia, South America, and Australia and Oceania were 12 000, 3570, and 3200 respectively. Existing data do, nevertheless, provide an effective basis for identifying the key features of the hydrology of the African continent and its rivers and for placing it within a broader global context.

Looking generally at the hydrology of Africa, its most distinctive feature is undoubtedly its *variability*. In a *spatial* context, the African continent arguably embraces greater contrasts than any other continent. According to the data presented by UNESCO (1978a,b), mean annual precipitation ranges from near zero over large areas of the Sahara to 9950 mm in the foothills of Mount Cameroon on the coast of the Gulf of Guinea. Likewise, mean annual evapotranspiration varies from a few millimetres in the Sahara and Namibian Deserts to in excess of 1300 mm in the central areas of the Congo Basin, and mean annual runoff ranges from essentially zero in the Sahara to >4000 mm in the coastal areas of Guinea and Sierra Leone. From a *temporal* perspective, variability must also be seen as a key feature of African hydrology. Such variability embraces both short-term year-to-year fluctuations in annual precipitation and runoff, particularly in the drier areas of the continent where the coefficient of variation of annual runoff frequently exceeds 0.5, and longer-term cycles of dry-and-wet periods, including the recent Sahelian droughts.

Comparisons of the hydrological characteristics of Africa with those of the other continents also serve to emphasize its unique features. It is often cited as the continent with the lowest average precipitation and runoff (mm year^{-1}) and the lowest runoff ratio (see Table 6.1). The recent work of McMahon *et al.* (1992), which compares a range of runoff characteristics for the individual continents, further emphasizes Africa's distinctive position. In their continental subdivision, these authors separate northern and southern Africa, but Africa's uniqueness is still clearly evident. Their results are necessarily limited by being based purely on a comparison of the statistical properties of the available flow records for the individual continents, rather than any

Table 6.1 The water balance of the continents

Continent	Precipitation (mm)	Evapotranspiration (mm)	Runoff (mm)	R/P
Africa	740	587	153	0.21
Europe	790	507	283	0.36
Asia	740	416	324	0.44
North America	756	418	339	0.45
South America	1600	910	685	0.43
Australia and Oceania	791	511	280	0.35

Source: Based on UNESCO (1978*a*).

spatial-weighting, but the rivers of Southern Africa stand out as evidencing the highest average coefficient of variation of mean annual runoff (Figure 6.1A), the highest average storage requirement for flow regulation (defined as the required storage for a constant draught of 80 per cent of the mean annual flow with a reliability of 95 per cent expressed as a ratio of the mean annual flow) (Figure 6.1B), the highest average flood variability (defined as the standard deviation of the logarithms of the annual peak discharges) (Figure 6.1D), and the highest average extreme flood index (defined as the ratio of the 100-year flood to the mean annual flood) (Figure 6.1E). Only in terms of the average specific mean annual flood do the rivers of Southern or Northern Africa fail to achieve the highest rank, but in this case African rivers again stand out, since Northern Africa occupies the lowest position (Figure 6.1C).

Turning to consider the main characteristics of the hydrology of the African continent and its rivers in more detail, it is clearly impossible to deal comprehensively with all relevant aspects within the constraints of a single short chapter and the scope of this brief review must of necessity be selective. Emphasis will be placed on continental-scale patterns and features. Attention will focus first upon the main components of the annual water balance (i.e. precipitation, evapotranspiration, and runoff) and their spatial patterns. Secondly, runoff and river behaviour will be considered more fully in terms of general regime characteristics and flood response. Thirdly, other aspects of the hydrological system, including groundwater, river-water quality and material transport by rivers will be briefly treated, and finally the impact of major water-resource development schemes and the implications of recent changes in the natural hydrological regime will be discussed.

Water-Balance Characteristics

The annual water balance exerts a fundamental control over the hydrological characteristics of an area and

Figure 6.2 presents information on the distribution of annual precipitation and evapotranspiration totals over the African continent. Particular attention can be drawn to two key features. The first is the very substantial range of annual precipitation receipt from near zero over parts of the eastern Sahara to nearly 10 000 mm on the coastal foothills of Mount Cameroon on the Gulf of Guinea, highlighted previously. The second is the relatively simple spatial pattern of annual precipitation totals. There is a clear latitudinal zonation, with the most arid belt lying between 20° and 30°N receiving an average of only 40 mm year^{-1}, and the zone of abundant rainfall located in the equatorial region between 10°N and 10°S receiving an average of 1350 mm year^{-1}. Superimposed on these general latitudinal gradients are the orographic effects associated with mountainous areas and some significant longitudinal gradients. In the portion of the continent north of 10°S precipitation totals show a general decline from west to east, whereas south of 10°S the situation is reversed and precipitation totals tend to reduce in a westward direction.

The pattern of mean annual evapotranspiration illustrated in Figure 6.2B primarily reflects the interplay of radiation receipt and moisture availability (i.e. precipitation inputs) and therefore again demonstrates clear latitudinal zonation. At the global scale, *potential* evapotranspiration (i.e. the evapotranspiration that would occur in the absence of any limitation on moisture supply and which is therefore closely controlled by radiation receipt) exhibits a general increase from the poles towards the Equator, in response to increasing solar radiation. However, the cloud associated with the intertropical convergence zone (ITCZ) causes a reduction in solar radiation receipt at the Equator itself, so that maximum levels of potential evapotranspiration occur at *c*.20°N and S. The effects of these controls on potential evapotranspiration rates over Africa are clearly shown by the inset to Figure 6.2B. In this region of the globe, potential evapotranspiration rates are also commonly higher in the northern hemisphere because

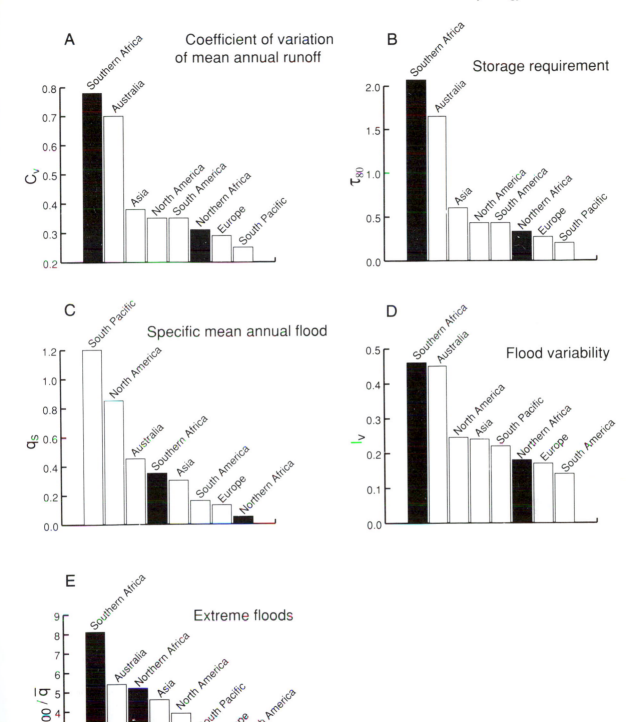

Fig. 6.1. A comparison of the runoff characteristics of the rivers of Northern and Southern Africa with those of other continents. (*Source:* data presented in McMahon *et al.*, 1992.)

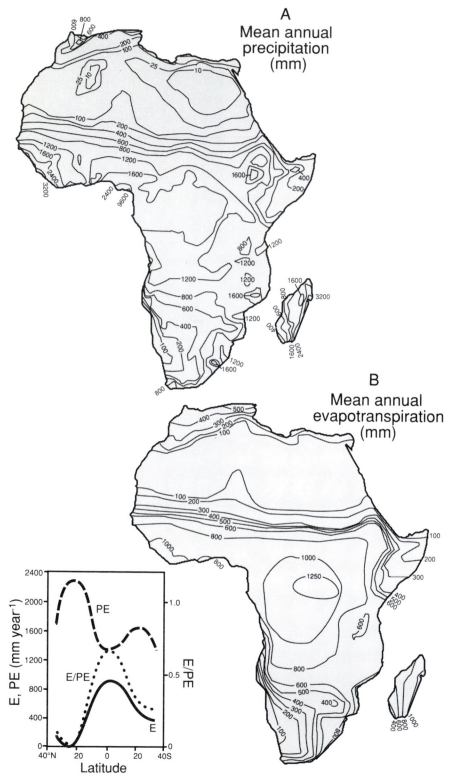

Fig. 6.2. The distribution of mean annual precipitation and evapotranspiration over the African continent. (*Source*: UNESCO, 1978*a*,*b*.)

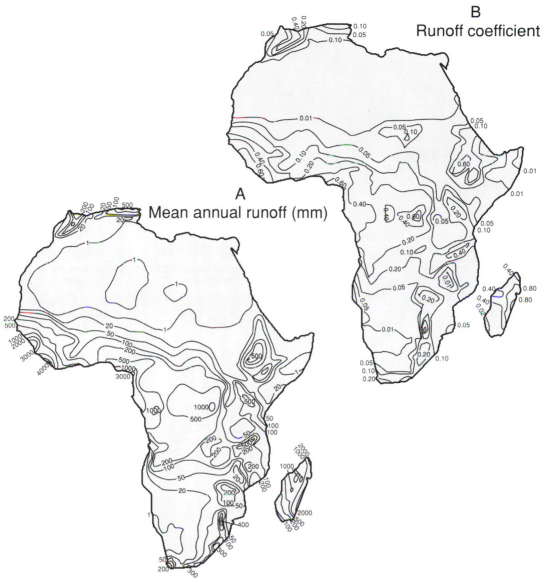

Fig. 6.3. The distribution of mean annual runoff and the runoff coefficient over the African continent. (*Source*: UNESCO, 1978*b*.)

of the more continental climate which results in reduced cloud cover and increased solar radiation receipt. Because of the higher precipitation inputs, *actual* evapotranspiration and the ratio of actual to potential evapotranspiration reach a maximum at the Equator, and, for a given latitude, values are generally lower north of the Equator due to the reduced moisture availability (see Figure 6.2B inset). Altitude also exerts a significant influence on the pattern of annual evapotranspiration losses depicted in

Figure 6.2B, since potential evapotranspiration rates have been shown to decrease by about 10–30 mm year^{-1} for an increase in elevation of 100 m.

Because it essentially represents the net balance between precipitation inputs and evapotranspiration losses, the map of mean annual runoff for Africa presented in Figure 6.3A reflects many of the features and controls associated with the patterns of precipitation and evapotranspiration over the continent discussed above.

Table 6.2 Mean annual runoff of the largest African rivers

River	Mean annual runoff		Contribution to total external runoff (%)
	mm	km³	
Congo	370	1414	38.0
Niger	128	268	7.2
Ogooue	732	149	4.0
Zambezi	80	106	2.8
Nile	25	73	2.0
Sanaga	503	68	1.8
Volta	103	41	1.1

Source: Based on UNESCO (1978*a*).

Table 6.3 Runoff losses in some major African river basins

River	Total basin runoff (mm)	Runoff at mouth	Loss (%)
Nile	70	25	64
Niger	153	130	15
Senegal	110	52	53
Zambezi	115	80	30
Orange	27	15	44
Congo	382	370	3

Source: Based on UNESCO (1978*a*).

Again, a wide range of values is involved, with annual runoff totals ranging from near zero in the Libyan Desert to >4000 mm on the coast of Liberia in West Africa. A clear latitudinal zonation is similarly evident, with runoff reaching a maximum in the equatorial zone, decreasing towards the tropics and increasing again on the northern and southern margins of the continent. Altitudinal effects are also important, with most of the highland areas in the east of the continent marked by areas of increased runoff, by virtue of the increased annual precipitation and reduced evapotranspiration losses. As demonstrated by Figure 6.3B, values of the runoff coefficient range from <0.01 in the desert areas of the Sahara and the Kalahari to >0.6 in the coastal areas of Sierra Leone and Liberia, and as high as 0.8 on the eastern coast of Madagascar.

According to UNESCO (1978*a*), the annual river flow reaching the oceans from the African continent amounts to 3720 km³, with an equivalent depth of 126 mm. Of this, 90 per cent flows to the Atlantic slope and 10 per cent to the Indian Ocean. In addition, there are substantial areas of internal drainage associated with the Sahara, the Logone and Chari basins draining to Lake Chad, the Okavango delta, the Awash River in Ethiopia, and Lake Rudolf in Kenya. These areas of internal drainage extend to *c*.9.6 × 10⁶ km², representing *c*.32 per cent of the total area of the continent, and receive an average annual runoff of *c*.158 km³ which is equivalent to about 4 per cent of the external runoff. An important feature of the runoff output from the African continent is that a small number of large rivers account for a major proportion of the total volume (see Table 6.2). A further distinctive characteristic of several of these large rivers is that they rise in humid upland areas and traverse arid desert areas or extensive areas of swamp before reaching the oceans (see Chapter 15, on wetlands). As a result, a considerable proportion of their runoff is lost to evaporation and seepage and flows decline towards their mouths (see Table 6.3).

The upper course of the Nile, and more particularly the White Nile and its tributaries, provides a classic example of greatly reduced downstream runoff due to high evapotranspiration losses from extensive areas of swamp. Within this area, between the Kenyan border and Malakal on the White Nile, the main tributaries of the White Nile pass through three major areas of swamp. These include, first, the Sudd or Bahr el Jebel swamps which permanently cover some 8300 km², but which according to Sutcliffe and Lazenby (1990) have extended over as much as 20 000–30 000 km² during recent periods of high runoff; secondly, the swamps of the Bahr el Ghazal which occupy *c*.15 000 km²; and, thirdly, the Machar swamps which cover about 6700 km². In the case of the Machar and the Bahr el Ghazal swamps, virtually all of the runoff entering the swamps is lost by evapotranspiration. Flows through the Sudd swamps are greater, but approximately 50 per cent of the inflow is still lost. According to data presented by Shahin (1985), the combined runoff loss for the three swamp areas is *c*.30 × 10⁹ m³ year⁻¹, which is approximately the same as the mean flow at Malakal downstream of the swamps. Actual evapotranspiration losses are even greater, since rainfall inputs to the swamps must also be taken into account (see Sutcliffe and Parkes, 1987), and these are estimated to be of the order of 5 mm day⁻¹. Other important areas of swamp or extensive seasonal inundation occur in the inland deltas of the River Niger and the Okavango River, in the lowland flood-plain areas of the Niger and Senegal Rivers, in the upper and middle Zambezi Basin (Barotse and Kafue swamps), in the Zaire basin (Kamulondo Depression), and in the basins of the Chari and Logone Rivers adjacent to Lake Chad. According to Blache (1964), the area of inundation associated with the floodplains of the Chari/Logone Basin can extend to >60 000 km² during the flood season.

River Regimes and Flood Response

With the wide range of climates existing over the continent and the associated variability in the seasonal distribution of precipitation and evapotranspiration, African rivers exhibit a wide range of flow patterns. In the case of the Upper Zambezi, for example, the flow pattern is dominated by the annual high water in March and a period of relatively low flows from August through to December. In contrast, on the Lower Congo the main flood season occurs during the period November–December, with a secondary peak in March–April (see Balek, 1983). Traditional flow regime analysis (see Pardé, 1955; Beckinsale, 1969) provides an effective means of characterizing this aspect of river behaviour and Figure 6.4 presents the results for the African continent of two studies which have attempted to generalize the flow regimes of the world's rivers. The map derived from the work of Beckinsale (1969) presented in Figure 6.4A was based primarily on an analysis of the influence of the major climate types on runoff patterns. This identifies five distinctive seasonal patterns, based on the number and timing of the high-water periods and the relative importance of low-water periods, which reflect the major climate types. Runoff-deficient desert areas are shown as an additional category. The approach used by Haines *et al.* (1988) to generate the map shown in Figure 6.4B was more objective and was based on analysis of existing river discharge records. Cluster analysis was used to distinguish major regime types in terms of the magnitude and timing of the high-water periods. In this case, nine separate regime types are identified, in addition to that representing essentially uniform flows throughout the year. These nine regime types include a wide variety of seasonal patterns, with peak-flows occurring in all four seasons. Rivers, such as the Nile which flow across several different regime zones clearly introduce further complexity into this pattern of regime behaviour by combining the regimes of their individual tributaries.

In common with the hydrological literature for other areas of the globe, there are many reports of high magnitude flood events (as distinct from regular seasonal floods) on African rivers which have caused serious damage or notable geomorphological effects. These include the Laingsburg flood which occurred in the southern Cape in 1981 (see Jordaan and Van Bladeren, 1990), the disastrous 1988 Orange River floods (see Bremner *et al.*, 1990), the exceptional flood which occurred on the Mejerda River in Tunisia in 1973 (see Claude and Loyer, 1977; Colombani *et al.*, 1984) and the severe floods which hit the Moulouya Basin in Morocco and the Zeroud Basin in Tunisia in 1963 and 1969 respectively

(see Heusch and Millies-Lacroix, 1971). However, no African rivers feature in the list of maximum floods in the world for a range of basin sizes compiled by Rodier and Roche (1984). This suggests that despite the incidence of a number of severe floods, neither the catchment characteristics nor the climatic and associated storm-generating conditions occurring in Africa are particularly conducive to record-breaking floods. This is further supported by Figure 6.1E, which shows that specific mean annual floods in both Southern and Northern Africa tend to be relatively low when compared to the other continents, although Figure 6.1D indicates that low frequency floods tend to be more extreme in terms of their ratio to the mean annual flood in both regions of Africa.

There are few areas with substantial snow cover where snowmelt floods are an important feature in Africa and rainfall provides the dominant source of flood runoff. Looking at rainfall-generated floods on African rivers, their underlying causes and their spatial distribution, in a little more detail, it is useful to refer to the attempt by Hayden (1988) to produce a classification of flood climates for the land surface of the globe. The results of this classification as they relate to Africa are presented in Figure 6.5. From this it can be seen that, with the exception of the northern margin, the flood climates of the continent reflect the existence of barotropic conditions in the troposphere (i.e. dominance of vertical motion and associated convective rainfall). Over much of the continent perennial barotropic conditions occur. Rising air associated with the Intertropical Convergence Zone (ITCZ) provides the dominant cause of flood-producing rainfall in the equatorial zone. However, the strength of this convergence is relatively weak over much of Eastern Africa, as compared to other regions of the globe, and there are few areas where highlands cause significant orographic enhancement. This may account for the relative lack of reports of severe high-magnitude floods in Equatorial Africa. Elsewhere within the perennial barotropic zone, convective rainfall associated either with local thunderstorms (unorganized) or with more widespread synoptic (organized) activity (e.g. easterly waves and tropical cyclones) provides the dominant cause of floods. In Southern Africa, conditions in the troposphere are characterized by seasonality, with barotropic conditions in summer being replaced by baroclinic conditions (vertical motions driven by solenoidal circulations) in winter. Figure 6.6 indicates that barotropic conditions still dominate the flood climates of this region, but the seasonality and juxtaposition of barotropy and baroclinicity would appear to result in increased potential for flood-producing precipitation, particularly

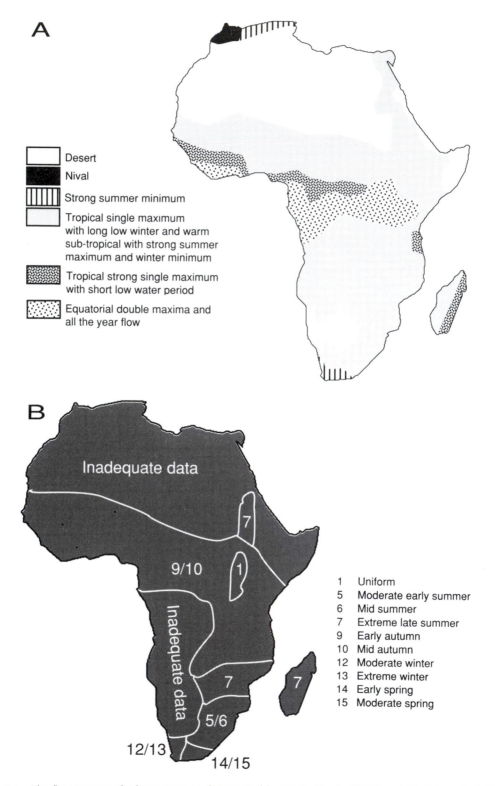

Fig. 6.4. The flow regimes of African rivers as characterized by (A) Beckinsale (1969), and (B) Haines *et al.* (1988).

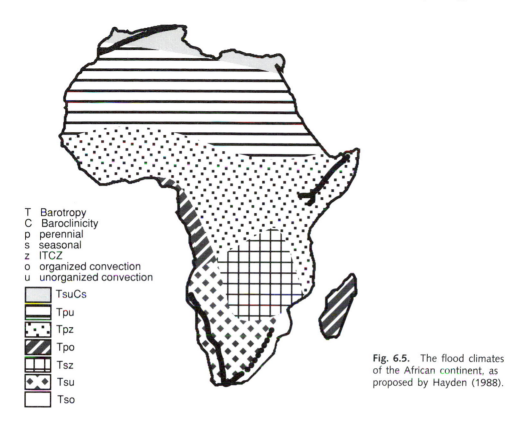

T   Barotropy
C   Baroclinicity
p   perennial
s   seasonal
z   ITCZ
o   organized convection
u   unorganized convection

- TsuCs
- Tpu
- Tpz
- Tpo
- Tsz
- Tsu
- Tso

Fig. 6.5. The flood climates of the African continent, as proposed by Hayden (1988).

in the eastern regions of South Africa, which are shown in Figure 6.5 as an area of seasonal barotropy with organized convection (Tso). Here, widespread severe flooding is relatively common and very severe floods occurred in this area in 1981, 1984, 1987, and 1988, with maximum storm-rainfall totals frequently exceeding 500 mm. Tropical cyclones, cold fronts, cut-off low-pressure systems, and tropical–temperate wave interactions are responsible for heavy rainfall. In the coastal areas of Northern Africa, the cyclones and frontal activity associated with the seasonal baroclinic conditions in winter supplement the local convective rainfall occurring during the summer to provide an additional cause of flood-producing storms.

Groundwater

Detailed information on groundwater and its behaviour is lacking for many areas of Africa and it is difficult to provide a quantitative assessment of the continent's groundwater resources. UNESCO (1978a) provide some generalized estimates of total groundwater storage (5.5×10^6 km³) and the average rate of annual recharge

(1.6×10^3 km³ year^{-1}), but these are of uncertain reliability. Because of the importance of local hydrogeological conditions, it is also more difficult to generalize about regional groundwater patterns in Africa than for other components of the hydrological cycle. However, in their reviews of African groundwater resources, Wright (1983) and Foster (1984) suggest that the continent can be conveniently subdivided into three geohydrological provinces, representing the basement shield, areas underlain by volcanic rocks, and sedimentary basins (see Figure 6.6). In addition, the main areas of fold mountains in the Maghreb region of Northern Africa and in South Africa are identified as being of limited importance for groundwater storage.

As shown in Figure 6.6, extensive regions of Africa are underlain by crystalline basement rocks formed primarily from major suites of Precambrian rocks. Granitic gneisses and lower grade metamorphic rocks are the dominant lithological types. Within these areas, some groundwater exists in the fractured bedrock, but the weathered mantle, which is commonly >10 m thick, provides the main aquifer storage. Transmissivities within these aquifer systems are generally low and their role

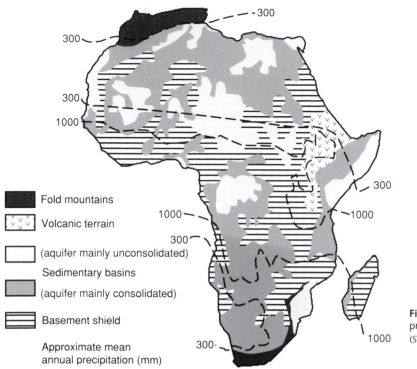

Fold mountains

Volcanic terrain

(aquifer mainly unconsolidated)

Sedimentary basins

(aquifer mainly consolidated)

Basement shield

Approximate mean
annual precipitation (mm)

Fig. 6.6.   The geohydrological
provinces of Africa.
(*Source*: Foster, 1984.)

as a runoff source and in supply is limited (see Wright, 1992). They do, however, provide important sources for domestic and livestock demands in rural areas. The areas of volcanic terrain shown on Figure 6.6 are of relatively limited extent and are restricted to parts of East Africa. Local variability in aquifer characteristics is frequently marked in these areas, and although substantial groundwater storage exists in some locations, in others storage is limited.

In contrast to the basement shield and the areas with volcanic rocks, the sedimentary basins, comprising both consolidated sedimentary strata and more recent unconsolidated deposits (see Figure 6.6), represent the most important and productive aquifers of the African continent and hold very large quantities of water in storage. Sandstones and limestone aquifers predominate in the older basins, which frequently contain marine sediments, and artesian conditions may be present over large areas. Unconsolidated terrestrial sediments characterize the younger basins, and here unconfined conditions are more common. Considerable potential for large-scale groundwater development exists in these sedimentary basins but careful scientific management will be important in many areas if over-exploitation is to be avoided. This is particularly important in the arid regions of the

continent where much of the groundwater storage will be fossil, reflecting recharge during past pluvial periods. Large groundwater reserves exist beneath the Sahara (see Ambroggi, 1966), but current rates of recharge are low and there is considerable scope for debate as to whether such reserves should be 'mined' to satisfy short-term developments (see Clark and Stoner, 1980). The Great Man-Made River Project currently under development in Libya (see Odone, 1984), which will transport vast quantities of water from Kufra in the southern Fezzan to the coast is clearly based on the mining of fossil water and as such can have only a limited life. Falkenmark (1989) estimates that by the year 2020 water use in Libya will amount to more than seven times the renewable supply.

River-Water Quality and Material Transport by Rivers

Although available data are still unsufficient to provide a rigorous and accurate assessment of material transport by African rivers, the work of Milliman and Meade (1983) and Meybeck (1979) provides a meaningful basis for several generalizations regarding their status relative to those of the other continents. Such data

Table 6.4 Material transport from the continents to the oceans

Continent	Specific suspended-sediment yield (t km^{-2} year^{-1})	Specific dissolved load (t km^{-2} year^{-1})	Overall denudation (mm 1000 years^{-1})[a]
Africa	35	13	17
Asia	229	57	102
Europe	50	92	42
North and Central America	84	43	43
South America	100	34	47
Oceania/Pacific Islands	589	56	241

[a] Assuming a 65 per cent denudation component for dissolved load and a rock density of 2.6 t m^{-3}.

Sources: Based on load estimates produced by Milliman and Meade (1983) and Meybeck (1979).

Table 6.5 Maximum and minimum mean annual suspended sediment yields recorded for African rivers draining basins of *c.*10 000 km^2

River	Basin area (km^2)	Mean annual sediment yield (t km^{-2} year^{-1})
Maximum		
Oued Ouerrha, Morocco	6183	3500
Oued M'Jara, Morocco	5190	2910
Oued Isser, Algeria	3595	2610
Oued Isser, Algeria	31 615	1712
Minimum		
Chari, Chad	193 000	0.93
Fafa, Central African Republic	6750	3.1
Bangoran, Central African Republic	2590	4.4
Gribingui, Central African Republic	5680	5.0
Bahr Sar, Chad	79 600	8.4
Ngoko, Cameroon	67 000	8.6

Sources: Based on data presented in Walling (1984) and Seyler *et al.* (1993).

summarized in Table 6.4 again emphasize the distinctive nature of African rivers, since the continent is characterized by the lowest specific suspended-sediment yield, the lowest specific dissolved load, and the lowest overall denudation rate as represented by river loads. This position reflects a number of important controls, including the large areas of desert with near-zero runoff which contribute little to the total river-borne flux from the land to the oceans, the absence of appreciable areas of contemporary tectonic activity, where rapid uplift promotes increased rates of mechanical and chemical denudation, and the extensive areas underlain by crystalline basement rocks which yield solute-poor runoff.

Existing data suggest that the mean annual suspended-sediment loads of African rivers draining basins of the order of 10 000 km^2 range between approximately 1.0 and 4000 t km^{-2} year^{-1}. Minimum reported values are associated with the Chari River and its tributaries draining to Lake Chad (see Table 6.5), but low specific suspended sediment yields of the order of 10 t km^{-2} year^{-1} have also been reported for other rivers in the Lake Chad basin, in the Central African Republic, and in Cameroon. Maximum reported values are for rivers draining the Maghreb region of Morocco, Algeria, and Tunisia (see Table 6.5). The highest specific suspended sediment yield for an African river reported to date is probably that for the 1310 km^2 Perkerra Basin in Kenya for which Dunne (1975) reports a sediment yield of 19 520 t km^{-2} year^{-1}. This drainage basin had, however, experienced severe overgrazing which had destroyed most of the natural vegetation cover.

To consider further the wide range of specific suspended-sediment yields noted above, there have been a number of attempts to produce generalized maps of the continental patterns involved which have in turn produced several inconsistencies and contradictions. Four such maps are presented in Figure 6.7 and it is readily evident that the patterns shown are quite different. The maps produced by Strakhov (1967) and Fournier (1960) inevitably suffer from having been produced at a time when few data were available and these authors placed considerable reliance on extrapolating data from other areas of the world. Fournier's map was based on a relationship between specific sediment yield and an index of precipitation amount and seasonality and there are clear similarities with the map of mean annual precipitation presented in Figure 6.2. The maps produced by Dedkov and Mozzherin (1984) and Walling (1984) are more reliable, since they are based on available measurements of sediment yield and show a reasonable degree of similarity. The broad patterns shown on these two maps can be accounted for in terms of the general influence of climate, relief, underlying geology, vegetation cover, and land use (see Walling, 1984). Thus, for example, the high sediment yields in the Maghreb region of North Africa can be accounted for in terms of the steep unstable terrain, the highly erodible rocks, particularly marls, the relatively high (500–1000 mm) and markedly seasonal annual precipitation, and vegetation removal by human activity. Conversely, the low sediment yields characterizing much of Central Africa reflect the lower relief, the greater density of the vegetation canopy, and in some cases the trapping of sediment in lakes, swamps, and floodplains along the

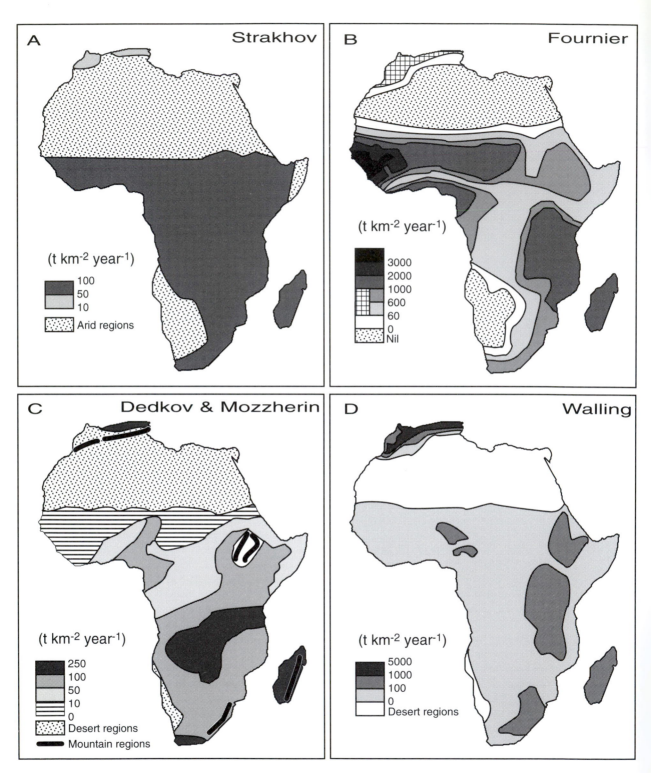

Fig. 6.7. The maps of specific suspended-sediment yield for the African continent produced by (A) Strakhov (1967); (B) Fournier (1960); (C) Dedkov and Mozzherin (1984); and (D) Walling (1984).

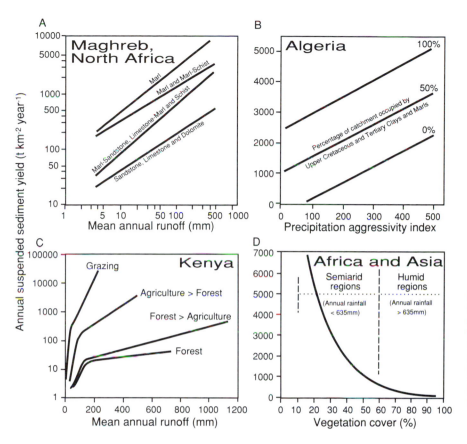

Fig. 6.8. Factors influencing the spatial distribution of specific suspended-sediment yield within particular regions of Africa.

(*Sources*: information presented by (A) Heusch and Millies-Lacroix (1971); (B) Demmak (1982); (C) Dunne (1979); and (D) Starmans (1970).)

Table 6.6 Some comparisons between measured suspended-sediment yields of African rivers and estimated rates of contemporary soil loss by water erosion depicted on the FAO map of Soil Degradation

River	Country	Basin area (km²)	Suspended-sediment yield (t km⁻² year⁻¹)	FAO estimate of soil loss (t km⁻² year⁻¹)
Watari	Nigeria	1450	483	1000–5000
Bunsuru	Nigeria	5900	438	1000–5000
Senegal	Mali	157 400	14.6	1000–5000
Faleme	Mali	15 000	40	1000–5000
Hamman	Algeria	485	198	1000–5000
Kebir Ouest	Algeria	1130	92	1000–5000
Mesanu	Ethiopia	150	1680	5000–20 000

Sources: Based on Walling (1987); FAO (1979).

river courses. The suspended-sediment yield from the 3.5×10^6 km² basin of the Congo River, which occupies nearly 12 per cent of the continent, is, for example only 13.8 t km⁻² year⁻¹ (see Nkounkou and Probst, 1987). Figure 6.8, based on the studies of Heusch and Millies-Lacroix (1971), Demmak (1982), Dunne (1979), and Starmans (1970), provides further information regarding factors accounting for variations of sediment yield within particular regions of Africa.

The low suspended-sediment yields characterizing much of the African continent can to some extent be seen as anomalous when compared to the high rates of soil erosion reported from many areas of the continent (see Chapter 18). Table 6.6 further emphasizes this contrast by comparing the measured suspended-sediment yields for a number of drainage basins with the estimates of contemporary soil-erosion rates within these basins depicted on the FAO map of current rates of soil degradation by water erosion in Africa north of the Sahara (FAO, 1979). In all cases, the estimates of soil-erosion rates are at least an order of magnitude greater than the reported sediment yields, and in several cases they are two orders of magnitude greater. Such data emphasize the low sediment-delivery ratios operating in many African drainage basins and the importance of sediment-conveyance processes in accounting for low sediment yields (see Walling, 1983). Although suspended-sediment load data are now available for a substantial number of African rivers, few measurements of bed-load transport are available. However, existing data (see Table 6.7) suggest that this component rarely exceeds 10 per cent of the long-term total sediment load.

As indicated in Table 6.4, the total dissolved loads of the rivers draining the African continent are generally low by world standards. This is primarily a result of the relatively low solute content of most of the rivers in the humid equatorial regions (e.g. the Congo) which account for the majority of the runoff and the overall flux of dissolved material (see Table 6.8). These low solute contents primarily reflect the solute-deficient crystalline basement rocks, the low relief, and the high rates of runoff. Solute concentrations are higher in rivers such as the Niger, Senegal, and the Zambezi which drain semi-arid areas and increase still further in rivers draining arid areas (e.g. the Nile and the Orange River), in line with the general inverse relationship between mean solute concentration and annual runoff reported by Walling and Webb (1983). However, the specific loads of such rivers draining arid areas are relatively low compared to those draining the more humid areas (see Table 6.8) because of their low runoff totals.

Detailed information on solute-load composition are now available for an increasing number of African rivers as a result of studies undertaken by French hydrologists and geochemists involved with ORSTOM projects in West and Central Africa. Table 6.9 provides information on the mean composition of the solute load of a selection of African rivers and of world average river water. These data emphasize the importance of SiO_2 and HCO_3 and also the dominance of Ca as the major cation, in the overall composition of the solute loads of rivers draining the humid equatorial regions. However, in the case of the Congo, the low total solute concentrations (see Table 6.8) emphasize the limited contribution from rock weathering and the relatively high Na and Cl concentrations push this river towards the 'precipitation dominance' category of Gibbs (1970), where atmospheric inputs account for a substantial proportion of the total solute load. The relative contribution of Cl and SO_4 concentrations increases in rivers draining arid areas (e.g. the Nile). Dissolved organic carbon (DOC) concentrations are also appreciable in many African rivers. Highest levels are found in the rivers draining the humid equatorial areas and values decline with increasing aridity.

Table 6.7 Estimates of the relative importance of suspended-sediment and bed-load transport to the total sediment loads of some African rivers

River	Basin area (km^2)	Sediment load (%)	
		Suspended	Bed load
Gwai, Zimbabwe[a]	14 400	90.8	9.2
Umsweswe, Zimbabwe[a]	1990	95.0	5.0
Limpopo, Zimbabwe[a]	196 000	90.9	9.1
Hunyani, Zimbabwe[a]	1510	100.0	0.0
Mellegue, Tunisia[b]	9000	80.9	19.1
Upper Niger, Nigeria[c]	730 000	93.5	6.5
Lower Niger, Nigeria[c]	—	95.0	5.0
Upper Benue, Nigeria[c]	107 000	94.0	6.0
Lower Benue, Nigeria[c]	304 000	95.0	5.0
Upper Tana, Kenya[d]	6668	94.0	6.0
Congo, Zaire[e]	3 500 000	93.0	7.0
Senegal, Mali[f]	—	95.0	5.0

Sources: Based on data presented by [a]Ward (1980); [b]Rodier *et al.* (1981); [c]Nedeco (1959); [d]Ongweny (1978); [e]Spronck (1941); [f]Gac and Kane (1986).

Table 6.8 Total dissolved solids transport by some major African rivers

River	Mean annual runoff (mm)	Mean concentration (mg l^{-1})	Total load (t x 10^6)	Specific load (t km^{-2} year^{-1})
Congo	338	28	36.6	9.9
Senegal	48	42	0.4	1.8
Niger	124	67	14.0	11.0
Zambezi	157	113	25.2	21.0
Orange	15	140	1.6	1.6
Nile	47	318	11.8	3.9

Source: Based on Martins and Probst (1991).

In the case of the Congo River, Probst *et al.* (1992) report that dissolved organic matter represents 30–40 per cent of the total dissolved solids.

In common with all continents, the human impact on the water quality of African rivers has increased significantly during the past few decades (see Dejoux, 1988). Lack of large-scale industrial development has prevented the excesses of water pollution reported in many rapidly industrializing developing countries (see Meybeck *et al.*, 1989) but such problems are inevitably increasing. For example Ajayi and Osibanjo (1981) indicate that all the major streams in the industrial sectors of Nigerian cities such as Lagos, Kano, and Kaduna are seriously polluted by industrial wastes and that all the streams flowing through the densely populated city of Ibadan are heavily polluted by wastes from domestic sources. Dejoux (1988) also reports the increasing presence of pesticides and related contaminants in African

Table 6.9 The mean solute composition of water from a selection of African rivers and of world-average river water

River	Element concentration (mg l^{-1})							
	SiO$_2$	Ca	Mg	Na	K	Cl	SO$_4$	HCO$_3$
Congo	9.7	2.2	0.9	2.0	1.3	2.7	2.0	7.1
Ouham	21.5	2.4	1.4	2.8	2.0	0.5	0.8	23.7
Ubangi	13.2	3.3	1.4	2.1	1.6	0.8	0.8	19.0
Gambia	8.5	3.9	1.6	1.6	1.4	0.9	—	24.2
Bandama	19.1	3.9	2.1	5.8	1.7	2.4	<0.8	39.7
Niger	16.2	6.1	2.2	2.8	1.6	0.9	1.4	35.9
Senegal	7.6	3.4	1.8	1.9	1.8	1.3	0.2	24.4
Chari	19.1	3.9	1.7	2.6	1.9	—	—	30.4
Nile	—	28.8	11.3	34.0	6.6	32.0	27.8	157.4
Orange	17.4	17.0	6.7	6.7	2.5	7.8	2.3	—
WORLD	11.6	14.6	3.8	5.1	1.3	5.3	8.5	57.7

Sources: Based on Nkounkou and Probst (1987); Gac and Kane (1986); and Probst *et al.* (1992).

Table 6.10 Reservoir capacity and baseflow augmentation in Africa as compared to the other continents

Continent	Annual runoff volume		Gross reservoir capacity			Baseflow augmentation	
	Total	natural baseflow (km^3 × 10^3)	Volume (km^3)	% of World	% of total runoff	Volume (km^3)	% of natural baseflow
Africa	4.23	1.5	1280	26.2	30.3	510	38.7
Europe	3.10	1.13	450	9.2	14.5	180	17.8
Asia	13.19	3.44	1770	36.3	13.4	710	23.2
N. America	5.95	1.9	975	20.0	16.4	390	23.1
S. America	10.38	3.74	340	7.0	3.3	140	4.0
Australia	1.97	0.46	65	1.3	3.3	30	6.5

Source: Based on Mahmood (1987).

rivers. Few African rivers remain in a truly natural state in terms of their water quality. The construction of dams and reservoirs and the development of irrigation schemes on many rivers has caused important changes. For example, Martins (1983) indicates that total dissolved-solids concentrations in the Niger have increased by about 10 per cent, as a result of the construction of the Kainji Reservoir and Martins and Probst (1991) suggest that further increases may have followed the construction of the Jebba Dam. Similarly, Kempe (1983) has estimated that total dissolved-solids concentrations in the River Nile at Cairo have increased by about 33 per cent as a result of the construction of the Aswan Dam and the development of irrigation and industrial activities. Construction of the Aswan Dam has also resulted in a major decrease in the suspended-sediment load transported by the lower Nile. Schamp (1983) estimates that the Nile carries $110-30 \times 10^6$ t year^{-1} of silt into Egypt, but only $1-2 \times 10^6$ t year^{-1} currently leave Lake Nasser, and much of this represents phytoplankton from the reservoir itself.

Changing River Behaviour

As noted above, the construction of dams on a number of African rivers has led to changes in their sediment and solute transport. Major changes in flow volumes and patterns have also occurred, due to both water abstraction for irrigation, and industrial and domestic use, and the attenuation and equalization of flows by reservoir storage (see also Chapter 7). Hydropower generation requires storage of water and its release at an essentially constant rate, in order to provide a stable supply of electricity, and the provision of reliable water supplies frequently necessitates long-term storage of runoff from wet years to sustain supplies in dry years. Dam construction has been an important feature of water-resource development in Africa and data presented by Mahmood (1987) again emphasize the distinctive position of the continent (Table 6.10). Comparing the individual continents, Africa has by far the highest proportion of its runoff impounded by reservoirs and by far the greatest augmentation of river baseflow by

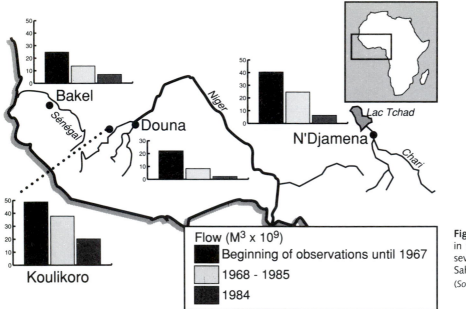

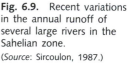

Fig. 6.9. Recent variations in the annual runoff of several large rivers in the Sahelian zone.
(*Source*: Sircoulon, 1987.)

reservoir storage. Current reservoir storage represents nearly one-third of the total runoff from the continent and baseflows are increased by nearly 40 per cent.

Obeng (1978) suggests that for Africa, the period between 1958 and 1968 can be designated the 'Decade of Large Dams'. During this period, four major dams were built to create enormous man-made lakes. These were Lake Kariba on the Zambezi, the Volta Lake on the Volta River, Lake Kainji on the River Niger, and Lake Nasser on the Nile. Together, these reservoirs have a surface area of more than 20 000 km^2 and a storage capacity in excess of 500 km^3 which alone is equivalent to about one-eighth of the total natural runoff from the land surface of the continent to the oceans. On the Zambezi, the construction of Lake Kariba was followed by that of Lake Cabora Bassa further downstream. The flows of this river are now highly regulated and its suspended-sediment load is of the order of 50 per cent of that before construction of the dams.

Returning to the theme of temporal variability in the hydrological behaviour of African rivers, there have been a number of reports in recent years which have emphasized the declining flows and reduced annual runoff of many African rivers (see Olivry, 1987; Mahé and Olivry, 1991; Olivry *et al.*, 1993; Sircoulon, 1987; Sutcliffe and Knott, 1987). Thus, for example, the mean annual flow of the Nile is reported to have fallen from 84 km^3 during the period 1900–59, to 72 km^3 during the period 1977–87, with an even more dramatic fall to <52 km^3

for the years between 1984 and 1987 (see Howell and Allan, 1990). Similarly, whereas the mean annual flow of the Senegal River at Bakel over the 84-year period to 1985 was 22.3 × 10^9 m^3, over the period 1968–85 the equivalent value was 13.7 × 10^9 m^3 (Sircoulon, 1987). The work of Hurst on the River Nile in the 1930s and 1940s has clearly demonstrated the tendency for that river to show runs of years with both above average and below average flows rather than a more random distribution (see Hurst, 1952) and it is well known that other African rivers exhibit similar behaviour. Some may therefore argue that the trends noted above are a reflection of the natural variability of river flows and their cyclic behaviour, but the persistence of the recent decline suggests that more fundamental shifts associated with climatic change may be involved.

Most attention has focused on rivers in the semi-arid Sahelian zone of tropical West Africa (Sircoulon, 1987). Here clear cycles of high and low flows are apparent, with droughts occurring during the periods 1910–20, 1940–9, and 1968 onwards and periods of high runoff between 1925–35 and 1950–65. However, there is currently no evidence of a reversal of the period of drought which commenced in 1968 and Figure 6.9 summarizes the evidence of declining flows provided by several rivers in this region. The Chari River represents the major source of water to Lake Chad and the declining flows of this river have had marked repercussions on the areal extent of the lake. After a period of high flows in the

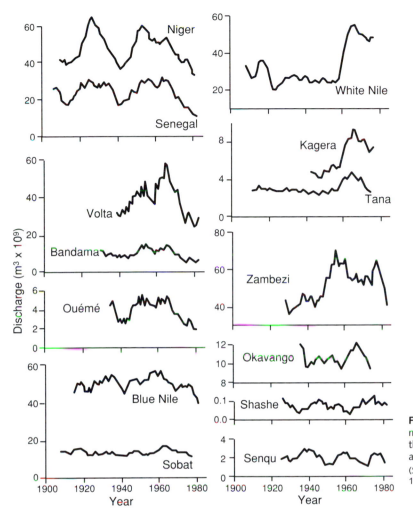

Fig. 6.10. Seven-year moving averages fitted to the annual discharge data for a selection of African rivers. (*Source*: Sutcliffe and Knott, 1987.)

early 1960s, Lake Chad occupied an area of 23 500 km², but after twenty years of reduced inflows the surface area was reduced to about 2000 km² and limited to a small area in the south of the original lake basin. In addition to reduced volumes of annual runoff and a reduction in the magnitude of the seasonal flood, Olivry (1987) has also demonstrated that flows on the Senegal River now recede more rapidly after floods in response to the reduced groundwater recharge.

More recently, attention has also been directed to the rivers of the humid tropical and equatorial regions of West and Central Africa, including the Congo, where flows also appear to have declined significantly since the 1980s. Olivry *et al.* (1993) report an analysis of flow data from the entire area of the Atlantic drainage of humid intertropical Africa which indicates that, relative to the

long-term (40-year) mean, flows during the period 1971–80 were reduced by 7 per cent, whereas during the period 1981–90 they were reduced by 16 per cent. For rivers draining the coastal area stretching from Guinea to Nigeria the equivalent values were as high as 18 and 32 per cent respectively. These reductions have caused significant problems in the operation of hydropower-generation equipment, particularly on the rivers of the Ivory Coast.

The trends noted above do not, however, apply to all areas of the continent. Sutcliffe and Knott (1987) present plots of the 7-year moving average for the annual flow series of a representative selection of African rivers and a number of these are reproduced in Figure 6.10. The marked reduction in the flows of the Senegal and Niger Rivers, noted above, are clearly evident in Figure

6.9, and the similar patterns evident for other West African rivers including the Volta, the Bandama in the Ivory Coast, and the Ouémé in Benin also confirm the trend of reducing flows in the Atlantic drainage of humid intertropical Africa. Evidence of recent droughts is also apparent in the flows of the Blue Nile. However, the flows of the White Nile present a rather different pattern which is dominated by the increased flows generated by the marked rise in water levels in Lake Victoria during the period 1961–4. The increased flows in the White Nile have clearly been important in balancing the reduced flows in the Blue Nile over the same period and thus reducing the overall reduction in the flows of the main Nile. Other rivers in East Africa, like the Kagera and Tana, and to some extent the Sobat, also show a similar pattern to the White Nile, although they are not sustained by lake storage. In further contrast, the flows of the Zambezi present a rather different pattern, with a period of increased runoff between 1950 and the late 1970s, and the flows of the Okavango, the Shashe in Botswana, and the Senqu in Lesotho show little evidence of clear trends. Future years will confirm whether these changes in the hydrology of many African rivers should be seen as a permanent shift induced by climatic change or as a marked, and therefore abnormal, perturbation in the general pattern of cyclic variability which characterizes the behaviour of African rivers.

References

Ajayi, S. O. and Osibanjo, O. (1981), 'Pollution Studies in Nigerian Rivers'. *Environmental Pollution* (B), 2: 87–95.

Ambroggi, R. P. (1966), 'Water under the Sahara', *Scientific American*, 214: 21–9.

Ball, J. B. (1939), *Contributions to the Geography of Egypt* (Cairo).

Balek, J. (1983), *Hydrology and Water Resources in Tropical Regions* (Amsterdam).

Beckinsale, R. P. (1969), 'River Regimes', in R. J. Chorley (ed.), *Water, Earth, and Man* (London), 455–71.

Blache, J. (1964), 'Les poissons du bassin du Tchad et du bassin adjacent du Mayo Kebbi', *Mémoires ORSTOM*, 4/2.

Bremner, J. M., Rogers, J., and Willis, J. P. (1990), 'Sedimentological Aspects of the 1988 Orange River Floods', *Transactions of the Royal Society of South Africa*, 47: 247–94.

Clark, L. and Stoner, R. F. (1980), 'Regional Groundwater Development in Temperate and Arid zones', in *Proceedings ICE Conference on Water Resources: A Changing Strategy (London)* (London), 85–92.

Claude, J. and Loyer, J.-Y. (1977), 'Les alluvions déposées par L'Oued Medjerdah lors de la crue exceptionnelle de mars 1973 en Tunisie: aspects quantitatif et qualitatif du transport et du dépôt', in *Erosion and Solid Matter Transport in Inland Waters*, IAHS pub. no. 122, 211–18.

Colombani, J., Olivry, J. C., and Kallel, R. (1984), Phénomènes exceptionnels d'érosion et de transport solide en Afrique aride et semi-aride, in *Challenges in African Hydrology and Water Resources*, IAHS pub. no. 144, 295–300.

Dedkov, A. P. and Mozzherin, V. I. (1984), *Erozia i Stok Nanasov na Zemle*, Izdatelstvo Kazanskogo Universiteta.

Dejoux, C. (1988), *La pollution des eaux continentales Africaines*, Orstom Travaux et Documents no. 213 (Paris).

Demmak, A. (1982), 'Contribution a l'étude de l'érosion et des transports solides en Algerie septentrionale', Docteur-Ingenieur thesis (Paris).

Dunne, T. (1975), 'Sediment Yields of Kenyan Rivers', unpublished report.

—— (1979) 'Sediment Yield and Land Use in Tropical Catchments', *Journal of Hydrology*, 42: 281–300.

Falkenmark, M. (1989), The Massive Water Scarcity now Threatening Africa: Why isn't it Being Addressed? *Ambio*, 18: 112–18.

FAO (1979), *A Provisional Method for Soil Degradation Assessment* (Rome).

Foster, S. S. D. (1984), 'African Groundwater Development', in *Challenges in African Hydrology and Water Resources*, IAHS pub. no. 144, 3–12.

Fournier, F. (1960), *Climat et erosion* (Paris).

Gac, J. Y. and Kane, A. (1986), Le Fleuve Sénégal: I. Bilan hydrologique et flux continentaux de matières particulaires a l'embouchure', *Bulletin Sciences Géologiques*, 39: 99–130.

Gibbs, R. (1970), 'Mechanisms Controlling World Water Chemistry', *Science*, 170: 1088–90.

Haines, A. T., Finlayson, B. L., and McMahon, T. A. (1988), 'A Global Classification of River Regimes', *Applied Geography*, 8: 255–72.

Hayden, B. P. (1988), 'Flood Climates', in V. R. Baker, R. C. Kochel, and P. C. Patton (eds.), *Flood Geomorphology* (New York), 13–26.

Heusch, B. and Millies-Lacroix, A. (1971) 'Un méthode pour estimer l'écoulement et l'érosion dans un bassin: Application au Maghreb', *Mines et Géologie*, 33: 21–39.

Howell, P. P. and Allan, J. A. (1990) (eds.), *The Nile: Resource Evaluation, Resource Management, Hydropolitics and Legal Issues* (London).

Hurst, H. E. (1952), *The Nile* (London).

Jordaan, J. M. and Van Bladeren, D. (1990), 'Hydraulic Phenomena Associated with Recent Disastrous Floods in South Africa', in W. R. White (ed.), *International Conference on River Flood Hydraulics* (Chichester), 219–28.

Kempe, S. (1983), 'Impact of Aswan High Dam on Water Chemistry of the Nile', in E. T. Degens, S. Kempe, and H. Soliman (eds.), *Transport of Carbon and Minerals in Major World Rivers*, Part 2 (Hamburg), 401–23.

Mahé, G. and Olivry, J. C. (1991), 'Changements climatiques et variations des écoulements en Afrique occidentale et central du mensuel a l'interannuel', in *Hydrology for the Water Managemement of Large River Basins*, IAHS pub. no. 201, 163–72.

Mahmood, K. (1987), *Reservoir Sedimentation: Impact, Extent and Mitigation* (Washington, DC).

Martins, O. (1983), 'Transport of Carbon in the Niger River', in E. T. Degens, S. Kempe, and H. Soliman (eds.), *Transport of Carbon and Minerals in Major World Rivers*, 435–49.

—— and Probst, J.-L. (1991), 'Biogeochemistry of Major African Rivers: Carbon and Mineral Transport', in E. T. Degens, S. Kempe, and J. E. Richey (eds.), *Biogeochemistry of Major World Rivers* (Chichester), 127–55.

McMahon, T. A., Finlayson, B. L., Haines, A., and Srikanthan, R. (1992), *Global Runoff: Continental Comparisons of Annual Flows and Peak Discharges* (Cremlingen-Destedt).

Meybeck, M. (1979), 'Concentrations des eaux fluviales en éléments majeurs et apports en solution aux oceans', *Revue de Géologie Dynamique et de Géographie Physique*, 21: 215–46.

—— Chapman, D. and Helmer, R. (1989), *Global Freshwater Quality: A First Assessment* (Oxford).

Milliman, J. D. and Meade, R. H. (1983), World-wide Delivery of River Sediment to the Oceans, *Journal of Geology*, 91: 1–21.

NEDECO (1959), *River Studies: Niger and Benue* (Amsterdam).

Nkounkou, R. R. and Probst, J.-L. (1987), 'Hydrology and Geochemistry of the Congo River System', in E. T. Degens, S. Kempe, and G. Weibin (eds.), *Transport of Carbon and Minerals in Major World Rivers*, Part 4 (Hamburg), 483–508.

Obeng, L. E. (1978), 'Environmental Impacts of Four African Impoundments', in C. G. Gunnerson and J. M. Kalbermatten (eds.), *Environmental Impacts of International Civil Engineering Projects and Practices* (New York), 29–43.

Odone, T. (1984), 'Man-made River Brings Water to the People', *Middle East Digest*, 10: 39–40.

Olivry, J. C. (1987), 'Les conséquences durables de la sécheresse actuelle sur l'écoulement du fleuve Sénégal et l'hypersalinisation de la basse Casamance', in *The Influence of Climatic Change and Climatic Variablility on the Hydrologic Regime and Water Resources*, IAHS pub. no. 168, 501–12.

—— Bricquet, J. P. and Mahé, G. (1993), 'Vers un appauvrissement durable des ressources en eau de L'Afrique humide', in *Hydrology of Warm Humid Regions*, IAHS pub. no. 216, 67–78.

Ongweny, G. S. O. (1978), 'Erosion and Sediment Transport in the Upper Tana Catchment', Ph.D. thesis (Nairobi).

Pardé, M. (1955), *Fleuves et rivières* (Paris).

Probst, J.-L., Nkounkou, R. R., Krempp, G., Bricquet, J. P., and Olivry, J. C. (1992), 'Dissolved Major Elements Exported by the Congo and the Ubangi Rivers During the Period 1987–1989', *Journal of Hydrology*, 135: 237–57.

Rodier, J. A., Colombani, J., Claude, J., and Kallel, R. (1981), *Le bassin de la Mejerdah*, Monographies Hydrologiques ORSTOM, no. 6.

—— and Roche, M. (1984), *World Catalogue of Maximum Observed Floods*, IAHS pub. no. 143.

Schamp, H. (1983), 'Sadd el-Ali, der Hochdamm von Assuan, I, II', *Geowiss. in unserer Zeit.*, 1/2: 51–59; 1/3: 73–85.

Seyler, P., Olivry, J. C. and Sigha Nkamdjou, L. (1993), 'Hydrogeochemistry of the Ngoko River, Cameroon: Chemical balances in a Rainforest Equatorial Basin', in *Hydrology of Warm Humid Regions*, IAHS pub. no. 216, 87–105.

Shahin, M. (1985), *Hydrology of the Nile Basin* (Amsterdam).

Sircoulon, J. H. A. (1987), 'Variation des débits des cours d'eau et des niveaux des lacs en Afrique de l'ouest depuis le début du 20ème siècle', in *The Influence of Climate Change and Climatic Variability on the Hydrologic Regime and Water Resources*, 13–25.

Spronck, R. (1941), 'Mesures hydrographiques effectuées dans la region divagante du bief maritime du fleuve Congo', *Mémoires de l'Institut Royal Colonial Belge*, 8: 3–56.

Strakhov, N. M. (1967), *Principles of Lithogenesis* (Edinburgh).

Starmans, G. A. N. (1970), 'Soil Erosion on Selected African and Asian Catchments', in *Proceedings International Water Erosion Symposium* (Prague).

Sutcliffe, J. V. and Parks, Y. P. (1987), 'Hydrological Modelling of the Sudd and Jonglei Canal', *Hydrological Sciences Journal*, 32: 143–59.

—— and Knott, D. G. (1987), 'Historical Variations in African Water Resources', in *The Influence of Climatic Change and Climatic Variability on the Hydrologic Regime and Water Resources*, IAHS pub. no. 168, 463–75.

—— and Lazenby, J. B. C. (1990), 'Hydrological Data Requirement for Planning Nile Management', in P. P. Howell, and J. A. Allan (eds.), *The Nile: Resource Evaluation, Resource Management, Hydropolitics and Legal Issues* (London), 107–36.

UNESCO (1978a), *World Water Balance and Water Resources of the Earth*, UNESCO Studies and Reports in Hydrology no. 25 (Paris).

—— (1978b), *Atlas of World Water Balance* (Paris).

Walling, D. E. (1983), 'The Sediment Delivery Problem', *Journal of Hydrology and Water Resources*, IAHS pub. no. 144, 69: 209–37.

—— (1984), 'The Sediment Yields of African Rivers', in *Challenges in African Hydrology*, 265–83.

—— (1987), 'Land Degradation and Sediment Yields in Rivers: A Background to Monitoring Strategies', in *Monitoring Systems for Environmental Control*, Report no. 13, SADCC Soil and Water Conservation and Land Utilization Programme Co-ordination Unit (Maseru, Lesotho).

Ward, P. R. B. (1980), 'Sediment Transport and a Reservoir Siltation Formula for Zimbabwe-Rhodesia', *Civil Engineer in South Africa* (Jan.), 9–15.

Wright, E. P. (1983), 'Groundwater Development', in *Proceedings ICE Conference on World Water 83* (London), 63–71.

—— (1992), *The Hydrology of Crystalline Basement Aquifers in Africa*, Geological Society Special Pub., no. 66, 1–28.

7 Lakes

William M. Adams

Introduction

Africa is rich in lakes, in both qualitative and quantitative terms. Figure 7.1 shows the largest lakes of the continent. The diversity of lake environments defies simple summary, for there is a vast literature on hydrology, physical limnology, biogeography, and climatic history (e.g. Grove *et al.*, 1975; and see Goudie and Nicholson, both this volume). There is also an enormous diversity of lakes, including deep tectonic lakes of the East African Rift, such as Lakes Malawi, Tanganyika, Albert, and Turkana (see Nyamweru, this volume), lakes created by volcanic action (e.g. Lake Kivu or Lake Chala in Kenya), shallow floodplain lakes (e.g. those in the Okavango Swamps, McCarthy *et al.*, 1993), soda lakes (notably in the East African Rift), multitudes of deflation basins or pans (as in the Kalahari and the Panlands of South Africa) and even some at high altitude of glacial origin (Livingstone and Melack, 1984). There are a number of excellent texts on tropical or African limnology, among them Beadle (1981), Serruya and Pollingher (1983), and Payne (1986).

Quantitatively, Africa's lakes are more easily described. Africa has twenty times the volume of lakes and reservoirs that occur in Latin America, in Lake Victoria it has the third largest lake in the world by area, and in Lake Tanganyika the third largest by volume (Bowen, 1982; Burgis and Morris, 1987). Lake Tanganyika is also the second deepest lake in the world, with 1 per cent of the total volume of freshwater on the surface of the planet (Bowen, 1982). The approximate contemporary dimensions of the largest lakes in Africa are given

in Table 7.1 (Balek 1983), although it should be noted that some of these have been subject to considerable variation in extent both on millenial and centennial time-scales, and also in more recent decades. This applies particularly to Lake Chad (Grove, 1978, 1985*a*). However, this table does demonstrate the enormous size of Africa's greatest lakes, and also the significance of those created artificially in the last three decades.

Lakes are classified as wetlands under the Ramsar Convention, and African lakes share with other freshwater wetland environments a particular importance for biodiversity, ecological productivity, and for resources for human use (Dugan, 1990; Hughes, this volume). Lakes in floodplain and inland delta areas are often particularly important economically, particularly in supporting fisheries (Welcomme, 1979). Anthropogenic impacts on lake environments have become increasingly significant features of their ecology, as indeed the creation of new lakes has been a significant impact on the freshwater ecosystems of rivers (Rzoska, 1976; Grove, 1985*b*; Davies, 1979).

The lakes of Africa have long attracted attention from outside explorers and researchers (Balek, 1977; Beadle, 1981). Scientific research programmes on lake ecology were in place before the Second World War, notably perhaps in the expeditions led by E. B. Worthington to the Great Lakes of Eastern Africa from 1925 to 1933 (Worthington, 1932, 1983; Worthington and Worthington, 1933). This research effort grew after that war, with for example the Belgian Exploration Hydrobiologique on Lake Tanganyika in 1946–7, and studies of the fisheries of colonial territories such as Nigeria (Welman, 1948). This work was enhanced by the establishment of African universities from the 1950s onwards (Beadle, 1981). Limnological research received a major boost

I would like to thank Dick Grove for supervising my first research on Africa, and for his continuing encouragement and support.

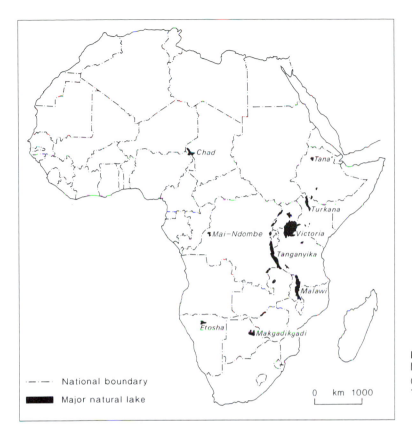

Fig. 7.1. Major natural lakes of Africa.

(*Source*: Livingstone and Melack, 1984.)

Legend:
- – · – National boundary
- ▆▆▆ Major natural lake

0   km   1000

Table 7.1 The dimensions of large natural and anthropogenic lakes in Africa

Lake	Area (km²)	Max depth (m)	Volume (km³)
Natural			
Victoria	66 400	92	2656
Tanganyika	32 890	1435	18 940
Malawi	30 800	706	7000
Chad	18 000	12	27
Turkana	8660	72	—
Albert	5500	57	64
Anthropogenic			
Volta	8480	70*	148
Kariba	5250	100*	160
Aswan	5120	95*	157
Cabora Bassa	2700	100*	66

* dam height.

Source: Balek, 1983.

with the research effort associated with the development of the first large anthropogenic lakes, for example the Kainji Biological Research Team (White, 1965; Ewer, 1966) and the Volta Basin Research Project (Lawson, 1963). The International Biological Programme Freshwater Section, active between 1963 and 1974, included among its bilateral projects joint programmes between France and the Republic of Chad, and UK and Uganda. There was a particular focus on the production ecology of Lake George, and an IBP meeting at Makerere University in Uganda in May 1968 (le Cren and Lowe-McConnell, 1980).

Physical Environment

Lakes in Africa are neither simple nor static environmentally. The dramatic changes in the spatial extent of the shallow waters of Lake Chad in the 1980s emphasize the importance of an understanding of hydrology. Lakes gain water from inflowing rivers, from rainfall, and from seepage from groundwater, and lose it from outflows, evaporation, or back to groundwater. The

Plate 7.1. (a) Lake Magadi, Kenya; (b) Lake Magadi Soda Works. The soda industry of Lake Magadi in Kenya is based on the collection of precipitated sodium carbonates. Lake Magadi is classified by Talling and Talling (1965) as a Class III lake (conductivities of 6000 to 160 000 μS cm^{-1}), with other closed basin lakes in arid belts of the Sahara and Namib, and elsewhere in the East African Rift (e.g. Lakes Zwai, Shala, and Abaya in Ethiopia, and Lakes Bogoria, Nakuru, Elementaita, Natron, and Manyara in Kenya and Tanzania) (photo: W. M. Adams).

balance of different inputs and outputs varies considerably between lakes. Those in closed basins tend to have low recharge rates, and to lose water primarily by evaporation. Other lakes, for example those in floodplains, gain most of their water from surface runoff. In Lake George, over 90 per cent of inputs come from river flows, while Lake Victoria, gets 83 per cent of its inflow from rainfall on the lake surface, and loses a similar proportion from evaporation. Discharge from Lake Victoria into the Nile system accounts for only 17 per cent of water loss (Payne, 1986). In lakes in areas with marked rainfall seasonality, the balance of inputs and outputs can change considerably on a seasonal time-scale, and so too can water levels and extent. There can also be changes in water balance on inter-annual and longer time-scales, even of the largest lakes. The level of Lake Malawi, for example rose in the 1920s by some 5 metres,

while that of Lake Victoria rose in the 1960s (Balek, 1977; Beadle, 1981). Such changes in lake-level are known to have occurred widely. The raised shorelines of lakes in closed basins have provided a particularly important indicator of palaeoclimates in Africa (Goudie, and Nicholson, this volume).

As might be expected from their diverse forms and origins, the water chemistry of African lakes is highly variable. However, the extent of ancient acid basement-complex rocks tends to lead to a dominance of sodium bicarbonates in many African lake waters. The soda industry of Lake Magadi in Kenya (Plates 7.1a,b) is based on the collection of precipitated sodium carbonate (Livingstone and Melack, 1984). A classification of African lakes was presented by Talling and Talling (1965), based on conductivity. Class I lakes have a conductivity of less than 600 μS cm^{-1}, and typically have open basins and significant inputs from river flows. Among the larger lakes of Africa, Class I lakes include Victoria (96 μS cm^{-1}), George (200 μS cm^{-1}), and Malawi (210 μS cm^{-1}) (Beadle, 1981). Class II lakes have conductivities of 600–6000 μS cm^{-1}, and in Africa range from Lake Tanganyika (610 μS cm^{-1}) to Lake Kivu (1240 μS cm^{-1}) and Lake Turkana (3300 μS cm^{-1}). Even Lake Turkana has a predominantly 'freshwater' fauna, but is salt to taste (Beadle, 1981). Class III lakes have conductivities of 6000 to 160 000 μS cm^{-1}, and are typically closed and lying within arid belts of the Sahara and Namib, and in the East African Rift. The saline East African lakes include a number in Ethiopia (lakes Zwai, Shala, and Abaya), and the famous lakes of Kenya and Tanzania, including Bogoria, Nakuru, Elementaita, Magadi, Natron, and Manyara. The conductivities of these lakes vary, for example Lake Elementaita at 43 750 μS cm^{-1}, Manyara at 94 000 μS cm^{-1} and Nakuru at 162 500 μS cm^{-1} (Payne, 1986). These lakes have a remarkable and specialized fauna, including the alga *Spirulina platensis*, food of the lesser flamingo *Phoeniconaias minor* whose breeding grounds on Lake Natron were only recorded in the 1950s (Brown, 1959).

Most African lakes do not experience major seasonal variations in temperature, although seasonal variations in incoming solar radiation do give a cool season of several months in many lakes, in January/February in the northern hemisphere and July/August in the south (Payne, 1986). Evaporative cooling and incoming rain are also major influences on the temperature of lake-surface waters (Livingstone and Melack, 1984). Although absolute temperature differences between surface and deep water can be relatively small, the response of water density to temperature change at the relatively high temperatures of African lakes can lead to considerable thermal stability. That stability is most likely to break down at the coolest season, and is greatly affected by wind. Thermal stability can be increased by salinity differentials (particularly in very deep lakes such as Malawi and Tanganyika), such that there can be semi-permanent stratification at depth, with a seasonal thermocline at lesser depth that is subject to mixing by wind action (Livingstone and Melack, 1984; Beadle, 1981).

Ecology

Primary production is dependent on insolation and light penetration (and hence on turbidity), temperature, and the availability of nutrients. Nutrient inputs to lakes come from inflowing rivers, rainfall, and fringing or otherwise linked swamp environments (Livingstone and Melack, 1984). Rates of primary production vary in

Table 7.2 Gross planktonic photosynthetic production of selected African lakes

| | Gross planktonic photosynthetic production | |
	per day (gC m^{-2})	per year (gC m^{-2})
Victoria	1.08–4.2	950
Tanganyika	0.8–1.1	
Kivu	1.8	
George	5.4	1980
Chad	0.7–2.7	
Turkana	0.25–6.2	
Nakuru	2.0–3.2	

Source: Beadle. 1981 (drawing on a range of sources).

different parts of lakes, and over time. Planktonic production is relatively easily assessed, and limited observations may be valid over extensive areas. Production by emergent and aquatic vegetation is more difficult to measure, but can be very important, particularly locally. There have been few studies that allow integrated estimates of primary production that are valid on an annual time-scale, particularly for the larger lakes. Beadle (1981) presents annual data from studies of Lake Victoria and Lake George, and daily estimates for other lakes (Table 7.2).

Among the most productive lakes in Africa are those that are shallow, such as Lake George, particularly where they are mixed by wind action. The production ecology of Lake George was studied extensively by the IBP project, and it is probably the best known of the African lakes in this respect (e.g. Burgis, 1974). Lake George is of medium size (250 km^2), and only about 2.4 m deep. It receives inflow from the north (1.9 × 10^9 m^3), and outflows to lake Edward (1.7 × 10^9 m^3). The turnover time of water is rapid (retention time 4.3 months). The lake has a diurnal cycle of stratification in terms of temperature, pH, and oxygen concentrations (Serruya and Poligher, 1983). The euphotic zone is shallow, and most of the primary production takes place in the top metre of water. There is a diurnal pattern of phytoplankton activity that matches the physical dynamics of the lake. Nutrient cycling is rapid and productivity is affected by the supply of nutrients in inflowing river water, but there are other complicating factors such as the influence of hippopotami on nutrient dynamics (Beadle, 1981). Some 95 per cent of the planktonic biomass of Lake George comprises phytoplankton, but the net production is small (only 28 per cent of gross production) and the level of secondary production is also small. There is, however, a diverse fish community, of which the largest species are commercially important (Beadle, 1981; Serruya and Poligher, 1983). Fish biomass has been

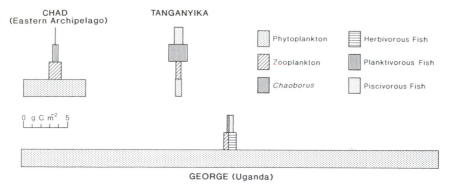

Fig. 7.2. Trophic structure of Lake Chad and Lake Tanganyika. (*Source*: Coulter, 1991.)

recorded at between 6.3 and 90 g live weight m^{-2}, with a mean of 23 g m^{-2} (Payne, 1986).

The Lake George ecosystem seems to be remarkably stable, both on an annual and inter-annual time-frame (data being available for six years). Other productive shallow lakes such as Lake Chad are much more variable because of much greater influence of nutrients from seasonal water inflows, and from the extensive area of land flooded annually. Indeed, Lowe-McConnell (1987) describes Lake Chad as a 'riverine lake', comparable to a vast littoral zone. It has a high benthic biomass (37 kg ha^{-1}, compared to 7.4 kg ha^{-1} for Lake George).

Lake Victoria presents a rather different and more complex picture. It is obviously very large (68 000 km^2), with a long shoreline (3440 km), and a particularly shallow western shore with numerous bays and islands with extensive areas of emergent and floating vegetation linked to swamps. The mean depth is 40 m, and the maximum depth 79 m (Serruya and Poligher, 1983). There are relatively small temperature differences with depth in Lake Victoria, but they are sufficient to create some vertical stratification. The depth of the euphotic zone is about 13 m. The phytoplankton flora is diverse and fairly constant through the year in terms of biomass, but there is a peak in primary production in July and August associated with mixing of water caused by south-east trade winds (Beadle, 1981). Patterns of primary and secondary production are highly variable in space (hence the range in the figures given in Table 7.1), and research on this and other major lakes is demanding in terms of resources.

Deeper lakes with clear water tend to have smaller standing crops of phytoplankton than shallow lakes, but rates of production can be high. Lake Tanganyika has remarkably clear water. There is a marked annual variation in primary production, which is highest in October–November. The trophic efficiency of the Lake Tanganyika ecosystem is very high, as is the yield of the fishery (Hecky, 1991). The high proportion of fish biomass to phytoplankton is unusual, as is the level of endemicity of the pelagic fish fauna. The trophic structure of the pelagic community is similar to marine environments. The planktonivorous fish are believed to be of marine origin (Hecky, 1991; Coulter, 1991a).

It is not, however, the productivity of the fisheries of the deeper Great Lakes of East Africa that has attracted the attention of biologists, but the diversity of their faunas. The fish faunas of Lakes Victoria, Tanganyika, and Malawi are dominated by cichlid species that exhibit astonishing patterns of speciation and endemism (Lowe-McConnell, 1987; Myers, this volume). There are 247 species of fish from nineteen families in Lake Tanganyika, 242 species from nine families in Lake Malawi, and more than 238 species from twelve families in Lake Victoria (Lowe-McConnell, 1987, table 2.2). Cichlids dominate these fauna, comprising no less than 84 per cent of the species in Lake Victoria, 82.6 per cent in Lake Malawi, and 55.1 per cent in Lake Tanganyika. In Lake Victoria some 150 species of the genus *Haplachromis* have adapted to exploit a wide range of different and quite specific food sources (Greenwood, 1974).

A great deal of research has been done on the endemic fish faunas of the Great Lakes of East Africa, particularly their origins and biogeographic affinities, and their evolution (Lowe-McConnell, 1975, 1987; Beadle, 1981; Coulter, 1991a). These processes obviously remain highly dynamic. Recent studies of changes in the levels of Lake Malawi between AD 1500 and 1850, and of mitochondrial DNA differentiation, suggest that the species flock of rocky-shore cichlid fish have evolved over only the last 200–300 years, in a manner linked to changing water levels and lake margins (Owen *et al.*, 1990).

Human Impacts on Lake Environments

Human impacts on lake environments are undoubtedly increasing in Africa. The major phenomenon of new impoundments is discussed in the next section, but

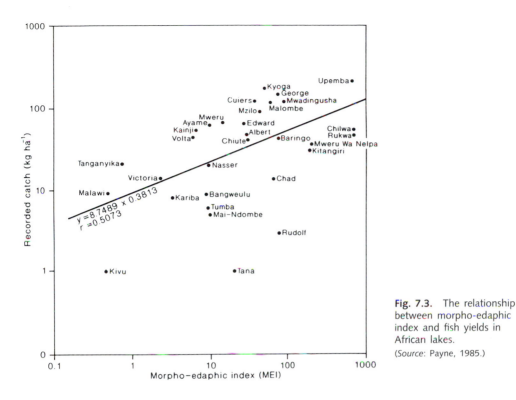

Fig. 7.3. The relationship between morpho-edaphic index and fish yields in African lakes.
(*Source*: Payne, 1985.)

hydrological impacts extend to even major natural lakes, notably in the case of Lake Victoria, whose level is controlled by the Owen Falls Dam. Other problems are still limited in extent and severity. Pollution is a relatively minor and localized problem in African lakes, partly because of the limited extent of urbanization and industrialization. Enrichment with sewage or eroded soils can be beneficial in the short run, although impacts can be serious. Soda lakes such as Lake Nakuru in Kenya are among those where sewage and waste water (and indeed the increased discharge from improved water-based sanitation) could have significant environmental impacts. A problem of eutrophication has developed since 1989 in Lake Victoria, with fish kills from deoxygenated water (Viner, 1992). There are particular fears for the effects of pollutants in deep lakes that are strongly stratified and have long water-residence times, for example Lake Malawi. The impacts of pulp-mill effluent or other pollution in such circumstances could be considerable and long-lasting (Payne, 1986).

The most widespread human use of lake environments in Africa is for fishing. Fish protein is an important element in diet for many people, and dried and smoked freshwater fish have long been traded over long distances. Fish production can be approximately estimated by using a function of the ratio of total dissolved solids to mean lake depth (the 'morphoedaphic index' (Payne, 1986) see Figure 7.3). In many lakes, fishing is low-technology and artisanal in form. On Lake Tanganyika, for example, various canoe-based methods have long been used, in particular involving the use of lights at night to attract clupeid fish. However, new methods including lift-nets, have been introduced within the artisanal fishery. More significant was the industrial fishery begun in 1953 using 15 m-long steel boats and purse-seine nets at night (Coulter, 1992b). Some sixty purse-seine boats operate, but are restricted to a few major ports and hence do not fish the entire lake. It is estimated that the annual yield of Lake Tanganyika might be between 380 000 and 460 000 tonnes, with yields of 300 kg ha^{-1} yr^{-1} possible (Coulter, 1991b).

In a number of lakes, fishing effort has tended to increase over time as new technologies have become available. Thus studies of catches of the tilapia *Oreochromis esculentus* in Lake Victoria, show that catches fell from 50–100 individuals per net in 1905 when gill nets were introduced to 6–7 fish per net in 1928 and only 1.6 in 1954. Mesh sizes were dropped in 1954, and again in 1966. The size of fish caught fell (Payne, 1986). Such declines in catches in the multi-species fishery typical of most African lakes can be associated with changes in the species composition of catches (e.g. on Lake

Fig. 7.4. Major anthropogenic reservoirs in Africa.
(*Source:* Bernacsec, 1984.)

Malawi), and associated changes in the economic value of the catch as particularly valuable species decline. The limited scientific knowledge about African lake fisheries, and the even more limited institutional capacity to control catch size and gear, suggests that effective management of lake-fish resources may be problematic.

In several lakes attempts have been made to improve fisheries by introducing new or exotic species. In some instances such interventions may be successful, for example in the case of the introduction of the pelagic fish *Limnothrissa miodon* to the artificial Lake Kariba from Lake Tanganyika in 1972 (discussed below). Four exotic tilapia (cichlidae) species were introduced into Lake Victoria in the 1950s to enhance the declining and over-exploited fishery, and have adapted fairly successfully to conditions (Viner, 1992). Other introductions have been more controversial. In the early 1960s, the Nile perch (*Lates niloticus*) was introduced to Lake Victoria from Lake Albert in an attempt to enhance the fishery. This has been achieved, with a dramatic increase from an annual yield of 40 000 tonnes to 500 000 tonnes over the next twenty years. However, there is evidence that the Nile perch is being overfished, with the yield of

filleted fish (320 000 tonnes per year) exceeding the calculated maximum sustainable yield. In addition, *Lates niloticus* is a large predator, and has had a role (with overfishing) in drastically reducing the population of pelagic *Haplochromis* species (Viner, 1992).

Artificial Lakes

Natural freshwater ecosystems have been transformed by human action on an extensive scale in Africa (Fig. 7.4). This intervention accelerated in the 1960s through the construction of large dams (Walling, this volume; Adams, 1992) (see Plates 7.2 and 7.3). In the 1960s alone there were some twenty-eight dams closed, impounding 18 000 km² (Bernacsek, 1984). The Aswan High Dam on the Nile was begun in 1960 and finished nine years later, and in the same decade the Akosombo Dam was built on the River Volta, the Kainji Dam on the Niger, and the Kariba Dam on the Zambezi (Hart, 1980; Mabogunje, 1973). The reservoirs created by these dams became the subject of considerable international research attention (Lowe-McConnell, 1966; Obeng, 1969; Rubin and Warren, 1968; Ackerman *et al.*, 1973).

Plate 7.2. Bakolori Dam, Sokoto River, Nigeria. The Bakolori Dam was built in the late 1970s on the Sokoto River to store seasonal flood water for year-round irrigation. It was originally conceived as part of an integrated river-basin development plan primarily to control flood peaks and provide a more moderate flow for extensive downstream development, but it was designed to serve a single purpose, to supply to a 22 000 ha irrigation scheme. The impacts of flow reduction in the downstream floodplain were considerable. Some 12 000 people were resettled from the reservoir area (photo: W. M. Adams).

Plate 7.3. Turkwel Reservoir, Kenya. The Turkwel River in north-west Kenya was dammed in the late 1980s to supply hydroelectric power. The reservoir has been very slow to fill, and will remain below design capacity until the end of the century. There has been great concern at the impact of reduced flood flows and evaporation losses on the recharge of the aquifer in Turkana District downstream of the dam, and particularly its implications for the survival of the riparian forest on which pastoralists depend, on pastoralists' shallow hand-dug wells, and on the supply of water to the District Headquarters at Lodwar (photo: W. M. Adams).

The ecological impacts of dams on river and flood-plain environments stimulated extensive research. In 1969, White argued that research effort devoted to the ecology of tropical impoundments had been on too small a scale and had started too late to provide 'either useful predictions or firm conclusions, about the course of biological events consequent upon impounding water in the Tropics' (White, 1969: 37). However, there is now a substantial body of knowledge on the ecological effects of river control (e.g. Baxter, 1977; Ward and Stanford, 1979).

The ecological successions in new anthropogenic lakes are complex. Organisms characteristic of flowing water (lotic ecosystems) are replaced by those of still water (lentic) ecosystems, and communities of planktonic and littoral species develop (Baxter, 1977). The trajectory of development of the reservoir ecosystem depends partly on the chemistry, turbidity, and temperature of inflowing waters, and partly on the nature of the substrate inundated. The Volta Lake in Ghana filled slowly over a period of seven years. No clearance of vegetation in the inundated area took place, and as a result there was initially considerable deoxygenation at depth as vegetation decayed. The lake became poor in nutrients with a low crop of phytoplankton (Baxter, 1977). At Kainji, filled a few years later in Nigeria, vegetation was cleared and burned before inundation. A diverse benthic fauna developed. Elsewhere, for example in Cabora Bassa or Kariba on the Zambezi system, inundation led to infestation with algae or macrophytes (Balon and Coche, 1974; Davies *et al.*, 1972). In 1962, 22 per cent of the surface area of Lake Kariba was covered with *Salvinia molesta*, although the area was reduced to very low levels through various control strategies by 1980 (Bernacsek, 1984). Floating plants such as *Salvinia*, or the Nile cabbage (*Pistia stratiotes*) or the water hyacinth (*Eichornia crassipes*) can create problems for hydroelectric turbines, and increase evapotranspiration losses from the reservoir surface.

The development of the fish fauna in tropical reservoirs is complex, and depends on the way in which the reservoir ecosystem develops (Lowe-McConnell, 1985; Bernacsek, 1984). There may be an initial peak of population as nutrients from flooded areas feed into the ecosystem, followed by a slump to a lower level (Petr, 1975). Aquatic macrophytes can provide substrates where fish can find food sources. It is possible to some extent to predict fish yields in reservoirs from data on morphometry, the morphoedaphic index, conductivity, and total dissolved solids (Marshall, 1984). Reservoir fish stocks will be at least in part pelagic, and may require very different fishing techniques from those previously used in river floodplains.

Some African reservoirs have developed productive economic fisheries (Marshall, 1984). There is substantial variation in the productivity of fish stocks and therefore the economic value of the fishery created (Welcomme, 1979; Lowe-McConnell, 1985; Bernacsek, 1984). On the Volta Lake in Ghana there were over 20 000 fishermen in 1976 (many of them not evacuees, but fishermen from elsewhere), and the annual catch was some 40 000 tons. This gave a catch of just under 5 kg per ha per year, more than four times that predicted before the dam was built. Volta lake fish is a major source of protein in Ghana, and indeed given the fixed low price at which the electricity generated by the dam is sold, this fishing industry rates as one of the most significant economic benefits of the dam. At Kainji, some of the 5000 fishermen displaced by the reservoir were able to change their fishing methods and continue to fish on

the new lake. Studies of seven man-made lakes in Africa done in the 1970s showed that catches varied from over 12 to under 0.8 kg per ha per year (Welcomme, 1979). The introduction of the pelagic fish *Limnothrissa miodon* to Lake Kariba from Lake Tanganyika in 1972 transformed a mixed fishery producing some 2000 tonnes per year. The new 'sardine' catch rose to 8000 tonnes in 1981, and a much enhanced total catch of 11 000 tonnes (Payne, 1986). There has been considerable research into the management of reservoirs to maximize fish production (Bernacsek, 1984).

The impacts of dam construction on running water ecosystems below new reservoirs are relatively well understood in temperate rivers, particularly where there are important sport fisheries. Knowledge of tropical rivers is less complete (but see Payne, 1986), although there is a considerable amount of data on freshwater fisheries in Africa (Hickling, 1961; Lowe-McConnell, 1975, 1985; Welcomme, 1979).

Passage through a reservoir has a number of effects on water quality (see also Chapter 6). Water released from low outlets in a dam is typically cool, often deoxygenated, and sometimes rich in hydrogen sulphide. Dams can affect patterns of invertebrate drift downstream, and river-bed degradation due to accelerated erosion downstream can cause significant changes to in-stream substrates. In tropical Africa, however, it is the impact of dams on natural flood regimes which is the most significant. Many fish exhibit fairly short 'lateral' migrations or longitudinal migrations of greater length in response to seasonal fluctuations in the river. As water spills out onto the floodplain it becomes enriched with organic matter from decaying vegetation, animal manure, and other materials. This creates a flush of algal, bacterial, and zooplanktonic growth, which forms a food source for invertebrates. Aquatic and emergent vegetation growth is also rapid and extensive. As a result there is abundant food for fish which follow the floodwaters out of the river channel, and in particular both food and shelter for young fish. Growth-rates are very rapid in these flood conditions. As the flood subsides, fish move back to the river channel, and in many cases eventually to the small and deoxygenated pools of largely dry river beds. This season sees high mortality, both from fish stranded in evaporating floodplain pools and those taken by birds, predatory fish, and human predators (Welcomme, 1979; Lowe-McConnell, 1985; Lowe-McConnell, 1987).

Dam construction has significant implications for the fish population of tropical floodplain rivers. Jubb (1972) records the failure of *Hydrocynus vittatus* and other species to spawn in the Pongola floodplain in South Africa

Plate 7.4. Fisherman using clap nets (Hausa: *Koma*) and calabash floats on the Sokoto River downstream of the Bakolori Dam, Nigeria. Dam construction has significant implications for the fish population of tropical floodplain rivers. On the Sokoto River, reductions were recorded in fish catches and fishing effort on the floodplain downstream of the Bakolori Dam, but a small tailrace fishery was established (photo: W. M. Adams).

following river control which reduced the annual flood to such an extent that floodplain pools remained isolated from the main channel. The Kainji Dam acts as a complete barrier to fish movement, and although catches at the foot of the dam are high, studies further downstream reported significant reductions in fish catches between 1967 and 1969 (Lelek and El-Zarka, 1973; Adeniyi, 1973; Lowe-McConnell, 1985), and associated reductions in fishing activity. Similar reductions in fish catches and fishing effort below dams were recorded elsewhere in Nigeria, for example on the Sokoto (Adams, 1992) (Plate 7.4). A more complex response was shown by the *Egeria* clam fishery of the lower Volta. Breeding in the clams is triggered by a rise in salinity following reduced river flow. During construction of the dam at Akosombo the critical salinity conditions moved 30–50 km inland, and once operating the constant flows pushed the fishery down to within 10 km of the river mouth (Chisholm and Grove, 1985).

The most obvious impact of dam construction is simply the loss of productive land beneath the reservoir, and the costs (economic and human) of resettlement. The area lost beneath reservoirs in Africa obviously varies a great deal, as does its quality. Among the largest reservoirs are the Volta Lake formed behind the Akosombo Dam, which covers 8500 km^2, flooding a substantial area of central Ghana. Kainji impounded 1200 km^2, including 15 000 ha of farmland. The Lagdo Dam on the Benue flooded 70 000 ha, including floodplain land stretching 2–5 km on both banks of the river.

The other major cost of reservoir construction is that of resettlement. Sometimes very large numbers of people are displaced. The Kossou Dam in the Ivory Coast displaced 85 000 people, the Volta Dam 84 000, Kariba 57 000, Kainji 50 000, and the Lagdo Dam on the Benue 35 000 people (Adams, 1985). The Aswan High Dam on the Nile displaced an even greater number of people, 120 000 Nubians, both in Egypt and Sudan. The two countries agreed on the construction of the High Dam in November 1959, as part of the negotiations over the Second Nile Waters Agreement. The reservoir eventually stretched 150 km inside Sudan, flooding twenty-seven villages and the town of Wadi Halfa. A total of 50 000 people were evacuated from Sudanese Nubia, and resettled far away on the Khashm el Girba irrigation scheme on the Atbara River (Dafalla, 1975).

The cost of resettlement needs to be looked at both in terms of the actual financial costs (surveying people and property, compensation or rebuilding of settlements and infrastructure, and actual translocation), and also the human cost of the stress caused by uprooting. The stress of resettlement is multidimensional, both psychological and socio-cultural. It is particularly grim that this stress is so often exacerbated by bad planning and inadequate provision for resettlement. Loss of assets, unfamiliar environments, unprepared resettlement sites, poor living conditions, and hopeless economic prospects are all elements in the human and economic costs of resettlement (Scudder and Colson, 1982).

Some African man-made lakes have been the subject

of very detailed resettlement planning exercises, most notably Volta Lake in Ghana and Kainji in Nigeria, both of them in the 1960s. However, even here, success was moderate. At Kainji, architect-designed concrete houses were judged uncomfortable and badly laid out, while at Volta Lake only 40 per cent of evacuees remained in resettlement villages in 1968. Resettlement has often not been successful. When construction of the Kariba Dam began, no resettlement studies had been carried out. The result was a 'poorly conceived and trauma-ridden crash programme to get people out of the lake basin before the dam was sealed' (Scudder and Habarad, 1991). While most evacuees were resettled nearby, in 1958 6000 Gwembe Tonga people were removed by force. In clashes nine people were killed and over thirty injured. Disputes about compensation following construction of the Bakolori Dam in Nigeria led to block-ades to stop construction work and an attack by riot police in 1980 which caused many deaths (Adams, 1988).

Where reservoir resettlement projects are poorly planned and under-financed, most evacuees are con-siderably worse-off in the short term. This impact may persist, as in the case of the Zambian Gwembe Tonga at Kariba, affected also by the economic decline in Zambia since 1974 (Scudder and Habarad, 1991). There can also be second-order problems for those living close to reservoirs, such as disease. River-blindness (onchocer-ciasis) is spread by biting flies whose larvae live in fast-flowing water such as that found in rapids and dam spillways. Bilharzia (schistosomiasis) spreads from an intermediate host in certain snails which like shallow water, and thrive on reservoir margins (Barrow, 1981).

The common failures of resettlement planning relate directly to its complexity and sensitivity (Adams, 1990). It needs to be done by sociologists or anthropologists, rather than by the hydrologists, engineers, and econo-mists who dominate the processes of dam design and construction. Like environmental impact assessment, it also needs to be done early in the project implementa-tion process, and it needs to be adequately funded. These conditions are too rarely met in Africa, and indeed elsewhere.

Conclusion

Lake environments in Africa are complex and, despite past research efforts, are still too little understood. Many of them support unusual assemblages of species of enormous intrinsic scientific interest, and that represent elements within the biodiversity of Africa that are of the greatest importance at both the continental and global scales (see Myers, this volume). Lakes also commonly provide resources for human use, and sustain both local populations and in some instances contribute signific-antly to national economies. However, as human pres-sure on lake ecosystems increases, through increased fishing, pollution of lake and inflowing waters, and species introductions, the capacity of these ecosystems to main-tain their biological functioning and existing species-diversity is threatened.

The level of research knowledge about lake environ-ments is still too slender to make reliable predictions about their ability to absorb the range and intensity of human demands now being put upon them. There is an urgent need for further research to provide an input to development planning. Such planning must be cau-tious, and must focus on ways to increase yields of use-ful products, and the streams of economic benefits that they imply, in a manner that sustains diversity and ecological productivity. Development must also take proper account of the needs and aspirations of local people, and should ideally be conceived, planned, and implemented by local communities themselves, em-powered with scientific knowledge and external finan-cial support.

References

Ackerman, W. C., White, G. F., and Worthington, E. B. (1973) (eds.), *Man-Made Lakes: Their Problems and Environmental Effects*, Geophysi-cal Monograph 17, American Geophysical Union (Washington, DC).

Adams, W. M. (1985), 'River Control in West Africa', in A. T. Grove (ed.), *The Niger and its Neighbours: Environment, History, Hydrobiology, Human Use and Health Hazards of the Major West African Rivers* (Rotterdam), 177–228.

—— (1988), 'Rural Protest, Land Policy and the Planning Process on the Bakolori Project, Nigeria', *Africa*, 58: 315–36.

—— (1990), *Green Development: Environment and Sustainability in the Third World* (London).

—— (1992), *Wasting the Rain: Rivers, People and Planning in Africa* (London).

Adeniyi, E. O. (1973), 'The Impact of Change in River Regime on Economic Activities below Kainji Dam', in A. L. Mabogunje (ed.), *Kainji Lake Studies*, ii, Nigerian Institute of Social and Economic Research (Ibadan), 169–77.

Balon, E. K. and Coche, A. G. (1974) (eds.), *Lake Kariba: A Man-Made Tropical Ecosystem in Central Africa*, Monographiae Biologicae 24 (The Hague).

Barrow, C. J. (1981), 'Health and Resettlement Consequences and Opportunities Created as a Result of River Impoundment in De-veloping Countries', *Water Supply and Management*, 5: 135–50.

Balek, J. (1977), *Hydrology and Water Resources in Tropical Africa*, Development in Water Science, vol. 8 (Amsterdam).

—— (1983), *Hydrology and Water Resources in Tropical Regions*, Development in Water Science, vol. 18 (Amsterdam).

Batisse, M. (1982), 'The Biosphere Reserve: A Tool for Environmental Conservation and Management', *Environmental Conservation*, 9: 101–11.

Baxter, R. M. (1977), 'Environmental Effects of Dams and Impoundments', *Annual Review of Ecology and Systematics*, 8: 255–84.

Beadle, L. C. (1981), *The Inland Waters of Tropical Africa: An Introduction to Tropical Limnology*, 2nd edn. (London).

Bernacsek, G. M. (1984), *Dam Design and Operation to Optimize Fish Production in Impounded River Basins*, CIFA Technical Paper 11 (Rome).

Bowen, R. (1982), *Surface Water* (London).

Brown, L. (1959), *The Mystery of the Flamingos* (Nairobi).

Burgis, M. and Morris, P. (1987), *The Natural History of Lakes* (Cambridge).

Chisholm, N. G. and Grove, J. M. (1985), 'The Lower Volta', in Grove (ed.), *The Niger and its Neighbours*, 229–50.

Coulter, G. W. (1991*a*), 'Zoogeography, Affinities and Evolution with Special Reference to Fish', in G. W. Coulter (ed.), *Lake Tanganyika and its Life* (London and Oxford), 275–305.

—— (1991*b*), 'Fisheries', in Coulter (ed.), *Lake Tanganyika and its Life*, 139–50.

Dafalla, H. (1975), *The Nubian Exodus* (London).

Davies, B. R., Hall, A., and Jackson, P. B. N. (1972), 'Some Ecological Aspects of the Cabora Bassa Dam', *Biological Conservation*, 8: 189–201.

Dugan, P. J. (1990) (ed.), *Wetland Conservation: A Review of Current Issues and Required Action* (Gland, Switzerland).

Ewer, D. W. (1966), 'Biological Investigations in the Volta Lake, May 1964 to May 1965', in R. H. Lowe-McConnell (ed.), *Man-Made Lakes* (New York), 21–31.

Greenwood, P. H. (1974), 'The Cichlid Fishes in Lake Victoria, East Africa: The Biology and Evolution of a Species Flock', *Bulletin of the British Museum, London, Zoological Supplement (Natural History)*, 6: 1–134.

Grove, A. T. (1978), 'Geographical Introduction to the Sahel', *Geographical Journal*, 144: 407–15.

—— (1985*a*), 'Water Characteristics of the Chari System and Lake Chad', in Grove (ed.), *The Niger and its Neighbours*, 61–76.

—— (1985*b*) (ed.), 'The Niger and its Neighbours: Environmental History and Hydrobiology, Human Use and Health Hazards of the Major West African Rivers (Rotterdam).

—— Street, F. A., and Goudie, A. S. (1975), 'Former Lake Levels and Climatic Change in the Rift Valley of Southern Ethiopia', *Geographical Journal*, 141: 177–219.

Hart, D. (1980), *The Volta River Project: A Case Study in Politics and Technology* (Edinburgh).

Hecky, R. E. (1991) 'The Pelagic Ecosystem', in Coulter (ed.), *Lake Tanganyika and its Life*, 90–110.

Hickling, C. F. (1961), *Tropical Inland Fisheries* (London).

Jubb, R. A. (1972), 'The J. G. Strydon Dam, Pongolo River, Northern Zululand: The Importance of Floodplain Pans below it', *Piscator*, 86: 104–9.

Lawson, G. W. (1963), 'Volta Basin Research Project', *Nature*, 199 (4896): 858–9.

le Cren, E. D. and Lowe-McConnell, R. H. (1980) (eds.), *The Functioning of Freshwater Ecosystems*, IBP, vol. 22 (Cambridge).

Lelek, A. and El-Zarka, S. (1973), 'Kainji Lake, Nigeria, in W. C. Ackerman, G. F. White and E. B. Worthington (eds.), *Man-Made Lakes: Their Problems and Environmental Effects*, 655–60.

Livingstone, D. A. and Melack, J. A. (1984), 'Some Lakes of Sub-Saharan Africa', in F. B. Taub (ed.), *Lakes and Reservoirs*, Ecosystems of the World, 23 (Amsterdam), 497.

Lowe-McConnell, R. H. (1966) (ed.), *Man-Made Lakes* (London).

—— (1975), *Fish Communities in Tropical Freshwaters: Their Distribution, Ecology and Evolution* (London).

—— (1985), 'The Biology of the River systems with Particular Reference to the Fishes', in Grove (ed.), *The Niger and its Neighbours*, 101–40.

—— (1987), *Ecological Studies in Tropical Fish Communities* (Cambridge).

Mabogunje, A. L. (1973) (ed.), *Kainji: A Nigerian Man-Made Lake*, Kainji Lake Studies, ii: *Socio-economic Conditions*, Nigerian Institute for Social and Economic Research (Ibadan).

McCarthy, T. S., Ellery, W. N., and Stanistreet, L. G. (1993), 'Lakes of the North-Eastern Region of the Okavango Swamps, Botswana', *Zeitschrift für Geomorphologie*, 37/3: 273–94.

Marshall, B. E. (1984), *Predicting Ecology and Fish Yields in African Reservoirs from Preimpoundment Physico-chemical Data*, CIFA Technical Paper 12 (Rome).

Obeng, L. E. (1969) (ed.), *International Symposium on Man-Made Lakes, Accra* (Accra).

Owen, R. B., Crossley, R., Johnson, T. C., Tweddle, D., Nornfeld, I., Davidson, S., Eccles D. H., and Engstrom, D. N. (1991), 'Major Low Levels of Lake Malawi and their Implications for Speciation Rates in Cichlid Fishes', *Proceedings of the Royal Society of London B.*, 240: 519–53.

Payne, A. I. (1986), *The Ecology of Tropical Lakes and Rivers* (Chichester).

Petr, T. (1975), 'On Some Factors Associated with High Fish Catches in New African Man-Made Lakes', *Archiv für Hydrobiologie*, 75: 32–49.

Rubin, N. and Warren, W. M. (1968) (eds.), *Dams in Africa: An Interdisciplinary Study of Man-made Lakes in Africa* (London).

Rzoska, J. (1976) (ed.), *The Nile: Biology of an Ancient River* (The Hague).

Scudder, T. and Colson, E. (1982), 'From Welfare to Development: A Conceptual Framework for the Analysis of Dislocated people', in A. Hansen and A. Oliver-Smith (eds.), *Involuntary Migration and Resettlement: The Problems and Responses of Dislocated People* (Boulder, Colo.), 267–87.

—— and Habarad, J. (1991), 'Local Responses to Involuntary Relocation and Development in the Zambian Portion of the Middle Zambezi Valley', in J. A. Mollet (ed.), *Migrants in Agricultural Development* (London), 178–205.

Serruya, C. and Pollingher, U. (1983), *Lakes of the Warm Belt* (Cambridge).

Talling, J. F. and Talling, I. B. (1965), 'The Chemical Composition of African Lake Waters', International *Revue der Gesamten Hydrobiologie (und Hydrographie)*, 50: 421–63.

Viner, A. (1992), 'Biodiversity, Fisheries and the Future of Lake Victoria', *IUCN Wetlands Programme Newsletter*, no. 6, Nov. 1992: 3–6.

Ward, J. V. and Stanford, J. A. (1979) (eds.), *The Ecology of Regulated Streams* (New York).

Welcomme, R. L. (1979), *Fisheries Ecology of Floodplain Rivers* (London).

Welman, J. B. (1948), *Preliminary Survey of the Freshwater Fisheries of Nigeria* (Lagos).

White, E. (1965) (ed.), *The First Scientific Report of the Kainji Biological Research Team* (Liverpool).

Worthington, E. B. (1932), *A Report on the Fisheries of Uganda Investigated by the Cambridge Expedition to the East African Lakes 1932–3* (London).

—— (1983), *The Ecological Century: A Personal Appraisal* (Oxford).

—— and Worthington, S. (1933), *Inland Waters of Africa* (London).

8 Soils

Olusegun Areola

Introduction

The soil is probably Africa's most important resource being the basis of the essentially agrarian economy of most countries. Therefore, the soil resources of Africa have received a great deal of attention, perhaps more than any of its other resource elements. Yet, overall knowledge of the soils is still rather scanty; there is a dearth of detailed studies as evident in the paucity of large-scale soil maps. The problem is partly historical; for long, soil studies in Africa suffered from sweeping generalizations and myths about the supposedly harsh and inhospitable nature of the African environment and the poverty of its resource-base including the soils. Such misconceptions led to an oversimplification of explanations of soil genesis and distribution on the continent.

This legacy of generalization and oversimplification persists in the present-day situation in which national and subregional soil maps are derived from small-scale soil maps of Africa or of the World rather than the other way round. The dearth of national soil maps which could be combined to produce regional or continent-wide maps is glaring evidence of the limited extent of detailed soil survey and mapping in African countries. The vast territories that have to be covered, the wide variety of landscapes and the complexity of the soil patterns have stretched the limited human and material resources available for soil survey and mapping in most countries. Indeed, much of what has been achieved in this regard, can be attributed largely to the technical-aid programmes enjoyed from the erstwhile colonial powers that ruled over the continent. In many cases, the reconnaissance and semi-detailed soil-survey and land-evaluation studies carried out by the technical-aid groups, such as, for example, the British Land Resources Development Centre (LRDC), provided the first, and still the only, systematic data and information on the environment of large sections of many African countries. Other groups included the French ORSTOM and the Belgian INEAC. In addition, United Nations agencies, particularly FAO and UNESCO, have provided tremendous resource-mapping and monitoring services to Africa in the areas of soil, land use, and forestry. The value of the land-evaluation studies is such that it is said that planners in some African countries would rather have them than strictly soil surveys that only result in the production of soil maps (see Woode, 1981). Information on the origin, morphology, and distribution of African soils has to be pieced together from the works of such resource-survey and land-research organizations, and also from the works of researchers in cognate fields especially geomorphology, agriculture, and soil science.

Soil Genesis

The discussion of African soils has almost always been, more or less, synonymous with the discussion of tropical soils. And, for a long time, the conception of these soils was unduly influenced by laterite (Kellogg, 1950). Although many tropical soils have no relations at all with true laterite, yet explanations of the origin, morphology, and distribution of laterite unwittingly came to dominate the discussions on African and tropical soils. According to Young (1976), there has probably been more written about laterite than about any other aspect of tropical soils.

However, an important lesson learnt from the obsession with laterite is that contemporary environmental conditions and processes alone cannot suffice to explain the nature and occurrence of tropical soils. Much present-day laterite is wholly or in part 'relict' (palaeosols) dating back to the Pleistocene or the Tertiary era. Soil genesis and distribution in Africa must be understood in the context of the history of landscape evolution, the past climatic changes, and the accompanying cycles of weathering and erosion that produced the sequence of erosional surfaces and landforms characteristic of the continent. In contrast to temperate zone soils which frequently go back only to the end of the Pleistocene, many tropical soils, with their great weathering horizons, are very much older, going back to the Tertiary era of about 60 million years ago.

Undoubtedly, it is in relation to his work on past climatic fluctuations and the reconstruction of past landscapes on the African continent that A. T. Grove's contribution to geomorphological research will be truly assessed (e.g. Grove, 1958, 1969, 1978; Grove and Pullan, 1964; Grove and Warren, 1968). Apart from soil erosion, his interest in African soils is mainly in relation to his exploration of the palaeoclimates and palaeoenvironments of the continent. He has shown keen interest in such 'relict' features as duricrusts and stonelines in soils and encouraged his students to research into them (e.g. Goudie, 1971). As is all too obvious in the works of Grove and others, there is much still to be unravelled or thoroughly understood about the evolution of Africa's landscapes, and even more so, about the formation of the soils. But, in this review, we shall attempt a simple synthesis of available knowledge on the soil-forming factors and processes in Africa.

Soil-Forming Factors

Climate

Although soil patterns in a number of places may be difficult to relate directly with present environmental conditions, there is a definable zonation in the distribution of the major soil groups which shows the unmistakable influence of climate. Langdale-Brown (1968) has noted that there is a marked similarity between the regional maps of the soils, vegetation, and rainfall of Africa. This shows that both the soils and vegetation of the continent are affected deeply by the climate. Since the landscape of most of Africa has remained structurally stable for millions of years, chemical weathering and soil development under hot and humid conditions have reached an advanced stage in many parts. Thus, Ahn (1970) has suggested that it is precisely in those characteristics of the soil which reflect climate and vegetation that (West) African soils are most likely to differ from those of other latitudes.

Rainfall and temperature are the two most important climatic influences on soils in Africa. To a large extent they determine the intensity of chemical and biochemical processes in the soil especially the 'richness of weathering' and the 'intensity of leaching'. As elsewhere, the balance between these two processes, weathering and leaching, broadly determines the extent of soil-profile development and the chemical and mineralogical composition of soils in different parts of the continent (Crompton, 1960).

However, there is a general agreement, that more than the regional climate, the factor that exercises the greater influence on soil-forming processes is the soil climate (see Mohr and van Baren, 1959). Unfortunately, soil climate has been little studied and analysed in Africa. It is true that some synoptic and agrometeorological weather stations keep records of soil temperature and soil moisture at different depths. But there has been very little attempt at soil climate–soil correlation studies.

Soil structural characteristics, especially the degree of particle aggregation and the proportion, size, and distribution of pore spaces, to a large extent determine soil climatic conditions and their influence on chemical processes and biological activity within the soils. Most tropical soils are believed to have a strong, stable structure which facilitates (1) the movement and diffusion of gases; (2) efficient heat flux; and (3) infiltration and free movement of water in the soil. These conditions promote high rates of microbial activity and efficient and complete decomposition and mineralization of organic matter. But, the same conditions also promote high rates of leaching in humid regions and seasons.

Young (1976) points to some broad correlations between climate and soil properties. For instance, topsoil organic matter and nitrogen contents increase with a fall in temperature and an increase in rainfall. This is because lower temperatures slow down the rate of loss of humus through oxidation while higher rainfall amounts slow down the rate of loss by the supply from the more abundant plant growth. There are inverse relations between rainfall and both pH and base saturation, since both properties are related to intensity of leaching.

The silt : clay ratio tends to decrease with increasing rainfall so that soils of rain-forest climates are frequently high in clay and very low in silt, whilst under the slower chemical weathering of semi-desert and desert soils, there are abundant sand- and silt-grade particles. There is also some relationship between clay-mineral synthesis and climate. Gibbsite is found only where the soil is subject to leaching throughout the year, that is in wet climates, whilst montmorillonite is absent in such conditions and is found only on freely drained sites where the soil

Plate 8.1. In the drier parts of Africa high levels of calcium carbonate or calcium sulphate may accumulate. In this example from southern Tunisia a columnar soil structure has developed in association with gypsum accumulation (photo: A. S. Goudie).

dries out seasonally. The absence of 2 : 1 lattice clays coupled with the presence of abundant free iron oxides is responsible for the very friable, crumb structure of the soils of the rain-forest zone (see Young, 1976: 8–9).

It is possible to recognize broad pedoclimatic zones in Africa in which broadly similar pedogenetic processes and closely related soil groups may be found. Such broad soil groups which show a marked correlation with climate and vegetation are referred to as zonal soils. Perhaps to illustrate the influence of climate and vegetation on soil formation and distribution in Africa, Vine (1966) grouped all the zonal soils of the tropics into two large orders: pedocals, and pedalfers.

The pedocals are the soils of the dry regions where the mean annual rainfall is generally less than 762 mm and the vegetation cover is sparse. The rainfall is so light that its leaching action is restricted to the top layers of the soil which are usually only a few centimetres thick. Thus, although weathering is not rich, calcareous materials ($CaCO_3$ and $CaSO_4$) tend to accumulate in the upper parts of the subsoils. (Plate 8.1).

The pedalfers are soils which have been formed under warm-to-hot and relatively humid climatic conditions and a substantial plant cover either of forest, woodland, or grass. They are generally intensely weathered and leached so that the more easily decomposed elements have long been broken down and washed away leaving the more inert and resistant compounds. For example, the soils have lost most of the combined silica leaving

a relative accumulation of iron and aluminium sesquioxides. The ratio of silica to sesquioxides in the soil matrix is sometimes used as an index of the balance of soil weathering and leaching. This ratio decreases with increasing rainfall intensity.

In spite of the obvious profound influence of climate on soil genesis, it is important to note, however, that parent materials, relief, and ground drainage play significant roles in determining the course and end-result of many soil-forming processes even where climate may be the dominant influence. Moreover, Africa has a wide variety of azonal and intrazonal soils which reflect local factors of relief, parent material, and ground drainage.

Parent Material

Unlike in the temperate region where it is often relatively easy to trace soils to their original parent rocks, it is not always easy to relate many tropical soils directly to the underlying rocks. This is due in part to the high degree of alteration of the soils relative to the underlying rocks. But, there is increasing mineralogical evidence to show that some of the soils were actually derived from rocks which previously overlay and differed from the present underlying solid rocks. Furthermore, and particularly in the humid tropics, a thick layer of regolith or weathered materials lies between the surface soils and the underlying rocks. Given the thickness of these weathered layers, it is conceivable that they are the result of the decomposition of rock layers which differed from the present underlying solid rocks. The chemical and mineralogical composition of these weathered layers cannot always be traced directly to the nature of the underlying rocks. Thus, as with the soils themselves, there is still a great deal to be learnt about soil parent materials in many parts of Africa.

Still, the influence of soil parent material on soil morphological characteristics and distribution is very glaring. Even, in respect of the zonal soils, it is usual to distinguish between acquired soil properties resulting from the climatic/ecological environment and the processes involved in soil formation and inherited or litho-logic properties imparted by the soil parent material. Such persistent inherited properties include stone content, texture, base content and, in certain circumstances, even soil colour and the type of clay minerals.

The importance of parent material as a fundamental soil-forming factor in Africa is clearly reflected in the fact that soil classification and mapping at the soil family or association level is based mainly on parent material lithology. Significant differences in soil morphology are related directly to differences in the mineralogical composition of the parent materials. The distinction is usually between quartzose acidic parent materials and more basic parent materials which are rich in the ferro-magnesian minerals.

The quartz-rich rocks are the most widespread parent materials and underlie the major zonal soil types of Africa. They include the granites, gneisses, migmatites, quartzites, sandstones, shale, and sand deposits of alluvial, aeolian, or marine origin. They give rise to soils with a high sand fraction and of relatively poor nutrient status, being deficient in calcium, magnesium, and iron and manganese. In such conditions of low base concentration in the soil solution, kaolinite is the major secondary mineral produced. The soils have little weatherable minerals, but rather, contain a high proportion of inert materials especially residual quartz. The poverty of these soils is further aggravated by the fact that, on the old erosion surfaces that characterize most parts of the continent, the parent materials probably have undergone several cycles of soil formation. The residues of such multiple pedogenesis cover large expanses of Africa and are among the poorest of parent materials (D'Hoore, 1964).

The basic rocks or rocks rich in the ferromagnesian minerals (e.g. amphibolite, gabbro, basalt) are much more localized in their occurrence and they have produced intrazonal soil types within the major climatic zones of Africa. Unlike in the case of the quartz-rich parent materials, the weathering environment of the basic rocks is rich in bases, thereby promoting the formation of 2 : 1 lattice clay minerals. The parent materials are relatively rich in weatherable minerals in the form of feldspars, hornblende, olivine, augite, and micas. The soil nutrient capital is continually replenished from these minerals, which explains the comparatively medium-to-high fertility status of the soils.

Between the basic and quartz-rich rocks, there are intermediate rock types with less silica content (55–66 per cent) and more of the ferromagnesian minerals than the granites, gneisses, and sandstones. Such rocks include schists, diorite, syenite, and biotite-hornblende gneisses. They give rise to soils of somewhat higher base status than the quartz-rich rocks in the same soil-forming environment.

Calcareous parent materials, including limestones, marble, and calcareous alluvium, also exist in certain areas where they have given rise to intrazonal soil types. The most important group of intrazonal soils developed on limestones are the tropical rendzinas (rendzic leptosols).

Relief

The significance of relief in understanding soil morphological characteristics and spatial distribution in Africa has long been recognized. In particular, the importance of the character and distribution of the various erosion surfaces has been highlighted by various authors including King (1951, 1962), D'Hoore (1964), Moss (1968), and Young (1976). Moss (1968) discussed their importance to pedogenesis from three perspectives: (1) the age of the soils in relation to the age of the surfaces; (2) the validity of the distinction made between surfaces of degradation and those of aggradation; and (3) the uniformity of soils on particular surfaces of like origin. Although some correlation may be made between soils and the surfaces on which they occur, field evidence suggests that this is not primarily a function of age and length of development. The surfaces are not of uniform age and since they take time to develop, the soils associated with them may be of widely differing ages. Most surfaces have undergone more than one cycle of erosion and present-day surfaces were not cut on virgin, unweathered rock but on materials which had been pre-weathered in previous cycles of erosion. Furthermore, most present-day surfaces are really not that old, in that they have been modified more recently by contemporary erosional processes.

However, where differences in age can be established, the sequences of erosion surfaces or of the development of an erosion surface may be useful in distinguishing between soils of differing intensities of weathering and leaching. Usually many different types of soils occur on the same erosion surface due to complexities introduced

Table 8.1 Suggested grouping of land surfaces according to general character

---

I Degradational surfaces
1. Surfaces with extensive development of hard laterite, e.g. Early Cenozoic (African) Surface. The occurrence of hard laterite implies other subsoil or parent materials in association with it.
2. Surfaces without extensive development of hard laterite, e.g. Late Cenozoic (post-African) Surface. Very variable, especially according to variations in parent rock, and past climates.

II Zones of transference
1. Montane zones: usually with high-level isolated fragments of older surfaces, sometimes with volcanic forms as well, e.g. Ethiopian Highlands, Kenya Highlands, Cameroon Highlands, Bauchi Plateau, etc.
2. Very dissected zones: with numerous river valleys, and a generally dense pattern of dissection, associated with generally relatively steep slopes, e.g. parts of the coastal area of West Africa, and also the area of Uganda immediately to the north of Lake Victoria.

III Aggradational surfaces
1. Surfaces with significant weathering and soil development, e.g. parts of the Eocene of the West African coastal zone.
2. Surfaces without significant modification:
   (a) Continental, e.g. the ergs and regs of the Sahara, and their fossil counterparts.
   (b) Littoral: Lagoon coasts, deltas, etc. e.g. Niger Delta, coasts of Dahomey and Togo, etc.
3. Inland swamps. Large areas of seasonally high groundwater away from the coast, e.g. inland delta of the Niger; raised swamps of Lake Victoria. The Kalahari salt pans might also be included in this group.
4. Floors of Rift Valleys. In East Africa only.

Source: R. P. Moss (1968).

by pre-weathering, variations in underlying geology, and the fact that the surface has developed over a period of time so that some parts are 'older' than others.

Moss also notes that the surfaces differ in character: there are zones of degradation (erosion) and zones of aggradation (deposition). But D'Hoore (1964) recognized a third type of surface when he noted that in Africa the relief consists of zones of departure (degradation), of transference (transportation), and of accumulation (aggradation). D'Hoore also noted that vast areas of tropical Africa which are flat or nearly flat do not belong to any of these three and they are covered with parent materials that have undergone more than one cycle of soil formation. The major areas of departure and transference are areas of marked relief, the mountains and the high plateaux and hill ranges. The major zones of aggradation are the lower erosion surfaces and the coastal or littoral belt of the continent. Moss (1968) has suggested a tentative classification scheme for these land surfaces according to their general character (Table 8.1).

Given the complex character of erosion surfaces, the multiple processes that have operated, the variety of underlying rocks, and the effect of pre-weathering, it should be clear that uniformity of soils on any particular surface is difficult to achieve. Hence, it is very difficult, except at a very general regional level, to make any direct correlation between the erosion surfaces and soil types. However, at the local level, it is easy to perceive direct relationships between soils and slopes. Milne's catena concept (1935) has stood the test of time such that, although it was initially developed as a mapping unit, it has since assumed a taxonomic status. This is an acknowledgement of the fact that there are distinct correlations between slope forms (and angles) and soil morphological characteristics (see Gerrard, 1981: 80–91). Soil formation is directly related to the processes of slope evolution or development.

The soil catena as a taxonomic unit received a big boost when the Australian CSIRO developed the concept of integrated surveys (Christian, 1959; Christian and Stewart, 1968). The land system and the *land facet*, both as conceptual and as landscape mapping units, are based on the assumed close relationship between slope processes and soil-forming processes and, consequently, soil morphology. The properties of the soil that are directly influenced by slope form and angle include soil depth, texture, stone content, moisture content, and occurrence of indurated layers or iron concretions. Slope may also influence other properties including colour, clay mineralogy, humus and nutrient status, and the content of weatherable minerals. The structural stability of the soil and, consequently, the risk of soil erosion also vary with slope.

Thus, slopes have been widely used in the rapid mapping of soils in Africa. However, the accuracy of such soil mapping depends on the ability to identify and demarcate distinct breaks of slope. Unfortunately, such distinct breaks of slope are very rare, especially on the low gentle and smooth relief which characterizes most of the continental interior of Africa, particularly the savanna lands. Because of the problem of boundary location in such terrains, in addition to inherent variability of soils, most slope units are really composite rather than simple homogeneous soil units. Although they often show marked homogeneity in soil physical properties, such soil-slope or pedomorphological units are highly variable in respect of the chemical characteristics (see Areola, 1982).

Still it is impossible to over-emphasize the value of slope form in understanding soil genesis, morphology, and spatial patterns in Africa. Slope determines the disposition of soil parent materials and hence the nature of soils at the local level. The vast areas of low, smooth relief generally have sedentary soil parent materials. But mass movement has resulted in a distinction being made between residual parent materials on the summits and brows of the interfluves and colluvial parent materials on the interfluve mid- and lower slopes. The soils on the summits are usually developed directly on the original parent rock or pre-weathered material and they are usually more clayey than the coarse, sometimes gravelly, slopewash soils on the side slopes. Movement of the soil solution is essentially vertical on the interfluve summits while lateral movement of the soil solution is a major process by which upper-slope soils are leached of their nutrients and colloids to enrich the soils lower down the slope. This lateral movement of the soil solution downslope, for instance, is crucial to the formation of the laterite horizon and the concentration and precipitation of iron concretions in the lower parts of the soil catena (see Prescott and Pendleton, 1952; Maignen, 1966).

Vegetation

The apparent correlation between the distribution of soils and vegetation in Africa is well documented. As Langdale-Brown (1968) has noted, the comparative geography of soils and vegetation is due both to the parallel effects of climate, parent materials, relief, and other living organisms and to interactions between soils and vegetation themselves. The vegetation creates a distinct microclimate at the soil surface within which soil-forming processes operate and are regulated.

But, the major influence of vegetation on soils relates to the supply of litter to form humus which transforms the layer of weathered rock minerals into a living soil. The supply of litter is largely dependent on the biomass of the vegetation. In this regard the tropical rain forest has the largest biomass, which is of the order of 300–900 tonnes per hectare (see Longman and Jenik, 1974). In comparison, for the woodland savanna it is about 60 tonnes per hectare; for the dry savanna it is 30 tonnes per hectare, while the semi-arid regions have less than 10 tonnes per hectare. The amounts of litter generated decrease correspondingly from the tropical rain forest to the arid regions. Litter production in the savanna depends very much on the proportion of wood species, which yield more than the herbaceous species. Unfortunately, bush burning annually destroys much plant biomass as well as litter which could have been incorporated in the soil.

Soil organic matter is comparatively higher under natural forest than under savanna vegetation. But much of the soil humus is concentrated in the surface layer and is rapidly exhausted once the forest is cleared. In the dry savanna, however, the soil organic matter is relatively more deeply and evenly distributed in the soil profile due to the deep and extensive root systems of the xeromorphic plants which decay *in situ* (Klinkenberg and Higgins, 1968).

Apart from the low litter supply, soils in the savanna may suffer another vegetation-related deprivation; certain types of grass vegetation tend to inhibit the mineralization of nitrogen and N-deficiency occurs in grain crops, particularly in the West African Guinea Savanna belt dominated by tall grass species, e.g. *Andropogon spp.* (Jones and Wild, 1975). On the other hand leguminous plants tend to enhance nitrogen-fixation, which considerably improves the fertility of the soils under them. *Acacia albida* and *Azadirachta indica* (neem) are two such leguminous species which are known markedly to improve soil fertility. In the forest and derived savanna regions, *Gliricidia sepium* appears to have a similar favourable effect on soil fertility.

Apart from the supply of litter and its soil protective function; vegetation plays other vital roles in soil formation. Plant roots play a major role in the movements of mineral elements up and down the soil profile and in the mechanical sorting of soil particles. Therefore, vegetation plays a major role in horizon differentiation. Root exudates provide acids and chelating agents which mobilize certain mineral compounds, for example, the sesquioxides of iron and aluminium.

Soil-Forming Processes

From the discussion of the soil-forming factors in Africa, it is perhaps easy to understand Moss's admonition that 'soil science in Africa has been bedevilled throughout its history by assumptions of uniformity, when an expectation of extreme complexity would be more appropriate' (Moss, 1968: 51). This complexity is also reflected in the soil-forming processes, most of which, are not yet quite fully understood. Again, discussions on soil formation in Africa were, for a long time, dominated by attempts at explaining the formation of laterite and the so-called 'lateritic soils'. This was particularly so in much of the French literature on the subject. The fact that the French territories extend mostly over the dry savannas, where laterite outcrops occur extensively on the surface, probably accounted for this obsession with laterite. Even so, there is still no agreement on the actual processes involved in the formation of laterite.

The major soil-forming regimes recognizable in Africa include ferrallitization, rubification, calcification, salinization, solonization, and gleying. The major difference between one and the other of these processes is in the relative strengths of weathering and leaching which determine the mineralogy and chemical composition of the resultant soils.

Ferrallitization is a soil process unique to the tropical environment and it is the dominant process in the formation of all the zonal soils of tropical Africa. It is promoted by the high temperatures throughout the year and the abundant, even if strongly seasonal, rainfall. Ferrallitization is basically an advanced stage of hydrolysis. Intense chemical weathering results in the complete breakdown of all weatherable primary minerals, except quartz, and the complete leaching of soluble salts and carbonates. The exchangeable bases also are strongly leached; most of the silica is also leached (referred to as 'desilication') and, even iron and aluminium oxides are partially leached. Ferrallitization leads to the gradual development of an indurated B horizon with abundant sesquioxides to an extent that the sesquioxide content may be up to 70–80 per cent of the total mass of the horizon (Knapp, 1979).

The process is accelerated by a number of conditions. First, a relatively low pH and low concentration of dissolved weathering products in the soil solution enhance desilication and enrichment of the soil with residual iron and aluminium. These conditions arise where plant roots and soil micro-organisms produce a lot of carbon dioxide and where there is a high rate of percolation of water through the soil. Secondly, although ferrallitization is a very slow process, the prolonged geomorphic stability of the old erosion surfaces of Africa, has provided sufficient time for the process to advance.

Finally, ferrallitization proceeds most strongly in basic parent materials which contain and release much iron and aluminium from easily weatherable minerals but which have little silica. Although kaolinite is the dominant clay mineral, gibbsite is also present. But, in acid parent materials, ferrallitization is much slower because there is more quartz and less weatherable minerals. In spite of the removal of silica through leaching (desilication), much still remains in the soil solution. Thus the silica combines with aluminium to form the 1 : 1 clay mineral, kaolinite (kaolinitization). Gibbsite is not normally present.

Hematite and goethite are two minerals that give the tropical soils their red and yellow colours. They are produced by the weathering of ferrihydrite ($Fe(OH)_3$). Hematite, which imparts the bright red colour to many tropical soils, is formed where (i) the iron concentration is high; (ii) organic matter content is low; (iii) soil pH is higher than 4; and (iv) temperatures are high thereby accelerating the dehydration of ferrihydrite and the decomposition of organic matter. Goethite, which imparts yellow or orange colours, is formed where one or more of these four conditions are not fully met.

Rubification (also referred to as ferrugination) is a form of *in situ* weathering of moderate to high intensity which produces the type of iron oxides that give the soils their brilliant red colour. The iron oxides form coatings on other soil constituents thereby imparting the reddish colour. Altogether the iron oxides stabilize the soil; and some may harden to form iron concretions. Rubification also results in the complete leaching of soluble salts and carbonates, moderate leaching of exchangeable bases, and partial leaching of silica. The dominant clay is also kaolinite but with some goethite and hematite. There may be small amounts of illite or montmorillonite but not gibbsite. According to Knapp, the ferruginous soils, may be thought of in many ways as the brown earths of the tropics because they are largely a response to the dominance of *in situ* weathering.

Calcification, which operates in the semi-arid areas, is characterized by weathering of moderate intensity and weak leaching. The soluble salts are completely leached but carbonates and bases are only moved from the upper horizons and reprecipitated in lower horizons, and distinct calcic horizons may be formed. There is little or no leaching of silica. The clay fraction contains kaolinite, illite, and montmorillonite. With increasing aridity, upward movement of the soil solution is encouraged by rapid evaporation on the soil surface. In such a situation, there is hardly any leaching; rather sodium and potassium salts are carried to the soil surface where they crystallize out. This is the process referred to as salinization. In extreme cases, sodium is so abundant that it replaces exchangeable calcium, a process known as alkalinization or solonization, which leads to the deflocculation and translocation of the clays. Although most of the soils in the arid lands of Africa are little more than layers of weathered materials with little or no real soil development (lithosols, regosols), there are pockets of soils dominated by the processes of calcification, salinization, and solonization.

Gleying is a soil-forming process associated with anaerobic conditions which is a feature of hydromorphic soils. It leads to the reduction and release of iron oxides and their translocation and reprecipitation as ferric iron oxides in the form of mottles and concretions.

Major Soil Groupings

A consequence of the diverse soil-survey and resource-mapping legacies that have been bequeathed to Africa is the multiplicity of soil classification schemes. The *Soil Map of Africa* produced by the Commission for Technical Cooperation in Africa (CCTA), was the first major effort to arrive at a comprehensive soil classification scheme and map legend for the whole continent by integrating

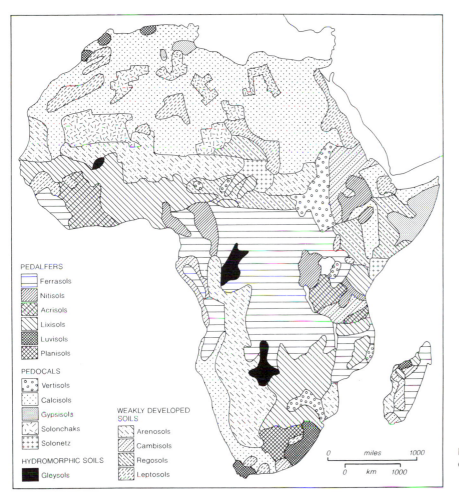

PEDALFERS
☐ Ferrasols
☐ Nitisols
☐ Acrisols
☐ Lixisols
☐ Luvisols
☐ Planisols

PEDOCALS
☐ Vertisols
☐ Calcisols
☐ Gypsisols
☐ Solonchaks
☐ Solonetz

HYDROMORPHIC SOILS
☐ Gleysols

WEAKLY DEVELOPED SOILS
☐ Arenosols
☐ Cambisols
☐ Regosols
☐ Leptosols

Fig. 8.1. The main soil types of Africa.

through 'successive approximations', the legends and mapping units from existing country and regional soil-survey organizations (D'Hoore, 1964). The soil legend and classification scheme of the CCTA Soil Map of Africa, is still perhaps the most widely known and used in Africa. But the FAO (FAO-UNESCO, 1974) has since introduced a more refined world soil-classification scheme which integrates African soil types. This FAO-UNESCO classification scheme was revised in 1988 and it is this revised scheme that will be applied in discussing the major types of soil in Africa.[1]

[1] For the description of the genesis, morphology and distribution of the soil types according to the FAO-UNESCO legend, the author is greatly indebted to Driesen, P. M. and Dudal, R. (1989) (eds.), *Lecture Notes on the Geography, Formation and Use of the Major Soils of the World*, co-produced by the Agricultural University, Wageningen, Netherlands, and the Faculty of Agricultural Sciences of the Catholic University of Leuven, Belgium.

Pedalfers

These, the largest group of soils in Africa, include the (i) ferrasols; (ii) nitisols; (iii) acrisols; (iv) lixisols; and (v) plinthosols in the FAO-UNESCO soil-classification scheme. They comprise with others, the group of soils referred to by Kellogg (1949) and Young (1976) as latosols, that is, all the 'red and yellow' soils of the tropics. Other pedalfers occur in the warm temperate and subtropical parts of Africa; they include the (vi) luvisols; (vii) planisols; and (viii) arenosols (discussed under 'weakly-developed soils'). Thus, with few exceptions, the pedalfers comprise all the zonal soils of Africa.

As shown in Figure 8.1, the ferralsols, occur in the central parts of the continent, traversing the equatorial forest/savanna transition, and savanna belts. They are red (hematite) and yellow (goethite) intensely weathered and leached soils with a high content of kaolinitic clay, iron, and aluminium oxides and hydroxides. These give the ferralsols a strong stable microstructure. They are found mostly developed on surfaces of low relief with deep pre-weathered materials derived from a wide variety of rocks.

Although the ferralsols have good physical properties they are chemically poor. Their low base content, virtual absence of weatherable minerals and very low cation-exchange capacity are major limitations. Liming and liberal applications of artificial chemical fertilizers are needed for productive arable farming.

Nitisols occur in areas with clearly defined wet-and-dry seasons although they occur also under more humid conditions. They occur in parts of south-eastern Nigeria, the Cameroon Highlands, and in upland areas of Ruanda–Burundi and Ethiopia. Nitisols, like ferralsols, are formed by the process of ferrallitization. But, the soil profile shows little horizon differentiation due to homogenization by termites, ants, worms, and other soil fauna. Clay translocation results in the development of a clay-enriched argic (Bt) horizon at depth. This process is probably responsible for the strong, angular blocky structure.

The clay fraction is dominated by kaolinite with small quantities of halloysite, illite, vermiculite, hematite, goethite, and gibbsite. The deep, uniformly red colour of these soils is due to the amorphous-as well as to the free-iron content. Compared to the other major soil groups, ferralsols, lixisols, and acrisols, nitisols have a relatively high cation-exchange capacity, and high organic matter content but variable base saturation.

Nitisols are among the more productive tropical soils. Their high porosity and deep solum promote deep rooting and their comparatively good chemical properties make them able to support both tree-crop and food-crop cultivation. They respond well to fertilizer applications. The strong, stable structure makes them less susceptible to erosion than most other soils.

Acrisols are the strongly weathered acid soil counterparts of the nitisols. They have an argic (Bt) horizon with low cation-exchange capacity and low base saturation. They are commonly developed on quartz-rich parent materials in areas of hilly or undulating topography and wet tropical and monsoonal climates. They occur mostly in West Africa (Ivory Coast, Liberia, Guinea) and the lakes region of East Africa.

The profile of most acrisols comprises a shallow A horizon with raw, acid organic matter overlying an incipient and yellowish E horizon and a strong and reddish coloured argic (Bt) horizon. Thus, although the dominant soil-forming process is ferrallitization, cheluviation is responsible for the subsequent redistribution of iron compounds to form the (E) and Bt horizons.

As to be expected, acrisols have a predominance of low activity clays (kaolinite with some gibbsite) and very low content of bases and weatherable minerals. The contents of iron, aluminium and titanium oxides are only slightly lower than those of the ferralsols. Acrisols have a weak microstructure and they are highly susceptible to erosion when the protective cover of vegetation is removed. The productivity of the soils depends on the maintenance of the thin surface layer and its organic matter. Modern agriculture on the soils requires intensive liming and application of chemical fertilizers.

Lixisols are the soils of the savanna and semi-arid regions. They are strongly weathered soils with a clay-enriched argic (Bt) horizon. Thus, they have mature ABtC profiles developed on the high plains with pre-weathered parent materials. It is believed that the lixisols are polygenetic soils with characteristics formed under a more humid climate in the past. One evidence for this is that relict plinthite and coarse reddish iron mottles or indurated iron nodules occur in the subsurface layers of some lixisols.

Lixisols cover vast areas in West Africa, East Africa and East-Central Africa (Malawi, Zimbabwe, Zambia, etc.). The soils generally have a reddish or, sometimes, yellowish subsoil, the product of rubification.

The soils have a predominance of kaolinite clays and high contents of iron, aluminium, and titanium oxides. Although the levels of available nutrients and mineral reserves are low, lixisols have generally better chemical characteristics than the ferralsols or acrisols because of their high soil pH and the absence of serious aluminium toxicity. The surface soils have a low aggregate stability and are prone to erosion if the protective cover of vegetation is removed. The *argic* horizon also may harden to form a claypan. But the soils are amenable to management.

Plinthosols (or soils with laterite) are soils having 25 per cent or more plinthite by volume in a distinct horizon that is at least 15 cm thick within 125 cm of the soil. Plinthite is formed in level to gently sloping terrains characterized by a fluctuating water-table. It is an indurated, mottled clayey material which hardens up upon exposure to the air (see Mohr and van Baren, 1959). A continuous ironstone layer (petroferric phase) is formed where the plinthite becomes exposed to the surface as in the old erosion surfaces of the interior savanna lands of Africa where many laterite-capped hills occur, especially on sandstone formations (see Buringh and Sombroek, 1971). In other areas hardened plinthite occurs in a concretionary layer that is not continuously cemented (skeletic phase); this is common in colluvial and alluvial materials at the footslopes and valley shoulders.

Plinthite or soft laterite occurs in the rain-forest areas, whilst hardened plinthite (petroferric and skeletic phases) occurs more commonly in the savanna regions with marked wet-and-dry seasons. The accumulation of the sesquioxides arises from the process of ferralitization (relative accumulation) and through enrichment from outside sources by both lateral and vertical seepage (absolute accumulation). The segregation of the iron mottles occurs at the zone of fluctuating water-table. The iron, in the ferrous form is mobilized and brought in place during the period of water saturation; this iron is precipitated as ferric oxide when the water-table falls. This ferric oxide will not be or, at best, is only partially, redissolved in the subsequent wet season.

Mineralogically, plinthite and ironstones have a high content of free iron oxides (lepidocrocite, goethite, and hematite), free aluminium oxides (gibbsite, boehmite) and quartz inherited from the parent material. The dominant clay mineral is kaolinite.

In the field, plinthite presents itself in various forms:

(a) The massive or continuous ironstone layer (petroferric phase) may be residual, where it is formed by the hardening of a continuous layer of plinthite; or recemented, where colluvial ironstone gravel, stones, and boulders are cemented together. Residual massive ironstone layers may be vesicular, with irregular cavities or ferruginized where the original rock structure is still visible.

(b) Discontinuous ironstone layers (skeletic phase) are either residual consisting of rounded or irregular nodules or concretions (pisolithes or pea iron); or colluvial, that is, ironstone gravel, stones, and boulders formed by the disintegration of massive ironstone layers.

The luvisols are soils in which clay has been washed down from the surface soil (lessivé) to accumulate in a distinct argic (Bt) horizon. They are found in the Mediterranean regions of North and South Africa on gently sloping lowlands abutting the highlands (Atlas Mts in the North and the Drakensberg Mts in the South). These are areas with distinct wet-and-dry seasons and a variable vegetation cover of deciduous forests, coniferous forests, and grasslands. Wide differences in parent material and environmental conditions are responsible for the considerable variation among luvisols.

Luvisols are generally only moderately weathered and leached and they have a relatively high base saturation. They contain less aluminium, iron, and titanium oxides than their tropical counterparts, the lixisols. They have SiO_2/Al_2O_3 ratio of more than 2. In the warmer environments, a calcic horizon may be present in the luvisol profile. The soils are generally fertile and have high

weatherable mineral reserves. Their favourable physical properties include the granular-to-crumbly structure of the surface soils which are also porous and well-aerated. A high silt content, however, may render luvisols prone to erosion especially on steep slopes and if the soils are tilled in wet condition and/or with heavy machinery.

Planosols occur mainly in the high veldt of South Africa, typically in seasonally waterlogged flat lands or gently undulating terrains. There are pockets of planosols also in the Sahel zone of West Africa. They have a characteristic eluvial E horizon overlying a distinct mottled and clay-enriched argic horizon. In very wet locations, an organic horizon may develop on the surface; but normally the surface layer is a mineral-soil layer with soft iron concretions.

The seasonal waterlogging of the soils causes them to undergo the process known as ferrolysis (Brinkman, 1979), an oxidation-reduction sequence set in motion by the chemical energy derived from the bacterial decomposition of soil organic matter. Ferrolysis leads to a sharp reduction in the clay content of the upper layers of the soils; these layers abruptly overlie a deeper horizon with considerably more clay. This abrupt change in clay content may also entail a change in the nature of the clay minerals.

Clay destruction through ferrolysis leads to lower cation-exchange capacity and a lower moisture retention capacity in the topsoil. The topsoil also has low structural stability. Finally, planosols are subject to waterlogging during the wet season when rain-water stagnates on level terrains. Planosols are capable of supporting land use of moderate intensity; indeed large areas are used for extensive grazing as in the high veldt of South Africa.

Pedocals

The pedocals of Africa include the (i) vertisols; (ii) calcisols; and (iii) solonchaks. While vertisols are mineral soils conditioned mainly by the parent material, the calcisols and solonchaks are conditioned mainly by the semi-arid-to-arid climate.

Vertisols are found in the Sahel zone at the southern borders of the Sahara Desert. They are associated with former lake floors and river floodplains with smectitic clays. About 8000–12 000 years BP climate was more humid than at present. Then, lakes in the Sahara, notably Lake Chad, as well as the inland delta of the Niger in Mali, and the alluvial plains of the Nile in present-day Sudan, extended over much wider areas. Later in the Holocene, about 5000 BP, the climate became drier again; lake-levels subsided and the river floodplains shrank (see Grove, 1959; Grove and Warren, 1968; Goudie, this volume). Alluvial and lacustrine deposits on these former lake floors and river floodplains form the parent materials of present-day vertisols.

Vertisols are heavy clay soils in which alternate swelling and shrinking of the clay produces deep cracks during the dry season, as well as slickensides and wedgelike structural elements in the subsoils. Furthermore, gilgai microrelief occurs in many areas with vertisols. Gilgai relief consists of low mounds and shallow depressions; this microrelief form is produced by soil heaving resulting from the swelling of the clay when wet. The commonest form is the normal or round gilgai found on level terrains; slightly sloping terrains produce wavy or linear gilgai while very slight slopes have the transitional form, the lattice gilgai.

Vertisols have a high cation-exchange capacity and a high base saturation. Soil reaction varies from weakly acid to weakly alkaline, pH values being from 6.0 to 8.0. Salinity develops in vertisols in the more arid areas and also in areas under irrigation. Sodicity of vertisols also occurs in the arid areas and in enclosed depressions in more humid areas. The sodium-clays arising from sodicity have greater tensile and shear strengths than calcium-clays; it is believed that clay dispersion is prevented by the seasonal saturation with water.

Areas of vertisols are cultivated under both rainfed and irrigated agriculture in some parts of Africa. Depressions with dark-coloured, heavy clay soils, known locally as *firki*, are commonly cultivated in the Chad Basin by the local farmers. Modern large-scale irrigation has been introduced also in the South Chad Irrigation Project in Nigeria. But, perhaps the best known irrigation project in areas of vertisols is the Ghezira–Manaqil project in the Sudan Republic. The level plains characteristic of the areas of vertisols make land reclamation and mechanical cultivation relatively easy. However, large areas of vertisols remain unused or only used for grazing and firewood exploitation. The local people find the highly clayey soils rather heavy and difficult to cultivate.

Calcisols occur in a wide range of landforms in arid regions, including pediments, alluvial fans, lake bottoms, and terraces. The parent materials are more often than not base-rich weathered materials. These soils are characterized by the presence, in their profiles, of a distinct layer of calcium carbonate accumulation (calcic horizon) within 125 cm of the soil surface. Some profiles have a clay-enriched argic B horizon pregnated with calcium carbonate. Precipitation of calcium carbonate in root channels, worm-holes, and cracks within the soils produces the pockets of soft powdery lime characteristic of some profiles. Still, in others, the calcium

carbonate accumulation is in the form of hard concretions referred to as calcrete (Goudie, 1973). Precipitation of calcrete is probably at the zone of fluctuating water-table or at the depth where the capillary rise of water from the water-table evaporates.

Calcisols occur extensively in the Sahara and Namibian Deserts and in the Mediterranean regions of North and South Africa. Many of these calcisols are old soils believed to be polygenetic with their formation taking different courses as climates changed from one geological period to another. The two critical factors involved in the dissolution of calcium carbonate and its subsequent accumulation in the calcic horizon are the carbon dioxide pressure of the soil air and the concentration of dissolved ions in the soil solution.

Calcisols are potentially fertile soils; although the soil organic level may be no more than 1 or 3 per cent, the soils are rich in nitrate and mineral nutrients. The exchange complex is usually completely saturated with bases with calcium and magnesium making up more than 90 per cent of all adsorbed cations.

Most of the areas of calcisols are used for grazing but there are also areas of rainfed arable cultivation. The full potential of the calcisols can be realized only under irrigation.

Solonchaks are the soils whose properties are dominated by the presence of free salts. They are intrazonal soils of depressions in arid regions which are either seasonally or permanently waterlogged. Thus, they are associated with inland river basins, ancient lake bottoms, and depressions in intermontane areas. They also occur in coastal areas elsewhere. The soils show gleyic properties at depth. The zone of salt accumulation depends on the depth of the water-table; low-lying areas with a shallow water-table have salt accumulation in the top few centimetres of the soil while areas of deep ground water-table have the greatest salt concentration at some depth below the surface.

The formation of solonchaks is due to the high rate of evaporation relative to precipitation. Salts dissolved in the soil solution crystallize after evaporation of the water and accumulate at the soil surface (external solonchaks) or at some depth (internal solonchaks). However, the bulk of the salts that accumulate in any particular area is imported from elsewhere, carried by rivers, surface runoff, or by seepage water. Seepage water is responsible also for bringing up salts from the underlying geological rock beds to the soil surface.

The high salt content strongly influences the structure of solonchaks. When wet, as in the early morning hours, the surface soil is a muddy mixture of salt and soil particles; but on drying out later in the day, the surface soil becomes a hard crust. The subsoil structure is often stronger especially in heavy clays. Strong peptization of the clays may make the soil virtually impermeable to water.

Solonchaks are characterized by the extremely high electrical conductivity. However, the quantities and the composition of the salts vary among solonchaks, giving rise to different types of saline soils. The major varieties are the sulphate, chloride, and soda types, and their different combinations. Salt accumulation affects plant growth directly and indirectly in three ways: (1) microbial activity is depressed and ceases entirely in soils with 3 per cent salt or more; (2) the composition of the soil solution is skewed thus upsetting the availability of vital nutrients to plants; and (3) high osmotic pressure of the soil moisture causes physiological drought. In severe situations the vegetation may be reduced to scattered stands of halophytic shrubs, herbs, and grasses. The imbalances in the nutrient composition of the soil have to be corrected before solonchaks can be successfully cultivated. This requires the addition of the missing elements (e.g. addition of gypsum) as irrigation cannot ordinarily desalinize the soil.

Hydromorphic Soils

The hydromorphic soils are represented in Africa by the gleysols and fluvisols.

The gleysols occur extensively in the inland Niger delta, the Niger Delta, the Congo Basin and interior parts of Angola. They are associated with depressed, low-lying areas with shallow groundwater. They are distinguished from fluvisols in the sense that they are not developed on alluvial parent materials but on recent unconsolidated colluvial deposits of variable mineralogy.

The dominant factor in the genesis and morphology of gleysols is the waterlogging at shallow depth in some period of the year or throughout the year. The presence of the dissolved products of organic matter decomposition in the water and the low redox conditions result in the reduction of ferric iron compounds to ferrous compounds, giving the subsoil layers of gleysols their grey, olive, or blue matrix colours. Subsequent oxidation of transferred ferrous compounds back to ferric iron oxides takes place in cracks, fissures, and dead-root channels in the soil. The subsoil thus develops a pattern of mottles around such air pockets. These characteristic gley colours occur in the zone of fluctuating water-table. Hence, these gleyic soil properties are associated strictly with the movements of the groundwater-table; the mottled horizons occur above a fully reduced subsoil which lies below the permanent water-table. These soils, thus differ from soils with 'stagnic properties' (e.g.

luvisols, planosols), that is soils affected by perched water-tables in which a reduced horizon occurs on top of an oxidized subsoil horizon.

Gleysols usually have a distinct litter layer overlying a dark, humic mineral surface soil. Below is a mottled grey or olive Bg horizon which grades into a fully gleyed grey, olive, or blue Cr horizon. Heavy clay gleysols develop prominent blocky structure, and, in the savanna, such heavy clay gleysols usually have very dark, humic surface layers.

The main constraint to the cultivation of gleysols is the waterlogging which necessitates the installation of a drainage system. But, in Africa such soils in the *fadama* and valley bottoms are used for wetland cultivation and to raise early crops of yam, vegetables, and also for the cultivation of rice and sugar-cane.

Fluvisols are alluvial soils which occur in seasonally flooded alluvial plains, valleys, and tidal marshes. They are developed on fluvial, lacustrine, or marine deposits mostly of recent origin. They occur in the Nile and Zambezi deltas, West African coasts, and the Lake Chad plains. Apart from a distinct humic top layer, there is little horizon differentiation. They are predominantly brown where the soils are well-aerated, or grey in water-logged areas. They vary in texture from coarse sand at river banks to heavy clays in the backswamps. Alternating reduction and oxidation conditions result in the mottling of the soil profiles.

Most fluvisols are fertile soils with neutral or near-neutral soil pH values which do not impede nutrient availability to plants. Some coastal fluvisols may be saturated with sodium and this may constitute a problem. But others usually have some calcium carbonate derived from shelly marine sediments and the exchange complex is saturated with base elements from the sea water.

A distinct group of fluvisols is that of the thionic fluvisols or acid sulphate soils, found in the coastal rice-growing areas of West Africa (Sierra Leone–Senegal). Thionic fluvisols differ from other fluvisols in that the parent materials have pyrite (FeS_2) present in them. A distinction is usually made between potential acid sulphate soils which are not yet oxidized but contain pyrite in the soil material; and actual acid sulphate soils which are oxidized and acidified. The thionic fluvisols of the West African coast are potential acid sulphate soils which occur mostly in areas of tidal flushing (see Isirimah and Ojanuga, 1987). Draining is crucial to the reclamation and cultivation of the potential acid sulphate soils. The acidity that results can be leached out with saline or brackish water as has been done successfully in Sierra Leone. This approach, though expensive, solves the problem once and for all. Much water is needed to achieve complete leaching. An environmental problem associated with the reclamation of potential acid sulphate soils is that the drain-water is highly acidic and may pollute the environment elsewhere.

Azonal Soils

Weakly developed or immature soils occupy large parts of Africa. They include the (i) arenosols; (ii) regosols; and (iii) leptosols.

Arenosols cover millions of hectares of land in the semi-arid zones of the southern Sahara and south-west Africa. These are areas of shifting sand dunes and soil formation in such areas has been impeded by the instability of the landscape; soil formation can only begin when sufficient vegetation has colonized and stabilized the sand dunes. The sand grains are often coated with iron oxides (goethite) giving the soils a deep red colour. In some places the coating is of brownish clay, carbonates, or gypsum.

Arenosols also occur in the humid tropical parts of Africa on beach ridges and coastal plains. The soils have a distinct top layer with humus and abundant iron oxides overlying a deep, uniformly grey subsoil showing little or no horizon differentiation other than mottling of the upper parts. These humid tropical arenosols are highly leached in consequence of which they are very deficient in base elements. Indeed, a distinct highly leached eluvial horizon (E horizon) may be found below the top humic layer of humid tropical arenosols. The soils are highly susceptible to erosion because of their very low structural stability.

Arenosols in the arid zones are used mainly for grazing but dry farming is also possible especially on those rich in gypsum and calcareous materials. However, the high rates of infiltration make irrigation virtually impracticable.

Regosols are developed on unconsolidated weathered materials, usually sand, and they have no diagnostic horizon other than a dark-coloured top layer with abundant iron oxides. They occur in level or rolling topography in the upland areas of the arid zones of Africa extending from West Africa to Ethiopia and Somalia.

Regosols are susceptible to erosion; whilst the lower water-holding capacity and high permeability also make them highly prone to drought. Their irrigation-water need is very high and this makes irrigation uneconomic in most places. The areas of regosols in Africa are used mainly for nomadic grazing.

Leptosols are weakly developed soils which are limited in depth by the presence of continuous hard rock (lithic leptosols) or highly calcareous material (rendzic leptosols) or a continuous cemented layer such as hard laterite (petroferric phase). Leptosols are usually associated with upland areas with strongly dissected or undulating topography such as in North Africa, the Sahara, South Africa, Central, and East Africa.

Leptosols are basically young soils with a distinct A horizon, with a weakly developed, or incipient (B) horizon between it and the parent material. But the profiles are better developed on limestones (rendzic leptosols) and they have generally better physical and chemical properties than the non-calcareous leptosols.

The non-calcareous leptosols vary widely in character depending on the nature of the parent material. They are generally shallow and stony and susceptible to erosion and drought. However, cultivation has been carried out successfully in some areas through terracing. Otherwise, the areas of leptosols are devoted to transhumance, forestry, and tourism.

Soil Use and Management

In universities, research institutes, and agricultural development projects all over Africa, efforts are being made to increase knowledge on the use and management of the different soils. Thus, there is a growing body of detailed, technical knowledge on the soils under use. It is true that the impact of these research efforts has not been felt as it should be, for (as noted by Moutappa, 1973) although there has been a great deal of research and experimental work on soil management and fertilizer use under sustained crop production in both East and West Africa, the results of these studies are seldom identifiable with specific soil and land characteristics. There is still some vestige of the past habit of making sweeping generalizations about African soils even in today's research reports on soil; from studies carried out in one or two spots, generalizations are made about 'African' or 'tropical' soils in general! But, there is no doubt that modern soil science in Africa has succeeded somewhat in changing the old prejudices about the presumed poverty of African soils (see Gourou, 1966; Manshard, 1974). African soils, like those of other continents, are amenable to management with a thorough understanding of their characteristics, their environmental settings and the cultural milieux in which they are being utilized (see Jones and Wild, 1975).

Soil research and agricultural promotional efforts are geared towards combating the related problems of soil fertility deterioration, soil desiccation or the irreversible drying out and hardening of soils; and accelerated soil erosion. Soil erosion is discussed in detail in Chapter 18, and wider issues of land degradation in Chapter 19. Unlike in the past, soil and land management systems are being developed on the understanding that peasant land-holdings and peasant husbandmen, and not Western-style large-scale holdings and land users, will continue to dominate the African rural economy for the foreseeable future.

Research on the maintenance of soil fertility and productivity focuses on the preservation of topsoils and their organic matter levels and buffering capacities, as well as on the integration of the application of artificial chemical fertilizers into traditional crop-management systems (see Landon, 1984; Mortimore, 1989). The need to minimize soil loss during cultivation has led to research into tillage methods, such as, for example, minimum tillage (Lal, 1974, 1980) and new cropping systems such as alley cropping. Cover crops, mulching, and supplementary irrigation techniques are being introduced not only to combat soil erosion but also to protect the soils against desiccation (Lal, 1975).

References

Ahn, P. M. (1970), *West African Soils* (London).

Areola, O. (1982), 'Soil Variability within Land Facets in Areas of Low, Smooth Relief: A Case Study on the Gwagwa Plains, Nigeria', *Soil Survey and Land Evaluation*, 2/1, 9–13.

Brinkman, R. (1979), 'Ferrolysis: A Soil-Forming Process in Hydromorphic Conditions', Ph.D. thesis (Wageningen, Neth.).

Christian, C. S. (1958), 'The Concept of Land Units and Land Systems', *Proceedings, 9th Pacific Science Congress*, 20, 74–81.

—— and Stewart, G. A. (1968), 'Methodology of Integrated Surveys', *UNESCO Natural Resources Research*, 6: 233–80.

Crompton, E. (1960), 'The Significance of the Weathering/Leaching Ratio in the Differentiation of Major Soil Groups, with Particular Reference to some Very Strongly Leached Brown Earths on the Hills of Britain', *Transactions, 7th International Congress of Soil Science*, 4: 406–11.

D'Hoore, J. (1964), *Soil Map of Africa: Explanatory Monograph*, CCTA Publication 93 (Lagos).

Driessen, P. M. and Duda, R. (1989) (eds.), *Lecture Notes on the Geography, Formation, Properties and Use of the Major Soils of the World* (Wageningen, Neth. and Leuven, Belg.).

FAO-UNESCO (1974), *FAO-UNESCO Soil Map of the World*, i. *Legend* (Paris).

Gerrard, A. J. (1981), *Soils and Landforms: An Integration of Geomorphology and Pedology* (London).

Goudie, A. S. (1971), 'Calcrete as a Component of Semi-Arid Landscapes', Ph.D. thesis (Cambridge).

—— (1973), *Duricrusts in Tropical and Subtropical Landscapes* (Oxford).

Gourou, P. (1966), *The Tropical World*, 4th edn. (London).

Grove, A. T. (1958), 'The Ancient Erg of Hausaland, and Similar Formations on the South Side of the Sahara', *Geographical Journal*, 124: 528–33.

—— (1959), 'A Note on the Former Extent of Lake Chad', *Geographical Journal*, 82: 465.

—— (1969), 'Landforms and Climatic Change in the Kalahari and Ngamiland', *Geographical Journal*, 135: 192–212.

—— (1978), 'Geographical Introduction to the Sahel', *Geographical Journal*, 144: 407–15.

—— and Pullan, R. A. (1964), 'Some Aspects of the Pleistocene Palaeogeography of the Chad Basin', *Samaru Miscellaneous Paper*, No. 3.

—— and Warren, A. (1968), 'Quaternary Landforms and Climate on the South Side of the Sahara', *Geographical Journal*, 134: 194–208.

Isirimah, N. O. and Ojanuga, A. G. (1987), 'Potential Acid Sulphate Peats of the Niger Delta Wetlands', *Soil Survey and Land Evaluation*, 7/3: 147–56.

Jones, M. J. and Wild, A. (1975), 'Soils of the West African Savanna', *Technical Communication, no. 55, Commonwealth Bureau of Soils*.

Kellogg, C. E. (1949), 'Preliminary Suggestions for the Classification and Nomenclature of Great Soil Groups in Tropical and Equatorial Regions', *Technical Communication, no. 46, Commonwealth Bureau of Soils*: 76–85.

—— (1950), 'Tropical Soils', *Transactions, 4th International Congress of Soil Science*, 266–76.

King, L. C. (1951), *South African Scenery* (Edinburgh).

—— (1962), *Morphology of the Earth* (Edinburgh).

Klinkenberg, K. and Higgins, G. M. (1968), 'An Outline of Northern Nigerian Soils', *Nigerian Journal of Science*, 2/2: 91–115.

Knapp, B. (1979), *Soil Processes* (London).

Lal, R. (1974), No-tillage Effects on Soil Properties and Maize Production in Western Nigeria', *Plant and Soil*, 40: 321–31.

—— (1980), 'Soil Conservation: Preventive and Control Measures', in R. P. C. Morgan (ed.), *Soil Conservation: Problems and Prospects* (Chichester), ch. 13, 175–81.

Landon, J. R. (1984) (ed.), *Booker Tropical Soil Manual* (New York).

Langdale-Brown, I. (1968), 'The Relationship between Soils and Vegetation', in R. P. Moss (ed.), *The Soil Resources of Tropical Africa* (Cambridge), ch. 3, 61–74.

Longman, K. A. and Jenik, J. (1974), *Tropical Forest and Its Environment* (London).

Manshard, W. (1974), *Tropical Agriculture* (trans. D. A. M. Naylong) (London).

Milne, G. (1935), Composite Units for the Mapping of Complex Soil Associations, *Transactions, 3rd International Congress of Soil Science*, 1: 345–7.

Mohr, E. C. and van Baren, F. A. (1959), *Tropical Soils* (The Hague and Bandung).

Mortimore, M. (1989), 'The Causes, Nature and Rate of Soil Degradation in the Northernmost States of Nigeria and An Assessment of the Role of Fertilizers in Counteracting the Processes of Degradation', World Bank Environment Working Paper No. 17.

Moss, R. P. (1968) (ed.), 'Soils, Slopes and Surfaces in Tropical Africa', in R. P. Moss (ed.), *The Soil Resources of Tropical Africa* (Cambridge), 29–60.

Moutappa, F. (1973), 'Soil Aspects in the Practice of Shifting Cultivation in Africa and the Need for a Common Approach to Soil and Land Resources Evaluation', *Draft Report on FAO/SIDA/ARCN Regional Seminar on Shifting Cultivation and Soil Conservation in Africa. Ibadan, 2–21 July 1973*, 53–64.

Naignen, R. (1966), 'Review of Research on Laterites', *Natural Resources Research* (UNESCO), 4.

Prescott, J. A. and Pendleton, R. L. (1952), *Laterite and Lateritic Soils* (Farnham Royal, Bucks.).

Vine, H. (1966), 'Tropical Soils', in: C. C. Webster and P. N. Wilson (eds.), *Agriculture in the Tropics* (London), ch. 2, 28–67.

Woode, P. R. (1981), 'We Don't Want Soil Maps. Just Give us Land Capability: The Role of Land Capability Surveys in Zambia', *Soil Survey and Land Evaluation*, 1/1: 2–5.

Young, A. (1976), *Tropical Soils and Soil Survey* (Cambridge).

9 The Geomorphology of the Seasonal Tropics

Andrew S. Goudie

Introduction

Various aspects of the geomorphology of the African continent are dealt with elsewhere in this book. Chapter 3, for example, discusses some of the geomorphological implications of long-term climatic changes, including changes in dune and lake environments. Chapter 1 looks at the geomorphology of Africa as a whole in terms of its tectonic and denudational history and draws attention to the origin of the great plainlands of Africa and some of the continent's unusual drainage patterns. Other chapters cover such matters as the geomorphology of African deserts, the nature and origin of the great rift structures of East Africa, the form of the African coastline, the development of African mountains and lakes, and the accelerated erosion of African soils. The purpose of this chapter is to look at some of the features and processes that contribute to the distinctive form of the great seasonally humid plains and basins that cover vast tracts of the African interior.

The features that will be considered are the plains themselves, inselbergs, deep weathering profiles and duricrusts, colluvial sediments and stone lines, dambos, and pans. The processes that will be considered are two that have a particular role in sediment mobilization—termites and fire. More general reviews of these forms and processes have recently been given by Thomas (1994).

Plains and Inselbergs

To many observers, plains from which rise isolated hills or groups of hills—inselbergs, bornhardts, tors, or koppjes—are one of the most striking and pervasive landform types in Africa. This is partly because of the widespread occurrence of rock types that are conducive to their formation—especially granitoid rocks—and to the long-term denudational history of the continent (see Chapter 1). The isolated hills or clusters of hills display an enormous range of forms and sizes from large mountains like the Brandberg of Namibia to small scatters of corestones. They have been the subject of very considerable controversy as to origin (see Thomas, 1974; Faniran and Jeye, 1983; and Pugh, 1966 for reviews).

One view is that deep weathering is crucial to inselberg formation (see, for example, Thomas, 1966). This was a proposal that was put forward as early as 1911 by Falconer (1911: 246), who worked in Nigeria:

A plane surface of granite and gneiss subjected to long continued weathering at base level would be decomposed to unequal depths, mainly according to the composition and texture of the various rocks. When elevation and erosion ensued the weathered crust would be removed, and an irregular surface would be produced from which the more resistant rocks would project . . . a repetition of these conditions of formation would give rise to the accentuation of earlier domes and kopjes and the formation of others at lower level.

This is a model that is similar to the ideas of Wayland (1934) on etchplain development and of the German School's idea of divergent weathering and erosion (Büdel, 1982). By contrast, King (1966: 97–8) maintained with vigour that weathering has little to do with inselberg formation, arguing that these features form during cycles of pedimentation or during dissection into hard rock. He contended that the great height of many inselbergs precludes a fundamental role for deep weathering penetration and that in some inselberg landscapes there is an absence of evidence for deep weathering:

. . . deep weathering as a pre-requisite is neither necessary nor in most cases likely. On the contrary, apart from a few minor

Plate 9.1. An inselberg in north-west Zimbabwe developed in a massive, sparsely jointed granite (photo: A. S. Goudie).

features such as perched summit blocks and rocking stones, the landforms in the major bornhardt fields of Central Africa and Brazil are determined not by deep weathering from a former summit plain, but by stream incision along joint lines between walls of impressively solid rock sometimes narrowly pedimented across rock at the base. Air photos show how remarkably the pattern of dissection by streams follows the joint system in the bed rock. In the field the relevant joints show in stream beds across bare rock, without pre-weathering.

The height of bornhardts, sometimes 1000–1500 feet consonant with the measure of relief between the earlier and current base-levels, surpasses any known depth of weathering in such rock types. Below 300 feet depth from surface, joint systems are normally found, e.g. in mining or in boring for water, to be tightly closed and cores recovered from such depths show the walls of joints to be ironstained but not significantly weathered . . .

Bornhardts are part of the lower cyclic landscape in which they are found, with the strength of new features, not the senility of old.

I have climbed and ruminated upon too many great bornhardts, in company with the leopard and the baboon, to believe that these most powerful of landforms, glorious in the sun and rain alike, ever originated foetally within the dark body of the earth. The leopard and the baboon don't believe so either.

It is not inconceivable that these landscapes can originate through both mechanisms (see Selby, 1977) and Brook (1978: 159) has proposed a compromise between the two theories:

Stream incision and scarp retreat in solid rock (pediplanation theory) and stream incision and scarp retreat or slope flattening in weathered rock (exhumation theory) are extremes of a wide range of conditions that exist in nature. It seems reasonable to suppose that if inselbergs can form under these extreme conditions they can also form by stream incision and slope evolution in any mixture of solid and weathered rock as long as some solid rock is present. Futhermore, by the exhumation theory chemical weathering is the dominant process in the molding of inselberg landscapes with subaerial pediplanation in rock playing a late and minor role. By the pediplanation theory, scarp retreat in solid rock is the dominant process and chemical weathering merely etches the final landsurface. Again these are extremes of a wide range of existing conditions all of which are probably capable of producing inselbergs.

What is certain is that lithological and structural controls are central in determining the form, distribution, and origin of inselbergs of all types. Jointing and fractures are significant, with inselbergs tending to be composed of massive and monolithic compartments of rock which are in compression (Plate 9.1) (Twidale and Bourne, 1978). Also important may be particular lithologies (see, for example, Jeje's 1973 work in Nigeria or Gibbons' 1981 work in Swaziland), while other workers have pointed to the preferential resistance afforded by potash metasomatism (e.g. Brook, 1978 on South Africa; Pye, Goudie, and Thomas, 1984 on Zimbabwe; and Pye, Goudie, and Watson, 1986 on Kenya).

Deep Weathering Profiles and Duricrusts

Low relief landsurfaces of considerable antiquity provide a geomorphological basis that favours deep penetration of weathering and thorough decomposition of rocks. Deep weathering is also probably promoted by high temperature and high rainfalls. Thus in many parts of Africa, particularly on granitic rocks, weathering profiles may extend to great depths below the ground surface. Although slight alteration of basement rocks has been found in the Zambian Copperbelt at depths of over 1000 m (Thomas, 1974), values of 30 to 60 m may be more normal, but because of local differences in jointing density and in rock composition these values vary greatly over short distances. The so-called 'weathering front' or 'basal surface of weathering' between sound rock and weathered rock (saprolite) may thus have a very irregular topography, often exhibiting a pattern of discrete basins and domical rises. As we have already seen, this characteristic may have considerable importance in understanding the inselberg or bornhardt landscapes of the continent.

Although the word 'duricrust' is by origin Australian (Woolnough, 1927), duricrusts (the 'cuirasses' of francophone scientists) are recognized as being important components of the surface geology and geomorphology of Africa. They can be defined as

A product of terrestrial processes within the zone of weathering in which either iron and aluminium sesquioxides (in the case of ferricretes and alcretes) or silica (in the case of silcrete) or calcium carbonate (in the case of calcrete) or other compounds in the case of magnesicrete and the like have dominantly accumulated in and/or replaced a pre-existing soil, rock, or weathered material, to give a substance which may ultimately develop into an indurated mass. (Goudie, 1973: 5)

Many duricrusts are associated with low-angle surfaces and this may account for their widespread development in Africa. Table 9.1 lists some recent studies of three main types of African duricrust. Calcretes tend to characterize more arid portions of the continent (with rainfalls below 850 mm), whereas silcretes, ferricretes, and alcretes (bauxites) may occur in more humid regions. This is brought out in the map of laterite and bauxite distribution shown in Figure 9.1. Many examples of laterite and bauxite may be relics related to past climatic conditions but many French workers see them as developing most powerfully in savanna environments (Tricart, 1972: 175). Many different processes are involved in duricrust formation (see Figure 9.2) and space precludes any detailed discussion here. Further information can be obtained in Goudie and Pye (1983).

Table 9.1 Examples of duricrust studies in Africa

Calcrete

South Africa	Netterberg (1980)
Namibia	Blümel and Vogt (1981)
Algeria	Horta (1979)
Tunisia	Bonvallot and Delhourie (1978)

Silcrete

Algeria	Smith and Whalley (1982)
Botswana	Summerfield (1982)
South Africa	Summerfield (1983)
Zaire	Veatch (1935)

Ferricrete and alcrete

Uganda	McFarlane (1983)
Guinea	Maignien (1958)
Zaire	Alexandre and Alexandre-Pyre (1987)

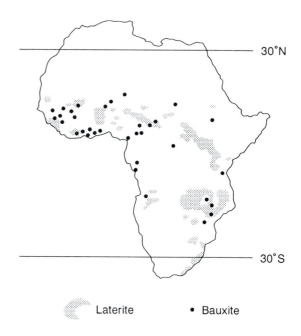

Fig. 9.1. The distribution of laterite and bauxite. (*Sources*: various sources in McFarlane, 1991.)

Duricrusts have a range of important geomorphological consequences: they act as caprocks to escarpments and terraces and preserve otherwise less durable landforms; they cap mesas and other residuals; they provide coarse resistant debris for stone lines and stone pavements; they cause relief inversion; they can be exposed to create infertile pavements (e.g. the laterite 'bowals' of West Africa); and they may be associated with pseudokarstic development (see, for example, Bowden, 1980). They can reach a great age and therefore their geomorphological impact can be felt over

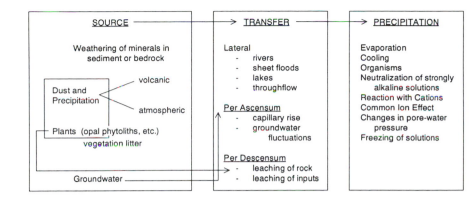

Fig. 9.2. Processes in duricrust evolution.

Plate 9.2. Colluvial aprons and associated gully (*donga*) development in central Swaziland (photo: A. S. Goudie).

extended periods of time. Tricart (1972) believes that their absence or presence is one of the fundamental controls of landscape evolution in the seasonal tropics.

Colluvial Sediments and Stonelines

Colluvium is one of the Cinderella's of geomorphology. It is little studied but probably hugely important. Nowhere is this more true than in Africa, where the significance of colluvial deposits (material that has been transported across and deposited on slopes as a result of wash and mass movement processes) is only now being recognized. These processes seem to operate where the amount of soil cover is limited (because of such factors as moisture deficiency and/or fire) and some may be relics of more extensive arid phases during portions of the Pleistocene.

Thomas and Thorp (1992) draw attention to relict hillslope deposits of possible mudflow origin in Ghana, Sierra Leone, and the Jos Plateau of Nigeria. Such deposits are also known from Zambia, where they partially infill the large limestone caves at Mumbwa.

The best evidence for these deposits and their age comes from Central and Southern Africa, where they mantle about 20 per cent of the land surface south of the Zambezi. They may display more than one phase of deposition and are often trenched at the present time by *dongas* (erosional gully systems of badland type) (Plate 9.2). Colluvial sequences and their associated dongas are to be seen in a wide variety of landscapes. The altitudinal range is from as low as 250 m above sea-level in the case of Skoteni in central Swaziland, to as high as 1600 m in the western Transvaal at Groot Marico. Rock types also vary from site to site. In Swaziland the

Table 9.2   Selected studies of stone lines in Africa

Country	Source
Zimbabwe	Stocking (1978)
Malawi	McFarlane and Pollard (1989)*
Cameroon	Embrechts and de Dapper (1989)*
Zaire	Alone et al. (1989)*
Sierra Leone	Teeuw (1989)*
Rwanda	Moeyersons (1989)*
Congo Republic	Stoops (1989)*
Nigeria	Burke and Durotye (1971)

* In Alexandre and Symoens (1989).

bedrock is generally either granite or granodiorite, whereas in Natal, Zululand, Lesotho, and the Orange Free State, as well as Zimbabwe, the rocks are of the Karoo Supergroup—shales and sandstones. In the northern Transvaal the bedrock comprises rocks of the Bushveld Igneous Complex, in Botswana it is granite and in the western Transvaal Precambrian volcanics and sedimentaries are the source of the colluvia (Watson et al., 1984).

Various attempts have been made to date phases of both colluvium deposition and donga formation, most of which seems to take place in areas which currently receive 600–800 mm of precipitation per annum (Price-Williams et al., 1982). In Swaziland there was a major phase of late Pleistocene colluviation which infilled dongas that had developed in an even earlier phase of colluviation and this was tentatively dated to between 30 000 and 10 000 years ago. Similar colluvial materials and dongas in central Zimbabwe are probably of a similar age (Shakesby and Whitlow, 1991), with the dongas resulting from excessive land use pressures in recent decades. In Natal, recent studies of Botha et al. (1992) indicate that thick colluvial deposits were laid down from 30 000 to 14 500 years ago when pollen analysis indicates that ground-cover was reduced and dry karroid-type vegetation was present.

Colluvial deposits may contain stone lines—phenomena that are widely encountered in natural sections, soil pits, and road cuttings alike. There is an extensive literature on stone lines in Africa (Table 9.2) and considerable debate as to origins (see Thomas, 1974: 130; Stocking, 1978; and Alexandre and Symoens, 1989). Hypotheses include the view that they are relict desert pavement layers, that they result from bioturbation by organisms such as termites and mole rats, that they are formed by downslope creep of resistant particles (e.g. vein quartz) derived from weathering profiles, and that they are residuals associated with a weathering front. There is in effect

controversy between those who hold allochthonist ideas (explaining the stone layer as an accumulation of transported material) and authochthonists (who suggest that the slope layer is formed by *in situ* weathering and transformation of the parent rock). In reality one is probably dealing with a portmanteau term and with a range of different phenomena that have the same name but differing origins.

Dambos

The word *dambo* is a Bantu one meaning 'meadow grazing' and has been used to describe grasslands in the catchment head-waters of Central Africa. The word has now been adopted by geomorphologists to describe shallow, predominantly linear depressions in the headward zones of rivers, which lack a marked stream channel. They tend to be grass-covered, to lack woodland vegetation and to be seasonally waterlogged. The term is probably synonymous with other local names such as *mbuga* (in East Africa), *matoro* (in Mashonaland), *vlei* (in South Africa), and, in some respects with the *fadama* of Nigeria and the *bolis* of Sierra Leone. Comparable European terms are *bas-fond* (French), and *Spültäl* (German). They are known from other parts of the tropics, but seem to display their widest extent (Figure 9.3) and greatest development in Africa (Table 9.3). They are particularly well developed in a broad latitudinal belt from Angola to Tanzania (Thomas and Goudie, 1985).

Dambos contain a wide range of soils. Although a sandy texture is most common, dambo soils can range from gley clays, through sandy clays, to peat (as on the high Nyika Plateau of Malawi) (Meadows, 1985). Compared to surrounding interfluve soils, dambo soils usually possess higher organic-matter levels, higher cation-exchange capacities and increased saturation levels with higher phosphate and phosphorous contents (Boast, 1990: 156). Ferricrete and calcrete layers may also outcrop along dambos.

The distinct lack of trees and shrubs within dambos results primarily from the closeness of the water-table to the surface, which causes waterlogging and thereby impedes growth. This relationship between vegetation and dambos is an important means by which they are identified on air photographs. Many dambos are now used extensively for grazing and cropping (see Chapter 15).

The pathways of water in dambos is the cause of some controversy (Boast, 1990: 162). Some see the main water supply as direct precipitation on the dambo surface while others believe that subsurface throughflow from interfluves is the main source. Some regard them

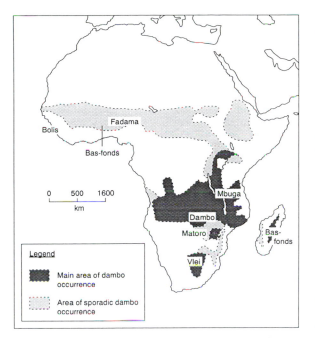

Fig. 9.3. The distribution of dambos and dambo-like forms in Africa.

(*Source*: Boast, 1990: fig. 1.)

Table 9.3 Some studies of dambos in Africa

Location	Source
Zambia	Mäckel (1974)
Malawi	Meadows (1985)
Sierra Leone	Millington *et al.* (1985)
Northern Nigeria	Turner (1985)
Zimbabwe	Whitlow (1984)
Madagascar	Raunet (1985)

500 mm per year. They appear to be developed on a wide range of rock types including unconsolidated Kalahari Sand, shales, quartzites, schists, gneisses, and granites, but their best development appears to be on the Basement Complex granitic and metamorphic rocks of Central Africa. This may be because the relatively sandy regolith or grus produced by the weathering of these rock types facilitates water infiltration and seepage, while the heavy textured subsoils and relatively impervious bedrock impede deep percolation (Boast, 1990: 165).

However, one of the prime controls on their distribution is geomorphological. They tend to be largely restricted to areas of low gradient, especially where associated with ancient erosion surfaces (e.g. the 'African' and 'post-African' erosion surfaces of King, 1962). Such surfaces are probably favourable to dambo formation as they allow the infiltration of surface water (reducing its erosive and channel-forming capabilities), permit an adequate hydraulic gradient to enable subsurface flow towards the valley bottoms, and allow the water-table to outcrop or be near to the surface over extensive areas.

The processes of dambo formation are also the subject of debate. One school sees them as simple extensions of the channelled-drainage network formed by headward erosion and then infilled by slope colluviation and channel alluviation. Surface wash by high-intensity rainfall at the start of the rainy season, when vegetation cover is least, moves surface sediment down-slope, and is, in this model, the prime geomorphological process that operates to mould dambos, though it may be assisted by suffosion (subsurface transport in pipes, etc). On the other hand McFarlane (1989) believes that the morphology of many dambos (which can, for example, link river valleys across interfluves) means that they have formed independently of the normal fluvial network and that chemical and biochemical corrosion (especially along fractures, faults, and joints) are the crucial processes in their development. Probably, as Boast (1990: 173) suggests . . . 'both fluvial incision coupled with colluviation and alluviation, and differential *in situ* weathering, have been important in the development of dambos at some time or other.'

Dambos then are intriguing and still imperfectly understood features, but they are as much a part of Africa as its magnificent bornhardts and its endless plains.

Pans

Dambos are not the only characteristic drainage features of large tracts of the African plainlands. Also important, and in some cases occurring in association with

as 'sponges' which store water and help to maintain dry-season flows, while others believe that they do not contribute significantly to dry-season baseflows of streams. Certain dambos have dry centres in the dry season either because they are areas of lower permeability clay-rich sediments or because they are areas where incision has caused water to drain out from the dambo bottom (McFarlane, 1989).

Dambos occur primarily where rainfall amounts range between 600 mm and 1500 mm per year and where there is a marked seasonal distribution, with drought lasting for six to eight months. However, the *bolis* of Sierra Leone are found where annual rainfall approaches 2500 mm, and some in Kenya occur where rainfall is less than

Table 9.4   Selected studies of African pans

Country	Source
Senegal	Tricart (1953)
Egypt	Said (1975)
Kenya	Ayeni (1977)
Zaire	de Ploey (1965)
Zambia	Williams (1982)
Mozambique	Tinley (1977)
Zimbabwe	Goudie and Thomas (1985)
Botswana	Lancaster (1978)
Namibia	Lancaster (1986)
South Africa	Goudie and Thomas (1985)

dambos (see, for example, Whitlow's 1985 discussion of Zimbabwe), are a series of small closed basins called 'pans'. These are widespread, especially in the drier parts of Africa (Table 9.4) and have a characteristic morphology that has been likened to a clam, a kidney, or a pork chop in planform (Goudie, 1991).

In Southern Africa it is evident (see Goudie and Thomas, 1985) that while widespread and numerous, pans are not ubiquitous. Most of them occur on the arid side of the 500 mm mean annual isohyet and the 1000 mm free surface evaporation loss isoline, with the notable exception of those in the Lake Chrissie area of the eastern Transvaal. Within the arid area there are some major concentrations of pans. In the western Orange Free State there is a belt that runs southwards from Kroonstad, through Wesselsbron, Boshof, and Dealsville, to south of Kimberley (Le Roux, 1978). In this area there may be more than 100 pans 100 km^{-2}. North of this belt is another one that is centred in the western Transvaal and north-eastern Cape Province. Three other main groups occur in Cape Province: those to the south of the Orange River between Kenhardt and Brandvlei, those running north-westwards from Upington across the Molopo Valley into Namibia, and those on the coastal plain near Cape Agulhas (Tinley, 1985).

There are also a large number of pans in Botswana, many of which are developed on unconsolidated Kalahari Beds, though they also occur on the Karoo and Precambrian rocks in the east of the country. In addition there are some major calcrete-rimmed pans, reaching a maximum area extent of about 8 km^2 (e.g. Ukwi and Tshane pans).

Pans are also widespread in Namibia. Notable examples occur near Aminuis in Hereroland and to the east of Keetmanshoop in the south of the country (Lancaster, 1986).

In central Mozambique oval pans reach a density of 200 per km^2 in the Gorongosa area (Tinley, 1977),

where annual precipitation levels reach 840–1000 mm. They occur partly as a result of the presence of susceptible materials (duplex sands) and partly because 'of reduced inflow into the dambos (shallow drainage lines) due to river capture and beheading of their catchments by other drainage' (Tinley, 1977: 70). They occur only in dambos, and relict links between the pans can be traced along the dambos.

The pans of Zimbabwe have received very little attention, being described in limited detail by Flint and Bond (1968) and Thomas (1982). They occur upon the Kalahari Sands of Matabeleland in the west of the country in an area of about 600 mm precipitation annually. They are particularly prolific in Hwange National Park where they attain a density of nearly 15 pans 100 km^{-2}.

Pans, (or 'Plains' as they are often called in Zambia) are also found upon the Kalahari Sands of western Zambia, and their distribution has been described by Williams (1982). They are restricted to an area east of the Southern Lueti and Zambezi Rivers, and predominantly occupy watershed locations.

The first attempts to explain pan development were made in the second half of the nineteenth century. Early workers gave some prominence to animal activities. The importance of this process was appreciated by H. Guillemard as early as 1877 (quoted in Hutchinson et al., 1932):

The idea of the pans being enlarged or deepened (I do not say originated) by the vast herds of game continually visiting them . . . was the natural result of constantly seeing the blesbok and springbok in the pans of the district of the Western Free State and Kalahari border. I don't think any naturalist could fail to come to this conclusion at that period. I don't think the modern could realize their myriads. Yet the old boers, even in those days considered that the game was finished . . . . At a place called Leuuw Pan I saw a herd of blesboks in 1877 which I and my friend estimated at over 8000.

This was the view supported in the context of South Africa by Alison (1899) and Passarge (1904). At much the same time the role of combined salt weathering and deflation was championed by Du Toit (1906: 257), who worked in the northern Cape.

Aridity (annual or seasonal) is an important predisposing factor for pan development. It contributes in a variety of ways: a limited vegetation cover in areas of low rainfall and high salinity permits deflational activity (Tricart, 1953); in areas of low rainfall animals tend to concentrate at any local source of water such as a pan would provide (Ayeni, 1977), thus creating trampling and overgrazing pressures on the ground surface and leading to physical removal of sediment in and on

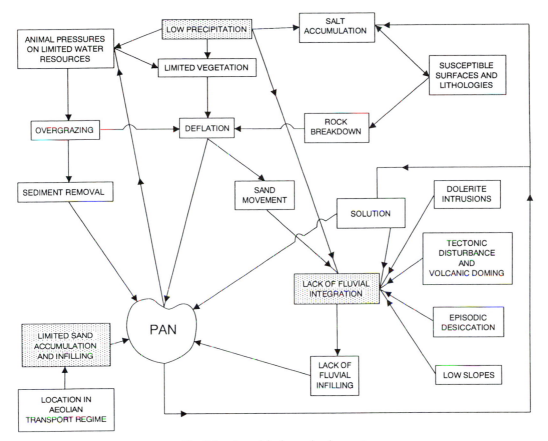

Fig. 9.4. A model of pan development.
(*Source*: Goudie, 1990: fig. 7.)

their bodies; aridity also favours salt accumulation and salt weathering; aridity can cause normal drainage disruption by aeolian blocking, leading to localized accumulation of water, salt development, animal activity, etc; unconsolidated aeolian materials can provide an ideal susceptible surface upon which deflation can occur.

Another predisposing condition for pan development and survival is that they should not be obliterated by the action of intergrated fluvial systems. Indeed, many pans develop along relict, non-functioning drainage lines.

For pans, however formed, to persist it is necessary that they should not be obliterated by aeolian infilling. Rates of sediment excavation have to exceed rates of external aeolian inputs.

Figure 9.4 illustrates a model of pan development developed in Goudie (1990: fig. 7). It incorporates the role of predisposing factors of aridity, lack of integrated fluvial systems, and lack of aeolian infilling. It also recognizes that there may be a range of factors that

initiate a depression (including wind excavation, animal wallowing, and karstic collapse), and a range of factors that encourage further enlargement (e.g. salt weathering, solution caused by focused water flow, and ongoing wind excavation). It also recognizes the importance of having a series of susceptible surface materials, be they soluble rocks, salt-laden shales, or unconsolidated sediments of aeolian, lacustrine, coastal, or fluvial origin. Many of the pans have probably been exposed to a range of past environmental conditions (see Chapter 3) and these will have contributed to their development.

In addition to dambos and pans there may be some other drainage forms that are related in their occurrence to rainfall seasonality and to larger term alternations of arid and humid conditions. These are the 'deranged drainages' of Garner (1974, chs. 4 and 7) or 'the great inland deltas' of McCarthy (1993). Such systems, which include the Nile, Niger, Chari, Okavango, and Zambezi may in large part be a consequence of the

Table 9.5 Density of termite mounds in Africa

Dominant species	Location	Source	Density (mounds/$^{ha-1}$)
Bellicositermes bellicosus rex	Central African Republic	Boyer (1973)	10–12
Bellicositermes natalensis	Central African Republic	Boyer (1973)	up to 30
Cubitermes exiguus	Zaire		0–652
Cubitermes fungitaber	Zaire	Lee and Wood (1971)	875
Cubitermes sankurensis	Zaire		8–850
Macrotermes spp	Ivory Coast	Bodot (1964)	86
Macrotermes bellicosus	Uganda	Pomeroy (1977)	0.75–9.0
Macrotermes bellicosus	Sierra Leone	Miedema and Van Vuure (1977)	5
Macrotermes subhyalinus	Uganda	Pomeroy (1977)	0.75–13.25
Ondontotermes spp	Kenya	Lee and Wood (1971)	5–7
Pseudoacanthotermes	Uganda	Pomeroy (1971)	0.37–7.32
Trinervitermes trinervoides	South Africa	Lee and Wood (1971)	5–7
Trinervitermes spp	Nigeria	Sands (1965)	74–148

formation of half-graben structures as a result of tectonic activity, but climatic factors contribute to their form and development.

Termites and Termitaria

Whether on the ground or inspecting air photos, the importance of termites in the African landscape is impressively evident (Smeathman, 1781). Drummond (1888: 146) reported that their heaps and mounds were 'so conspicuous that they may be seen for miles, and so numerous are they and so useful as cover to the sportsman, that without them in certain districts hunting would be impossible'. He went on to suggest (p. 154) that 'the soil of the tropics is in a state of perpetual motion. There is so to speak, a constant circulation of earth in the tropics, a ploughing and harrowing.' He postulated (p. 158) that while, to paraphrase Herodotus, Egypt was the gift of the Nile, that great river's silt load resulted from 'the labours of the humble termites in the forest slopes about Victoria Nyanza'. More recently, Tricart (1972: 189) remarked that 'the African savannas are the domain of giant termitaries and of the most evolved and active termites.'

Termites occur in great numbers in certain parts of Africa (2–3 m ha^{-1} in Senegal and 9.1 m ha^{-1} in the Ivory Coast have been reported in UNESCO, UNEP, FAO, 1979), and one common genus, *Macrotermes*, is particularly large with a length of around 20 mm and a wing span of 90 mm. They create structures with a variety of forms, some of which attain heights in excess of 9 m (see Pullan, 1979 for a discussion of their distribution and form). Undoubtedly soil characteristics are an important control of mound formation. Mounds tend to be rare on sands (where there is insufficient binding

material), on deeply cracking vertisols (which are unstable), or on shallow soils (where there is a shortage of building material). Moreover, the particular form of mound produced by any individual species may vary according to environmental conditions (see, for example, Darlington's 1984, discussion of *Macrotermes subhyalinus* mounds in Kenya).

The density of termite mounds per unit area also varies with species, with large structures such a those being produced by *Macrotermes* being less dense than those produced by species such as *Trinervitermes* (Table 9.5).

Whether termites live in the conspicuous constructions just discussed (Plate 9.3) or, as most do, in hidden subsoil chambers, they have a significant effect on soils because of both their mechanical activities and their feeding habits. Many mounds are augmented with calcium in comparison with surrounding soils, to the extent that in some cases carbonate nodules may develop (Watson, 1974). Whether or not termites contribute to laterite formation is a rather more contentious matter (compare Erhart, 1951 with Tricart, 1957). Grassé and Noirot (1959) dismissed the importance of termites for laterite formation in West Africa on various grounds: the volume of individual masses of vesicular lateritic ironstone is often much greater than that of termitaria; the vesicles of laterite are not arranged in ways that even vaguely resemble any part of termitaria; and neither termitaria nor surrounding gallery networks have a form resembling that of lateritic crusts. Soils inhabited by termites do, however, contain complex systems of passages and storage pits which considerably modify soil structure. Darlington's (1982) study of a Kenyan mound built by *Macrotermes michaelseni* revealed the presence of 6 km of passage and 72 000 storage pits in an area of just 8000 m^2. The soil surface is also

Plate 9.3. A large termite mound in northern Kenya (photo: A. S. Goudie).

modified because of the extremely efficient way in which termites remove dead wood and litter (see Buxton, 1981; Collins, 1981). Finally, mounds are frequently characterized by higher clay contents than surrounding soils.

It is possible that termites contribute to soil erosion, colluviation, stone-line formation, and stream-sediment loads in three main ways: by removing the plant cover (as just described); by digesting or removing organic matter which would otherwise be incorporated into the soil, and thus making the soil more susceptible to erosion; and by bringing to the surface fine-grained materials for subsequent wash and creep action. Abandoned mounds may be subject to quite rapid rates of erosion (Lepage, 1984), sometimes accelerated by wild and domestic animals, and they become surrounded by large 'wash pediments' that can be up to 60 m in diameter.

Thus termites, which occur in huge numbers and have a very high biomass, create major structures, move major quantities of soil in the process of making these, consume very appreciable amounts of litter and wood

debris, create new soil horizons, cover sands and stone lines, and probably exercise a major control on the rate of operation of such important physical processes as infiltration, surface wash, rain-splash detachment, and soil creep, especially in the savanna lands of Africa (Goudie, 1988: 188).

Fire as a Geomorphological Factor

As the chapters on the various major African biomes repeatedly show (see, for example, Chapter 17), fire is an important component of the African environment (Phillips, 1974) and one that is in some ways complementary to that of termites and other organisms (Plate 9.4). As such it may have a range of geomorphological implications which include the baring of the ground surface so that it is exposed to erosion by the first rains of the wet season, changes in the structure of soils in terms of such characteristics as organic content and hydrophobicity which may relate to their erodibility, and the direct weathering of rock outcrops by fire spalling.

Biomass burning appears to be especially significant in the tropical environments of Africa in comparison with other tropical areas (Table 9.6). This is primarily because of the great extent of savanna which is subjected to regular burning. As much as 75 per cent of African savanna areas may be burned each year (Andreae, 1991). It is probably an ancient phenomenon in the African landscape antedating the appearance of humans (Menaut *et al.*, 1991), but humans have greatly increased the role of fires in the continent and may have used it for over 1.4 million years (Gowlett *et al.*, 1981).

Conclusion

This brief chapter has sought to outline some of the main processes and forms that characterize the great tropical plainlands of Africa and which render them so distinctive. Of the processes, termites and fire are especially important in Africa, and with humans play a major role in determining rates of sediment translocation. The origin of the plainlands themselves (be they pediplains, peneplains, or etchplains) is still the subject of controversy, as is the evolution of inselbergs. Deep-weathering profiles and duricrusts have developed on the plains and translocated eluvium has contributed to the formation of colluvial aprons and stone lines. Pervasive low-angle surfaces have also contributed (along with highly seasonal climates) to the development of two drainage forms of considerable importance—dambos and pans.

Plate 9.4. Fire in the highveld of Swaziland (photo: A. S. Goudie).

Table 9.6 Biomass burning in the tropical regions

Region	Biomass (Tg dm/yr)				REGION TOTALS	
	Forest	Savanna	Fuelwood	Agricultural waste	Tg dm/yr	Tg C/yr
America	590	770	170	200	1730	780
Africa	390	2430	240	160	3210	1450
Asia	280	70	850	990	2190	980
Oceania	—	420	8	17	450	200
TOTAL TROPICS	1260	3690	1260	1360	7580	3410

Source: Andreae, 1991, table 1.3.

References

Alexandre, J. and Alexandre-Pyre, S. (1987), 'La reconstitution à l'aide des cuirasses latéritiques de l'histoire géomorphologie du Haut-Shaba', *Zeitschrift für Geomorphologie Supplementband*, 64: 119–32.

—— and Symoens, J.-J. (1989) (eds.), 'Stone lines', *Geo-Eco-Trop*, 11: 1–237.

Alison, M. S. (1899), 'On the Origin and Formation of Pans', *Transactions, Geological Society of South Africa*, 4: 159–61.

Andreae, M. O. (1991), 'Biomass Burning: Its History, Use and Distribution and its Impact on Environmental Quality and Global Climate', in J. S. Levine (ed.), *Global Biomass Burning* (Cambridge, Mass.) 3–21.

Ayeni, J. S. O. (1977), 'Waterholes in Tsavo National Park, Kenya', *Journal of Applied Ecology*, 14: 369–78.

Blümel, W. D. and Vogt, T. (1981), 'Croûtes calcaires de Namibie: problèmes géomorphologiques et études micromorphologiques', *Récherches Géographiques de Strasbourg*, 12: 55–68.

Boast, R. (1990), 'Dambos: A Review', *Progress in Physical Geography*, 14: 153–77.

Bodot, P. (1964), 'Études écologiques et biologiques des termites dans les savanes de Basse Côte d'Ivoire', in A. Bouillon (ed.), *Études sur les termites africains* (Paris), 251–62.

Bonvallot, J. and Delhourie, J.-P. (1978), 'Étude de différentes accumulations carbonatées d'une toposéquence du centre Tunisien (Djebel Semmama)', *103 Congrès national des sociétés savantes, Nancy, Sciences*, iv. 281–9.

Botha, G. A., Scott, L., Vogel, J. C. and Von Brunn, V. (1992), 'Palaeosols and Palaeoenvironments during the Late Pleistocene Hypothermal in Northern Natal', *South African Journal of Science*, 88: 508–11.

Bowden, S. J. (1980), 'Sub-Laterite Cave Systems and Other Pseudokarst Phenomena in the Humid Tropics: The Example of the Kasewe Hills, Sierra Leone', *Zeitschrift für Geomorphologie Supplementband*, NF 24: 77–90.

Boyer, P. (1973), 'Action de certains termites constructeurs sur l'évolution des sols tropicaux', *Annales des Sciences Naturelles, Zoologie*, 15: 329–498.

Brook, G. A. (1978), 'A New Approach to the Study of Inselberg Landscapes', *Zeitschrift für Geomorphologie Supplementband*, 31: 138–60.

Büdel, J. (1982), *Climatic Geomorphology* (Princeton).

Burke, K. and Durotye, B. (1971), 'Geomorphology and Superficial Deposits Related to Late Quaternary Climatic Variation in South Western Nigeria', *Zeitschrift für Geomorphologie*, 15: 430–44.

Buxton, R. D. (1981), 'Termites and the Turnover of Dead Wood in an Arid Tropical Environment', *Oecologia*, 51: 379–84.

Collins, N. M. (1981), 'The Role of Termites in the Decomposition of Wood and Leaf Litter in the Southern Guinea Savanna of Nigeria', *Oecologia*, 51: 389–99.

Darlington, J. P. E. C. (1982), 'The Underground Passages and Storage Pits Used in Foraging by a Nest of the Termite *Macrotermes michaelseni* in Kajiado, Kenya', *Journal of Zoology*, 198: 237–47.

—— (1984), 'Two Types of Mound Built by the Termite *Macrotermes subhyalinus* in Kenya', *Insect Science Applications*, 5: 481–92.

de Ploey, J. (1965), 'Position géomorphologique, génèse et chronologie de certains depôts superficiels au Congo occidental', *Quaternaria*, 7: 131–54.

Drummond, H. (1888), *Tropical Africa* (London).

Du Toit, A. L. (1906), 'Geological Survey of Portions of the Divisions of Vryburg and Mafeking', *10th Annual Report Geological Commission of the Cape of Good Hope*, 205–58.

Erhart, H. (1951), 'Sur le rôle des cuirasses termitiques dans la géographie des regions tropicales', *Compte rendu de L'Academie des sciences*, 233: 804–6.

Falconer, J. D. (1911), *The Geology and Geography of Northern Nigeria* (London).

Faniran, A. and Jeje, L. K. (1983), *Humid Tropical Geomorphology* (London).

Flint, R. F. and Bond, G. (1968), 'Pleistocene Sand Ridges and Pans in Western Rhodesia', *Bulletin Geological Society of America*, 79: 299–314.

Garner, H. F. (1974), *The Origin of Landscapes: A Synthesis of Geomorphology* (New York).

Gibbons, C. L. M. H. (1981), 'Tors in Swaziland', *Geographical Journal*, 147: 72–8.

Goudie, A. S. (1973), *Duricrusts in Tropical and Sub-tropical Landscapes* (Oxford).

—— (1988), 'The Geomorphological Role of Termites and Earthworms in the Tropics', in H. A. Viles (ed.) *Biogeomorphology* (Oxford), 166–92.

—— (1991), 'Pans', *Progress in Physical Geography*, 15: 221–37.

—— amd Pye, K. (1983) (eds.), *Chemical Sediments and Geomorphology* (London).

—— and Thomas, D. S. G. (1985), 'Pans in Southern Africa with Particular Reference to South Africa and Zimbabwe', *Zeitschrift für Geomorphologie*, 29: 1–19.

Gowlett, J. A., Harris, J. W. K., Walton, D., and Wood, B. A. (1981), 'Early Archaeological Sites, Hominid Remains and Traces of Fire from Chesowanja, Kenya', *Nature*, 284: 125–9.

Grassé, P.-P. and Noirot, C. (1959), 'Rapports des termites avec les sols tropicaux', *Revue de Géomorphologie Dynamique*, 10: 35–40.

Horta, O. S. (1979), 'Les encroûtements calcaires et les encroûtements gypseux en géotechnique routière', *B.E.T. Lab. Mechanique Sols, Alger*, Mémoire Technique 1.

Hutchinson, G. E., Pickford, G. E. and Schuurman, J. F. M. (1932), 'A Contribution to the Hydrobiology of Pans and Other Inland Waters of South Africa', *Archiv für Hydrobiologie*, 24: 1–136.

Jeje, L. K. (1973), 'Inselberg's Evolution in a Humid Tropical Environment: The Example of South-Western Nigeria', *Zeitschrift für Geomorphologie*, 17: 194–225.

King, L. C. (1962), *South African Scenery* (Edinburgh).

—— (1966), 'The Origin of Bornhardts', *Zeitschrift für Geomorphologie*, 10: 97–8.

Lancaster, N. (1986), 'Pans in the Southwestern Kalahari: A Preliminary Report', *Palaeoecology of Africa*, 17: 59–67.

Le Roux, J. S. (1978), 'The Origin and Distribution of Pans in the Orange Free State'. *South African Geographer*, 6: 167–76.

Lee, K. E. and Wood, T. G. (1971), *Termites and Soils* (London and New York).

Lepage, M. (1984), 'Distribution, Density and Evoution of *Macrotermes*

bellicosus Nests (Isoptera: Macrotermitinae) in the North-east of the Ivory Coast', *Journal of Animal Ecology*, 53: 107–17.

McCarthy, T. S. (1993), 'The Great Inland Deltas of Africa', *Journal of African Earth Sciences*, 17: 275–91.

McFarlane, M. J. (1983), 'Laterites', in A. S. Goudie and K. Pye (eds.), *Chemical Sediments and Geomorphology* (London), 7–58.

—— (1989), 'Dambos: Their Characteristics and Geomorphological Evolution in Parts of Malawi and Zimbabwe, with Particular Reference to Their Role in the Hydrological Regime of Surviving Areas of African Surface', *Commonwealth Science Council*, i, session 3: 254–308.

—— (1991), 'Some Sedimentary Aspects of Lateritic Weathering Profile Development in the Major Bioclimatic Zones of Tropical Africa', *Journal of African Earth Sciences*, 12: 267–82.

Mäckel, R. (1974), 'Dambos: A Study in Morphodynamic Activity of Plateau Regions of Zambia', *Catena*, 1: 327–65.

Maignien, R. (1958), *Le cuirassement des sols en Guinée, Afrique occidentale*, Mémoires du Service de la Carte Géologique D'Alsace et de Lorraine, no. 16: 239.

Meadows, M. E. (1985), 'Dambos and Environmental Change in Malawi, Central Africa'. *Zeitschrift für Geomorphologie Supplementband*, 52: 147–69.

Menaut, J. C., Abbadie, L., Lavenu, F., Loudjani, P., and Podaire, A. (1991), 'Biomass Burning in West Africa Savannas', in J. S. Levine (ed.), *Global Biomass Burning* (Cambridge, Mass.), 133–42.

Miedema, R. and Van Vuure, W. (1977), 'The Morphological, Physical and Chemical Properties of Two Mounds of *Macrotermes bellicosus* (Smeathman) Compared with Surrounding Soils in Sierra Leone', *Journal of Soil Science*, 28: 112–24.

Millington, A. C., Helmisch, F. and Riebergen, G. J. (1985), 'Inland Valley Swamps and Bolis in Sierra Leone: Hydrological and Pedological Considerations for Agricultural Development', *Zeitschrift für Geomorphologie Supplementband*, 52: 201–22.

Netterberg, F. (1980), 'Geology of Southern African Calcretes: 1. Terminology, Description, Macrofeatures and Classification', *Transactions Geological Society of South Africa*, 83: 255–83.

Passarge, S. (1904), *Die Kalahari* (Berlin).

Phillips, J. (1974), 'Effects of Fire in Forest and Savanna Ecosystems of Sub-Saharan Africa', in T. Kozlowski and C. E. Ahlgren (eds.), *Fire and Ecosystems* (New York), 435–81.

Pomeroy, D. E. (1977), 'The Distribution and Abundance of Large Termite Mounds in Uganda', *Journal of Applied Ecology*, 14: 465–75.

Price-Williams, D. Watson, A., and Goudie, A. S. (1982), 'Quaternary Colluvial Stratigraphy, Archaeological Sequences and Palaeoenvironment in Swaziland, Southern Africa', *Geographical Journal*, 148: 50–67.

Pugh, J. C. (1966), 'The Landforms of Low Latitudes', in G. H. Dury (ed.), *Essays in Geomorphology* (London), 121–38.

Pullan, R. A. (1979), 'Termite Hills in Africa: Their Characteristics and Evolution', *Catena*, 6: 267–91.

Pye, K., Goudie, A. S. and Thomas, D. S. G. (1984), 'A Test of Petrological Control in the Development of Bornhardts and Koppies on the Matopos Batholith, Zimbabwe'. *Earth Surface Processes and Landforms*, 9: 455–67.

—— and Watson, A. (1986), 'Petrological Influence on Differential Weathering and Inselberg Development in the Kora Area of Central Kenya', *Earth Surface Processes and Landforms*, 11: 41–52.

Raunet, M. (1985), 'Les bas-fonds en Afrique et à Madagascar— Géochimie—Pédologie—Hydrologie'. *Zeitschrift für Geomorphologie Supplementband*, 52: 25–62.

Said, R. (1975), 'Some Observations on the Geomorphological Evolution of the South-Western Desert of Egypt and its Relation to the Origin of Ground Water', *Annals, Geological Survey of Egypt*, 5: 61–70.

Sands, W. A. (1965), 'Termite Distribution in Man-Modified Habitats in West Africa, with Special Reference to Species Segregation in the Genus *Trinervitermes*', *Journal of Animal Ecology*, 34: 557–71.

Selby, M. J. (1977), 'Bornhardts of the Namib Desert', *Zeitschrift für Geomorphologie Supplementband*, 18: 121–43.

Shakesby, R. A. and Whitlow, R. (1991), 'Perspectives on Prehistoric and Recent Gullying in Central Zimbabwe'. *Geojournal*, 23: 49–58.

Smeathman, H. (1781), 'Some Account of the Termites which are Found in Africa and Other Hot Climates', *Philosophical Transactions of the Royal Society of London*, 71: 139–92.

Smith, B. J. and Whalley, W. B. (1982), 'Observations on the Composition and Minerology of an Algerian Duricrust Complex', *Geoderma*, 28: 285–311.

Stocking, M. A. (1978), 'Interpretation of Stonelines', *South African Geographical Journal*, 60: 121–34.

Summerfield, M. A. (1982), 'Distribution, Nature and Genesis of Silcrete in Arid and Semi-arid Southern Africa', *Catena*, suppl. 1: 37–65.

—— (1983), 'Silcrete a Palaeoclimatic Indicator: Evidence From Southern Africa'. *Palaeogeography, Palaeoclimatology, Palaeoecology*, 41: 65–79.

Thomas, D. S. G. (1982), 'Evidence of Quaternary Palaeoclimates in Western Zimbabwe: A Preliminary Assessment', paper of the Southern African Conference of the Commonwealth Geographical Bureau, Lusaka, 9–15 June 1982.

Thomas, M. F. (1966), 'Some Geomorphological Implications of Deep Weathering Patterns in Crystalline Rocks in Nigeria', *Transactions of the Institute of British Geographers*, 40: 173–93.

—— (1974), *Tropical Geomorphology* (London).

—— (1994), *Geomorphology in the Tropics* (Chichester).

—— and Goudie, A. S. (1985) (eds.), 'Dambos: Small Channelless Valleys in the Tropics', *Zeitschrift für Geomorphologie Supplementband*, 52: 1–222.

—— and Thorp, M. B. (1992), 'Landscape Dynamics and Surface Deposits Arising from Late Quaternary Fluctuations in the Forest-Savanna Boundary', in P. A. Furley, J. Proctor, and J. A. Ratter (eds.), 1992, *Nature and Dynamics of Forest–Savanna Boundaries* (London), 213–53.

Tinley, K. L. (1977), 'Framework of the Garongosa Ecosystem', unpublished D.Sc. thesis, University of Pretoria.

—— (1985), 'Coastal Dunes of South Africa', *South African National Scientific Programmes Report*, 109.

Tricart, J. (1953), 'Influence des sols salés sur la deflation éolienne en basse Mauritanie et dans la Delta du Sénégal', *Revue de Géomorphologie Dynamique*, 23: 145–58.

—— (1957), 'Observations sur le rôle ameublisseur des termites', *Revue de Géomorphologie dynamique*, 6: 170–2, 179.

—— (1972) *Landforms of the Humid Tropics, Forests and Savannas* (London).

Turner, B. (1985), 'The Classification and Distribution of Fadamas in Central North Nigeria', *Zeitschrift für Geomorphologie Supplementband*, 52: 87–113.

Twidale, C. R. and Bourne, J. A. (1978), 'Bornhardts', *Zeitschrift für Geomorphologie Supplementband*, 31: 111–37.

UNESCO/UNEP/FAO (1979), *Tropical Grazing Land Ecosystems* (Paris).

Veatch, A. C. (1935), 'Evolution of the Congo Basin'. *Memoirs, Geological Society of America*, no. 3.

Watson, A., Price-Williams, D., and Goudie, A. S. (1984), 'The Palaeoenvironmental Interpretation of Colluvial Sediments and Palaeosols of the Late Pleistocene Hypothermal in Southern Africa', *Palaeoclimatology, Palaeogeography, Palaeoecology*, 45: 225–49.

Watson, J. P. (1974), 'Calcium Carbonate in Termite Mounds', *Nature*, 247: 74.

Wayland, E. J. (1934), 'Peneplains and Some Other Erosion Platforms', *Annual Report for 1934, Bulletin, Uganda Geological Survey Department*: 77–9.

Whitlow, J. R. (1984), 'A Survey of Dambos in Zimbabwe', *Zimbabwe Agriculture Journal*, 81: 129–38.

Williams, G. J. (1982), 'A Preliminary Landsat Interpretation of the Relict Landforms of Western Zambia', paper presented at Southern African Conference of the Commonwealth Geographical Bureau, Lusaka, 9–15 June 1982.

Woolnough, W. G. (1927), 'The Duricrust of Australia', *Journal of Proceedings of the Royal Society of New South Wales*, 61: 24–53.

10 Biogeography

Michael E. Meadows

Introduction

Africa is a fascinating and even puzzling continent to a biogeographer. The spectrum of environments which exist there spans entire moisture and temperature gradients, from perhaps the most arid to among the most well-watered places on earth, from the coolness of the Cape to the furnace that is the Sahara (see Chapter 3). This environmental diversity is mirrored in the proliferation of its fauna and flora, for Africa has seemingly every conceivable combination of climatological, geological, and pedological factors; the plant and animal communities have evolved over time to reflect this heterogeneity. Moreover, it is an ancient continent that has provided a cradle for a wide range of taxonomic groups, from among the very first prokaryotic life-forms which show up in the Precambrian rocks of South Africa (Knoll and Barghoorn, 1977), to the first primates, ancestors of humans (Szalay and Delson, 1979) and, indeed, the first members of our own genus and species (Stringer, 1989). Thus, the biogeographical image of Africa as a continent of rain forests and savannas and deserts is one that has been sketched not only by the impact of its contemporary environmental features but also by the forces of time—evolutionary, geological, and climatic change. Any biogeographical analysis, therefore, of the pattern of plants and animal communities of Africa must begin with an examination of history. It is only through an understanding of development and change through time that we can begin to appreciate the present-day phytogeographical, zoogeographical, and ecological characteristics of the continent and its islands. In so doing, we can in turn begin to comprehend that these patterns of life are today precariously balanced: the biology of much of Africa appears to be threatened by the activities of one of its very own species.

This chapter examines Africa's historical biogeography as a background to the contemporary plant and animal distribution patterns which are elaborated upon in some of the following chapters. These present-day patterns, classified into a number of biomes, are then described in terms of their major biogeographical and ecological features, mainly on the basis of their vegetation characteristics. These include biological diversity, levels of endemism (Plate 10.1), and important structural and functional characteristics of the ecology of each major biome. The chapter concludes with some comments on the potential future of Africa's rich biogeographical heritage.

The Historical Biogeography of Africa

In considering the evolutionary development of Africa's biota, the process of continental drift, with its underlying mechanism of plate tectonics, emerges as the major driving force. The shifting continent, in relation to the geography of the earth itself and in its juxtaposition with respect to the other major continents, explains much of the historical biogeography and, indeed, contemporary distribution patterns of Africa. Africa has, at various stages, connected to and disconnected from the other Gondwanan continents of the southern hemisphere and has undergone several phases of contact with and subsequent separation from the Laurasian continents of the northern hemisphere. Madagascar probably parted company with Africa in the Jurassic, but has remained close enough to it to retain significant biogeographical ties (Briggs, 1987). Land plants occupied the earth's terrestrial habitats from about the Silurian onwards, and we witness a progressive diversification thereafter, an

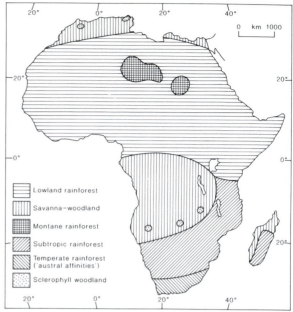

Fig. 10.1. Inferred distribution of vegetation: late Cretaceous–Palaeocene.

1. Lowland rain forest; 2. Savanna woodland; 3. Montane rain forest; 4. Subtropic rain forest; 5. Temperate rain forest ('austral affinities'); 6. Sclerophyll woodland. (*Source*: Axelrod and Raven, 1978.)

Plate 10.1. The baobab, *Adansonia digitata* is one of the most characteristic trees of Africa. A baobab, locally called a boab, also occurs in north-western Australia (*Adansonia gregorii*) but the island of Madagascar has a much richer endemic baobab flora (photo: A. S. Goudie).

adaptive radiation which reached its fullest expression in the Carboniferous swamps. These tropical wetlands, globally widespread at the time, were characterized by tall trees, in particular of the genera *Lepidodendron* and *Sigillaria* (Ingrouille, 1992), but appear to be less well developed in the southern continents of Gondwana, where cooler, more seasonal climates seem rather to have supported a lower-growing vegetation (White, 1986). The end of the Palaeozoic in Africa is characterized by widespread glaciation and the development of a Gondwanan flora which was depauperate relative to the forests thriving in more tropical climates in the northern hemisphere (Polunin, 1960).

While the first gymnosperms were probably initially small trees of the understorey, they rose to full prominence in the Mesozoic. All the southern continents of Gondwana are characterized during the Triassic and Jurassic by the *Glossopteris* seed-fern flora. The vegetation consisted of shrubs and trees up to 6 m tall, with pteridosperms, cycads, and other ancestors of the modern coniferous trees (Ingrouille, 1992). Thus, at this early stage in the evolution of its flora and flora, Africa appears to be biogeographically similar to its Gondwanan sister-continents.

Axelrod and Raven's (1978) reconstruction of the historical biogeography of Africa from the Cretaceous onwards points to the importance of two major factors in determining its distinctive course of evolution thereafter, namely the fragmentation of Gondwana leading to the isolation of Africa, and accompanying topographic and climatic changes. While the fossil record is incomplete, the evolutionary record here onwards points to the differentiation of Africa itself into several major ecosystems, or biomes, which provide a useful framework on which to base a description of the historical biogeography of the continent (Figures 10.1–10.4).

During the late Cretaceous (Figure 10.1) Africa was still situated to the south of its present position astride the Equator, and tropical conditions prevailed mainly in the northern part of the continent. Rain forest dominated the region up to 15° north and south of the

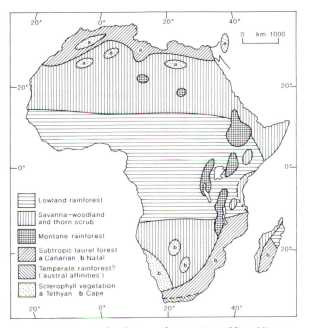

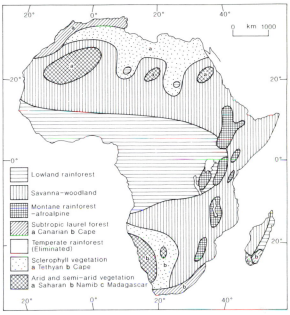

Fig. 10.2. Inferred distribution of vegetation: Oligo–Miocene.
1. Lowland rain forest; 2. Savanna woodland and thorn scrub; 3. Montane rain forest; 4. Subtropic laurel forest, a. Canarian, b. Natal; 5. Temperate rain forest? ('austral affinities'); 6. Sclerophyll vegetation, a. Tethyan, b. Cape. (*Source*: Axelrod and Raven, 1978.)

Fig. 10.3. Inferred distribution of vegetation: late Miocene–early Pliocene.
1. Lowland rain forest; 2. Savanna woodland; 3. Montane rain forest-afroalpine; 4. Subtropic laurel forest, a. Canarian, b. Cape; 5. (eliminated); 6. Sclerophyll vegetation, a. Tethyan, b. Cape. (*Source*: Axelrod and Raven, 1971.)

Equator, a tropical humid-climate vegetation formation which illustrates that already the true flowering plants (angiosperms) had replaced the gymnospermous *Glossopteris* flora in its entirety (Axelrod and Raven, 1978). In so far as the East African mountains had still to be formed, we can envisage this rain forest extending from coast to coast, although it retained only limited connections with the South American tropical flora. In practice, the floras of tropical South America and Africa are sufficiently distinct to suggest that the two continents must largely have been separated prior to the development of the world's seed flora (Smith, 1973; Thorne, 1973; Briggs, 1987). Other plant communities in Africa at this time include temperate rain forest, abundant indications of which are found in all the Cretaceous-age fossil-bearing deposits of the southern land-masses, although in Southern Africa it appears to have been species-poor and lacking the characteristic *Nothofagus* (southern beech) element (Van Steenis, 1972), so typical of the rest of Gondwana at the time. More seasonal climates also existed at this time, and these promoted the widespread occurrence of savanna woodland and sclerophyllous scrub (Figure 10.1), although severe aridity was still to develop in Africa.

By the mid-Tertiary (Figure 10.2), Africa was located at approximately its contemporary latitude. The major changes this induced relate to the 'migration' of the rain forest flora southwards to occupy Central Africa, still spanning the west and east coasts. The emergence of the East African mountains led to the evolution of montane elements along the spine which approximates the present-day rift valley. Sclerophyll-dominated vegetation is found for the first time in the south-western parts of Africa, as evidenced by the fossil Banke flora at Arnot, in Namaqualand (Scholtz, 1985). This is important, for it documents the arrival in that part of Africa of a flora that has subsequently diversified to become one of the most species-rich in the continent, if not in the world (Linder *et al.*, 1992). As far as the fauna is concerned, this appears to have been a time of dramatic upheaval, and twenty-nine new families and seventy-nine new genera make their appearance in the late Oligocene/early Miocene (Briggs, 1987) and a further eighteen families by the late Miocene.

By the late Tertiary (Figure 10.3) aridity was becoming a dominant environmental factor in parts of North Africa (the proto-Sahara) and in Southern Africa. The Southern African arid and semi-arid areas developed as a consequence of the isolation of Antarctica (Kennett, 1980), the growth of an Antarctic ice sheet and the

Fig. 10.4. Distribution of tropical rain forest in present-day Africa compared to inferred distribution in refugia during Pleistocene cool-arid periods.
(*Source*: Mayr and O'Hare, 1986.)

consequent global drop in sea-level (Siesser and Dingle, 1981), and the establishment of the cold Benguela current (Siesser, 1980). At the same time, the closing of the Tethys Sea in the north opened up further opportunities for exchanges of biota between Africa and Eurasia (Axelrod and Raven, 1978). The trend to cooler and drier climates reduced the area of rain forest such that savannas became progressively more widely distributed in Africa, a vegetation change paralleled by an increase in the diversity of grazing mammals (Axelrod and Raven, 1978). This development virtually eliminated the temperate rain forest flora from the continent. The contemporary biomes of Africa therefore have a character which, with the exception of the equatorial rain forests, is controlled to a greater or lesser degree by seasonality with respect to precipitation. The trend to more xeric conditions has waxed and waned during major climatic fluctuations of the Quaternary (Goudie, 1992), at once favouring the expansion of tropical forests and at other times forcing its contraction (Bonnefille and Riollet, 1988) (Figure 10.4). The degree to which such changes may have influenced the diversity of Africa's biota and its levels of endemism is, perhaps, difficult to estimate. The relatively depauperate nature of the African tropical forests by comparison with those of either South America or Malesia (Flenley, 1979), taken together with

the existence of 'islands' of rain forest with relatively high levels of endemism (Hamilton, 1976) certainly points to the significance of more recent environmental changes and evolutionary events in shaping the contemporary biogeography of the continent. Environmental change has typified the Quaternary and fluctuations in both temperature and rainfall have been documented throughout Africa; such changes were undoubtedly instrumental in promoting major shifts in the ranges of its biota.

The situation of Madagascar (see Battistini and Richard-Vindard, 1972), which has been isolated for a long period of geological time is somewhat distinct from that of the continental mainland. Zoologically, this island is well known for a fauna which, while relatively depauperate in terms of total numbers of species, is rich in strange and unique forms (Briggs, 1987). Botanically, the flora possesses high levels of species-richness and endemism and indicates perhaps what might have prevailed in Africa had not the ravages of declining temperatures and increasing aridity repeatedly perturbed the vegetation on the mainland.

In concluding this brief review of the historical biogeography of Africa, it is apparent that, despite the fact that it is geologically one of the oldest continents and has a fossil biota stretching back to the origin of life itself, the major contemporary distribution patterns seem largely to have been determined by the dramatic events of only the last 65 million years or so. In the light of this, the discussion now turns to the biogeography and ecology of the contemporary African flora and fauna.

The Contemporary Biogeography and Ecology of Africa: A Descriptive Framework

The Biomes of Africa

In attempting to describe the major features of the biogeography and ecology of a continent the size of Africa two decisions must be made regarding a choice of a descriptive framework. First, a solution to the problem of whether to use plants or animals as the fundamental classification basis must be sought. In this case (and in most others), the decision is made to concentrate on vegetation, mainly because of the fact that vegetation is the entry-point of energy into ecosystems (all other organisms are effectively dependent on the carbon-fixing plants) and because vegetation is such a reliable indicator of other environmental variables. Vegetation distribution, therefore, indicates more about the environment than the mere distribution of its constituent species, it is a

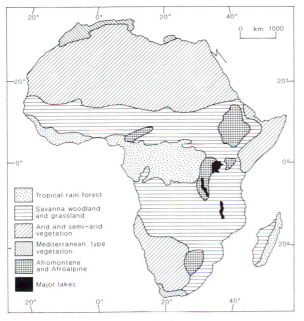

Fig. 10.5. The major biomes of Africa.

land with a Mediterranean-type climate, and the montane regions (Figure 10.5), thus producing basically five major biomes. Compared to the twenty phytochorological units of White's (1983) *Vegetation of Africa*, this is a much simplified and generalized classification, although it shows strong concordance with White's eight 'formations of regional extent'.

The Influence of Climate

At the continental scale, the overriding importance of climate in determining the characteristic pattern of contemporary biotic distributions becomes immediately apparent. While climate is clearly only one of a myriad of environmental variables which influence biotic distributions, it is widely regarded as dominant at an appropriately large scale. Polunin (1960) describes climate as 'the master' (p. 9) controlling environmental variable. Walter and Lieth's (1967) *Klimadiagram* approach to ecology acknowledges the paramount importance of climate, in particular the balance between moisture inputs, in the form of precipitation, and outputs, determined largely by temperature- and humidity-controlled evapotranspiration. Despite the obvious logical over-simplication in so doing, the comparison between vegetation, mapped as major biomes, and mean annual precipitation shows striking similarities and conformities. Given the clear relationships between climate and the distribution of Africa's biota, the possible changes of climate in the future brought about by anthropogenic increases in the concentration of so-called greenhouse gases have the potential to shift markedly the ranges of many of its plant and animal species (Smith *et al.*, 1992).

Having decided that a logical biogeographical and ecological framework for Africa can be constructed on the basis of biomes, which effectively integrate climate and vegetation with other environmental variables, it is now possible to proceed to a description of the individual biomes themselves.

Tropical Rain Forest

This formation, also known as equatorial forest, tropical lowland forest, or tropical evergreen forest (Richards, 1952) has been referred to as 'the apex of creation' (Whitmore, 1990: 9) on account of its unsurpassed luxuriance and profusion of biological diversity. The biodiversity of African forests is discussed by Myers (Chapter 20). In Africa, rain forests are distributed in Central and West Africa and eastern Madagascar, the largest continuous area being the 1.8×10^6 km^2 of the Zaire basin (Whitmore, 1990) associated with annual precipitation values

clue to a wide range of habitat factors and, moreover, an indication of the dependent animal species. Having made this decision, the second problem revolves around choosing between an analysis of the vegetation on a purely phytogeographical basis (using phytochorological mapping units) and an analysis made on the basis of *ecological* relationships, incorporating other aspects of the environment into the framework. White's (1983) description of the vegetation of Africa employs a phytochorological or floristic type of analysis and, while this provides us with a meaningful division of the continent into some twenty major mapping units, the system makes limited reference to either the physical environment or physiognomic characteristics of the vegetation, both of which are useful and important components of the broader understanding of why and how different parts of Africa support different vegetation communities. The framework adopted here, then, is more ecological in its approach, and follows the logic of Moreau (1966) and Axelrod and Raven (1978) in employing the 'biome' as the fundamental mapping unit. These large-scale mapping units represent reasonably homogeneous areas of the earth's land surface and illustrate the close ties that exist between the biota, in particular the vegetation, and general environmental conditions. Africa can be categorized into the following major biomes: tropical rain forest, tropical savannas (with or without an overstorey of trees), semi-arid and arid formations, areas of shrub-

in excess of 1400 mm and consistently warm temperatures. The tropical rain forest biome of mainland Africa (discussed in further detail in Chapter 11) closely approximates the distribution of White's (1983) Guinea–Congolian regional centre of endemism. Structurally, the rain forests are complex and multilayered, with the tallest 'emergent' stratum consisting of trees up to 50 m in height. A more complete canopy exists in the vicinity of 20–30 m, with various other strata dependent upon the age of the formation and degree of human disturbance. A great diversity of life-forms (synusiae) is associated with the rain forest; apart from trees (dominant), shrubs, bamboo, and herbaceous elements, there may be many climbers, epiphytes, saprophytes, and even parasites. Several variants exist, depending on mean annual precipitation, substrate (especially nutrient status), and drainage characteristics. Indeed, Whitmore (1990) identifies at least eight major formations, all of which occur within Africa, classified on the basis of moisture inputs, soil nutrient status, and elevation. The most widely occurring category in the Zaire basin is referred to by Whitmore (1990) as semi-evergreen rain forest, since it includes a number of deciduous species occurring in the canopy.

Considering the huge area occupied by the forests, rather little is known about their detailed ecology, particularly with respect to ecosystem energy flow characteristics such as biomass, primary productivity, and secondary productivity. Several 'myths' have emerged over the years, perhaps partly in consequence of the awe that these forests undoubtedly inspire, but claims for prolific biomass, extreme high values for productivity and, paradoxically, nutrient-deficient soils are generalizations which have, to a greater or lesser degree, all been questioned in recent years. One of the problems is the lack of detailed ecological analysis, and here the African rain forest appears to have been even less well served than those found in Amazonia and Malesia. Hall and Swaine's (1981) monograph on the forest vegetation of Ghana provides us with a notable, and welcome, exception; indeed the West African forests are probably the best known of all the African rain forest formations, as studies on those of Ghana (Hall and Swaine, 1976, 1981); Ivory Coast (Bernhard-Reversat et al., 1978; Alexandre, 1984); and Nigeria (Jones, 1955–6; Hall, 1977) demonstrate. Primary productivity rates are commonly thought of as being very high, given the favourable moisture conditions and year-round warm temperatures, but the potentially elevated rate of photosynthesis promoted by large quantities of chlorophyll biomass is balanced by losses through high rates of respiration. Net primary productivity, thus, may be

in the range of 10 t to 20t ha^{-1} yr^{-1}. While this may be regarded as high, it is not significantly higher than for some temperate forests where respiration losses are lower (Collinson, 1988).

In respect of nutrient cycling, tropical forests were initially thought to survive on universally nutrient-poor soils and their high biomass was explained by rapid and efficient nutrient cycling through the litter layer. As a result, it was postulated that tree roots should be concentrated in the surface 30 cm or so, because the bulk of the soil contained minimal nutrition and was thought to be irrelevant from the viewpoint of biogeochemical cycling. Indeed, the roots of many forest trees *are* confined to the superficial layers of the soil (Bernhard-Reversat et al., 1978), but the soils themselves turn out to be highly variable in their nutrient content (Jordan, 1985; Proctor, 1986) and not all nutrients are necessarily concentrated in the above-ground biomass. Other sources of nutrients, for example aerosols (see Talbot et al., 1990) may also be significant, particularly on the ancient planation surfaces of the African continent where soils are indeed weathered and much less fertile.

The typical rain forests of Africa are undeniably species-rich, although rather less so than comparative areas of the Amazon basin and Malesia. Richards (1973) has gone so far as to call Africa 'the odd man out' because of its relatively depauperate rain-forest flora. This phenomenon is the most challenging biogeographical problem in the African forests. Flenley (1979) notes other unanswered questions, including the search for an explanation for West African–East African disjunctions and the existence of three geographical regions with rather higher levels of both plant species diversity and levels of endemism. The impoverished nature of the flora is illustrated by the distribution of tropical plant families: there are seventeen endemic to tropical Africa compared to forty-seven in the Neotropical region (Briggs, 1987). Thorne (1973) points to the separation of Africa from the rest of Gondwanaland before the development of the seed flora, which would explain the limited phytogeographic relationships between Africa and South America, but does not account for impoverishment *per se*. A more extreme environmental history may, however, have been important in producing extinction events, and the aridity of many parts of tropical Africa during the late Quaternary could certainly have promoted the retreat of rain forest to refugia (Figure 10.4) (Hamilton, 1976; Mayr and O'Hara, 1986; Tallis, 1991). As an explanation of biogeographic pattern in a range of rain-forest taxa, the Pleistocene refuge hypothesis so far seems to have stood the test of time in Africa, although it has been much more critically debated for South America (see Bush *et al.*, 1990). The islands of richness are usually coincident between different groups of plants and animals. Nevertheless, some disparities occur; for example, the African rain forests are relatively poor in plant species numbers but are especially rich in terms of primates, and there may be up to fourteen sympatric primate herbivore species (Whitmore, 1990).

The complexity of the African rain forests with respect to both their ecological and biogeographical characteristics remains an enigmatic research problem and, perhaps, yet another argument for their conservation in the face of increasing levels of human utilization.

Tropical Savannas

Savanna vegetation in Africa covers extensive areas between the equatorial forests and the mid-latitude desert to the north and south, and indeed occupies up to 65 per cent of the continent (Figure 10.5) (Cole, 1986). Savanna prevails in strongly seasonal climates where there is a marked dry season, varying in length from a few weeks to eight or nine months at the arid end of the gradient. The vegetation consists of a mosaic of formations, the most important diagnostic character being the presence of a continuous stratum of grasses at the ground layer. Trees may vary in abundance from a continuous closed-canopy woodland, to more open park-like savannas studded with microphyllous trees, to grasslands with scattered trees or bushclumps, to treeless grasslands (Cole, 1986). In other words, the range of vegetation formations which fall under the general term 'savanna' is enormous, such that generalizations concerning their biogeography and ecology are likely to be somewhat simplistic. Broadly speaking, savanna woodlands, parklands, and grasslands coincide with White's (1983) Zambezian and Sudanian regional centres of endemism, although there are also savanna formations in the Somalia–Masai centre of endemism and the regional transition zones named Guinea–Congolia/Zambezia, Guinea–Congolia/Sudania, Kalahari–Highveld, Sahel, and the mosaic around Lake Victoria. The classification, ecology, and human use of savannas is discussed in detail in Chapter 12.

Controlling environmental variables, that is those factors that determine the relative contributions of woodland and the size of constituent trees, are complex, including precipitation and length of dry season, soil nutrient status (in turn usually a function of the age of the landscape surface), soil texture, slope, and drainage. Cole's (1986) scheme, which applies well to the African savannas, sees the major factors as precipitation, texture, and soil-moisture relationships, although

woodlands, with a tree stratum of between 15 and 20 m high, are invariably associated in south Central Africa with Tertiary planation surfaces, so that geomorphological history, nutrient status, and the presence of laterite must also be considered important determinants (Cole, 1963, 1986).

The floristic composition and structure of savanna vegetation varies from region to region. In general, the African savannas differ from those of South America and Australia in their relatively restricted development of grasslands (Cole, 1986). The more typical image, therefore, of an African savanna is one dominated by an over-storey of deciduous trees. In Southern and south Central Africa, the trees are characteristically of the genera *Brachystegia*, *Isoberlinia*, and *Julbernardia*, all members of the same family (Fabaceae), which achieves a similar level of importance in the landscape to Myrtaceae (typified by members of the genus *Eucalyptus*) in Australian savannas. The so-called *miombo* woodlands are widespread in Zimbabwe, Zambia, Angola, Mozambique, and even north of the Equator. The West African savannas, on the other hand, are less diverse (Cole, 1986) and more seriously impacted by human activities. The miombo elements, especially *Brachystegia*, are absent from the savanna woodlands of West Africa, and its biogeographical relationships show greater affinity to the Guinea–Congolian rain forests.

Compared to the tropical rain forests of Africa, the savannas are ecologically and biogeographically considerably better studied. There have been two detailed analyses of energy throughput in African savanna ecosystems, namely the *lamto* savanna study from Ivory coast (Menaut and Ceser, 1982), and the Nylsvlei savanna study from South Africa (Huntley 1982). These studies indicate that biomass and productivity are highly variable in both space and time in savannas, depending on soil conditions, precipitation inputs, fire history, and grazing. Biomass is typically in the range 10 to 80 t ha^{-1} and productivity, dominated at Nylsvlei at least by the grass layer, in the range 1 to 5 t ha^{-1} yr^{-1}. Human-induced burning is an especially important factor in energy flow and nutrient cycling and, as pointed out by Robertson and Rosswall (1986), increasing disturbance under human-population pressure may produce marked increases in nitrogen loss in an ecosystem that would otherwise operate sustainably.

The development of savannas in Africa is a relatively recent phenomenon and can be related in particular to the increasing aridity and decreasing temperatures of the mid- to later Tertiary (Axelrod and Raven, 1979). This again supports the contention that geologically recent climatic changes of the later Tertiary and

Quaternary must have been especially important in adjusting the ranges of species and, accordingly, contemporary biogeographic patterns. The co-evolution of an exceedingly rich mammalian fauna based on the grazing food chain paralleled the spread of savannas as seasonal aridity developed, such that Africa now supports a diversity of grazing herbivores, and associated carnivores, second to none in the world. Seasonal aridity is also conducive to fire on a regular basis and this obviously has become an important ecological factor in the savannas of Africa. It is, however, a tool easily used, and abused, by human populations. Fire management, taken together with clearance of savanna vegetation for agricultural purposes, means that the savanna biome, like its rain-forest neighbour, is a highly threatened natural resource.

Semi-Arid and Arid Biomes

Towards the polar margins of the savanna woodland and grasslands of Africa, mean annual precipitation declines below 400 mm and the length of the dry season becomes a major ecophysiological constraint. It is here that we encounter the semi-arid and arid biomes of the continent (Figure 10.5) with more continuous cover of vegetation at the wetter end of the moisture gradient (areas sometimes referred to as 'steppe' or arid savannas) and the true desert at its drier end (see Chapter 13). Such regions coincide approximately with White's (1983) Sahara, Somalia–Masai, and Karoo–Namib regional centres of endemism and the Sahel and Kalahari–Highveld regional transition zones. The two major core arid regions are the Sahara and Namib Deserts where, although their respective causes of aridity are climatologically quite different, precipitation inputs are sporadic and generally below 100 mm per year. The terms arid, semi-arid, desert, and semi-desert tend to be used rather loosely, but areas in Africa with between 100 mm and 400 mm per annum precipitation are taken here to be semi-arid, and include the Sahel region to the immediate south of the Sahara and the Karoo and Kalahari 'deserts' of Southern Africa (Le Houérou, 1979; Leistner, 1979), the former being essentially a shrubland and the latter regarded more as an area of arid savanna (Thomas and Shaw, 1991).

The environmental conditions in such arid and semi-arid regions, with low and unreliably distributed rainfall, low humidities, and extremes of temperature, have prompted a fantastic array of adaptational responses in both plants and animals. These may involve avoidance of the problem environmental phenomena through, in the case of animals, behavioural responses, or in the case of plants, dormancy, either of the whole plant (for example annuals or ephemerals avoid the dry period by surviving it only as seed) or of individual organs. Other desert organisms *tolerate* the aridity through the employment of physiological adaptations (for example in plants the reduction of transpiring area, moisture storage, and the development of crassulacean acid metabolism for photosynthesis). The study of these complex, sometimes even bizarre, evolutionary adaptations has been a rich area for ecophysiological research (Louw and Seely, 1982). Among the more remarkable is the fog-basking behaviour of the Namib tenebrionid beetle, *Onymacris unguicularis*, which ascends to dune crests during nocturnal fogs (a reasonably common occurrence in the coastal regions as moisture-laden winds blow onshore from the cold Benguela current of the Atlantic Ocean to the west), promptly does a beetle version of a handstand and imbibes the moisture that condenses on the dorsal surface and runs down to the mouthparts (Hamilton and Seely, 1976).

From an ecological perspective, the deserts and semi-arid parts of Africa have been the subject of several major studies, with work by Kassas (1952, 1953; Kassas and Imam, 1954, 1959; Kassas and Girgis, 1964) on the Egyptian deserts; that of Bille and Poupon (1972) in Senegal; Seely (1990) in Namibia; and Cowling et al., (1986) and Cowling and Roux (1987) in the Karoo of South Africa. Clearly, under the constraints imposed by the physical environment, biomass and net primary productivity are limited. Biomass, concentrated mainly below ground for much of the year, rarely exceeds a few tonnes per hectare, and productivity, often concentrated in the short period after rains, would normally not exceed 0.5 t ha^{-1} yr^{-1}, and is often much less. Productivity is essentially a precipitation 'event-driven' phenomenon and may respond quickly to positive anomalies in the mean annual amounts.

Despite the obvious environmental constraints, the species-richness of arid and semi-arid regions of Africa is remarkably high, and probably underestimated owing to the remoteness of some of the regions. White (1983) alludes to the complexity of desert vegetation, the diversity of growth forms combining with habitat heterogeneity to produce a high degree of spatial variation. North African semi-arid and desert zones support more than 3000 plant species, about 10 per cent of which are endemic (Le Houérou, 1986). The Namib, Kalahari, and Karoo of Southern Africa are perhaps even richer in plant species and all these arid and semi-arid regions are surprisingly diverse. There are rather few phytogeographical connections between the different African desert regions, especially at the generic and specific level

(Shmida, 1985). Indeed, there are only eight desert species shared between the Sahara and the Namib, a phenomenon which Shmida (1985) suggests is a consequence of distinctive contemporary environments as well as differences in historical biogeography. The age of deserts and the antiquity of the desert flora has been the object of considerable biogeographical debate, although Shmida (1985) feels that the balance of the evidence suggests an ancient flora which expanded and diversified during the aridification of the Cenozoic. In Southern Africa, however, the development of the Benguela current in the middle Miocene was crucial to the evolution of aridity there, meaning that the Namib Desert, and its associated flora, is correspondingly young (Ward and Corbett, 1990). All the more remarkable, then, that the Southern African arid and semi-arid flora is so rich; Hilton-Taylor and Le Roux (1989) document the diversity of succulents in southern Africa as '. . . unparalleled elsewhere in the world' (p. 215), with over 2000 succulent taxa, and perhaps as many as 7000 plant species associated with the Karoo–Namib region as a whole.

Much has been made of the dynamic nature of arid and semi-arid environments in Africa, particularly in response to both natural climatic variability and the influences of human activities. 'Desertification', the spread of desert-like conditions into formally more biologically productive areas at the desert margins, has been portrayed as an inexorable and largely irreversible process in many semi-arid areas of Africa. The underlying causes of such a process have come under scrutiny in recent years (see Chapter 19). Complex socio-economic factors seem to be at work, in addition to a postulated aridification of climate acting in concert with so-called overgrazing and other associated inappropriate land-management practices. In fact, the assumed widespread occurrence of the process of desertification has itself been questioned (Hellden, 1991). There is little doubt, however, that a combination of unreliable precipitation, sparse and unproductive vegetation, and increasing numbers of people with few socio-economic options other than direct exploitation of the land, produces a potentially unstable ecological situation in many semi-arid parts of Africa, particularly at the southern margins of the Sahara.

Mediterranean-Type Shrublands of North Africa and the South-Western Cape

The distinctive climate of regions with a warm, dry summer alternating with a cool, rainy period is characteristic of the Mediterranean regions of North Africa and of the south-western corner of Southern Africa (see Chapter 17). The dominant vegetation formation in these regions is equally distinctive, varying proportions of evergreen sclerophyllous shrubs and trees producing a range of structural vegetation types, from open shrublands to closed-canopy woodland. These regions correspond broadly to White's (1983) Mediterranean and Cape regional centres of endemism and the Mediterranean–Sahara regional transition zone.

The vegetation of Mediterranean North Africa and the south-western Cape appears to exhibit convergent evolution. The fact of the season of the year in which temperatures are most suitable for growth being coincident with the period of minimum moisture availability makes for strong selection pressures. Added to this is the low nitrogen and phosphorus status of many of the associated soils (Specht and Moll, 1983) and these phenomena taken together appear to have favoured the evolution of sclerophylly. Canopy height and percentage cover are variable and nutrient-related, varying from closed-canopy oak forest on moderately leached soils in the Mediterranean basin, to the open shrubland of Southern Africa's *fynbos* communities developed either on highly leached and acidic, sandstone-derived soils, or on calcium-rich, high-pH soils on limestones (Specht and Moll, 1983). The lack of development of true woodland in the *fynbos* of the south-western Cape has intrigued biogeographers. The 'vacant tree niche' (Moll *et al.*, 1980) certainly sets it apart from the Mediterranean Basin, where oak-dominated woodlands on calcareous soils, and even gymnosperm forests on leached soils, occur widely. Fire, recurring on a frequency of from once every few years to once in several decades, is an additional ecosystem constraint in the shrublands although it is probably the major soil-mineralizing agent.

Fynbos in the south-western Cape Province of South Africa has been subject to much greater intensity of study than its counterpart in North Africa, perhaps in part because of its prolific plant-species richness (Hilton-Taylor and Le Roux, 1989). Much of the ecological and biogeographical research on *fynbos* has recently been reviewed (Cowling, 1992). The prevalence of sclerophylly, in particular among members of the Proteaceae and Restionaceae families, has been attributed to its adaptive function in regard to drought, to an improved nutrient-use efficiency in the low-nutrient environments and to the reduction in herbivory that it promotes (Stock *et al.*, 1992). Their occurrence mainly on nutrient-poor substrates means that many *fynbos* plants are carbon-enriched relative to nitrogen and other plant nutrients and this excess carbon is utilized in the development of, for example, secondary compounds which further restrict herbivory. Under the constraints imposed by dry summers, nutrient-depleted soils, and fire, it is perhaps hardly surprising that biomass and productivity values in these ecosystems are low. Phytomass rarely exceeds a few tonnes per hectare, and is obviously much lower after fire.

By far the most striking biogeographical aspect of the vegetation at the south-western tip of Africa is its plant-species richness. There are some 8500 plant species packed into what amounts, at least on a continental scale, to a very small geographical area (90 000 km²). The level of endemism (68 per cent of the species, Bond and Goldblatt, 1984) has rendered the flora sufficiently distinctive as to be accorded the status of Floristic Kingdom (Good, 1974). The diversity, which is notable at the species level, is all the more remarkable given relatively recent (Tertiary to Quaternary) development of summer aridity in this part of the world (Linder *et al.*, 1992).

Human impact on the Mediterranean-type shrublands has been more significant than in many other African biomes, perhaps because of the long history of human occupation of these landscapes. The shrubland vegetation of the Mediterranean Basin may, indeed, itself be a product of the human-induced degeneration of forests (Goudie, 1990). The fauna appears to have been subjected to especially cruel exploitation in the Mediterranean Basin, and the lust for blood that passed as entertainment during the height of the Roman Empire plundered elephant, rhinoceros, and zebra to such an extent that they were extinct in North Africa by the early centuries AD (Ponting, 1991). These landscapes have also been extensively altered by the invasion of introduced alien plant species. In the Cape *fynbos*, the apparently vacant tree niche has provided an opportunity for a number of introduced Australian members of the genus *Acacia* to expand their ranges to the degree that they represent arguably the most significant threat to the conservation of the region's species diversity (Breytenbach, 1986).

The Afromontane Biome

The mountains of Africa (see also Chapter 16) represent an archipelago-like chain of islands of distinctive biogeography (White, 1983). The zonation of vegetation on the African uplands illustrates classically the interrelationship of climate and biogeography, changing conditions with altitude being mirrored in the associated plant and animal communities (Hedberg, 1955). Of particular significance in the African uplands are

Afromontane forests (up to 3000 m in the equatorial zone), the 'ericaceous' belt dominated by giant heaths and heath-like plants, and the Afroalpine high altitude communities with their giant senecios and lobelias (Lind and Morrison, 1974). These regions correspond with White's (1983) Afromontane centre of endemism and the Afroalpine centre of extreme floristic impoverishment.

Biogeographical and ecological research in high-montane Africa has been confined to a few, isolated case studies, either of individual mountains and mountain ranges, or of individual species. The African mountains represent, however, a fascinating biogeographical arena, places where bizarre plant forms appear to have evolved, places where the normal annual climatic procession of winter and summer is repeated diurnally: at the summits, each day is a summertime and winter occurs every night. The Afroalpine flora associated with the summit areas is truly remarkable. Only a limited number of vascular plants survive the constraints of this environment and it is therefore an impoverished flora (Hedberg, 1986). That the selection pressures provided by such conditions are severe is suggested by the convergence illustrated by this and high-altitude floras from the other continents (Hedberg and Hedberg, 1979). Plants exhibiting giant rosettes, tussocks, and cushion-like forms are the most significant structural elements. Levels of endemism are high and many species, for example the grass, *Poa kilimanjarica*, are restricted to individual mountains (Markgraf-Dannenberg, in Hedberg, 1986).

At more moderate altitudes on the African mountains are found the upland evergreen forests, often in conjunction with grassland and heathland in a kind of mosaic. Approximately 4000 plant species are associated with this Afromontane flora and a degree of endemism (75 per cent) which points to its isolation for a considerable period (White, 1983). The origin of the grassland element in this flora has been the subject of particular controversy, White (1981) arguing its recent and anthropogenic origin, grassland having spread at the expense of forest in response to human-induced burning during the last 1000 years. Climatic changes during the Quaternary must, however, have produced major altitudinal shifts in the distribution of the various zonation belts of vegetation on these mountains, with cooler and more arid climates favouring grassland, long before the development of anthropogenic forcing. There is some palaeoecological evidence to support the idea of a more ancient origin for at least some of the Afromontane grasslands (Meadows and Linder, 1993). As with the other African biomes, however, the most significant biogeographical features appear to have become established only in the rather recent geological past.

The Biogeography of Africa: Past, Present, and Future

Africa, then, is an ancient continent, and yet one for which contemporary phytogeographical and zoogeographical patterns have essentially developed in the context of changing climates and continental configurations mainly during the last 65 million years or so. The Quaternary in particular seems to have been a crucial influence, as changes in climate (more particularly of precipitation) have provoked marked contractions and expansions of the ranges of taxa occupying the various African biomes (see Chapter 3). The Quaternary has also, of course, provided the temporal stage for the evolution and development of the human genus and species, an influence which, perhaps, more than any other, seems set to play the crucial role in the future evolution of the African flora and fauna.

The scale of human impact on the biogeography of the continent continues to grow in tandem with the exponential growth in human population numbers that has characterized the recent past. There is a long history of human interaction with the African environment, however, and some vegetation communities may themselves by a direct product of that interaction, for example the savanna grasslands (Singh *et al.*, 1983). Thus, although the biogeography of Africa is a product of a diverse set of influences—the making and breaking of land connections and changes in climate and geomorphology—the continued existence of many of its plant and animal species may now depend more on social than on physical environmental factors. That the biogeographical legacy of the continent is exceedingly rich is beyond question, it remains to be seen, however, if this wealth can survive the transformation and exploitation that all too often accompanies socio-economic development (see Chapter 20).

References

Alexandre, D. Y. (1984), 'Strata in Tropical Rain-Forest at Taî (Ivory Coast)', in A. C. Chadwick and S. L. Lutton (eds.), *Tropical Rainforest: The Leeds Symposium* (Leeds).

Axelrod, D. I. and Raven, P. M. (1978), 'Late Cretaceous and Tertiary Vegetation History of Africa', in M. J. A. Werger (ed.) *Biogeography and Ecology of Southern Africa*, (The Hague). 77–130.

Battistini, R. and Richard-Vindard, G. (1972) (eds.), *Biogeography and Ecology in Madagascar* (The Hague).

Bernhard-Reversat, F., Huttel, C., and Lemée, G. (1978), 'Structure and Functioning of Evergreen Rainforest Ecosystems of the Ivory Coast', in *Tropical Forest Ecosystems* (Paris), 557–79.

Bille, J. C. and Poupon, H. (1972), 'Recherches écologiques sur une savane sahélienne du Ferlo septentrional, Sénégal: Biomasse végétate et production primaire nette', *Terre et Vie*, 26: 366–82.

Bond, P. and Goldblatt, P. (1984), 'Plants of the Cape Flora: A Descriptive Catalogue', *Journal of South African Botany*, suppl. vol. 13: 1–455.

Bonnefille, R. and Riollet, G. (1988), 'The Kashiru Pollen Sequence (Burundi): Palaeoclimatic Implications for the Last 40 000 years BP in Tropical Africa', *Quaternary Research*, 30: 19–35.

Breytenbach, G. J. (1986), 'Impacts of Alien Organisms on Terrestrial Communities with Emphasis on Communities of the Southwestern Cape', in I. A. Macdonald, E. J. Kruger, and A. A. Ferrar (eds.), *The Ecology and Management of Biological Invasions in Southern Africa*. (Cape Town), 229–38.

Briggs, J. C. (1987), *Biogeography and Plate Tectonics* (Amsterdam).

Bush, M. B., Colinvaux, P. A., Wiemann, M. C., Piperno, D. R., and Liu, K. B. (1990), 'Late Pleistocene Temperature Depression and Vegetation Change in Ecuadorian Amazonia', *Quaternary Research*, 34: 330–45.

Cole, M. M. (1963), 'Vegetation and Geomorphology in Northern Rhodesia: An Aspect of the Distribution of the Savanna of Central Africa', *Geographical Journal*, 129: 290–310.

—— (1986), *The Savannas: Biogeography and Geobotany* (London).

Collinson, A. S. (1988), *Introduction to World Vegetation*, 2nd edn. (London).

Cowling, R. M. (1992) (ed.), *The Ecology of Fynbos: Nutrients, Fire and Diversity* (Cape Town).

—— and Roux, P. W. (1987), 'The Karoo Biome: A Preliminary Synthesis. Part 2: Vegetation and History', *South African National Scientific Programmes Report* (Pretoria), 142.

—— and Pieterse, A. J. H. (1986), 'The Karoo Biome: A Preliminary Synthesis. Part I: Physical Environment', *South African National Scientific Programmes Report* (Pretoria), 124.

Flenley, J. (1979), *The Equatorial Rain Forest: A Geological History* (London).

Good, R. (1974), *The Geography of the Flowering Plants*, 4th edn. (London).

Goudie, A. S. (1990), *The Human Impact on the Natural Environment*, 3rd edn. (Oxford).

—— (1992), *Environmental Change*, 3rd edn. (Oxford).

Grove, A. T. (1967), *Africa South of the Sahara* (Oxford).

—— (1989), *The Changing Geography of Africa* (Oxford).

Hall, J. B. (1977), 'Forest Types in Nigeria: An Analysis of Pre-exploitation Forest Enumeration Data', *Journal of Ecology*, 65: 187–99.

—— and Swaine, M. D. (1976), 'Classification and Ecology of Closed-Canopy Forest in Ghana', *Journal of Ecology*, 64: 913–51.

—— —— (1981), *Distribution and Ecology of Vascular Plants in a Tropical Rain Forest* (The Hague).

Hamilton, A. C. (1976), 'The Significance of Patterns of Distribution Shown by Forest Plants and Animals in Tropical Africa for the Reconstruction of Upper Pleistocene Palaeoenvironments: A Review', *Palaeoecology of Africa*, 9: 63–97.

Hamilton, W. J. and Seely, M. K. (1976), 'Fog Basking by the Namib Desert Beetle, *Onymacris unguicularis*', *Nature*, 262: 284–5.

Hedberg, I. and Hedberg, O. (1979), 'Tropical-Alpine Life-forms of Vascular Plants', *Oikos*, 33: 247–307.

Hedberg, O. (1955), 'Altitudinal Zonation of the Vegetation of the East African Mountains', *Proceedings of the Linnaean Society of London* 165: 134–50.

—— (1986), 'Origins of the Afroalpine Flora', in F. Vuilleumier and M. Monasterio (eds.), *High Altitude Tropical Biogeography* (New York and Oxford), 443–68.

Hellden, U. (1991), 'Desertification: Time for an Assessment', *Ambio*, 20: 372–83.

Hilton-Taylor, C. and Le Roux, A. (1989), 'Conservation Status of the Fynbos and Karoo Biomes', in B. J. Huntley (ed.), *Biotic Diversity in Southern Africa* (Cape Town), 202–23.

Huntley, B. J. (1982), 'Southern African Savannas', in B. J. Huntley, and B. H. Walker (eds.), *Ecology of Tropical Savannas* (Berlin).

Ingrouille, M. (1992), *Diversity and Evolution of Land Plants* (London).

Jordan, C. F. (1985), *Nutrient Cycling in Tropical Forest Ecosystems* (Chichester).

Kassas, M. (1952), 'Habitat and Plant Communities in the Egyptian Desert, 1: Introduction, *Journal of Ecology*, 40: 342–51.

—— (1953), 'Habitat and Plant Communities in the Egyptian Desert, 2: The Features of a Desert Community', *Journal of Ecology*, 41: 248–56.

—— and Imam, M. (1954), 'Habitat and Plant Communities in the Egyptian Desert, 3: The Wadi-bed Ecosystem', *Journal of Ecology*, 42: 424–41.

—— —— (1959), 'Habitat and Plant Communities in the Egyptian Desert, 4: The Gravel Desert', *Journal of Ecology*, 47: 289–310.

—— and Girgis, W. A. (1964), 'Habitat and Plant Communities of the Egyptian Desert, 5: The Limestone Plateau', *Journal of Ecology*, 52: 107–19.

Kennett, J. P. (1980), 'Palaeoceanographic and Biogeographic Evolution of the Southern Ocean during the Cenozoic, and Cenozoic Microfossil Datums', *Palaeogeography, Palaeoclimatology, Palaeoecology*, 31: 123–52.

Knoll, A. H. and Barghoorn, E. S. (1977), 'Archaean Microfossils Showing Cell Division from the Swaziland System of South Africa', *Science*, 198: 396–8.

Le Houérou, H. N. (1979), 'North Africa', in D. W. Goodall and R. A. Perry (eds.), *Arid Land Ecosystems: Structure, Functioning and Management* (Cambridge), 83–108.

—— (1986), 'The Desert and Arid Zones of Northern Africa', in M. Evenari, I. Noy-Meir, and D. W. Goodall (eds.), *Hot Deserts and Arid Shrublands*, vol. B (Amsterdam), 101–47.

Leistner, O. A. (1979), 'South Africa', in Goodall and Perry (eds.), *Arid Land Ecosystems*, 109–44.

Lind, E. M. and Morrison, M. E. S. (1974), *East African Vegetation* (London).

Linder, H. P., Meadows, M. E., and Cowling, R. M. (1992), 'History of the Cape Flora', in R. M. Cowling (ed.), *The Ecology of Fynbos*, 113–34.

Louw, G. and Seely, M. (1982), *Ecology of Desert Organisms* (London).

Mayr, E. and O'Hara, R. J. (1986), 'The Biogeographic Evidence Supporting the Pleistocene Refuge Hypothesis', *Evolution*, 40: 55–69.

Meadows, M. E. and Linder, H. P. (1993), 'A Palaeoecological Perspective on the Origin of Afromontane Grasslands, *Journal of Biogeography*, 20: 345–55.

Menaut, J. C. and Cesar, J. (1982), 'The Structure and Dynamics of a West African Savanna', in B. J. Huntley and B. H. Walker (eds.), *Ecology of Tropical Savannas* (Berlin).

Moll, E. J., McKenzie, B., and McLachlan, D. (1980), 'A Possible Explanation for the Lack of Trees in the Fynbos, Cape Province, South Africa', *Biological Conservation*, 17: 221–8.

Moreau, R. E. (1966), *The Bird Faunas of Africa and its Islands* (London).

Polunin, N. (1960), *Introduction to Plant Geography and Some Related Sciences* (London).

Ponting, C. (1991), *A Green History of the World* (London).

Proctor, J. (1989) (ed.), *Mineral Nutrients in Tropical Forest and Savanna Ecosystems* (Oxford).

Richards, P. W. (1952), *Tropical Rain Forest* (Cambridge).

—— (1993), 'Africa the "Odd Man Out"', in B. J. Meggers, E. S. Ayensu, and W. D. Duckworth (eds.), *Tropical Forest Ecosystems in*

Africa and South America: A Comparative Review (Washington, DC). 21–6.

Robertson, G. P. and Rosswall, T. (1986), 'Nitrogen in West Africa: The Regional Cycle', *Ecological Studies*, 42: 80–100.

Scholtz, A. (1983), 'The Palynology of the Upper Lacustrine Sediments of the Arnot Pipe, Banke, Namaqualand', *Annals of the South African Museum*, 95: 1–109.

Seely, M. K. (1990) (ed.), *'Namib Ecology: 25 Years of Namib Research'* (Pretoria).

Shmida, A. (1985), 'Biogeography of the Desert Flora', in M. Evenari, I. Noy-Meir, and D. W. Goodall (eds.), *Hot Deserts and Arid Shrublands* (Amsterdam). 23–77.

Siesser, W. G. (1980), 'Late Miocene Origin of the Benguela Upwelling System of Northern Namibia', *Science*, 208: 83–96.

Singh, J. S., Lauenroth, W. K., and Milchunas, D. G. (1983), 'Geography of Grassland Ecosystems', *Progress in Physical Geography*, 7: 46–80.

Smith, A. C. (1973), 'Angiosperm Evolution and the Relationship of the Floras of Africa and America', in B. J. Meggers, E. S. Ayensu, and W. D. Duckworth (eds.), *Tropical Forest Ecosystems in Africa and South America: A Comparative Review* (Washington, DC), 49–61.

Smith, T. M., Shugart, M. M., Bonan, G. B., and Smith, J. B. (1992), 'Modelling the Potential Response of Vegetation to Global Climate Change', in F. I. Woodward (ed.), *The Ecological Consequences of Global Climate Change (Advances in Ecological Research, Volume 22)* (London), 93–116.

Specht, R. L. and Moll, E. J. (1983), 'Mediterranean-type Heathlands and Sclerophyllous Shrublands of the World: An Overview', in F. J. Kruger, D. T. Mitchell, and J. V. M. Jarvis (eds.), *Mediterranean-type Ecosystems: The Role of Nutrients*, (Berlin), 41–65.

Stock, W. D., van der Heyden, F., and Lewis, O. A. M. (1992), 'Plant Structure and Function', in Cowling (ed.), *The Ecology of Fynbos*, 226–40.

Stringer, C. B. (1989), 'Documenting the Origin of Modern Humans', in E. Trinkhaus (ed.), *The Emergence of Modern Humans* (Cambridge), 67–96.

Szalay, F. S. and Delson, E. (1979) *'Evolutionary History of the Primates'*, (New York).

Talbot, R. W., Andreae, M. O., Andreae, T. W., and Harris, R. C. (1990), 'Regional Aerosol Chemistry of the Amazon Basin During the Dry Season', *Journal of Geophysical Research*, 93: 1499–508.

Tallis, J. H. (1991), *Plant Community History* (London).

Thorne, R. F. (1973), 'Floristic Relationships between Tropical Africa and Tropical America', in: Meggers, Ayensu, and Duckworth (eds.), *Tropical Rain Forest Ecosystems in Africa and South America*, 27–47.

Thomas, D. S. G. and Shaw, P. A. (1991), *The Kalahari Environment* (Cambridge).

Van Steenis, C. G. G. J. (1972), '*Nothofagus*, Key Genus to Plant Geography', in D. M. Valentine (ed.), *Taxonomy, Phytogeography and Evolution* (New York), 275–88.

Walter, H. and Lieth, M. (1967), *Klimadiagram-Weltatlas* (Jena).

Ward, S. D. and Corbett, I. (1990), 'Towards an Age for the Namib', in Seely (ed.), *Namib Ecology*, 17–26.

White, F. (1981), 'The History of the Afromontane Archipelago and the Scientific Need for its Conservation', *African Journal of Ecology*, 19: 33–54.

—— (1983), *The Vegetation of Africa: A Descriptive Memoir* (Paris).

White, M. E. (1986), *The Greening of Gondwana* (Sydney).

Whitmore, T. C. (1990), *An Introduction to Tropical Rain Forests* (Oxford).

11 Forest Environments

Alan Grainger

Introduction

Forest does not dominate the vegetation cover of Africa as it does other tropical continents. Only 7 per cent of the continent is presently covered by closed forest compared with 37 per cent of South America. This is partly explained by the low precipitation received by large parts of Africa, which even under purely natural conditions would result in a complex mosaic of closed forests, open woodlands, and grasslands. But equally important is that the primal forest cover has been extensively cleared or degraded by a long history of human activity. So great has this influence been that we can still only guess what the original forest cover was like. Even in areas where forests appear to be untouched by human hands, closer examination often reveals evidence of disturbance long ago.

In Africa, therefore, probably more than in any other part of the tropics, present forest distribution cannot be understood apart from that of human settlement. Over vast areas humans and trees coexist, but in a relationship greatly biased to the human side. Only in limited areas of sparse population do great closed forests remain, yet even here they are in a state of symbiosis with tribes of forest dwellers, and both are threatened by external pressures for deforestation.

A conventional biogeographical description of the ecology and climax distribution of the main types of forest ecosystems in Africa and how their distribution is related to environmental factors, therefore seems inappropriate. Given the complexity of the distribution it would also be very long (White, 1983). It is more meaningful instead to examine forest distribution and composition as they are today, not as they possibly once were, for the biogeography of Africa is now predominantly cultural, not natural. Thus, the large area of derived savanna woodland, though an aberration from the climax vegetation cover, has to be accepted as present reality, just like the largely agricultural landscape of Britain, which long ago was also heavily forested. This chapter therefore views Africa's forests from a cultural biogeographical perspective, examining how they have been moulded by, and have accommodated to, human pressures, experienced for longer here than anywhere else on the planet.

The chapter has three main parts. The first describes the main types of natural forest ecosystems, how their climax distribution is linked to climate and landforms, and where examples of each type still remain today. Part two examines the various human processes which have modified and transformed the natural distribution, structure, and composition of the forests and led to their present state. Part three analyses the present forest distribution from a quantitative point of view and how it has changed in recent history. It begins by looking at the areal distribution of African forests, and shows how estimates have gradually improved owing to the transition from expert assessments to measurement-based estimates and then to maps. It then assesses progress in monitoring the two main types of changes in forest distribution: deforestation and degradation, which includes two aspects of environmental change treated in more detail elsewhere: desertification (Chapter 19) and biodiversity decline (Chapter 20).

Natural Forest Ecosystems

Vegetation Classification

Systems used to classify vegetation on continental and global scales have traditionally used a physiognomic approach and taken account of links between vegetation and climate, so maps of the assumed distribution of climatic climax vegetation types have been based on the distribution of climatic zones. The wide range of ecosystems present in the world are grouped for convenience into a small number of major types, called biome types, each of which is represented by distinctive biomes in different continents (Eyre, 1968). Africa's biomes include Mediterranean vegetation, desert, tropical savanna grassland and scrub, tropical rain forest, tropical seasonal forest, and montane vegetation, shown in the inset to Figure 11.1. The main map, produced by Eyre, is more detailed since some of the biomes have been subdivided.

Floristic composition varies greatly within each biome, so the next level in hierarchical classifications commonly divides biomes into plant communities with distinct flora, often related to local soil, slope, and drainage conditions. Other systems classify continental vegetation into large phytogeographical zones. White (1983) divided Africa into twenty zones; ten were centres of endemism and the rest transitional, mosaic, or impoverished areas (Figure 11.2).

White co-ordinated production of the UNESCO vegetation map of Africa (Figure 11.3) (UNESCO-AETFAT-UNSO, 1983). Its classification system differed from previous systems, and even from UNESCO's own international vegetation classification system (UNESCO, 1975). Vegetation was classed mainly independently of climate and other environmental factors, with sixteen principal physiognomic types in five groups (Table 11.1).

White claimed that the first two groups were more representative of actual vegetation cover than previous typologies, and helped to overcome confusion between the various types of 'savannas'. But physiognomic types were considered to be inadequate by themselves, and so they were included within the overall framework of twenty phytogeographical zones, which was helped by coincidence between the physiognomic and floristic boundaries (White, 1983). The map was composed of eighty mapping units, defined using the different physiognomic types found in each phytogeographical zone, transitions between them, and mosaics of different combinations. The mosaics and transition zones gave a more realistic coverage of vegetation distribution than in traditional maps, which assumed definite boundaries between biomes.

Since the complete set of UNESCO mapping units is rather unwieldy, this chapter uses the simpler and traditional climate-based approach to classification, making reference to the UNESCO approach as appropriate.

Table 11.1 Synopsis of the main vegetation types in the UNESCO-AETFAT-UNSO Vegetation Map of Africa

Formations of regional extent
1. Forest. A continuous stand of trees at least 10 m tall, their crowns interlocking.
2. Woodland. An open stand at least 8 m tall with a canopy cover of at least 40 per cent. The field layer is usually dominated by grasses.
3a. Bushland. An open stand of bushes usually 3–7 m tall with a canopy cover of at least 40 per cent.
3b. Thicket. A closed stand of bushes usually 3–7 m tall.
4. Shrubland. An open or closed stand of shrubs up to 2 m tall.
5. Grassland. Land covered with grasses and other herbs, either without woody plants or the latter not covering more than 10 per cent of the ground.
6. Wooded grassland. Land covered with grasses and other herbs, with woody plants covering 10–40 per cent of the ground.
7. Desert. Arid landscapes with a sparse plant cover.
8. Afroalpine vegetation. Physiognomically mixed vegetation on high mountains where night frosts are liable to occur throughout the year.

Transitional formations of local extent
9. Scrub forest. Intermediate between forest and bushland or thicket.
10. Transition woodland. Intermediate between forest and woodland.
11. Scrub woodland. Stunted woodland less than 8 m tall or vegetation intermediate between woodland and bushland.

Edaphic formations
12. Mangrove. Open or closed stands of trees or bushes occurring on shores between high and low-water mark.
13. Herbaceous freshwater swamp and aquatic vegetation.
14. Halophytic vegetation. Saline and brackish water swamp.

Formation of distinct physiognomy but restricted distribution
15. Bamboo

Unnatural vegetation
16. Anthropic landscapes. Areas where natural vegetation has been entirely eliminated.

Source: Based on White (1983).

Vegetation and Climate
Closed and Open Forests

There are two main physiognomic types of tropical forests, closed and open. Most forests in the humid tropics are closed, with an almost continuous closed canopy covering a large proportion of the land beneath it. Forests become increasingly deciduous as the climate becomes drier and more seasonal, until in semi-arid areas rainfall is usually insufficient, and too seasonally concentrated, to support dense tree cover throughout the year. This results in forests that contain mixtures of trees, bushes, and shrubs scattered over grassy plains, and because of the open canopy they are known as open (or

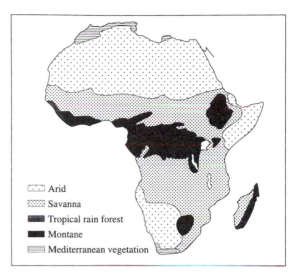

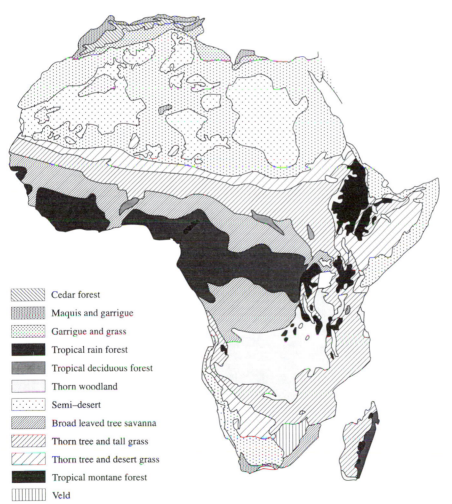

Arid

Savanna

Tropical rain forest

Montane

Mediterranean vegetation

Cedar forest

Maquis and garrigue

Garrigue and grass

Tropical rain forest

Tropical deciduous forest

Thorn woodland

Semi–desert

Broad leaved tree savanna

Thorn tree and tall grass

Thorn tree and desert grass

Tropical montane forest

Veld

Fig. 11.1. The potential distribution of vegetation in Africa. Inset shows global biome types.
(*Source:* Based on Eyre, 1968.)

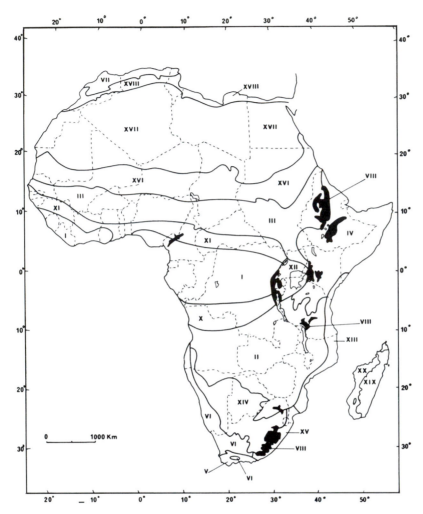

Fig. 11.2. The major phytogeographical zones of Africa.

I. Guineo–Congolian regional centre of endemism. II. Zambezian regional centre of endemism. III. Sudanian regional centre of endemism. IV. Somalia–Masai regional centre of endemism. V. Cape regional centre of endemism. VI. Karoo–Namib regional centre of endemism. VII. Mediterranean regional centre of endemism. VIII. Afromontane archipelago-like regional centre of endemism, including IX, Afroalpine archipelago-like region of extreme floristic impoverishment (not shown separately). X. Guinea–Congolia/Zambezia regional transition zone. XI. Guinea–Congolia/Sudania regional transition zone. XII. Lake Victoria regional mosaic. XIII. Zanzibar–Inhambane regional mosaic. XIV. Kalahari–Highveld regional transition zone. XV. Tongaland–Pondoland regional mosaic. XVI. Sahel regional transition zone. XVII. Sahara regional transition zone. XVIII. Mediterranean/Sahara regional transition zone. XIX. East Malagasy regional centre of endemism. XX. West Malagasy regional centre of endemism. (*Source*: White, 1983.)

savanna) woodlands. One of the peculiar features of African vegetation is that open woodlands are widespread in humid areas which on the basis of climatic considerations should be covered by closed forest.

Climax and Actual Vegetation Cover

Five main African forest types are considered in this chapter: tropical rain forest, tropical wetland forest, tropical montane forest, tropical dry deciduous forest, and open savanna woodland. Two others, thorn scrub, and Mediterranean vegetation, are only of limited extent or are found in areas peripheral to the main focus of the chapter. The *almost* ideal distribution of major climax vegetation types, or biomes, is shown in Figure 11.1. Eyre acknowledged that because of extensive human modification it was difficult to be sure what the original climax cover was in some areas, so he included savanna woodland and other present non-climax vegetation types,

with queries attached to them. The map of African vegetation in Grove (1978) distinguished what at that time was a standard representation of tropical rain forest distribution from its postulated climax distribution based on rainfall considerations (Figure 11.4), showing present savanna areas which originally may have been covered by tropical rain forest.

Closed Forests and Seasonality in the Humid Tropics

The core of Africa's closed forest area is in the humid tropical zone—that part of the wider tropics near the Equator and characterized by high humidity and moderately high temperature. More precisely, precipitation exceeds potential evapotranspiration, which is basically satisfied if mean rainfall is over 75 mm per month or 900 mm per annum (Fosberg *et al.*, 1961).

Mean rainfall and the degree of seasonality vary

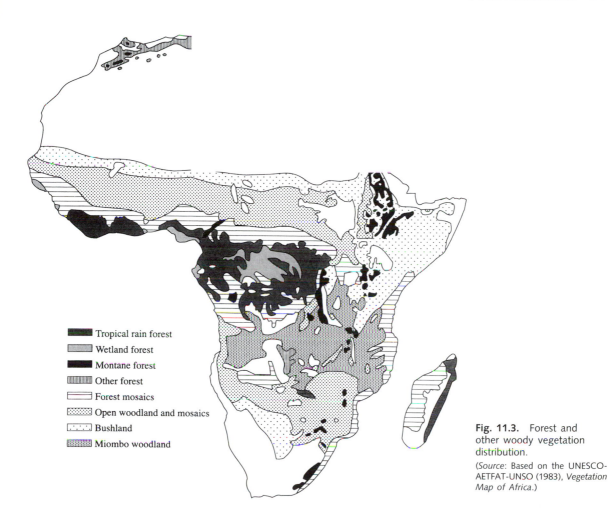

Tropical rain forest
Wetland forest
Montane forest
Other forest
Forest mosaics
Open woodland and mosaics
Bushland
Miombo woodland

Fig. 11.3. Forest and other woody vegetation distribution.

(*Source*: Based on the UNESCO-AETFAT-UNSO (1983), *Vegetation Map of Africa.*)

greatly in Africa. In the humid tropics of West and Central Africa, mean annual rainfall is generally between 1500 and 2500 mm, rising to 3000–4000 mm (and above) in the most westerly areas (Liberia and Sierra Leone) and on the border between West and Central Africa, in Nigeria and Cameroon. There are one or two dry seasons, lasting from 3 to 5 months each year. Permanently humid areas, with no distinct dry season, do exist, e.g. in northern Congo, but are of only a limited extent. Generally, mean annual rainfall near the Equator is rather lower than in South America and much lower than in South-east Asia, where there is an extensive permanently humid zone.

Forests in the humid tropics are referred to collectively as tropical moist forests, a term devised initially to cope with the poor resolution of tropical forest statistics

(Sommer, 1976), though it is also useful when the forest terminologies of different tropical regions are incompatible. It comprises two main types: tropical rain forest and tropical moist deciduous forest. The richest tropical moist forest of all is tropical evergreen lowland rain forest, which requires permanently humid conditions: low seasonality, not high rainfall, is the critical factor (Whitmore, 1984). Where there is a marked dry season in Asia and Latin America, tropical moist deciduous forest (the 'monsoon forest' of Schimper (1898)) is found. Its canopy trees are evergreen, but the taller emergent trees which penetrate the canopy are deciduous and lose their leaves at the same time in the dry season (Whitmore, 1984).

Whitmore (1984) believed that, owing to its generally lower rainfall and pronounced seasonality, Africa

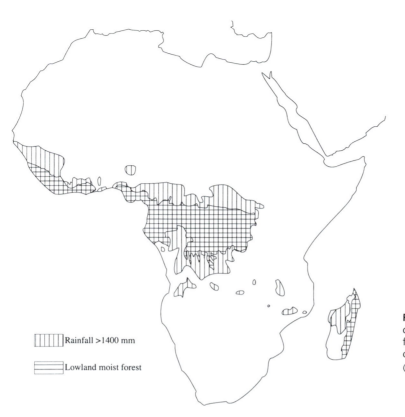

Rainfall >1400 mm

Lowland moist forest

Fig. 11.4. 'Actual' distribution of tropical rain forest and climax area based on rainfall data. (*Source*: Grove, 1978.)

lacked tropical evergreen rain forest of the kind found in Asia, and that the richest form of this biome type (Table 11.2) was semi-evergreen rain forest, which has a proportion of deciduous trees. But limited areas of true evergreen rain forest are thought to exist by other authorities (Richards, 1952; Longman and Jeník, 1987). Tropical moist deciduous forest, on the other hand, which is a common transitional vegetation cover on the fringes of tropical rain forest in Asia and, to a lesser extent Latin America, does not have the same role in Africa, where the boundary between tropical rain forest and savanna is usually quite sharp.

The overall degree of deciduousness has long been used as an indicator to distinguish major forest types (Eyre, 1968), but leads to various problems of particular relevance to Africa. One arises simply because of a multiplicity of terms: some classifications use 'semi-evergreen' and others 'semi-deciduous', and the difference between them is often obscure. Another stems from an inconsistent use of terms in different countries and continents, with the same terms being used to refer to different types of forest. Thus, in Asia, tropical evergreen rain forest is composed wholly of evergreen trees, and

if some deciduous trees are present it is called semi-evergreen rain forest. But while 'evergreen rain forest' or 'evergreen forest' is often mentioned in the African literature it may not necessarily correspond to its equivalent in Asia. In a review of this issue, Letouzey (1978) quoted Van Steenis's view that the wettest tropical moist deciduous forest in Asia may be equivalent to what is called evergreen rain forest in Africa.

Two terms have been used to describe African tropical moist forest with a deciduous component: evergreen seasonal forest and semi-deciduous (or mixed deciduous) forest. These correspond to the wet-evergreen and dry-evergreen forest classes of Jones and Keay, quoted by Richards (1952). Evergreen seasonal forest contains deciduous emergent trees, while semi-deciduous forest has deciduous trees in the main canopy as well (Longman and Jeník, 1987). Semi-deciduous forest has more deciduous trees, though according to Richards (1952) it still does not have as many as a typical tropical moist deciduous (monsoon) forest in continental Asia. In structure and physiognomy, both types are closer to the rain forests of Asia and Latin America. The canopy is lower and more open, but emergents can grow up to 40 m.

Table 11.2 Types of tropical moist forest ecosystems

Site type	Forest type requirements	Specific site or climate
Dryland	Lowland evergreen rain forest	Ever-wet climate
	Lowland semi-evergreen rain forest	Short dry season
	Evergreen seasonal forest	
	Mixed deciduous forest	
	Moist deciduous forests (various types)	Pronounced dry season
	Heath forest	Podzolized sandy soils
	Limestone forest	Over limestone
	Ultrabasic forest	Over ultrabasic rocks
Wetland	Freshwater swamp forest	
	Permanent swamp forest	Permanently wet
	Seasonal swamp forest	Periodically wet
	Peat swamp forest	Coastal or inland
	Saltwater swamp forest	
	Mangrove forest	Coastal
	Brackish water swamp forest	Coastal
	Beach forest	Coastal
Montane	Lower montane rain forest	Transition height varies
	Upper montane rain forest	Transition height varies
	Sub-alpine forest	Transition height varies

Source: Adapted from Whitmore (1990).

The transition between the two types is not easy to spot in the field. Apart from the different degrees of deciduousness, the main distinction is floristic (Richards, 1952). Many tree species are found in both types of forest but, for example, *Lophira procera* grows in tropical rain forest and evergreen seasonal forest, though not in semi-deciduous forest, and *Triplochiton scleroxylon* and *Chlorophora excelsa* occur only in semi-deciduous forest. *Khaya ivorensis* and *K. anthotheca* are found in evergreen seasonal forest and *K. grandiflora* in semi-deciduous forest. In Ghana, Hall and Swaine (1981) divided the two forest types further, distinguishing between wet evergreen, moist evergreen, moist semi-deciduous, and dry semi-deciduous forests, the latter being found on the border with savanna (Figure 11.5).

Given the difficulties of applying standard global vegetation classifications to Africa, the tropical moist forest terminology in this chapter is based on Whitmore's (1990) classification of tropical rain forest sub-types or formations (Table 11.2), which uses Asian forests as points of reference, though it has been amended here to include evergreen seasonal forest and semi-deciduous forest as sub-categories of tropical semi-evergreen rain forest.

The Transition from Moist to Dry Forest

The general trend towards a drier and more seasonal climate with increasing latitude leads to a distinct 'banding' of vegetation types to the north and south of the core area of tropical rain forest along the southern coast of West Africa (the Gulf of Guinea) and in the Congo Basin (Figure 11.1). Many countries therefore contain a range of humid and dry biomes. Liberia is the only country in West Africa wholly within the tropical rain forest biome boundary. The pattern is complicated by local climatic variations due to ocean currents and the angle presented by the coast to rain-bearing monsoon winds. Thus, the narrow coastal plain of the Congo is covered by a mosaic of savanna and forest, and the West African block, which extends as far west as Guinea Bissau, is divided into two by the 'Dahomey Gap', a wedge of land with a more seasonal climate in Togo, Benin, and eastern Ghana which is covered by savanna. The species compositions of the two West African sub-blocks differ.

On the northern fringes of the rain-forest belt in West and Central Africa there is usually a sharp transition from tropical rain forest to open savanna woodland (discussed in Chapter 12). Savanna woodland has various sub-types, which in Eyre's classification pass from broad-leaved tree savanna to thorn-tree tall grass savanna and thorn-tree desert grass savanna as mean annual rainfall declines from 1500 mm to as low as 200 mm (a thorny habit reduces transpiration losses as rainfall declines).

Substantial areas of dry closed forest also occur. Immediately to the south of the tropical rain forest zone in Central Africa is a large area of broad-leaved tree

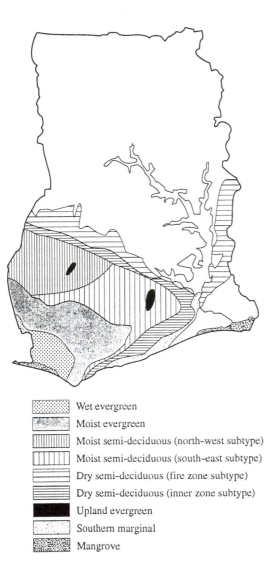

Fig. 11.5. Distribution of forest types in Ghana.
(*Source:* Hall and Swaine, 1981.)

savanna that gives way to an area of tropical dry deciduous closed forest. Then, as the climate becomes even
drier and more seasonal with increasing latitude, the
sequence of open forest types follows that in the north.

Vegetation and Landforms

Africa's most typical landforms are plains or low hills,
so lowland forests are widespread in both moist and dry
areas. Mountains are mainly found in the east of the continent, where fracturing led to rift valley formation and
volcanic activity. The overall pattern is complicated

by the elevation, depression, and slight folding of the
old shield into a series of plains at differing heights up
to 2600 m above sea-level. But essentially Africa may
be divided into two parts: (a) the north and west, consisting of low plains ranging from 150–600 m; and (b)
the south and east with high plains over 1000 m (Grove,
1978). Tropical rain forest grows both in the lowlands
near the coast and on higher plateaux: for example, the
Congo Basin, 250–480 m above sea-level, is a broad,
shallow depression in the surface of one plateau. In contrast, montane forest has only a limited extent, being
found mainly in a band in eastern Africa, running from
northern Ethiopia to Malawi, skirting the eastern edge
of the Congo Basin.

Tropical Rain Forest

Tropical rain forest is the first of five principal forest
types identified in this chapter. All five are now briefly
described but there is no space to refer to all the different types of tropical moist forest in Table 11.2. Detailed
descriptions of these are given in Whitmore (1990) and
Eyre (1968).

Tropical Evergreen Lowland Rain Forest

The richest type of tropical rain forest is tropical evergreen lowland rain forest (Table 11.2). Its dense assem-
blage of trees, with a closed canopy 30–40 m above the
ground, pierced by taller 'emergent' trees, is an efficient
collector of solar energy and has the highest biomass
density and net primary productivity of any ecosystem
in the world. There has been a long debate on the presence of distinct layers of vegetation (Richards, 1952),
but now the ecosystem is usually recognized as having
just (a) an emergent layer; (b) a canopy, together with
the trees immediately below it and for most purposes
coincident with it; (c) an intermediate layer of small fruit
trees and adolescent seedlings; and (d) a sparse herb
layer, as little light penetrates the canopy and reaches
the forest floor (Whitmore, 1984). The marked gradation of temperature and light from canopy to ground
provides many animal and plant niches, leading to
the high species-diversity which is another rain forest
characteristic.

Tropical rain forest trees usually have tough leathery
leaves to withstand the high temperatures experienced
in the canopy. Water also drains rapidly off them, though
there is debate about the effectiveness of their distinctive 'drip tips' in this regard (Longman and Jeník, 1987).
Many trees in lowland tropical rain forests also have
large buttresses, up to 3–4 m in height, whose purpose is thought to be to provide additional mechanical
support, since many (but not all) tree roots lie just

Legend:
- Wet evergreen
- Moist evergreen
- Moist semi-deciduous (north-west subtype)
- Moist semi-deciduous (south–east subtype)
- Dry semi-deciduous (fire zone subtype)
- Dry semi-deciduous (inner zone subtype)
- Upland evergreen
- Southern marginal
- Mangrove

underneath the forest floor so as to capture nutrients released from decaying dead organic matter. The abundance of vegetation has misled many people into thinking that the underlying soils are very fertile. Instead, there exists an intricate set of nutrient-cycling processes (Proctor, 1983, 1989), though they are largely lost when forests are cleared and cultivated, exposing the soils and often leading to rapid fertility depletion.

Tropical rain forests contain the highest density of woody dicotyledons in the world, but palms are also very prevalent, and attached in various ways to the trees are woody or herbaceous climbing plants, called lianes, which reach up to just below the canopy and are rooted in the ground. Many of them are members of the genus *Ficus*, and are often exploited as sources of valuable fibres. The top of the rain-forest canopy is usually covered with epiphytes, such as orchids, which are also physically attached to host trees but not rooted in the ground. Instead their roots hang freely, or rest in hollows in tree trunks, collecting nutrients in various ways. Evergreen rain forest would occur naturally only in very limited areas of Africa, including southern Liberia, a small coastal area in Nigeria, the south-west and south-east tips of Ivory Coast, coastal Cameroon, and parts of the Congo Basin (Sayer *et al.*, 1992).

Tropical Semi-Evergreen Lowland Rain Forest

Semi-evergreen lowland rain forest is the most common form of tropical rain forest in Africa but not necessarily equivalent to forest of the same name in Asia and Latin America, which has fewer emergent trees than evergreen rain forest and up to a third of all trees are deciduous, though they do not shed their leaves simultaneously (Taylor, 1960). In Africa, there are two main sub-types: (1) *Evergreen seasonal forest*, which contains deciduous emergent trees that shed their leaves in the dry season, though not all trees do so simultaneously; (2) *Semi-deciduous forest*, in which most emergent trees are quite drought resistant and shed their leaves in the dry season, as do some canopy trees. It has herbaceous climbers but almost no epiphytes, and tree leaves lack drip tips (Longman and Jeník, 1987).

The UNESCO Vegetation Map of Africa divides rain forests into two main categories, 'wetter types' and 'drier types', and distinguishes forests in the Guineo–Congolian phytogeographical zone from those in the Malagasy zone. Drier types are described as 'semi-evergreen', but seem to correspond to the semi-deciduous category above (White, 1983), so wetter types include both evergreen seasonal forest and evergreen rain forest, as defined above.

Bands of forest in West Africa typically follow a north–south climate gradient. For example, in Ivory Coast, tropical evergreen rain forest in the south-east and south-west extremities gives way to evergreen seasonal forest and then farther north to semi-deciduous forest, though little of this forest remains. A similar sequence occurs in Ghana (Figure 11.5) (Hall and Swaine, 1981). Rain forests in Central Africa are slightly taller than in West Africa, but even in the Congo Basin, which straddles the Equator, most rain forest is semi-evergreen, and evergreen seasonal forest gives way to semi-deciduous forest on the block's northern fringe in the south of the Central African Republic. Semi-evergreen lowland rain forest also occurs naturally in Uganda, a small area of Kenya, and eastern Madagascar (Sayer *et al.*, 1992).

Tropical Wetland Forests

A second group of rain forests (Table 11.2) is found in wetland areas. It can be divided into the two main subgroups of freshwater and saltwater swamp forests (Whitmore, 1984).

Among saltwater swamp forests, mangroves are low (up to 10 m high) bushy forests of limited species diversity situated along coasts and in the lower reaches of river estuaries. Once very prevalent in Africa, clearance for rice cultivation has been extensive so today they occur mainly in Gambia, Guinea, Guinea-Bissau, the Niger Delta in Nigeria, Cameroon, Gabon, and Tanzania (Sayer *et al.*, 1992; Orme, this volume). Some mangroves (e.g. *Rhizophora* spp.) have stilt roots; some (e.g. *Avicennia* spp.) have aerial tap roots or pneumatophores; some have neither. Mangroves form valuable barriers between the sea and farmlands, and by trapping silt around their tangled stilt roots or pneumatophores they are an important natural means for reclaiming land from the sea.

Freshwater swamp forests are mainly found adjacent to rivers. Those closest to rivers are usually permanently under water, and quite stunted with intermingling branches and roots. Farther away are those swamp forests whose habitat is flooded only in the rainy season. These have a much higher canopy and often contain valuable timber trees. Peat swamp forests grow in a variety of conditions, from lowlands to mountains, where anaerobic conditions favour peat-bog development. Freshwater swamp forests are still widespread, but mainly concentrated in the central Congo Basin in Congo and Zaire (Figure 11.3), along other rivers in these two countries, above the tidal limit in the Niger Delta, and in Tanzania (Sayer *et al.*, 1992).

Tropical Montane Forests

Montane forests are found as part of a distinct pattern of vegetation zones on mountains (Figure 11.6), linked to

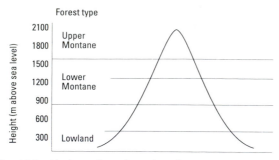

Fig. 11.6. Ideal zonation of montane forests. (*Source*: Based on Whitmore, 1984.)

the trend toward cooler and moister climates and greater daily temperature variation as altitude increases. Only some can be described as rain forests, so a general category is used here. In the humid tropics lowland rain forest normally merges into *lower montane rain forest* (also called sub-montane forest in some classifications (Eyre, 1968)) at about 1000 m, the exact altitude depending on the height of the mountain (Whitmore, 1984). The canopy of lower montane rain forest is only slightly lower than that of lowland rain forest.

The transition to *upper montane forest* (also called montane forest) is sharper, since the latter has a flat rather than undulating canopy, only half as high as that of lower montane rain forest (typically 15–25 m). Upper montane rain forest is usually above the cloud line so its trees tend to be covered with mosses and filmy ferns, hence the popular names of 'mossy forest' and 'cloud forest'. Both lower and upper montane forest may contain coniferous trees and tree species characteristic of temperate forests. Closer to the summit is *sub-alpine forest*, with an even lower canopy, and gnarled and stunted trees.

The ideal sequence from lowland rain forest to upper montane forest is not as common in Africa as in other parts of the humid tropics, but it does occur, for example, on the eastern fringe of the Congo Basin and in the Adamawa Highlands of north-west Cameroon, where lower montane forest begins at about 1200 m and gives way to upper montane forest at 1800 m (Gartlan *et al.*, 1992). Most of the montane forests have been cleared in the small and largely highland countries of Burundi and Rwanda, which adjoin the mountain chain in eastern Zaire.

Patterns in East Africa proper are more complex. Most of the tropical rain forest remaining there is montane forest and its distribution is highly fragmented. A large part of Ethiopia consists of high plateaux, and on the lower slopes of the Central Ethiopian Plateau in the south-west of the country, moist broadleaved forests,

containing such genera as *Albizia*, give way to coniferous forests of *Juniperus* and *Podocarpus* with increasing altitude. On the high plateau at 2500–3200 m are montane woodlands with mixtures of coniferous (*Juniperus*) and broadleaved (*Acacia*) species, giving way to montane savannas at even higher altitudes.

Most mountains in Kenya, Tanzania, and Uganda rise out of relatively dry, elevated plateaux covered by savanna. But with increasing altitude conditions become moister and montane forests occur with genera such as *Ocotoea*, *Podocarpus*, and (in drier areas) *Juniperus*, e.g. between 2000 and 3500 m on Mount Kenya and Mount Elgon. But as these conditions are also highly attractive for agriculture, forest clearance has been extensive (Lewis and Berry, 1988; Lanly, 1981). Montane rain forest is also found in the extreme south of Sudan.

Tropical Dry Deciduous Forest

Tropical dry deciduous forest is found in places on the northern rain-forest fringe in West and Central Africa, e.g. in the Central African Republic, Congo, and Ghana, and extensively in south Central Africa, where it is called 'miombo woodland'. It has a much lower canopy (20 m) than rain forest and a distinct grassy layer that dies off in the 4–7 month dry season (Eyre, 1968). It would be more aptly described as dense savanna woodland rather than closed forest, since the trees have spreading, flat tops and there is good visibility through the forest, though canopy cover is still greater than in most open savanna woodlands.

Open Woodlands

A large part of Africa, including many areas where climatic considerations would predict the occurrence of closed moist forest as the natural vegetation (Figure 11.4), is covered by open (savanna) woodlands. These are various combinations of trees, bushes, shrubs, and grasslands, defined in Table 11.1, the sub-types being differentiated more by the relative heights of the woody plants and grasses than by woody plant density. The principal sub-types were named by Eyre (1968) as: 'high-grass, low-tree savanna', which contains tree species similar to those in the dry deciduous forest; 'acacia, tall-grass savanna', where the trees are much higher; and 'acacia, desert-grass savanna', which is found in drier areas where the grassy component is purely nominal. Only in parts of East Africa does savanna woodland give way in drier areas to a dense thorn-tree scrub (Eyre, 1968).

The origin of savannas has long been controversial. Schimper (1898) regarded them as climatically determined, but the general view later swung to treat them

Table 11.3.   Land uses which replace forest after deforestation

Agricultural land uses
1. Shifting agriculture
    (a) Traditional long-rotation shifting cultivation
    (b) Short-rotation shifting cultivation
    (c) Encroaching cultivation
    (d) Nomadic pastoralism
2. Permanent agriculture
    (a) Permanent staple crop cultivation
    (b) Tree crop and other cash crop plantations
    (c) Government sponsored resettlement schemes
    (d) Cattle ranching
Other permanent land uses
3. Mining
4. Hydroelectric schemes
Other temporary land uses
5. Fuelwood cutting
6. Narcotic cultivation

Source: Modified from Grainger (1993*a*).

as the result of frequent human disturbance and burning. In recent decades the influence of edaphic limitations has been argued for some areas, e.g. where soils cannot support forest growth since they become too waterlogged in rainy seasons and too moisture-deficient in dry seasons (Cole, 1986; Moss, 1982). Large areas of savanna woodland undoubtedly result from prolonged human disturbance which continues to this day, and savannas, or open woodlands, are at the heart of the complexity of natural vegetation distribution in Africa: half of the eighty mapping units in the UNESCO Vegetation Map of Africa involve woodlands, bushlands, scrub forest, or grassland.

Human Impacts on Natural Forests

Owing to extensive human modification and transformation of natural vegetation, the actual distribution of forests in Africa today is very different from that in Figure 11.1, even though the savanna woodlands shown there were ostensibly the contemporary vegetation cover at the time the map was produced. This section examines the three main human impacts—deforestation, impeded regeneration, and forest degradation—which have affected both open and closed forests and have formed the basis of Africa's modern cultural landscape.

Deforestation

The most drastic vegetation modification of all is deforestation, defined here as 'the temporary or permanent clearance of forest for agriculture or other purposes' (Grainger, 1986, 1993*a*). The key word here is clearance: if forest is not cleared, then deforestation does not

take place. The usual motive for deforestation is to replace forest by another land use. A wide range of land uses are involved, and for convenience they may be classified into types of shifting agriculture, permanent agriculture, and non-agricultural land uses (Table 11.3).

Shifting Agriculture

One distinction between the effects of different land uses is whether or not the resulting deforestation is permanent or temporary. In traditional shifting cultivation, after forest has been cleared the patch of land is typically cultivated for 1–2 years and then abandoned to allow forest and soil fertility to regenerate. Farmers return 15–25 years later to repeat the process, but the long-term vegetation cover on that patch is still forest. Over a wider area this leads to a mosaic of clearances and areas of forest at different stages of regeneration. The impact changes as rotation periods decline.

Permanent Agriculture and Other Land Uses

On the other hand, the introduction of permanent agriculture, whether this involves cultivating field crops or tree crops, or grazing livestock on permanent pastures, requires the permanent replacement of forest. So the vegetation cover is not merely modified or converted, but transformed into a new state that remains as long as human intervention persists (Poore, 1976). The same holds true when forest is cleared to make way for reservoirs for hydroelectric power stations, or for opencast (surface) mines, though forest may be replanted when mining ends.

Fuelwood Cutting

Fuelwood cutting is a major cause of deforestation and degradation only in dry (open and closed) forests. Fuelwood is still the leading energy source in many African countries, accounting for over 90 per cent of all energy needs in countries like Ethiopia, Tanzania, and Mali (Foley, 1987). In 1989, 88 per cent of the 485 million cubic metres of roundwood (i.e. log) production in Africa was for fuelwood or charcoal, and only 12 per cent for industrial wood for conversion into sawnwood and plywood, etc. (FAO, 1992).

Fuelwood is traditionally gathered from dead wood on farms and common lands, but to supply the growing urban market, for domestic and industrial uses, gangs of fuelwood cutters travel about the countryside, felling live trees and transporting the wood back to cities in trucks, often converting it to charcoal when the market is far away (Grainger, 1990). Areas of open woodlands are either cleared of trees or suffer a severe cut in tree density. Fuelwood production in Africa rose by a third between the late 1970s and late 1980s, almost twice as fast as industrial wood production, and while there is debate about the scale of the problem and its causes, there is no doubt that extensive deforestation and rising demand for fuelwood have led to supply problems in many countries (FAO, 1981; Floor and Gorse, 1987).

Impeded Forest Regeneration

Deforestation may be prolonged by subsequent burning and/or livestock grazing that impedes forest regeneration. The earliest Africans burnt open and closed forests to flush out game, and their successors today burn savanna woodlands every year to promote the growth of fresh grasses for their livestock. Frequent fires, combined with intensive animal browsing of trees and shrubs, have had major effects on the continent's vegetation, preventing the regeneration of trees and shrubs that would otherwise produce a substantial woody cover, and resulting in a vast area of plagioclimax savanna woodland (Pears, 1985). This has resulted in a very different kind of landscape from that present before human beings began forest destruction. Those trees which remain scattered over the grasslands are, like the baobab tree (*Adansonia digitata*)—that curious chunky protuberance commonly found in savannas, those which are physiologically adapted to survive fires unharmed.

Burning also impedes forest regeneration on shifting cultivation plots. If clearing and burning become too frequent, fire-resistant weeds can invade and prevent forest regeneration once they establish a foothold, resulting in the large areas of weedy grasslands now found in the humid tropics. A similar effect can result from the spread of encroaching cultivation, in which landless people move into an area, usually by road, clearing forest in vast swathes on either side. Instead of moving elsewhere after a year or so, they stay until the land's fertility is depleted and then move on to clear more forest, leaving a scrubby, weedy wasteland behind them.

Degradation

Deforestation and Degradation

Deforestation refers to both the human activity of forest clearance and its physical effect in changing vegetation cover from forest to non-forest. But it is an inadequate term to describe impacts that modify forest cover but do not clear it, or involve temporary clearance followed by the growth or planting of new woody vegetation. The term now often used for this purpose is degradation (Serna, 1986). Vegetation degradation is defined here as a 'temporary or permanent reduction in the density or structure of vegetation cover or its species composition'. Deforestation is an extreme form of this, reducing the density of vegetation cover temporarily to zero. There are two main types of degradation in closed forests: replacing forest by an inferior woody cover, and modifying its composition by extractive uses. In open woodlands in dry areas, clearance and degradation of tree cover is a major contributor to desertification.

Replacing Forest by an Inferior Woody Cover

Sometimes deforestation is only temporary, as with shifting cultivation, though if rotations are short, e.g. six years or even shorter as is now common, this only allows time for a short bushy type of forest, called forest fallow (Lanly, 1981), to regenerate before it is cultivated again. Forest fallow, not forest, therefore effectively becomes the long-term vegetation cover. It is clearly degraded relative to mature tropical rain forest, as it has a lower biomass density and canopy height, an inferior structure, and a different species composition.

Even if deforestation is permanent, forest may be replaced by an agroecosystem with a woody component, e.g. an oil-palm plantation. Since it is inferior to forest in height and biomass density, the overall impact of the change may also be referred to as forest degradation.

Agroecosystems containing a woody component are often grouped under the heading of *agroforestry systems*, in which trees, field crops and/or livestock raising are integrated together in the same or adjacent locations (Nair, 1989). Their multi-storeyed nature is well suited to the humid tropics, which has an abundance of trees capable of yielding multiple products, e.g. food, fodder, wood, oils, etc. The mixtures of different crop plants can partially replicate the structure of natural ecosystems and so are fairly efficient collectors of solar energy.

There are many types of agroforestry systems, including intercropping in tree-crop plantations, in which different tree crops, like rubber and cocoa, are grown together (Alvim and Nair, 1986); alley cropping, in which field crops like maize are grown in corridors between rows of trees such as *Leucaena leucocephala* (Getahun *et al.*, 1982); and agroforestry gardens, in which farmers plant mixtures of commercial and subsistence tree crops and timber trees in 'buffer zones' between cleared land and forest reserves (Kio and Ekwebalan, 1987).

Agroforestry is an important intermediate stage between forest and cleared land. On the one hand, it provides significant economic benefits to farmers and contributes to sustainable land use; but on the other, it makes mapping vegetation cover in humid tropical areas difficult, as the wide range of combinations of field crops, tree crops, and herbaceous vegetation is matched by a great diversity of vertical and horizontal structures of the resulting agroecosystems.

Extractive Forest Uses

Degradation also occurs as a result of extractive forest uses, like selective logging and the collection of minor forest products, which modify forests even though they do not clear them.

Selective logging. Most high value industrial roundwood is logged in tropical rain forests, not dry forests. But only a few of the many thousands of tree species in tropical rain forests are in commercial demand on national and international markets. Light and medium-density woods converted into plywood, veneers, and sawnwood include okoume (*Aucomea klaineana*), obeche (*Triplochiton scleroxylon*), and limba (*Terminalia superba*). The highest-quality decorative veneers and sawnwood are made from kokrodua (*Pericopsis elata*), African mahogany (*Khaya* spp.), and sapele (*Entandophragma cylindricum*).

The limited number of commercial species has led to the widespread use of selective logging systems, in which typically only 2–10 large commercial trees out of a total of up to 300 are felled and removed per hectare. The commercial cut per hectare is usually about 10–30 m³, compared with an average total of 258 m³ of all timber (Lanly, 1981). Left behind in the forest are non-commercial trees and smaller trees of commercial species that will be ready for felling in another 25–40 years.

Selective logging relies on the natural regeneration of forest with little human intervention. Unlike clear-felling, it does not clear forest and cause deforestation, except for that needed for logging roads and trails. Forest structure and composition are degraded relative to mature forest when tall trees are felled and removed. But if logged forest is protected from deforestation and illegal logging, the degradation is only temporary and forest should regenerate naturally. Other aspects of the degradation caused by logging are the fragmentation of forest cover by clearance for logging roads, and soil erosion and compaction on logging roads and skid trails.

If logging is not carried out skilfully, many more trees can be killed or damaged when commercial trees are felled, resulting in even more degradation. Government forestry departments usually prescribe the number and minimum size of trees which may be extracted, so that enough small trees remain for the second harvest. Minimum cutting limits are 100 cm dbh for prime species in Liberia and Ghana, 80–100 cm dbh for Cameroon, 70 cm in Congo and Gabon, and 60 cm dbh in Ivory Coast (Rietbergen, 1989). But unscrupulous private operators often take out more trees than this and prejudice the sustainability of management. Sustainability is also affected if timber poachers enter the forest after the loggers have left, or if the original operators themselves come back and undertake a second logging before the end of the harvest cycle, which has been a fairly common practice in African forests since the 1950s.

Various types of tropical silviculture have been tested in Africa. These are forest management systems that require more intensive human intervention than selective logging in order to ensure a high proportion of commercial species in the regenerated forest. For example, shelterwood systems, in which the forest canopy is opened and undergrowth, climbers, and small trees removed, so seeds of commercially desirable species can germinate under the shelter of larger trees (Nwoboshi, 1987). But after forty years of trials in Nigeria and Ghana, these systems were abandoned in the mid-1960s as growth rates were poor and high labour costs made the practices uneconomic (Reitbergen, 1989). Consequently, selective logging and artificial forest plantations became the most widely used forest management practices.

Collection of minor forest products. Timber is not the only forest product. A wide range of other 'minor forest products' or 'non-timber forest products', including gums, resins, oils, latexes, fibres, fruits, and medicines, are also important to local people and on a wider basis (Grainger, 1993a). Some are collected for subsistence purposes by indigenous peoples living in tropical rain forests, others are sold on local, national, and world markets. Collection of minor forest products does not, under normal circumstances, cause undue stress on the forest, though sometimes whole trees may be felled just to obtain fruits or orchids from their upper branches.

Some products eventually become so important that the source plants are domesticated and bred as major commercial crops. Oil from the kernels of the oil palm (*Elaeis guineensis*) is used to make margarine and detergents. It grows wild in the rain forests of West Africa and for a long time kernels were simply collected from wild trees for export. Plantations were first established in Zaire just after the turn of this century, and are now found all over the world, though even today large quantities of kernels are still collected from wild trees in Nigeria.

The least degrading extractive forest use is hunting and gathering by indigenous forest peoples, like the Pygmies of the Congo Basin, who depend for their subsistence on hunting game and gathering everything else they need, from medicines to essential materials, in the forest. But their society is so closely linked with the forest that were the forest to disappear, so would they.

The survival of hunter-gathering peoples, and of minor forest product extraction by local people, is threatened when governments allocate to private concessionaires the rights to exploit large areas of forest for timber or clear them for agriculture. Protests by rubber tappers in Amazonia have led to the establishment of specific 'extractive reserves' there, but whether similar reserves will be set up in Africa remains to be seen (Grainger, 1993a).

Desertification

Desertification, the degradation of land in dry areas, has two main components, the degradation of soil and the degradation of the vegetation covering that soil. It is not, as still widely imagined, simply the conversion into desert of land on the desert fringe, but a more wide-ranging phenomenon, in both space and degree (see Chapter 19). Land can suffer from desertification even if it is a long way from a natural desert, and there is also a whole spectrum of desertification, from the very mild to the most severe. Deforestation, by severely degrading the overall vegetation cover of an area, is a form of desertification but is often the first step in a longer-term process (Mabbutt, 1986), as it reduces the soil's protection from erosion. If this is followed by overgrazing of the resulting grasslands, causing their density and species composition to become increasingly inferior, this not only increases the degree of vegetation degradation but also the likelihood of soil erosion (Grainger, 1990).

Cultural Forest Resources Outside the Closed-Forest Zone

Degradation also has a spatial dimension. For example, an entire area can be said to have a degraded vegetation cover if the latter consists of a mosaic of clearances and scrubby forests. Because of the extensive human impact on forests in Africa, neat distinctions between 'forested areas' and 'settled areas', though superficially attractive, are unrealistic. It is equally simplistic to imagine that trees and forest cover are lacking from predominantly agricultural areas which at first sight seem to contain either heavily degraded forest or no forest at all. The tree cover which remains may be highly dispersed, and as such may have been overlooked from the point of view of mapping or management. But whether they are of natural or human origin, these types of forest, called here cultural forest resources, often have significant value to local people and are now receiving increased attention from development agencies.

Agroforestry Resources

Trees play a significant productive role even in predominantly agricultural areas, usually mixed with farming practices in agroforestry systems. Some of these systems, such as bush fallowing in dry areas, are long established. Typically, a piece of land is cropped for 4–5 years before being left fallow for 5 years or more, when nitrogen-fixing shrubs like *Acacia* are allowed to invade. But with the growth of population and other pressures, fallow periods decline, the systems deteriorate, and the land becomes more susceptible to soil erosion (Grainger, 1990).

Another type of agroforestry system involves planting trees on farms to protect crops from damage and dehydration by winds. A classic example of a large-scale wind-break scheme is found in Niger's semi-arid Majjia Valley, 500 km north-east of the capital Niamey. Crops there were regularly damaged by strong harmattan winds before 325 km of wind-breaks were planted by farmers between 1975 and 1985. Besides reducing crop damage the wind-breaks have also led to higher millet yields (Kerkhof, 1990).

Social Forestry Resources

Increasing tree cover in predominantly agricultural areas is not easy, especially where trees are planted communally on village lands for joint benefits. Social forestry projects have not been as successful in Africa as they have in Asia, though the two major constraints, villagers' distrust of government foresters, and unwillingness to risk time and money in new and uncertain enterprises, are universal.

The most successful forms of social forestry in dry Africa have involved planting shelter-belts around towns and villages and stabilizing coastal sand dunes. 'Greenbelts' of drought-resistant trees protecting towns from fierce winds, sand storms, and encroaching sand dunes are now common in the Sahel. Ouagadougou, the capital of Burkina Faso, for example, is surrounded by a 500 ha greenbelt, and its trees can also be harvested for fuelwood, poles, and fodder. In one of many dune-stabilization projects along the Somalian coast, local people banded together in 1982 to stabilize a 30 ha section of dune that threatened to cover the road between Mogadishu and the port of Brava (Grainger, 1990). In both cases, new cultural forest resources were established only because they satisfied immediate social and environmental needs.

A New Awareness of Open Forests as Cultural Resources

The dependence on tropical rain forests of indigenous peoples, like the Pygmies, has been acknowledged for a long time, but only recently has it become appreciated just how much people living in open-forest areas depend on the trees for a variety of products. Part of the problem was that tree-based foods, like shea-nut butter, did not fit into conventional food categories dominated by cultivated crops like millet. This ignorance led to costly mistakes in the design of forestry projects which cleared supposedly 'degraded' open forests and replaced them with plantations to supply what local people were assumed to need, i.e. fuelwood. Unfortunately, no one bothered to ask their views, and many projects failed.

The need to avoid further failures led in the 1980s to new attempts to understand people's dependence on open forest resources. The extent of gathering was shown to be very wide indeed. Thus, the leaves of the baobab tree are used to make sauces, and it is also a source of fodder, medicine, fibres, dyes, and soap. According to Williams (1984):

If you ask a woman about all the things consumed during the course of a day and the ingredients of each meal you may learn for example that she ate some shea-nut fruits (*Butyrospermum parkii*) in the morning when she was working in her field and brought home the nuts to cook shea-nut butter, a much utilized cooking fat. Perhaps also she had some dol, or local beer, that was fermented from the fruit of a local tree, such as *Sclerocarya birrea*. By asking about the ingredients that were used to make the tao and gumbo sauce you can learn that more than just cultivated millet and gumbos are used. The tao (a thick porridge made with millet) is often cooked with an acid to make it more digestible; commonly used acids are lemon juice, tamarind juice, tamarind fruit, tamarind leaves, or leaves of other tree species. The sauce probably contains, in addition to the gumbos, shea nut butter, soumbala (a condiment made from the fermented seeds of the néré tree *Parkia biglobosa*), salt, hot peppers and other seasonings.

Findings like these have led to a major shift in the attitude of development agencies like the World Bank towards open forests and it is time for them to be re-evaluated by biogeographers too. In the conventional view, non-climax open forests would be regarded just as heavily degraded vegetation. But this is only half of the picture —or even less than half, for there is much still to be be learned about the physical characteristics of degraded forests too. It is also vital to acknowledge that the human impacts that modified forest cover not only had a purpose, but probably led to a continuing set of dependent relationships, like those described above. Local people show their reliance on baobab trees, for example, by protecting them from felling.

Regenerating and Managing Open Forests

Faced with the failure of many plantation projects, various development agencies began to ask themselves in the 1980s whether natural open forests, which they had previously regarded as worthless degraded lands, could not instead be brought under some form of communal management for wood production and other economic and environmental benefits. In a pioneering experiment in natural woodland management, the Forest Service in Niger started to restore the Guesselbodi National Forest Reserve, near Niamey, in partnership with local people. The reserve had suffered from severe vegetation and soil degradation, and there were too few government foresters to protect it from further degradation and restore it themselves.

When the project began in 1981, foresters talked with people from a number of nearby villages to obtain their suggestions, and information on past and present forest uses. Villagers were allowed to use a 30 ha area of the reserve for farming, as long as they planted crops between protective wind-breaks of exotic drought-resistant trees, like *Prosopis juliflora* and *Acacia holocerica*. Existing trees in the reserve are now being coppiced, and forest cover restored by both planting and natural regeneration of indigenous trees. In time these will be harvested for firewood to be sold in Niamey. Grazing is forbidden for three years after planting, but local people can cut hay for their animals and gather minor forest products as before. When it resumes, grazing is strictly regulated to prevent tree damage (USAID, 1979; Heermans, 1986).

This is just one of a number of experiments in co-operative natural woodland management now being conducted all over the tropics. It is a logical development of the new emphasis on community forestry that began in the mid-1970s, showing that even degraded forests are still valuable cultural resources, and by restoring natural forests instead of establishing plantations, it builds on an existing human–forest interface instead of trying to create a new one.

Present Forest Distribution

A major task for biogeographers today is to map the extent of present continental and global vegetation distributions, assess how they have changed under human influence from those shown in ideal biome maps, and monitor current patterns of change. This section reviews recent estimates of Africa's forest area and their dependence on remote-sensing data.

Expert Assessments and Measurements

For a long time estimates of tropical forest distribution relied on subjective expert assessments (Grainger, 1993b). This was inevitable given the poor availability of data based on actual measurements, owing to the great difficulty of undertaking regular aerial photographic surveys of national forest cover. The launch in 1972 of the first earth-resources satellite, Landsat 1, enabled large

Table 11.4 Areas of tropical closed forests, tropical moist closed forests, tropical dry open woodlands, and all tropical forest in Africa in 1980 (million ha)

	Closed forest		Open woodland[a]	All tropical forest
	All[a]	Moist[b]		
West Sahel	1	0	43	44
West Humid	18	18	38	56
Central Africa	173	173	163	336
East and Insular	25	14	192	217
South Africa	0	0	51	51
TOTALS	217	205	486	703

Sources: [a]Lanly (1981); [b]Grainger (1983), based on Lanly (1981).

Table 11.5 Areas of all tropical forest (closed and open) in Africa 1980 and 1990 (million ha)

	1980 original[a]	1980 revised[b]	1990[b]
West Sahel	44	44	41
West Humid	56	62	56
Central Africa	336	216	204
East and Insular	217	89	81
South Africa	51	159	146
TOTALS	703	569	528

Notes: Differences between the original and revised 1980 figures are due to improved estimates in 1993, different regional groupings (particularly affecting West Sahel/West Humid, and South/East and Insular divisions), and the inclusion of only 37 countries in the original 1980 estimate compared with 40 in revised one.

Sources: [a]Lanly (1981); [b]FAO (1993).

areas of forest to be surveyed far more easily than before (Grainger, 1984). But though some forest services have used satellite imagery for national forest inventories for many years, it has taken two decades for it to become the major data source for continental forest assessments in the tropics. Although there were technical problems, such as high cloud cover in the humid tropics and high ground reflectance in dry areas, the main constraints were organizational, and the slow acceptance of the need for measurement.

Total Forest Area

1980 FAO-UNEP Estimate

Africa contained 703 million ha of tropical forest in 1980, according to a joint assessment by the UN Food and Agriculture Organization (FAO) and UN Environment Programme (UNEP) (Lanly, 1981). Less than one-third of this, 217 million ha, was closed forest, and most of that was in West and Central Africa, and Madagascar. The other two-thirds, 486 million ha, was open forest, of which three-quarters was in East and Central Africa (Table 11.4). A major limitation of this study was that it was only partially based on remote sensing measurements; subjective assessments and official statistics of uncertain reliability were also used.

1990 FAO Estimate

According to FAO's 1990 Forest Resources Assessment, the total forest area in 1990 was only 528 million ha. FAO attributed the large drop to better data, rather than a huge forest clearance in the 1980s. It also revised its 1980 estimate to just 569 million ha (Table 11.5). The area of closed forest declined only slightly to 205 million ha (though this strictly referred just to closed broadleaved forest as FAO did not give formal estimates for 'closed forest' and 'open forest'). This left the area of

open forest as only 323 million ha: previous estimates were known to be very inaccurate so the revision was understandable (FAO, 1993).

Distribution of Tropical Moist Forest

Since tropical moist forest accounts for most closed forest in Africa its distribution is of major importance. In 1980 it was estimated to cover 205 million ha, based on FAO-UNEP data (Grainger, 1983), though only half of all countries and 17 per cent of the total forest area had been surveyed by any form of remote sensing in the 1970s (Grainger, 1984). By the late 1980s, according to the World Conservation Union (IUCN) the area was at least 183 million ha (Table 11.6), of which 165 million ha was in Central Africa. This estimate made greater use of satellite data (Sayer *et al.*, 1992).

FAO's 1990 Forest Resources Assessment estimated the area of tropical rain forest in Africa in 1990 as only 87 million ha, but this was clearly too low. Because of classification problems it excluded tropical rain forest in montane areas and a lot of tropical rain forest which was listed as tropical moist deciduous forest. The 'tropical moist deciduous forest' class included both open and closed forests, making it difficult to separate seasonal tropical rain forest from open savanna woodland in moist areas (Grainger, 1995a). This author tried to overcome the lack of comparability of estimates in FAO (1993) by using the FAO data to produce his own rough estimate of 215 million ha for the tropical moist forest area in 1990, but this is not very reliable (Table 11.6) (Grainger 1995b).

Central Africa contains about 85 per cent of the remaining tropical moist forest, with 8 per cent in West Africa and 7 per cent in East and Insular Africa. The

Table 11.6 Estimates of tropical moist forest areas in Africa 1980–1990 (million ha)

	1980	Late 1980s		1990
	Grainger (1983)[a]	IUCN (1992)	Myers (1989)	Grainger (1995)[f]
West Africa	18	14	3[c]	17
Central Africa	173	165	145	184
East and Insular[e]	14	8[b]	2[c]	15
TOTALS	205	—	152[d]	215

[a] Based on data in Lanly (1981).

[b] Totals by this author from IUCN data for illustration only; IUCN did not give totals for these regions due to poor data and an African Total is therefore not appropriate.

[c] Very incomplete regional data.

[d] Myers' own regional total: the global total for all tropical moist forest was adjusted by him to account for countries omitted, but not the regional totals.

[e] East and Insular includes South regional grouping in FAO (1993).

[f] Based on data in FAO (1993).

Sources: IUCN (Sayer *et al.*, 1992); Myers (1989); Grainger (1995*b*).

Table 11.7 Estimates of tropical moist forest areas in Africa in 1980, 1988, and 1989 (million ha)

	Grainger (1983) 1980	IUCN 1988	Myers 1989
Angola	2.9	nd	nd
Benin	0.05	0.04	nd
Burundi	nd	0.04	nd
Cameroon	17.9	15.5	16.4
Central African Republic	3.6	5.2	nd
Congo	21.3	nd	9.0
Ethiopia	nd	4.8	nd
Equatorial Guinea	1.3	1.7	nd
Gabon	20.5	22.7	20.0
Gambia	nd	0.05	nd
Ghana	1.7	1.6	nd
Guinea	2.1	0.8	nd
Guinea-Bissau	0.7	nd	nd
Ivory Coast	4.5	2.7	1.6
Kenya	0.7	nd	nd
Liberia	2.0	4.1	nd
Madagascar	10.3	4.2	2.4
Nigeria	6.0	3.9	2.8
Rwanda	nd	0.2	nd
Senegal	0.2	0.2	nd
Sierra Leone	0.7	0.5	nd
Sudan	nd	nd	nd
Tanzania	1.4	nd	nd
Togo	0.3	0.1	nd
Uganda	0.8	0.7	nd
Zaire	105.7	119	100.0

Note: *nd* = no data listed.

Sources: Grainger (1983); Sayer *et al.* (1992) (IUCN); Myers (1989).

forest in Central Africa is concentrated in four countries Zaire, Cameroon, Congo, and Gabon (Table 11.7). Zaire contains one tenth of all tropical moist forest in the world. By contrast, tropical moist forest in West Africa has been heavily cleared, and in East Africa its distribution is highly fragmented. Apart from lowland rain forest in Uganda, the forest in East Africa is mainly restricted to montane areas in Kenya, Tanzania, Ethiopia, Rwanda, and Burundi, and the islands of Mauritius, Reunion, the Seychelles, and Madagascar, all of which have suffered major forest loss (Figure 11.8) (Polhill, 1989; Lanly, 1981).

Maps of Forest Distribution

There are two main types of vegetation maps of Africa: those like Figure 11.1 that portray potential vegetation distribution, and those that show the vegetation as it really is. It took fifteen years to produce the UNESCO Vegetation Map of Africa and then another three years to publish it: work began in 1965 and the map was only released in 1983. Although the map is very detailed, for large areas it now only really shows potential distribution (Figure 11.3). The map is an advance over Figure 11.1 as it represents transitions between, and mosaics of, different vegetation covers. The full map (not the simplified version in Figure 11.3) also shows areas that have suffered severe human disturbance, e.g. in northern Africa, while lesser disturbance is implicitly recognized in mosaic areas. However, there are significant differences between the distributions of tropical rain

forest in Figures 11.1 and 11.7 because, besides the shortage of data, the long time taken to produce the map militated against taking full account of changes between 1965 and 1983 (UNESCO-AETFAT-UNSO, 1983).

It has been possible to use satellite imagery to produce maps of Africa's actual forest distribution for over twenty years, but such maps only became available in the late 1980s. FAO-ICIV (1987) produced the first map of present vegetation by combining Landsat multi-spectral scanner (MSS) images and NOAA-AVHRR images. Then a forest map of West Africa was produced by a UNEP team in the late 1980s (Figure 11.7) (Päivinen and Witt, 1989) and included in IUCN's atlas of Africa's tropical forests (Sayer *et al.*, 1992). It showed only fragments of tropical moist forest remaining, mostly in Liberia, western Ivory Coast near the Liberian border, and southeastern Nigeria. All the rest had been converted to mosaics of farmland and degraded forest, the aftermath of a strong frontier expansion of agriculture for much

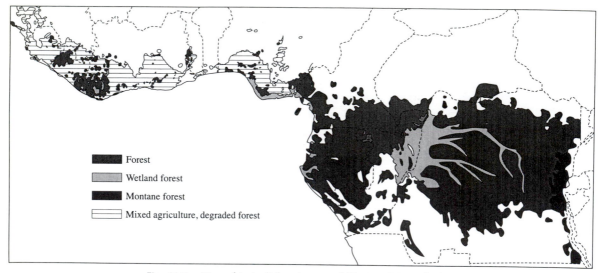

Fig. 11.7. Map of 'actual' forest cover of West and Central Africa.
(*Sources*: Based on Sayer *et al.*, 1992; Päivinen and Witt, 1989; and Laporte *et al.*, 1985.)

of this century, pushing south into the northern edge of the rain-forest belt.

The IUCN atlas also included a new map for Central Africa, though it made only partial use of satellite data, combining an AVHRR-derived map of Zaire in 1988, produced by a NASA team (Laporte *et al.*, 1995), with 'reasonably up-to-date' maps from the mid-1980s for Cameroon and the Central African Republic, and maps for the Congo and Gabon of less certain provenance (Figure 11.7). It showed tropical moist forest as concentrated in Zaire, Congo, and Gabon, extending into southern and western Cameroon and southwestern Central African Republic. Large areas of forest had been cleared in southern Zaire and south-eastern Congo.

These maps represent an advance, but are still only a first step, given the partial use of satellite imagery in Central Africa, and reliance on images collected by the AVHRR sensor, which despite its name has a much lower resolution (1–4 km) than sensors on the Landsat and SPOT satellites (20–79 m). When the first continental maps of actual forest distribution appeared, old norms about preparing global forest statistics had to be reconsidered. Biogeography will be affected too: for a long time it has tended to focus on studying either relatively undisturbed vegetation or its development in the absence of human influence. But the growing availability of large-area maps of actual vegetation cover should lead to increasing research into human-induced vegetation change.

Maps of Forest Attribute Distribution

The rapid expansion of research into global environmental change has meant that data are required not only on changes in forest areas, but also on changes in different forest attributes, such as biomass density, biodiversity, commercial timber cut per hectare, albedo, etc. Since these show great spatial variation, gaining a more accurate understanding of the consequences of deforestation will depend on knowing which forest is cleared and what its attributes are.

Geographic Information Systems (GIS) techniques (Burrough, 1986) can be used to produce maps of the distribution of any number of attributes attached to a base-map of forest area. Such work is still at an early stage, as with forest area maps, but should progress quickly as more empirical measurements of the distribution of forest attributes are made. Initial work is concentrating on mapping biomass density, but eventually there should be maps for a wide range of physical attributes, and also derived socio-economic attributes. For example, a map of the commercial timber volume per hectare attribute could be converted, using a simple extraction-cost function into a map of unit logging costs (Grainger, 1986).

Changes in Forest Distribution

Tropical forest distribution in Africa is still changing rapidly. As well as the decline in area, forest composition

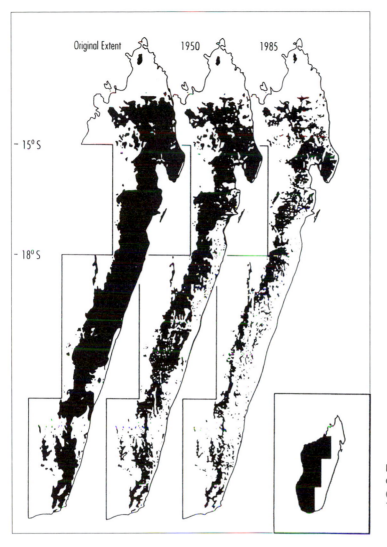

Fig. 11.8. Changes in forest cover in Madagascar. (*Source:* Green and Sussman, 1990.)

and structure are also being degraded, with consequent changes in the attributes of remaining forest. Each type of change has two key indicators: (a) the annual rate, e.g. the annual deforestation rate; and (b) the overall reduction from a nominal maximum value, e.g. the long-term regression of forest cover from its climax area. Estimates of both these deforestation indicators are still quite inaccurate. Estimating the rate and degree of degradation is even more difficult, but growing concern about desertification and the decline in biodiversity and biomass should lead to improved monitoring methods.

Contemporary Rates of Deforestation

In the late 1970s Africa lost 3.6 million ha of tropical forest a year, comprising 2.3 million ha of open forest

and 1.3 million ha of closed forest. Of the latter, 1.2 million ha was tropical moist forest (Tables 11.8 and 11.9) (Lanly, 1981; Grainger, 1983).

By the late 1980s Myers (1989) estimated that the deforestation rate for tropical moist forest had risen to 1.6 million ha per annum. FAO (1993) estimated a mean deforestation rate for all forest in the 1980s of 4.1 million ha per annum, although this referred to more countries than the earlier study. A rough calculation by this author suggests that this total included 1.4 million ha (Table 11.8), of tropical moist forest, 0.1 ha of tropical dry closed forest and 2.6 million ha of open forest (Grainger, 1995b). Again the estimates in FAO (1993) were not very comparable with previous estimates.

Table 11.8 Deforestation rates in Africa in the 1970s and 1980s

	All tropical closed forest		Tropical moist forest		
	1976–80[a]	1980s[b]	1976–80[c]	1980s	
				Myers[d]	Grainger[e]
West Africa	0.725	0.886	0.675	0.700	0.162
Central Africa	0.332	1.140	0.292	0.700	1.068
East, South, and Insular Africa	0.276	2.075	0.196	0.200	0.181
TOTALS	1.333	4.100	1.204	1.600	1.411

Sources: [a]Lanly (1981), [b]FAO(1993); [c]Grainger (1984); [d]Myers (1989); [e]Grainger (1995*b*).

Table 11.9 Estimates of deforestation rates for tropical moist forest in Africa, 1976–80 (thousand ha per annum)

	1976–80[a]
WEST AFRICA	676
Benin	2
Gambia	nd
Ghana	15
Guinea	nd
Guinea Bissau	15
Ivory Coast	310
Liberia	41
Nigeria	285
Senegal	nd
Sierra Leone	6
Togo	2
CENTRAL AFRICA	332
Angola	40
Cameroon	80
Central African Republic	5
Congo	nd
Equatorial Guinea	15
Gabon	27
Zaire	165
EAST, SOUTH, AND INSULAR AFRICA	196
Burundi	nd
Ethiopia	nd
Kenya	11
Madagascar	165
Rwanda	nd
Sudan	nd
Tanzania	10
Uganda	10

Note: nd = no data available.

Sources: [a]Grainger (1984), based on Lanly (1981).

Estimates of deforestation rates tend to be less accurate than those of forest areas since satellite images have been used less frequently for measuring national deforestation rates than for surveying forest areas (Grainger, 1984).

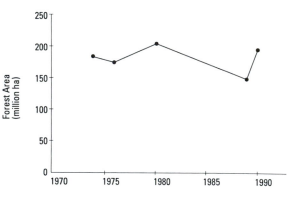

Fig. 11.9. Supposed trends in African tropical moist forest area, based on a time-series plot of various estimates 1970–1990.

The latest FAO study actually estimated deforestation rates by comparing national forest areas in 1981 and 1990. But it relied heavily on the use of mathematical models to project forest area forward from the survey year to 1990. For thirty-five of the forty African countries listed only one national forest survey was available to FAO, and for twenty-three of these the survey dated from before 1981, so a model was needed to estimate both the 1981 and 1990 areas (FAO, 1993).

Long-Term Trends in Forest Area

It is generally agreed that Africa's forest area has declined in recent decades, but no consistent downward trend is apparent from the sequence of estimates since 1970 (Figure 11.9). On the contrary, tropical moist forest area actually seemed to *increase* from a base of 175 million ha in the early 1970s (Sommer, 1976) to 205 million ha in 1980 (Grainger, 1983). Myers's (1989) estimate of 152 million ha for 1989 conflicts with this author's rough revised estimate of 215 million ha for 1990, based on data in FAO (1993) (Table 11.6). Such confusion can be explained by inaccuracies in the national estimates on which the regional figures were based.

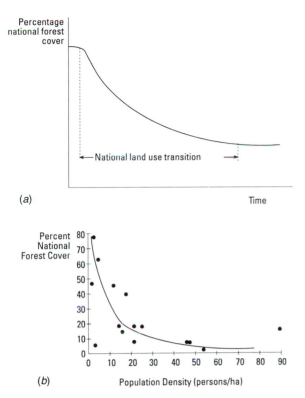

(a)

(b)

Fig. 11.10. (a) Idealized form of national land use transition; (b) Cross-sectional plot of national forest cover versus population density in Africa.
(Sources: Grainger, 1986, 1995c.)

Despite this uncertainty, it is clear that forest loss has not been uniformly spread. Sommer (1976) estimated that the potential climax area of tropical moist forest was 362 million ha. Half of this had been lost by the mid-1970s. West and East Africa had each lost 72 per cent but Central Africa only 45 per cent. IUCN used the new maps of tropical moist forest to estimate that by 1988 regression in West Africa had risen to 89 per cent (Sayer *et al.*, 1992). Its estimate of regression in Central Africa (41 per cent) was similar to Sommer's.

Tracing Patterns in Long-Term Forest Decline

Understanding the reasons for long-term trends in forest area is essential if there is to be progress in controlling deforestation and projecting possible future trends. Many tropical countries seem to show a consistent decline in forest area as population rises and forest is cleared so that agriculture and settlements can spread. But this decline, or national land use transition (Grainger, 1986, 1995c) (Figure 11.10A) should come to an end if farm productivity rises faster than population so that no new farmland is needed to supply rising food production or if government intervenes to protect remaining forest.

This pattern is evident in the African humid tropics when the percentage of national land area under forest in countries in humid tropical zone is plotted against population density (Figure 11.10B). The loss rate declines at high population density, by which time a lot of forest has been cleared. Countries which had already reached an advanced stage of deforestation in 1980 included Ghana (with a national forest cover of 8 per cent), Nigeria (7 per cent), and Uganda (4 per cent).

One of the conditions for controlling deforestation at an early stage, the development of highly productive agriculture, has not yet been achieved in many African countries. Much of the growth in farm production since 1945 has come by expanding farmland area, not yield per hectare. Future control of deforestation will depend on increasing agricultural productivity on the best lands, protecting remaining forests better, and improving the sustainability of natural forest management (Grainger, 1993a).

Trends in Degradation

Forest degradation is another important characteristic of the cultural modification of vegetation cover, but a lot more difficult to monitor than deforestation since changes in vegetation density, structure, and composition may not lead to easily identifiable changes in the forest canopy. Infra-red and visible satellite sensors respond to the overall leaf area of vegetation and so can easily confuse mature and young secondary forest. So data on spatio-temporal trends in biomass density in Africa are very limited. However, microwave satellite sensors, e.g. on the European Space Agency's ERS-1 satellite, are height-sensitive and should lead to a new era of biogeographical studies. There is great potential, for example, to study the increasing fragmentation of primary forest in Central Africa into mosaics of mature forest, forest fallow, and older secondary forest. As natural landscapes become more culturally transformed, the mosaics become more complex and the sizes of individual blocks smaller, making image interpretation difficult. Fragmentation is very apparent in the original UNEP map of West Africa (Sayer *et al.*, 1992) but less so in Figure 11.7 owing to difficulties in reproduction.

Conclusions

Africa's present forest cover differs greatly from its climax distribution and is constantly changing even today. Mounting human pressures mean that the trend is usually towards less forest, though there is some cause for hope that deforestation can be brought under control. Whether this happens or not, studying contemporary changes in vegetation in space and time provides a tremendous challenge for biogeographers, particularly as this field has been so neglected in the past. But the changes can only be understood by reference to the full range of human impacts associated with them, impacts which spread in space and time in response to underlying social, economic, and political pressures.

At least as important as monitoring changes in forest area is the monitoring of degradation, and the accompanying trends in forest attributes, such as biomass density and biodiversity, many of which are linked with major environmental impacts like global climate change. We cannot be certain if or when vegetation zones and species composition will shift as a result of the greenhouse effect, but mapping the changes as they occur will be a key task for biogeographers of the future. The principal task today is to map the present distribution of Africa's forests and their species composition as best we can, to provide a baseline to assist future generations in planning the effective conservation of Africa's still considerable wealth of flora and fauna.

Africa's forests may be more limited in extent and structure than in other parts of the tropics but they are of huge importance, not only for their unique biodiversity but also for how they have accommodated to human pressures. The balance between natural wealth and human pressures in these forests is clearly extremely delicate, yet at the same time is a rich area for biogeographical study. Hopefully the same pressures that have modified Africa's forests so heavily will also contain the seeds of their survival.

References

Adefolalu, D. K. (1983), 'Desertification of the Sahel', in Ooi Jin Bee (ed.), *Natural Resources in Tropical Countries* (Singapore), 402–38.

Alvim, R. and Nair, P. K. R. (1986), 'Combination of Cacao with Other Plantation Crops: An Agroforestry System in Southeast Bahia, Brazil', *Agroforestry Systems*, 4: 3–15.

Brown, S. and Lugo, A. E. (1984), 'Biomass of Tropical Forests: A New Estimate Based on Forest Volumes', *Science*, 223: 1290–3.

Burrough, P. A. (1986), *Principles of Geographic Information Systems for Land Resources Assessment* (Oxford).

Charney, J. (1975), 'Dynamics of Deserts and Drought in the Sahel', *Quarterly Journal of the Royal Meteorological Society*, 101: 193–202.

Cole, M. M. (1986), *The Savannas: Biogeography and Geobotany* (London).

Eyre, S. R. (1968), *Vegetation and Soils: A World Picture* (London).

FAO (UN Food and Agriculture Organization) (1981), 'Map of the Fuelwood Situation in the Developing Countries' (Rome).

—— (1992), *Forest Products Yearbook 1990* (Rome).

—— (1993), *Forest Resource Assessment 1990: Tropical Countries*, FAO Forestry Paper no. 112 (Rome).

FAO-ICIV (Institut de la Carte Internationale de la Vegetation) (1987), 'Digital Map of the Vegetation of Africa' (Toulouse and Rome).

Floor, W. and Gorse, J. (1987), 'Household Energy Issues in West Africa' (Washington, DC) Ms.

Foley, G. (1987), *The Energy Question* (London).

Fosberg, F. R. Garnier, B. J., and Küchler, A. W. (1961), 'Delimitation of the Humid Tropics', *Geographical Review*, 51, 333–47.

Getahun, A., Wilson, G. F., and Kang, B. T. (1982), 'The Role of Trees in Farming Systems in the Humid Tropics', in L. H. MacDonald (ed.), *Agro-forestry in the African Humid Tropics* (Tokyo), 28–35.

Grainger, A. (1983), 'Improving the Monitoring of Deforestation in the Humid Tropics', in S. L. Sutton, T. C. Whitmore, and A. C. Chadwick (eds.), *Tropical Rain Forest Ecology and Management* (Oxford), 387–95.

—— (1984), 'Quantifying Changes in Forest Cover in the Humid Tropics: Overcoming Current Limitations', *Journal of World Forest Resource Management*, 1: 3–63.

—— (1986), 'The Future Role of the Tropical Rain Forests in the World Forest Economy', D.Phil. thesis (Oxford).

—— (1990), *The Threatening Desert: Controlling Desertification* (London).

—— (1993a), *Controlling Tropical Deforestation* (London).

—— (1993b), 'Rates of Deforestation in the Humid Tropics: Estimates and Measurements', *Geographical Journal*, 159, 33–44.

—— (1995a), 'An evaluation of the FAO Tropical Forest Resource Assessment 1990', *Geographical Journal*, in press.

—— (1995b), 'Forest Areas and Deforestation Rates in the Humid Tropics: A Reassessment', Working Paper no. 95/14, School of Geography University of Leeds.

—— (1995c), 'The Forest Transition: An Alternative Approach', *Area*, 27, 242–51.

Green, G. and Sussman, R. (1990), 'Deforestation History of the Eastern Rain Forests of Madagascar', *Science*, 248: 212–15.

Grove, A. T. (1978), *Africa* (Oxford).

Hall, J. B. and Swaine, M. D. (1981), *Distribution and Ecology of Vascular Plants in a Tropical Rain Forest* (The Hague).

Heermans, J. (1986), 'The Guesselbodi Experiment with Improved Management of Brushland in Niger', *Development Anthropology Network*, 4/1: 11–15.

Kerkhof, P. (1990), *Agroforestry in Africa* (G. Foley and G. Barnard (eds.)) (London).

Kio, P. R. O. and Ekwebelan, S. A. (1987) 'Plantations Versus Natural Forests for Meeting Nigeria's Wood Needs', in F. Mergen and J. R. Vincent (eds.) *Natural Management of Tropical Moist Forests* (New Haven), 149–76.

Laporte, N., Justice, C., and Kendall, J. (1995), 'Mapping the Dense Humid Forest of Cameroon and Zaire using AVHRR Satellite Data', *International Journal of Remote Sensing*, 16: 1127–45.

Laval, K. (1986), 'General Circulation Model Experiments with Surface Albedo Changes', *Climatic Change*, 9: 91–102.

Lanly, J. P (1981) (ed.), *Tropical Forest Resources Assessment Project (GEMS): Tropical Africa, Tropical Asia, Tropical America*, 4 vols. (Rome).

Letouzey, P. (1978), 'Floristic Composition and Typology', in UNESCO–UNEP–FAO, *Tropical Forest Ecosystems: A State of Knowledge Report* (Paris), 91–111.

Lewis, L. A. and Berry, L. (1988), *African Environments and Resources* (London).

Longman, K. A. and Jeník, J. (1987), *Tropical Forest and its Environment*, 2nd edn. (London).

Mabbutt J. A. (1986), 'Desertification Indicators', *Climatic Change*, 9: 113–22.

Moss, R. P. (1982), 'Reflections on the Relation Between Forest and Savanna in Tropical West Africa', *Discussion Papers in Geography*, no. 23, Department of Geography, University of Salford.

Myers, N. (1989), *Deforestation Rates in Tropical Forests and Their Climatic Implications* (London).

Nair, P. K. R. (1989), 'Classification of Agroforestry Systems', in P. K. R. Nair (ed.), *Agroforestry Systems in the Tropics* (Dordrecht, Neths.), 39–52.

Nwoboshi, L. C. (1987), 'Regeneration Success of Natural Management, Enrichment Planting, and Plantations of Native Species in West Africa', in F. Mergen and J. R. Vincent (eds.), *Natural Management of Tropical Moist Forests* (New Haven), 71–91.

Päivinen, R. and Witt, R. (1989), 'The Methodology Development Project for Tropical Forest Cover Assessment in West Africa', unpublished report (UNEP–Global Resource Information Database: Geneva).

Polhill, R. M. (1989), 'East Africa', in D. G. Campbell and H. D. Hammond (eds.), *Floristic Inventory of Tropical Countries* (New York), 5–30.

Poore, M. E. D. (1976), *Ecological Guidelines for Development in Tropical Forest Areas.* (Morges, Switz.).

Proctor, J. (1983), 'Mineral Nutrients in Tropical Forests', *Progress in Physical Geography*, 7: 422–31.

—— (1989) (ed.), *Mineral Nutrients in Tropical Forest and Savanna Ecosystems* (Oxford).

Richards, P. W. (1952), *The Tropical Rain Forest* (Cambridge).

Rietbergen, S. (1989), 'Africa', in M. E. D. Poore (ed.), *No Timber Without Trees* (London), 40–73.

Sayer, J. A., Harcourt, C. S., and Collins, N. M. (1992), *The Conservation Atlas of Tropical Forests: Africa* (London).

Schimper, A. F. W. (1898), *Plant Geography Upon a Physiological Basis*, English trans. by W. R. Fisher, G. Groom, and I. B. Balfour (1903) (Oxford).

Serna, C. (1986), 'Degradation of Forest Resources: Special Study on Forest Management, Afforestation and Utilization of Forest Resources in the Developing Regions, Asia Pacific Region' (Rome).

Sommer, A. (1976), 'Attempt at an Assessment of the World's Tropical Forests', *Unasylva*, 28: 5–25.

Taylor, C. J. (1960), *Synecology and Silviculture in Ghana* (London).

Tucker, C. J. and Choudhury, B. J. (1987), 'Satellite Remote Sensing of Drought Conditions', *Remote Sensing of Environment*, 23: 243–51.

Tueller, P. T. (1987), 'Remote Sensing Science Applications in Arid Environments', *Remote Sensing of Environment*, 23: 143–54.

UNEP (UN Environment Programme) (1992), *World Atlas of Desertification* (London).

UNESCO (1978), *Tropical Forest Ecosystems: A State of Knowledge Report* (Paris).

UNESCO–AETFAT–UNSO (UN Educational, Scientific and Cultural Organization, Association pour l'étude taxonomique de la flore d'Afrique Tropicale, UN Sudano-Sahelian Office) (1983), *Vegetation Map of Africa* (Paris).

USAID (US Agency for International Development) (1979), 'Niger Forestry and Land Use Planning Project', (Washington, DC).

Verstraete, M. M. (1989), 'Land Surface Processes in Climate Models', in A. Berger, S. Schneider, and J. Cl. Duplessy (eds.) *Climate and Geo Sciences* (Dordrecht, Neths.), 321–40.

White, F. (1983), *The Vegetation of Africa: A Descriptive Memoir* (Paris).

Whitmore, T. C. (1984), *Tropical Rain Forests of the Far East*, 2nd edn. (Oxford).

—— (1990), *An Introduction to Tropical Rain Forests* (Oxford).

Williams, P. (1984), 'A "Minor" Forest Product?' (Hanover, NH).

12 Savanna Environments

Martin E. Adams

Definition and Classification

The African savanna ecosystem is remarkable for its enormous extent. About two-thirds of the continent are classified as savanna. This broad zone contains a large and rapidly growing human population as well as the majority of the continent's livestock and remaining large mammal biomass. The threats to African savanna habitats, believed to be the centre of origin of the human species, have received much less attention than those confronting tropical rain forests. Yet, increasingly large areas of savanna vegetation are being exposed to stress and disturbance through grazing, fuelwood, and timber collection, and land clearing for arable crops. However, in the last decade or so, knowledge of the physiognomy and the dynamic nature of savanna vegetation has been greatly extended. Much of the research has been conducted in East and Southern Africa and Australia, with the objective of improving savanna conservation and management. The work has been complemented by policy-related studies of wildlife and range management which seek to apply important research findings to the resolution of land-use problems in Africa's savanna areas.

Definitions of Savanna

There has been a lack of agreement among ecologists and biogeographers about the use of the term 'savanna' to describe an ecosystem. According to Beard (1953), the word originates from an Amerindian word used in Haiti and Cuba for a treeless plain and was incorporated in sixteenth century Spanish, as *savana*. Beard describes savanna as a plant community with a dominant stratum of herbaceous plants, sometimes with scattered shrubs, tress, and palms. Pratt *et al.* (1966) rejected its use in East Africa in the classification of vegetation,

because it had been used in so many different ways. However, according to Werner *et al.* (1991), quoting Frost *et al.* (1986), there is now general accord that the term can be broadly used to describe 'all tropical and subtropical ecosystems characterized by a continuous herbaceous cover of (heliophilous) C_4 grasses that show seasonality related to water, and in which woody species are significant but do not form a closed canopy or a continuous cover' (p. xi). Dominance of heliophilous grasses of the Megatherm C_4 group of plants (one of the five major plant groups, classified according to temperature response) is considered to be a diagnostic feature of tropical savanna. Plants with the C_4 photosynthetic pathway are favoured by high light intensity, high temperature (optimum temperatures 30 to 32 °C, lower temperature threshold 10 °C, upper temperature threshold 46 °C) and high evaporation rates, characteristic of the seasonally humid/arid tropics (Nix, 1983).

Classification and Extent of Savanna Types in Africa

There are numerous differences between the fauna and flora of the northern and southern savannas of Africa, despite the fact that the broad belt of tropical rain forest has seldom, if ever, completely bisected the savanna from the Atlantic to the Indian Ocean. General patterns of African biomes are discussed by Meadows (Chapter 10). The southern and eastern savannas are much more varied and floristically complex and support a much wider variety of animal species. In terms of the diversity of plant species and their physiognomy, the core area is East Africa, comprising much of Tanzania and Kenya. In terms of the numbers of mammals represented, especially large ones, the eastern savannas are richer than anywhere else on earth. This derives from the fertility

Table 12.1 Terminology and classification of African savannas

Category	Vegetation association	Keay et al. (Yangambi classification)
Savanna Woodland Deciduous and semi-deciduous woodland of tall trees (>8 m high) and tall mesophytic grasses (>0.8 m high); spacing of the trees more than diameter of the canopy.	*Brachystegia-Julbernardia* woodlands. *Cryptosepalum* low forest and woodland. *Baikiaea plurijuga* woodlands. *Colophospermum mopane* woodlands.	Woodlands, savannas (and steppes). Woodlands, savannas (and steppes). Dry deciduous forest and savanna. Woodlands, savannas (and steppes).
Savanna Parkland Tall mesophytic grassland (grasses 0.4–0.8 m high) with scattered deciduous trees (<8 m high).	*Acacia-Terminalia-Piliostigma-Combretum* grasslands.	Woodlands, savannas (and steppes).
Savanna Grassland Tall tropical grassland without trees or shrubs.	*Hyparrhenia-Themeda-Setaria-Echinochloa* grasslands. *Trichoptery* grasslands.	Swamps.
Low Tree and Shrub Savanna Communities of widely spaced low-growing perennial grasses (<0.8 m high) with abundant annuals and studded with widely spaced, low-growing trees and shrubs (often <2 m high).	*Chrysopogon-Aristida-Cenchrus* grasslands with *Acacia* and *Commiphora* spp. *Stipogrostis uniplumis* grasslands with *Acacia-Grewia* spp.	Wooded steppe with abundant *Acacia* and *Commiphora*.
Thicket Communities of trees and shrubs without stratification.	*Acacia-Commiphora* thickets. *Colophospermum mopane* scrub.	Thicket.
Intermediate types	*Burkea africana* woodland, *Terminalia sericea* woodland. *Terminalia prunioides* woodland, *Combretum imberbe woodland.*	Woodlands, savannas (and steppe).

Source: Based on Cole (1986).

of the region's volcanic soils, the diversity of its climate and relief as well as from its evolutionary history. Kafue National Park in the southern savannas has the highest number of antelope species.

Because of the varied form and composition of savanna vegetation, including a range of conditions from open grassland to open woodland, world-wide classifications devised by early plant geographers have been found unsatisfactory for differentiating savanna vegetation, especially in East Africa. For example, the typology of the Vegetation Map of Africa (Keay, 1959), based on the classification approved by the Yangambi Conference in West Africa (Conseil Scientifique pour l'Afrique, 1956), is considered problematic because it subdivides 'savanna' and 'steppe' on the basis of the height of the herbaceous layer, that is 80 cm (Pratt *et al.*, 1966; Menaut, 1983; and Cole, 1986).

To be meaningful, a classification has to take into account floristic, ecological, and physiognomic features, which are difficult to integrate in a single system and even more difficult to portray on a map. The simple scheme proposed by Cole (1963 and 1986) which

classifies savanna ecosystems into five broad categories according to physiognomy and floristic composition is adopted here. It provides the most satisfactory basis for describing the distribution and extent of the main vegetation categories of savanna in Africa. Within this broad scheme (see Table 12.1 and Figure 12.1) are included, for example, the 'Steppe' and 'Dry Tropical Scrub and Thorn Forest' of Preston James and others, which are shown in *The Times Atlas of the World* (Bartholomew, 1992).

Savanna Woodland

This biome, with tall trees (15–20 m) and a well-defined stratum of tall perennial grasses, occupies the plateau bordering the tropical rain forests of the Zaïre Basin (see Figure 12.1). It occurs where rainfall averages 600–1500 mm yr^{-1}, although on shallow, sandy soils with a low moisture-holding capacity it is found in areas with much higher rainfall. The rain forest penetrates deeply into the savanna woodland, often as gallery forests along river banks and sometimes as forest remnants on the plateau, forming a forest–savanna mosaic. However,

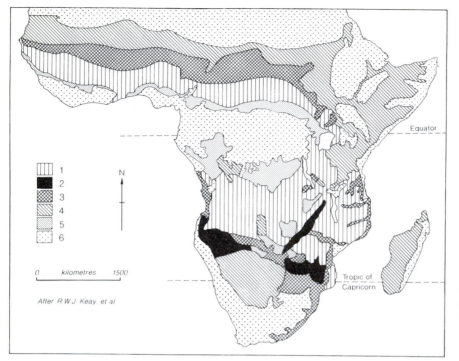

1. Savanna Woodlands predominantly of *Brachystegia*, *Isoberlinia* and *Julbernardia* spp trees
2. Savanna Woodlands dominated by *Colophospermum mopane* trees
3. Savanna Parklands and associated low tree and shrub and low savanna woodlands dominated by *Acacia* tree and shrub species
4. Low Tree and Shrub Savannas dominated by *Acacia* tree and shrub species and *Aristida*, *Stipagrostis* and *Chrysopogon* grass species
5. Mosaic distribution of Savanna woodlands, savanna grasslands and swamp communities
6. Other types of vegetation

Fig. 12.1. Distribution of major categories of savanna vegetation of Africa. (*Source*: Cole, 1986: 111.)

owing to the contrasting floristic and physiognomic features of the two biomes, they are usually separated by a sharp boundary, which marks the limit of penetration by grassland fires.

The commonly occurring *Brachystegia-Julbernardia* savanna woodland extends over a very large part of south-central Africa and includes the so-called *miombo* forest of Tanzania, the plateau woodland of Zambia and Malawi, and the *sandveld* of Zimbabwe. These coincide with high-level planation surfaces and the associated infertile lateritic soils (oxisols) derived from *in situ* weathering of Precambrian rocks.

Cole (1986) argues that these communities, found on Mesozoic planation surfaces, which pre-date the opening-up of the South Atlantic 100–130 million years ago (see Summerfield, this volume), represent the core area of savanna vegetation on the African continent. They are comparable to those of the *campos cerrados* on the Brazilian plateau. Proteaceous tree species within the savanna woodland of Africa are assumed to be relics of the original Gondwanan vegetation. Members of the *Proteacae* family are also found in the *campos cerrados* and in the Australian savannas.

Where post-Tertiary dissection has taken place and the soils have higher levels of exchangeable bases, variation in the composition of the savanna woodland is greater. Such discrete vegetation associations, reflecting local differences in geology and geomorphology, occur along the fault lines followed by the wildlife-rich Luangwa and Zambezi river valleys. On these more fertile soils, biomass of mammals may be higher, although species diversity may be lower than on the more ancient erosion surfaces.

In southern Angola and northern Namibia, and along the Limpopo valley in northern Transvaal and southern Zimbabwe, *Baikiaea plurijuga* and *Colophospermum mopane* are the dominant woodland species. Both members of the Cassia subfamily, the former is also associated with open deciduous woodland on deep Kalahari sand with lower rainfall (500–750 mm) and the latter with hotter and drier river basins where areas of predominantly base-rich clay soils, often alkaline, are derived from the erosion of late-Karoo volcanics (e.g. Stormberg basaltic lavas) of Triassic age. *Colophospermum mopane* woodland reaches its southernmost limit in Africa on the Olifants river in the Kruger National Park, beyond which it is susceptible to damage by frost. North of the Equator, *Acacia seyal-Balanites aegyptiaca* woodland occupies similar sites to those occupied by *Colophospermum mopane* woodland in the southern savannas. It is associated with extensive areas of vertisols developed on fluvial deposits of Quaternary age derived largely from the basic volcanics of the Ethiopian Highlands deposited by the Blue Nile and Sobat drainage systems.

Savanna Parkland and Low Tree and Shrub Savanna

These two categories cover vast areas of the Sudano–Sahelian zone and the Horn of Africa. They also occur, on a smaller scale, in Namibia, Botswana, and adjacent parts of Transvaal and northern Cape Province in South Africa. Savanna parkland occupies an intermediate position on the rainfall gradient between the savanna woodland and the low-tree and shrub savanna, with rainfall ranging from about 400 to 800 mm yr^{-1} (see Figure 12.1). The transformation from tree to shrub dominance is gradual and coincides with the area where mean annual rainfall is 350–400 mm yr^{-1}.

Savanna parkland and low-tree and shrub savanna are found on erosion surfaces which are younger than those occupied by savanna woodland. Large expanses of low-tree and shrub savanna are associated with wind-blown deposits, both in the Sudano-Sahelian savanna belt and in Southern Africa. Composition varies, but the dominant genus is usually *Acacia*. On the drier margins, a savanna–desert mosaic occurs as a result of different soil conditions and thus no sharp boundary between the two biomes can be drawn. With declining rainfall, low-tree and shrub savanna is gradually replaced by dwarf shrubs and desert grass communities.

Thicket

Cole (1986) includes thicket and scrub in her classification, but has little to say about the category. It is anomalous in so far as the herbaceous layer is absent. Thus it falls outside the savanna, however defined. Yet, as explained below, thickets can be replaced by savanna grassland and vice versa. Thicket is defined by Menaut (1983) as shrubby vegetation, evergreen or deciduous, frequently impenetrable by large animals, often in clumps with a grass layer absent or discontinuous. Thickets occur in a large range of environments and are associated with secondary formations which follow exposure to heavy grazing and an associated reduction in grass fires. For example, *Dichrostachys cinerea* thickets occur in the heavily grazed rangeland of the Kalahari (South Africa, Botswana, and Namibia) but, as rainfall decreases, they are replaced by *Acacia mellifera* thickets. A similar pattern is repeated in east-central Sudan and the coastal districts of Kenya, a situation which has prompted the local pastoralists to replace most of their cattle with goats in, for example, Kitui District over the last four decades. Thicket is the most common formation in much of Somalia, particularly the inter-riverine

area between the Juba and the Shebele, where caprines greatly outnumber bovines. Heavy browsing without fire can result in dense regeneration thickets. Thickets are often found at the foot of escarpments and inselbergs where eluvial sand is deposited on *in situ* clays.

Mosaic of Savanna Woodland, Savanna Grassland and Swamp Communities

Savanna mosaic describes the complex distribution of wooded and open savanna which results from 'contrasting edaphic conditions caused by the juxtaposition of different combinations of relief and drainage, bedrock and superficial geology variously related to the geomorphic evolution of the landscape and to both long-term and short-term climate changes' (Cole, 1986: 254). Typical savanna mosaics occur in the Rift Valley region of East Africa, the Upper Nile Basin of Sudan, the Okavango Delta in Botswana, and many other localities with changes in relief. East Africa, with its complex pattern of volcanic mountains, rift valleys, rivers, lakes, and seasonal swamps, provides the most impressive and extensive example. Here, the vigorous mix of wooded and open savanna on volcanic soils supports a high biomass.

Although tall savanna grassland does not appear as a separate category on the sketch map (see Figure 12.1), it none the less occurs throughout the savanna woodland where seasonal inundation is common. Cole (1982 and 1986) notes that, compared with the treeless savannas of the eastern Orinoco plains (i.e. the *llanos*) of South America or the Great Artesian Basin in Australia, the savanna grassland biome is not well represented in Africa. Cole reports that the most extensive grassland areas in Africa are those on the Kafue Flats and the Busango plains along the Kafue river in Zambia and on the Springbok Flats in the Bushveld Basin in South Africa. She does not mention the extensive grasslands bordering the Sudd in the White Nile and the Sobat basin in southern Sudan, the largest swamp in the world according to White (1983). Admittedly, within this flood region, there are areas of very slightly higher ground which are only flooded to shallow depth and carry *Acacia seyal-Balanites aegyptiaca* woodland with up to 1000 mm yr^{-1} of rain (Harrison and Jackson, 1958; Wickens, 1991). The savanna grassland in southern Sudan would nevertheless appear to be much more extensive than in Zambia. Indeed, Lebon (1965) has estimated that the 'seasonally wet grassland' in the White Nile Basin covers about 160 000 km^{-2}. On these extensive level plains, vertisols (i.e. black, cracking montmorillonitic clays) cannot absorb rainfall of more than about 700 mm yr^{-1} without flooding and tree species give way to open grass plains. These are dominated by *Hyparrhenia rufa* or *Setaria incrassata* grasslands. The former, bordering the permanent swamps, are known as the *toich* in southern Sudan. They are burned regularly by the Nilotic pastoralists, a process which further inhibits the development of trees due to the intensity of fires in the tall, rank grassland.

Functional Adaptations of Savanna Organisms

Savanna flora and fauna have developed a wide range of morphological, physiological, and behavioural adaptations. These responses to short-wet and long-dry seasons set them apart from species in other terrestrial environments. The rapid and innovative changes, which often characterize plants and animals in xeric environments, have been described for flowering plants by Stebbins (1952, 1974), for ruminants by Webb (1977, 1978), and by Leakey and Lewin (1992) and Kingdon (1993) for hominids. Stebbins has argued that during the increasing aridity of the middle to late Cenozoic, savanna-like formations acted as a 'species pump' whereby phyla adapted first to more xeric and then back to more mesic conditions.

Flora

The savanna flora comprise the continuum of short annual grasslands, wooded perennial grasslands, and grassy woodlands which lie between the mid-latitude deserts and the tropical rain forest. At the wetter end of the spectrum, grasses, trees, and shrubs are more tolerant of fierce fires which rage through the tall, dry, rank grassland during the dry season. Towards the drier end of the gradient, weather conditions tend to become more and more unpredictable, with large annual variations in the amount and timing of rainfall and the occurrence of long periods of drought. Fire tolerance is less essential for plants in the lower rainfall areas because there is less combustible material in the dry season. This overall gradient is locally interrupted by changes in vegetation structure which can be explained by reference to changes in soil nutrient and drainage conditions, the influence of grazing animals, and man's land-use practices.

Although savanna trees and shrubs show contrasting forms in Africa, South America, and Australia, they generally exhibit anatomical modifications of leaves and roots which give greater water-storage capacity and reduce loss through transpiration. In the drier parts of Africa's savanna zone, the dominant *Acacias* have microphyllous pinnate leaves. In savanna woodland, the leaves

of typical trees (e.g. *Julbernardia* and *Brachystegia* spp.) are also pinnate, but somewhat larger (i.e. mesophyllous). R. H. V. Bell (personal communication) argues that the slightly offset pinnate leaflets of *Julbernardia* in *miombo* woodland assist laminar airflow, evapotranspiration, and gaseous exchange and are an adaptation which favours biomass production on infertile soils. Trees with thick, shiny, leathery (i.e. sclerophyllous) leaves, common in South American savannas, are rare. On the other hand, fire tolerance and spinescence of shrubs and trees are more strongly developed in Africa. Spines are believed to be an adaptation to reduce evapotranspiration (Menaut, 1983), as well as a means of repelling browsing animals (Cumming, 1982). The characteristics of the dominant heliophilous C_4 grasses are broadly similar throughout the world's tropical savannas. Aspartic acid-forming C_4 grasses predominate in the drier areas where available soil nitrogen is higher. Malate-forming C_4 grasses predominate in the moister areas.

Acacia mellifera, a commonly occurring shrub on rangelands throughout low-tree and shrub savanna in West, East and Southern Africa, provides a typical example of the *Leguminosae*, or pod-bearing family, to which many African savanna tree species belong. *A. mellifera* often forms dense thickets of even age, 2–3 m high and tens of metres across. It comes into leaf towards the end of the dry season before the rains when the maximum temperature and the largest diurnal temperature ranges occur. The individual bushes are hemispherical. Low, sweeping branches are covered in paired, sharp, backward-curving prickles, in which man and beast can become hopelessly ensnared. The grass and herb layer beneath and between the bushes is usually poorly developed. In the clay soils of east-central Sudan, a shallow but extensive root system radiates from the root crown, many of the roots extending 8–15 m parallel to the surface at a depth of 25 cm, enabling the plant to exploit soil moisture and nutrients in a large volume of soil. However, roots rarely penetrate deeper than a metre, which is the approximate depth of annual water penetration in vertisols, receiving about 400 mm rain yr^{-1}. In the dry season, a grass fire may scorch the leaves and the ends of the branches, but will not reach the main stem. In post-mature forms, when the protective lower branches are lost, the hemispherical shape is replaced by an inverted cone and the grass and herb layer extends to the stem. *A. mellifera* thickets thus become prone to fire damage and may be rapidly replaced by grassland (Adams, 1967).

In adapting to climatic stress, savanna trees and shrubs display a wide variety of responses which increase their chance of survival: temporal patterns of growth, flowering, and fruiting; the shedding of leaves and root-lets for periods which vary with the length of the dry season; the development of extensive vertical root systems where groundwater is accessible (e.g. *Acacia albida, A. tortilis*); the rapid growth of seedling root systems with the onset of the rains; the funnelling of rainfall to the root zone by the canopy and fluted trunk (e.g. *Balanites aegyptiaca*) (Glover *et al.*, 1962); the storing of water in the bole of the trunk (e.g. *Adansonia digitata*—the baobab); the development of ligno tubers or xylopodia which allow regeneration after destruction of aerial parts by fire or drought; seeds capable of long dormancy; the development of thick leathery casings able to withstand desiccation (e.g. the pods of *Acacia erioloba*); and the dispersal of seed by wind (e.g. the winged fruit of *Combretaceae*).

In African savannas, as elsewhere, perennial grasses often have a bunchy or tussock form that is produced by seasonal aerial shoots arising from rhizomes close to the surface. Growing points (buds) are protected from fire and herbivores by the dense and tight bases of the tillers and the leaf sheaths which form the tussocks. An adaptation of grasses and herbs is the export of hairy and/or barbed seeds which lodge in animals' hooves or get entangled in their hair (and in the socks of bipeds), e.g. the troublesome *Cenchrus biflorus* and *Tribulus terrestris*. In Southern Africa, the latter is an important indicator of former occupation by pastoralists (Kinahan, 1991). Like the winged fruit of savanna trees, this clinging property of seeds is an important adaptation in an open savanna environment. Another co-evolutionary relationship is the development of poisonous secondary compounds by plants exposed to browsing (Owen-Smith, 1993). For example, *Acacias* subject to browsing by antelopes produce leaf tannin in lethal quantities and emit ethylene which stimulates other trees 50 m downwind to step up their production of leaf tannin within five to ten minutes, which explains why giraffe, for example, browse only one *Acacia* in ten, avoiding trees which are downwind (Van Hoven, 1991). Savanna woodland species on infertile soils tend to have a higher level of tannin and hydrocyanin, as they can less afford the loss of metabolic tissues to herbivores (Bell, 1982).

Fauna

Adaptations of savanna fauna include the ability to withstand long periods without access to water by aestivation under the ground during the dry season. For example, catfish of the family *Mormyridae* are endemic to Africa and are particularly well adapted to exploit ephemeral pools during the summer rains. They migrate into burrows as the surface water dries up, appearing again at the onset of the next rains. Most savanna invertebrates and smaller vertebrates are *r*-strategists, with the capacity to reproduce rapidly to use up the available resources of a habitat before other competing species can exploit them. They develop rapidly in response to the favourable impact of rains on primary production and food supply. For example, outbreaks of the African army worm (*Spodoptera exempta*), and both the migratory and the red locust, which are particularly damaging to grasses and cereal crops, are associated with years of good rainfall.

Migration and mobility in search of food have been important features of savanna vertebrates. Migration by large herds of ungulates has been severely attenuated by fences and extensive human habitation. A surviving example is the spectacular migration of vast herds of white-eared kob in Equatoria Province, Sudan, which spend the dry season in Boma National Park and move in an elliptical orbit south-westwards in the early rains (July) and return north-eastward in September. About one million animals move as a single unit (Fryxell, 1983). A better known and equally impressive migration is the seasonal movement of wildebeest in the Serengeti/Maasai Mara ecosystem.

Environmental Determinants Governing Savanna Physiognomy

Over Africa, there is a well-recognized positive correlation between plant biomass and rainfall. In general, the length of the wet season constrains the length of the growing season and available biomass, but there are exceptions which reflect local topographic and edaphic variations, the effects of fire, wildlife, man, and his animals. In the last three decades, information has been gathered on soils and vegetation, on numbers, densities and seasonal occupance of animals, and the effects on physiognomy of excluding animals and of controlling fires, especially in national parks and wildlife reserves in South and East Africa. The savanna ecosystem is now recognized to be more dynamic than was previously realized. Disturbance of dominant species is often found to lead to far-reaching chain reactions in other system components.

The role of environmental factors in determining the physiognomic character of savanna vegetation, both in space and time, remains the subject of a continuing debate. Alternative models have been devised which allow tropical savannas to be compared to one another. Older models were based on two or three environmental variables (e.g. temperature, rainfall, soils), but more recently they include a greater number (Belsky,

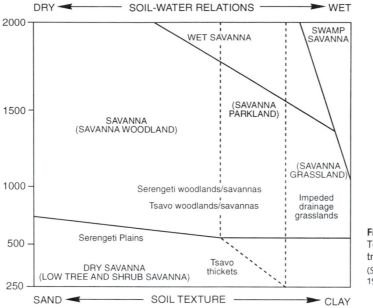

Fig. 12.2. The Johnson-Tothill (1985) model of tropical savannas.
(*Source*: adapted by Belsky, 1991.)

1991). The factors behind changes in tree : grass ratios in savannas have been debated since Hancock (Sudan Government, 1944) formulated the hypothesis of the grass–*Acacia* cycle. He suggested that there were 'grass' and 'tree' periods, and that, following plagues of grasshoppers or dry years, grasses may set no seed, thus rendering the land free of grass fires, which favoured the establishment of trees. A related observation by Jackson (1955) was that bushes and trees tended to be about the same age over large areas. The relative importance of factors determining tree : grass ratios continues to be debated, particularly by wildlife managers and livestock producers who seek to maintain a balance of plant species which is consistent with their production objectives.

Alternative Models

A simple model for the Sudan savanna belt where soil and slope conditions remain homogeneous over vast areas was developed by Smith (1949) as an aid to agricultural land classification. He found that the distribution of tree species (e.g. *Acacia seyal, A. mellifera*, and *A. tortilis*) could be explained by rainfall and soil texture and that these factors were interdependent in a fairly simple way. He showed how level or 'datum' sites with clay soils (i.e. vertisols) required one and a-half times as much rainfall for a species to grow satisfactorily as was required by the same species on sandy soils. This reflects the capacity of clay soils to hold water at a suction greater than sands and greater than that exerted by the

roots of the observed species of trees, which, with no recorded exceptions, can be shown to progress through their rainfall span, via the same sequence of site/soil types.

Johnson and Tothill's (1985) model of tropical savannas, adapted to elucidate East African savannas (Figure 12.2), illustrates the role of rainfall and soil texture in determining savanna structure. In this model, savanna grasslands typically occur under the wettest conditions (600–2000 mm annual rainfall) and on clay soils. Walker (1987) developed a model that incorporates all important environmental factors influencing savanna physiognomy and which shows the complexity of ways in which environmental factors interact (Figure 12.3). The large arrows illustrate that East African savannas are most strongly influenced by rainfall, affecting species composition through its effect on soil moisture and nutrients; by elephant and fire which reduce tree cover and prevent its regeneration; and by ungulates which affect grassland composition. A model by Frost *et al.* (1986) allows for the more accurate comparison of environmental factors (Figure 12.4). It combines available nutrients (AN), and plant available moisture (PAM), which, in turn, integrates rainfall, water infiltration, evapotranspiration, soil texture, and hydrologic regime into a single function.

The Effect of Fire

Grass fires, ignited both by lightning and by man, have played a key role in the evolution of savanna

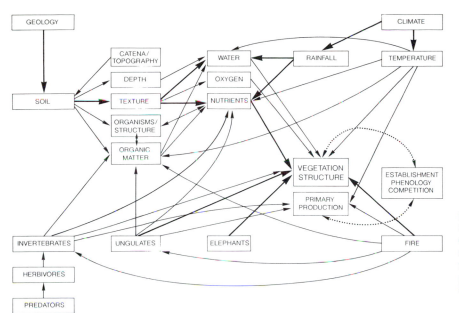

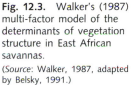

Fig. 12.3. Walker's (1987) multi-factor model of the determinants of vegetation structure in East African savannas.

(*Source*: Walker, 1987, adapted by Belsky, 1991.)

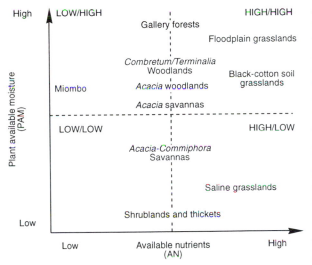

Fig. 12.4. Frost *et al.* (1986) functional classification of East African savannas.

(*Source*: adapted by Belsky, 1991.)

vegetation. Fires occur most frequently under conditions in which seasonal rainfall is sufficient to produce the fuel loads necessary for fire to carry. The effect of fire depends on the composition of the plant communities, their phenology, and the size and shape of woody species, the intensity of herbivory, the time of the year, and the frequency of burning (Adams, 1967; Trollope, 1982; Gillon, 1983). Fire generally favours the development

and maintenance of a predominantly grassland vegetation by destroying the shrub and tree seedlings, preventing them from reaching a more fire-resistant stage. Although grass is burnt over large areas of savanna almost every dry season, it does not burn uniformly and some areas are spared. If an area escapes fire in this way over several dry seasons in succession, seedlings attain sufficient height to withstand grass fires. This partly explains how woodland and thicket can be entirely composed of young, or mature, or of entirely moribund forms. Episodic rainfall events are another important factor in the periodic establishment and survival of woody species, especially when rainstorms coincide with a favourable seed-production year.

The Effect of Soils and Herbivory

Bell, a wildlife ecologist who carried out pioneering research defining grazing ecosystems in the Serengeti National Park, Tanzania (see Bell, 1971), and conducted numerous low-level aerial counts of wildlife in East Africa, developed a hypothesis to account for the fact that, despite its high plant biomass, the broad-leaved *Brachystegia* woodland, which extends over a very large part of south-central Africa, supports a much lower animal biomass (made up of fewer small herbivores, yet more elephant and buffalo) than the fine-leaved woodland and short to medium grasslands of East Africa's volcanic soils. Bell (1982) postulated that standing plant biomass in savanna communities decreases with increasing soil nutrients due to increased herbivory. He explained the higher proportion of elephant and buffalo (in relation to other species) in *miombo* forest by their greater tolerance of poor food quality. However, Belsky (1991) argued that, in the case of the drier parts of the Serengeti Plains, the lack of trees is due to high soil alkalinity and impermeability due to calcrete. She found that trees did not grow on these plains, even when protected from fire and herbivores.

The Effect of Herbivores

On a continental scale, wild herbivore effects are insignificant compared to those of soils and climate, but within a particular ecosystem they may determine savanna structure. The complexity of these interactions can be demonstrated by reference to the Serengeti ecosystem. Tribal traditions, archaeological evidence, and written records suggest that the area occupied by the Serengeti National Park in Tanzania and the adjacent Mara Game Reserve in Kenya was open and tsetse-free and extensively grazed by cattle a century or more ago (Dublin, 1986). The great 1880s rinderpest pandemic caused spectacular losses of both domestic and wild grazers. Woodland became established in the areas of higher rainfall, that is in the central and northern Serengeti Park and the Mara. This provided tsetse habitat, which prevented reoccupation by pastoralists and resulted in the abandoned area being included within the wildlife reserves. Not until the 1960s, after elephant invaded the Park, was the woodland once again opened up for grazing animals. Giraffe have significantly decreased the proportion of mature trees by maintaining browsing pressure on recruitment-sized *Acacia tortilis*. This prevents trees maturing to larger size classes and renders them vulnerable to fires (Ruess and Halter, 1990).

In the early 1960s, Serengeti's migratory wildebeest population was estimated at 0.19 million, but in 1989 the Serengeti Ecological Monitoring Programme estimated it to be 1.69 million. Meanwhile, populations of many other herbivores have remained stable or have shown a concurrent increase. There is now a pronounced annual migration of ungulates between the Mara and the southern Serengeti plains. Further, in some years the former westward movement towards Lake Victoria does not take place. Instead the animals migrate directly northwards to the Mara and once again maintain the open grassland environment which existed one hundred years previously.

The changes to the savanna physiognomy caused by herbivores over the last forty years in the Serengeti ecosystem provide some indication of the impact of mammals on the savanna mosaic before human activities became the dominant influence. Fluctuating populations would have kept the entire ecosystem in a state of flux.

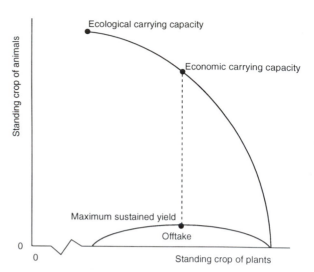

Fig. 12.5. The relationship between plant and herbivore populations.
(*Source*: adapted from Caughley, 1979, and Bell, 1985.)

Contemporary animal biomass of 18 000 kg km^{-2} on volcanic soils in East Africa has been quoted by Coe *et al.* (1976). Such production would have been matched during favourable periods 100 000 years ago.

Modelling Plant and Animal Biomass

Figure 12.5, originally presented by Caughley (1979) and elaborated by Bell (1985) and by Behnke and Scoones (1993), provides a schematic overview of the relationship between plant and wild herbivore populations in which the former is inversely related to the latter. The top curve marks combinations of plant and animal densities in a hypothetical system. At the far end of the horizontal axis, the curve shows the situation when there is a small animal population and a large reserve of forage. As the herbivore population increases, the plant biomass declines. In an undisturbed system, the increase in animal numbers will eventually be checked by the declining availability of forage. This will occur when the production of edible plants equals the rate of their consumption and the animal population ceases to grow because limited feed-supplies negatively affect reproductive performance and/or produce death rates equal to birth rates. This point of equilibrium is termed the 'ecological carrying capacity' or 'K'. There is no surplus production either of animals or biomass. Animals may be plentiful, but they will not be in good condition; neither will the vegetation be as dense, nor comprise the same species as it would in their absence.

The offtake curve in Figure 12.5 indicates the different offtake needed to maintain combinations of plant

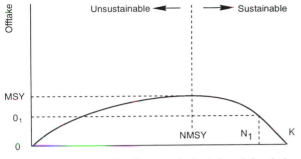

Fig. 12.6. The relationship between offtake and population stability.
(*Source*: adapted from Bell, 1985.)

and animal densities other than those occurring at ecological carrying capacity. Initially, the offtake curve rises from zero at very low stocking rates and increases with increasing size of the herbivore population. The point of 'maximum sustainable yield' (MSY) usually lies at about a half to two-thirds of the stocking density at ecological carrying capacity. This is termed 'economic carrying capacity' (NMSY) by Caughley (1979). As the animal population grows beyond this point, the offtake rate begins to fall and ultimately returns to zero as increasingly high rates of mortality and falling birth rates remove the opportunity for further offtake. These various thresholds will vary from year to year according to rainfall conditions, disease incidence, etc. If the effects of different levels of offtake are observed over time, it is possible to build up an offtake curve such as that shown in Figure 12.6. This shows the levels of offtake at which a population can stabilize. For example, a population subjected to an offtake O_1 will fall to N_1 and then stabilize. If offtake is increased to MSY, the population will decline to NMSY before stabilizing. An offtake in excess of MSY will cause the population to decline to extinction.

This model has several practical applications. By applying an offtake of known quantity to a known population (e.g. wildebeest, for which the shape of the curve is generally known), the size of any illegal offtake can be estimated. Also, knowledge of the relationship between the herbivore population and its food supply allows the prediction of the effect on MSY and NMSY of land, within a particular wildlife reserve, being converted to agriculture and settlement.

Human Adaptations and Impacts

On the basis of available fossil evidence, the wooded and open savannas of East Africa appear to have the longest

history of human habitation, extending from the Pliocene, that is 3.0–2.0 million years ago (Stringer, 1992). Australopithecine hominid use of East African savanna landscapes (e.g. Tabarin in Kenya, and Olduvai and Laetoli in the Serengeti-Ngorongoro area of Tanzania, and Omo in Ethiopia) dates back further, 3.0–4.0 million years or more. Leakey and Lewin (1992) and Kingdon (1993) argue that the colonization of African savanna by protohominids was the initial and decisive step taken in the history of the human species. To the west of the Great Rift Valley, about 10 million years ago, luxuriant forests provided ideal habitats for quadrupedal apes, but for bipedal apes (the hominid family) the savanna mosaic habitats created by the rift system and the changing climate were crucial, providing a centre of endemism for the hominids. Their primary evolutionary adaptation to bipedal locomotion, which took place about 7.5 million years ago in a period of cooling and drying, was a response to the need to forage for food in a more open savanna environment. While savannas provided a rich variety of foods, including many fruiting and seeding plants, these resources were more widely dispersed than in tropical rain forest and their availability was much more subject to seasonal fluctuations.

Studies of pollen grains and the nature of sediments at Olduvai and Laetoli have allowed the description of the environment of these early hominids, 2.5 million years ago. Pleistocene palaeontologists paint a picture of a dynamic mosaic of savanna grassland and woodland with a fauna which resembled that of modern Serengeti woodland habitats (Andrews, 1989, 1992). The savanna zone of East Africa is also postulated by Kingdon (1993) as the centre of origin of modern humans (*Homo sapiens*) during the later Pleistocene (0.4–0.2 million years BP). However, Bell (personal communication) argues that the distribution of hominid sites is entirely determined by fossil deposits of the right ages. Most of these are in the Rift Valley, but where suitable sites are found elsewhere (e.g. Transvaal) hominids are also found. It cannot be assumed therefore that hominids were largely confined to the Rift Valley. Whatever the resolutions of this issue, it would seem that the basic climatic adaptation of the human species is to hot and seasonally dry conditions, namely the savanna (Baker, 1992).

Hunting

In the savanna zone of sub-Saharan Africa, prehistory covers the period before the fifteenth century to the nineteenth century, depending on locality. It covers the Stone Age in which small groups pursued an environmentally opportunistic life-style of foraging, or hunting and gathering, using stone implements. Dry-season movement was tethered to a few perennial water sources.

Wherever fresh water was to be found, game animals were usually available too. The prolific rock art at many of these sites in Zimbabwe, Botswana, and Namibia reflects the rich ritual life and social order of the hunters (Thomas and Shaw, 1991).

There has been a long debate as to whether savanna communities are a natural climatic climax or whether they are primarily induced by humans through burning. As emphasized in the discussion of environmental factors influencing savanna physiognomy, fire is just one of these factors, although often a very important one. While most of the fire-tolerant plants (principally grasses and fire-tolerant trees such as *Acacia*, *Combretum*, and *Terminalia* species) evolved long before humans learned to kindle fire, their pre-eminence since that time (and long before the development of agriculture) is a by-product of the subsistence hunter's burning techniques. Animal populations were also fundamentally influenced by human activities. The regular 'cleaning' by fire of vast tracts of land by human hunters, put leaf- and herb-eaters at a disadvantage. At the same time, the grazers (e.g. wildebeest, kob, reedbuck, and zebra) were greatly favoured by regular dry-season burning of the vegetation. Available fossil evidence confirms that grass-eaters, in relation to browsers, greatly increased in importance towards the end of the Pleistocene. Selective hunting techniques also had consequences for populations of certain species, especially when maintained for decades or more to the point where offtake by humans exceeded the maximum sustainable yield (MSY). Resident animal species dependent on fixed watering points would have been more easily depleted than migratory types. Those which were unable to cope with fire and hunting died out whilst surviving animals declined in body weight (Kingdon, 1993). Bell (1985) explains the resilience of migratory herds on the basis of their large population size and the physical difficulty that hunters face in exceeding MSY.

Present-day ethnological observations of the foraging life-style of the San, no longer with stone tools, have been used as analogues of Later Stone Age behaviour (Deacon, 1984). The study of the use of fire by Australian Aborigines prior to the advent of Europeans also supports the hypothesis that burning was carefully planned and skilfully managed by hunter-gatherers, not only for the purposes of hunting, but also to increase the supply of early successional types of food plants (Harrington *et al.*, 1984). In Australian, as in African savannas, animals are attracted to patches regenerating after fire, where they are more easily hunted. Thus the impact of prehistoric humans on savanna ecosystems was far-reaching. As Kingdon (1993) observes,

African savanna ecosystems have evolved in tandem with *Homo*. On the one hand, the animals and plants evolved some defences against human depredation. On the other, a complex of endemic diseases, competition, predators and social and environmental constraints probably combined to keep human populations in some sort of balance. The African Eden was no paradise, but it was where hominids had been part of the landscape for at least 5 or 6 million years. (p. 71)

Pastoralism

Since prehistory, societies have moved from hunter-gatherer to pastoralist to agro-pastoralist and back again to pastoralist and hunter (Haaland, 1972; Denbow and Wilmsen, 1986). The few surviving hunters and gatherers have minimal impact on savanna fauna and flora, compared to post-Stone Age peoples (Silberbauer, 1972) now engaged in multiple land uses, relying on increasing levels of technological sophistication at each stage.

The Iron Age saw the development of pastoralism and agriculture accompanied by the growth of large and socially complex settlements. The low-tree and shrub savanna of the Sudan–Sahel and the Horn of Africa has been subject to pastoral influences (i.e. cattle) for at least 7000 years (Epstein, 1971). The earliest remains of domestic livestock (i.e. sheep and goats) in East Africa date to 3500 years BP (Robertshaw and Collet, 1983). The level of impact of humans and their grazing animals on the savanna ecosystem has varied greatly. The distribution of *trypanosomiasis* and tsetse fly (*Glossina morsitans*) has been a major limiting factor on human incursions (Ford, 1971). Homewood and Rodgers (1991) show how present-day savanna vegetation and wildlife populations in northern Tanzania have evolved alongside and under the influence of Maasai pastoralists for the last 2000 years. They argue that there have been no serious environmental changes attributable to pastoralism in the Ngorongoro Conservation Area, nor have the wildlife populations suffered from the presence of domestic stock. However, the Maasai of Ngorongoro are probably exceptional. The advent of human beings and their animals to African savannas, especially in the last one hundred years or so, has generally constituted an unprecedented ecological invasion, a dominance which is likely to expand (Cumming, 1982; Kingdon, 1993).

Over the past four decades, far-reaching technological changes in the savanna zone of Africa have affected the practice of extensive long-range seasonal transhumant herding of the Sudan-Sahel. The introduction of boreholes and pumps, and range enclosure have been major factors in the sedentarization of pastoralists (Adams, 1982, 1993; Tourè, 1990). In Southern Africa, the installation of boreholes, with or without the subdivision of the range by fencing, has transformed the vegetation. The only areas left untouched during the dry season are those where exploitable groundwater is unavailable and where the provision of surface water is too costly. Whether this heavy stocking results in 'degradation', representing an unsustainable process, is a matter for debate (Abel and Blaikie, 1989; Abel, 1990; Biot, 1990; White, 1992; de Queiroz, 1993; see Warren, this volume). For example, perennial grasses may be replaced by annuals, but such changes are not necessarily synonymous with degradation (Westoby *et al.*, 1990). In wet periods, vegetation is in surplus. In the dry season, when grazing pressures are high, it is less prone to damage. The length and severity of dry periods prevent animal populations from building up to the point that they challenge vegetation in the growing season. However, there is little doubt that bush encroachment by sickle bush (*Dichrostachys cinerea*) and blackthorn (*Acacia mellifera*) has become a major problem for pastoralists in the low-tree and shrub savanna rangelands of Southern Africa, one which is very costly to reverse in the short term.

Behnke and Scoones (1992) have used Caughley and Bell's model (see Figures 12.5 and 12.6) in an analysis of range-management strategies and policies in African savannas. They point out that most livestock owners find it profitable to hold their livestock populations somewhere short of ecological carrying capacity. However, what constitutes an economically optimal stocking rate varies according to a producer's husbandry practices and management objectives. Part of the alarm of colonial and post-colonial administrators over the 'overstocking' and 'degradation' of communal grazing land arises from the false assumption that economically profitable stocking rates for commercial ranchers are the ones which are biologically sustainable and that pastoralists' stocking rates are not. However, for a variety of reasons, African pastoralists are able to profitably sustain higher stocking rates than commercial beef ranchers. The importance of assumptions about overstocking to ideas about desertification are discussed by Warren (this volume).

Drawing on observations of fluctuating animal and plant biomass and associated episodic events (flood, drought, fire, etc.), which keep the savanna ecosystem in a state of flux (Westoby *et al.*, 1989), Behnke and Scoones argue that grazing systems are also likely to be in a state of constant disequilibrium. Thus the concept of a single, safe livestock-carrying capacity based on classical succession theory (Clements, 1936), applicable in temperate climates, is not appropriate to the management of savanna grazing systems where there is need to 'track' spatial and temporal fluctuations in feed supply by a flexible response to unpredictable events (see also Warren, this volume).

Forestry

Many of the early difficulties experienced by modern forestry in savanna environments arose from a failure to take adequate account of existing environmental limitations (e.g. drought, fire, and the depredations of livestock) and potentials (e.g. episodic rainfall events leading to widespread seed germination and seedling establishment). As with 'modern' livestock and crop production, preoccupation with the forestry systems and methods introduced from temperate regions has been overtaken by interest in indigenous management systems. This interest has been spurred by the increasing realization that wood fuel will continue indefinitely to be in demand as the primary source of energy for Africa and that savanna woodlands are under threat.

The removal of biomass from the savanna ecosystem for burning depletes the soil of nutrients as well as its capacity to store water and to replenish the vegetation cover. The first step is the removal of timber, either for firewood or conversion into charcoal, often as a by-product of clearing communal land for cultivation. Over vast areas of the African savanna, the scale of this extraction has been modest and continues to be within sustainable limits. However, in the last two decades, the clearing of forest solely for charcoal by urban contractors has been an accelerating problem, wherever feeder roads provide access to savanna woodland.

Plantations of exotic species have proved less successful than originally expected, particularly on the dry side of the 800 mm annual isohyet. However, most indigenous woody species continue to grow, even when browsed, burned, or lopped. Protecting them from livestock and from grass fires leads to much increased productivity. Helping more tree seedlings to survive until they are well established is much less costly and more effective than attempts to establish plantations or village woodlots. Thus, local methods of managing that natural regeneration of savanna woodland, linked to animal and crop production, are the subject of growing interest, together with the resolution of related land-tenure issues (Shepherd, 1992).

Cultivation

Up to a century or so ago, various local systems of land-rotation cultivation had no very serious impact on savanna ecology, defined in terms of an irreversible decline in output from cultivation. This was because the soil and vegetation were given adequate time to regenerate between periods of cropping. The circle *citemene* system provides an example. This was practised on the infertile oxisols of the Zaïre–Zambezi watershed. Trees were lopped over a wide area of *Brachystegia-Julbernardia* (*miombo*) woodland and the branches piled or burnt on a small garden enriched by nutrients carried from a surrounding area, ten times the size. A variant, along the margins of the broad *dambo* system of the upper Saisi River, involved the hoeing of *Hyparrhenia* grass into large circular mounds where it was left to rot and subsequently cultivated to maize, finger millet, and pulses (Allan, 1965). Increasing population and commercialization of production have inevitably led to the breakdown of many of these practices in the savanna zone. Nye and Greenland (1960) have described how tropical rain forest and savanna woodland, particularly in West Africa, have been transformed into less-productive savanna grassland by land clearing for cultivation, aided by fire.

Yet, modern agriculture in areas cleared of savanna woodland has proved even less sustainable than more traditional systems of crop farming, because of the low inherent fertility of ancient soils and/or soil erosion, weed infestation, and the unpredictability of rainfall. Over the last three decades, evidence has steadily been accumulated of the unsustainable systems of mechanized agriculture introduced by Europeans to the African savanna: in Nigeria (Baldwin, 1957), in Tanzania (Allan, 1965), and in Sudan (Simpson and Simpson, 1991). For example, the ill-fated Groundnut Scheme, in what was Tanganyika, was located at Urambo and Nachingwea on land cleared from *miombo* woodland. At Urambo, in Tabora District, an area of advanced planation and impeded drainage, soils proved to be of low fertility and mechanized groundnut production uneconomic. At Nachingwea, in Mtwara District, the environment was more complex and the soils more fertile as a result of recent elevation following a period of advanced planation. *Brachystegia-Julbernardia* woodland was confined to the ridges and intervening sites carried *Combretum* species and the valuable *Pterocarpus angolensis*. However, the reddish-brown sandy loams were sticky when wet, but hard and massive when dry, with a common tendency to cap after heavy rain. Large-scale commercial mechanized production was abandoned. Following the extraction of the more valuable timber species, these areas have remained sparsely settled and extensively used for shifting hoe cultivation, with the widespread distribution of the tsetse fly being both the cause and the effect of sparse settlement.

Conclusions

Compared with the rain forests, Africa's savanna ecosystem has been a source of less public concern, despite its much greater geographical area, its contribution to global photosynthesis, and its role as the primary habitat for human beings. There is little systematic information about its current state. This is partly because changes in savanna mosaics are difficult to quantify, unlike inroads into closed-canopy rain forest. Physiognomic change is, in any case, a characteristic of the savanna. None the less, the destructive impact of long human occupation of the ecosystem has been profound. Early hunters were fewer in number and their technology primitive. Their ecological impact has been dwarfed by the massive degradation caused by mechanized farming in, for example, the clay plains of east-central Sudan. Yet, it is important to recognize that these differences are primarily ones of scale—of human numbers and of technology. Man's tendency to over-exploit the basis of his subsistence is endemic.

Africa's savanna habitats are being rapidly degraded, primarily as a result of demographic pressure. Satellite images reveal blisters of devegetation surrounding all major settlements, especially in the Sudano–Sahelian zone. Because of their poverty, isolation, and poor communications, trade is insignificant. People are therefore unable to make the economic substitutions that would ameliorate or even forestall degradation. Rising poverty both causes and arises from environmental problems. Countries falling within the savanna zone, both within the Sahel, East and Southern Africa, suffer periodic drought, crop failure, famine, and related insecurity. Indeed, large areas of savanna in Africa now, once again, lie outside effective national territories.

If traditional pastoralism and fuelwood collection are still relatively benign influences, then extensive arable cultivation (manual and mechanized) which results in the permanent removal of the vegetation cover is not. In the absence of technical advances which would allow the intensification of arable production, larger areas of savanna will inevitably come under the plough, both as a result of falling yields per unit area and because of population increases.

Much of the research work on savanna ecosystems has been conducted in relatively small, protected areas in East and Southern Africa, but these are under increasing threat. For example, while the government of Tanzania has not been averse to setting aside large areas of potentially cultivable land for conservation and wildlife, opposition to this policy has come from local people as well as from commercial interests. This opposition is sure to grow. Much of the Serengeti National Park (especially within the northern extension and the western corridor) and the highlands of Ngorongoro are suitable for mixed farming. The drier areas are highly valued by local people for grazing domestic stock. The Maasai-Mara National Reserve, over the frontier in Kenya, is threatened by commercial wheat production, which has greatly decreased the area available to migratory wildlife outside the reserve. Following majority rule, similar pressures are building up in Namibia, and the same will undoubtedly follow in South Africa, where national parks have essentially been the playground of whites. While combining pastoralism and wildlife conservation may be feasible under special circumstances in East Africa, joint use of land for arable crops and wildlife is impossible. As farmers on the boundaries of African national parks well know, the benefits of wildlife conservation are usually enjoyed by a very different group from that which bears the brunt of its costs. The remaining savanna environments in Africa are expected to diminish further until the benefits from the wildlife industry are shared more equitably with local communities.

References

Abel, N. O. J. (1990), 'De-stocking Communal Pastures in Southern Africa: Is it Worth it?', *Technical Meeting on Savanna Development and Pasture Production* (London).

—— and Blaikie, P. M. (1989), 'Land Degradation, Stocking Rates and Conservation Policies in the Communal Rangelands of Botswana and Zimbabwe', *Land Degradation and Rehabilitation*, 1: 101–23.

Adams, M. E. (1967), 'A Study of the Ecology of *Acacia mellifera, A. seyal* and *Balanites aegyptiaca* in Relation to Land Clearing', *Journal of Applied Ecology*, 4: 221–37.

—— (1982), 'The Baggara Problem: Attempts at Modern Change in Southern Darfur and Southern Kordofan', *Development and Change*, 13: 259–89.

—— (1993), 'Options for Land Reform in Namibia', *Land Use Policy*, 10: 191–6.

Allan, W. (1965), *The African Husbandman* (London).

Andrews, P. J. (1989), 'Palaeoecology of Laetoli', *Journal of Human Evolution*, 18: 173–81.

—— (1992), 'Reconstructing Past Environments', in S. Jones, R. Martin, and D. Pilbeam (eds.), *Cambridge Encyclopedia of Human Evolution* (Cambridge), 191–5.

Baker, P. T. (1992), 'Human Adaptations to the Physical Environment', in Jones, Martin, and Pilbeam (eds.), *Cambridge Encyclopedia of Human Evolution*, 46–51.

Baldwin, K. D. S. (1957), *The Niger Agricultural Project* (Oxford).

Bartholomew (1992), *The Times Atlas of the World* (London).

Beard, J. S. (1953), 'The Savanna Vegetation of Northern Tropical America', *Ecological Monographs*, 23: 149–215.

Behnke, R. H. and Scoones, I. (1992), 'Rethinking Range Ecology: Implications for Range Management in Africa', in R. H. Behnke Jr., I. Scoones, and C. Kerven (eds.), *Range Ecology at Disequilibrium: New Models of Natural Variability and Pastoral Adaptation in African Savannas* (London), 1–30.

Bell, R. H. V. (1971), 'A Grazing System in the Serengeti', *Scientific American*, 224: 86–93.

—— (1982), 'The Effect of Soil Nutrient Availability on Community Structure in African Ecosystems', in B. J. Huntley and B. H. Walker (eds.), *Ecology of Tropical Savannas* (Berlin), 193–216.

—— (1985), 'Carrying Capacity and Offtake Quotas', in R. H. V. Bell and E. McShane Caluzi (eds.), *Conservation and Wildlife Management in Africa* (Washington, DC).

Belsky, A. J. (1991), 'Tree/Grass Ratios in East African Savannas: A Comparison of Existing Models', in P. A. Werner (ed.), *Savanna Ecology and Management: Australian Perspectives and International Comparisons* (Oxford), 139–45.

Biot, Y. (1990), 'How Long Can High Stocking Densities be Sustained?', *Technical Meeting on Savanna Development and Pasture Production* (London).

Caughley, G., 'What is This Thing Called Carrying Capacity?', in M. S. Boyce and L. D. Hayden-Wing (eds.), *North American Elk: Ecology, Behaviour and Management* (Laramie, Wyo.).

Clements, F. E. (1936), 'Nature and Structure of the Climax', *Journal of Ecology*, 24: 252–84.

Coe, M. J., Cumming, D. M., and Phillipson, J. (1976), 'Biomass and Production of Large African Herbivores in Relation to Rainfall and Primary Production', *Oecologia*, 22: 341–51.

Cole, M. M. (1963), 'Vegetation Nomenclature and Classification with Particular Reference to the Savannas', *South African Geographical Journal*, 55: 3–14.

—— (1982), 'The Influence of Soils, Geomorphology and Geology on the Distribution of Plant Communities in Savanna Ecosystems', in Huntley and Walker (eds.), *Ecology of Tropical Savannas*, 145–74.

—— (1986), *The Savannas: Biogeography and Geobotany* (London).

Conseil Scientifique pour l'Afrique (1956), Specialist Meeting on Phytogeography (Yangambi, Zaire).

Cumming, D. H. S. (1982), 'The Influence of Large Herbivores on Savanna Structure in Africa', in Huntley and Walker (eds.), *Ecology of Tropical Savannas*, 217–45.

Deacon, J. (1984), 'Later Stone Age People and Their Descendants in Southern Africa', in R. Klein, *Southern African Prehistory and Paleoenvironments* (Rotterdam), 221–328.

Denbow, J. and Wilmsen, E. N. (1986), 'Advent and Course of Pastoralism in the Kalahari', *Science*, 234: 1509–15.

de Queiroz, J. S. (1993), 'Range Degradation in Botswana: Myth or Reality? *ODI Pastoral Development Network Paper*, 35b (London).

Dublin, H. T. (1986), 'Decline of the Mara Woodlands: The Role of Fire and Elephants', Ph.D. thesis (University of British Colombia).

Epstein, H. (1971), *The Origin of the Domestic Animals of Africa*, 2 vols. (New York).

Ford, J. (1971), *The Role of Trypanosomiasis in African Ecology: A Study of the Tsetse Fly Problem* (Oxford).

Frost, P., Medina, E., Menaut, J. C., Solbrig, O., Swift, M., and Walker, B. (1986), *Responses of Savannas to Stress and Disturbance: A Proposal for a Collaborative Programme of Research* (Biology International Special Issue, 10) (Paris).

Fryxell, J. (1983), 'Wildlife Migration in Boma National Park', *Swara*, 6: 12–15.

Gillon, D. (1983), 'The Fire Problem in Tropical Savannas', in F. Bourlière (ed.), *Tropical Savannas, Ecosystems of the World*, 13: 617–41.

Glover, P. E., Glover, J., and Gwynne, M. D. (1962), 'Light Rainfall and Plant Survival in East Africa: II. Dry Grassland Vegetation', *Journal of Ecology*, 50: 199–206.

Haaland, G. (1972), 'Nomadization as an Economic Career among Sedentaries in the Sudan Savanna Belt', in I. Cunnison and W. James (eds.), *Essays in Sudan Ethnography* (London), 148–72.

Harrington, G. N., Wilson, A. D., and Young, M. D. (1984), *Management of Australia's Rangelands* (East Melbourne).

Harrison, M. N. and Jackson, J. K. (1958), *Ecological Classification of the Vegetation of the Sudan* (Khartoum).

Homewood, K. M. and Rodgers, W. A. (1991), *Maasailand Ecology; Pastoralist Development and Wildlife in Ngorongoro, Tanzania* (Cambridge).

Jackson, J. K. (1955), *Memoirs of the Forestry Division* (Khartoum).

Johnson R. W. and Tothill, J. C. (1985), 'Definitions and Broad Geographic Outline of Savanna Lands', in J. C. Tothill and J. G. Mott (eds.), *Ecology and Management of the World's Savannas* (Canberra), 1–15.

Keay, R. W. J. (1959), *Vegetation Map of Africa South of the Tropic of Cancer* (Oxford).

Kinahan, J. (1991), *Pastoral Nomads of the Central Namib Desert; The People History Forgot* (Windhoek; Namibia).

Kingdon, J. (1993), *Self-Made Man and his Undoing* (London).

Leakey, R. and Lewin, R. (1992), *Origins Reconsidered* (London).

Lebon, J. H. G. (1965), *Land Use in the Sudan* (Bude).

Menaut, J. C. (1983), 'The Vegetation of African Savannas', in Bourlière (ed.), *Tropical Savannas, Ecosystems of the World*, 13, 109–8.

Nix, H. A. (1983), 'Climate of Tropical Savannas', in Bourlière (ed.), *Tropical Savannas, Ecosystems of the World*, 13, 37–62.

Nye, P. H. and Greenland, D. J. (1960), *The Soil under Shifting Cultivation* (Harpenden).

Owen-Smith, N. (1993), 'Woody Plants, Browsers and Tannins in Southern African Savannas', *South African Journal of Science*, 89: 505–10.

Pratt, D. J., Greenway, P. J., and Gwynne, M. D. (1966), 'A Classification of East African Rangeland, with an Appendix on Terminology', *Journal of Applied Ecology*, 3: 369–82.

Robertshaw, P. and Collet, D. (1983), 'A New Framework for the Study of Early Pastoral Communities in East Africa', *Journal of African History*, 24: 289–301.

Ruess, R. W. and Halter, F. L. (1990), 'The Impact of Large Herbivores on the Seronera Woodlands, Serengeti National Park, Tanzania', *African Journal of Ecology*, 28: 259–275.

Shepherd, G. (1992), *Managing Africa's Tropical Dry Forests: A Review of Indigenous Methods*, ODI Occasional Paper 14 (London).

Silberbauer, G. B. (1972), 'The G/wi Bushman', in M. Bicceri (ed.), *Hunters and Gatherers Today* (New York), 271–335.

Simpson, I. G. and Simpson, M. C. (1991), 'Systems of Agricultural Production in Central Sudan and Khartoum Province', in G. M. Craig (ed.), *Agriculture of the Sudan* (Oxford), 252–79.

Smith, J. (1949), *Distribution of Tree Species in the Sudan in Relation to Soil Texture and Rainfall* (Khartoum).

Stebbins, G. L. (1952), 'Aridity as a Stimulus to Plant Evolution', *American Naturalist*, 86: 33–44.

—— (1974), *Flowering Plants: Evolution Above the Species Level* (London).

Stringer, C. B. (1992), 'Evolution of Early Humans', in Jones, Martin, and Pilbeam (eds.), *Cambridge Encyclopedia of Human Evolution*, 241–51.

Sudan Government (1944), *Report of the Soil Conservation Committee* (Khartoum).

Tourè, O. (1990), 'Where Herders Don't Herd Anymore; Experience From the Ferlo, Northern Senegal', Drylands Programme Research Paper, International Institute for Environment and Development (London).

Thomas, D. S. G. and Shaw, P. A. (1991), The Kalahari Environment (Cambridge).

Trollope, W. S. W. (1982), 'Ecological Effects of Fire in South African Savannas', in Huntley and Walker (eds.), *Ecology of Tropical Savannas*, 292–306.

Van Hoven, W. (1991), 'Mortalities in Kudu Populations Related to Chemical Defence in Trees', *Revue zoologique africaine*, 105, 141–5.

Walker, B. H. (1987), 'A General Model of Savanna Structure and Function', in B. Walker (ed.), *Determinants of Savannas*, (Oxford), 1–12.

Webb, S. D. (1977), 'A History of Savanna Vertebrates in the New World: Part I, North America', *Annual Review of Ecology and Systematics*, 8: 355–80.

—— (1978), 'A History of Savanna Vertebrates in the New World: Part II, South America and the Great Interchange', *Annual Review of Ecology and Systematics*, 9: 393–426.

Werner, P. A., Walker, B. H., and Stott, P. A., 'Introduction', in Werner (ed.), *Savanna Ecology and Management*, pp. xi–xii,.

Westoby, M. B., Walker, B. H. and Noy-meri, I. (1989), 'Opportunistic Management for Rangelands Not at Equilibrium', *Journal of Range Management*, 42: 266–74.

White, F. (1983), *The Vegetation of Africa; A Descriptive Memoir to Accompany the UNESCO/AETFAT/UNSO Vegetation Map of Africa* (Paris).

White, R. (1992), *Livestock Development and Pastoral Production on Communal Rangeland in Botswana* (London).

Wickens, G. E., 'Natural Vegetation', in Craig (ed.), *Agriculture of the Sudan*, 54–67.

13 Desert Environments

Nicholas Lancaster

Introduction

Regions classified as arid cover some 60 per cent of the African continent and include the Earth's most extensive desert region, the Sahara, as well as the Kalahari, Namib, and Karoo regions of Southern Africa (Figure 13.1). Arid regions are also important in the Horn of Africa. Based on the aridity index (the ratio between precipitation (P) and potential evapotranspiration (PET)), the Sahara and Namib are hyper-arid (P/PET <0.03) to arid (0.03 <P/PET <0.20) whereas the Horn of Africa, the Kalahari, and the Karoo are semi-arid (0.20 <P/PET <0.50) to arid (0.03 <P/PET <0.20) (UNESCO, 1979).

The present environments of all the deserts of Africa reflect both a long history of geologic and geomorphic evolution as well as the effects of Quaternary climatic changes. The origins of the present-day aridity of large parts of the African continent are directly linked to the development of the 'modern' climatic patterns of the globe during the late Tertiary period (Goudie, this volume).

In this chapter, following a brief introduction to the characteristics of each desert region, I discuss landforms and geomorphic processes which reflect the distinctive nature of the African deserts and their contribution to our understanding of the fundamentals of landform development in deserts. For a comprehensive physical geography of the African deserts, the reader is referred to Williams and Faure (1980) and Cloudsley-Thompson (1984) for the Sahara, and Seely (1990), Thomas and Shaw (1991), and Lovegrove (1993) for the southern African deserts.

The African deserts have played an important role in the development of paradigms of desert research. This is especially true of aeolian processes and landforms. Early ideas about the role of wind action were

Fig. 13.1. The African continent showing the location and extent of the major desert areas.

developed in Southern Africa by Passarge and others (Passarge, 1904; Walther, 1924). Pioneering studies of sand transport by wind were undertaken in the Western Desert of Egypt by R. A. Bagnold (Bagnold, 1941), while modern dune studies owe much to the support of the Desert Ecological Research Unit in the Namib, which has also played a major role in developing modern concepts of the ecology of desert reptiles and invertebrates. The wealth of palaeoenvironmental evidence from the Sahara has been instrumental in understanding the

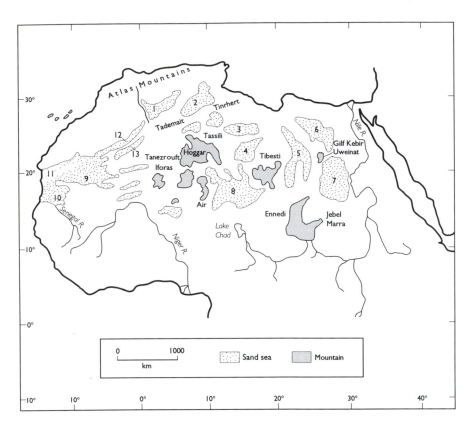

Fig. 13.2. The Sahara: major physiographic features. Major sand seas: 1. Grand Erg Occidental; 2. Grand Erg Oriental; 3. Ubari; 4. Murzuk; 5. Calanscio; 6. Great Sand Sea; 7. Selima; 8. Fachi-Bilma and Ténéré; 9. Majabat al Koubra; 10. Aouker; 11. Akchar; 12. Iguidi; 13. Chech.

response of low-latitude areas to changes in ice volume in polar regions (e.g. Williams, 1975; Petit-Maire, 1986), and Holocene climate changes in the region may provide good analogs for the response of deserts to anthropogenically induced global warming (Petit-Maire, 1991). However, the geographic coverage of modern research on African deserts is extremely variable: the Western Desert of Egypt, and the deserts of Tunisia, the Kalahari, and the Namib are relatively well known, whereas large areas of the Sahara have received scant attention since the pioneering efforts of French investigators, and the deserts of the Horn of Africa are scarcely known to modern workers.

Characteristics of the Major Desert Regions

The Sahara

The Sahara is the world's largest desert region and covers an area of approximately 7 million km², as large as the continental United States and Alaska combined (Figure 13.2). It stretches from the Atlantic coast to

the Red Sea, and from the Mediterranean to the Niger River and the margins of the Ethiopian Highlands. The northern boundary of the Sahara lies on the southern piedmont of the Atlas Mountains and the Mediterranean Sea coast. Its southern boundary is much less distinct, and the desert grades into the semi-arid Sahel zone around 15°N, or the latitude of the Niger bend and Lake Chad.

The Sahara is characterized by generally low relief (Mainguet, 1983). The only major mountain areas occur in the central parts of the desert, and include the Hoggar, Tibesti, Aïr, and Ennedi Massifs which reach elevations of as much as 3300 m; as well as the Gilf Kebir and Uweinat of Egypt. Ephemeral drainages extend from these mountain masses to the surrounding basins. Elsewhere there are extensive plateaux (e.g. the Tanezrouft, Tassili, Tademaït, and Tinrhert) developed on flat-lying or gently tilted Palaeozoic and Mesozoic sedimentary rocks and bounded by prominent escarpments. The surfaces of these plateaux are characterized by bare rock (*hammada*) or gravel (*reg*) surfaces. Intervening basins developed after Tertiary and Quaternary uplift of the central Sahara as well as by differential erosion of sedimentary rocks (Grove, 1980). Many of these basins are occupied by extensive sand-seas or *ergs* that cover approximately 28 per cent of the area of the Sahara.

The climate of the Sahara is arid to hyper-arid, with large areas in the central and eastern parts of the desert receiving less that 100 mm of rain a year (Smith, 1984). The extreme aridity of parts of the eastern Sahara has encouraged its use as an analog for surface processes on Mars (e.g. El Baz *et al.*, 1979; Breed *et al.*, 1987). The Sahara is also one of the hottest regions of the world, with average maximum temperatures exceeding 35 °C over wide areas, and over 40 °C in the southern part of the region. The Sahara forms part of a larger zone of extreme aridity that includes the deserts of the Arabian peninsula, southern Iran, and Pakistan. The primary cause of the aridity is the persistence of subtropical anticyclonic cells and dry descending air masses, but their effect is accentuated by the great width of the African continent at this latitude and the consequent lack of oceanic influences. The anticyclonic cells shift their position seasonally by 5–15° of latitude. In winter this allows depressions passing through the Mediterranean to affect the northern Sahara, which receives most of its rainfall during this season. In summer, the high pressure weakens somewhat and moves north, so that moist air from the Gulf of Guinea can invade the western and southern Sahara behind the intertropical convergence zone (ITCZ) which reaches its northern limits at around 18°N.

Large areas of the northern Sahara have significant groundwater resources in major regional aquifers. Most of the Algerian and Tunisian Sahara, as well as areas of Morocco and Libya, are underlain by the artesian Bas Sahara aquifer system (Gischler, 1979). Important areas of groundwater discharge occur in the *chotts* of Tunisia (Millington *et al.*, 1987; Roberts and Mitchell, 1987). The Nubian aquifer system forms another large regional aquifer in the Egyptian and Libyan Sahara and the Chad Basin of the northern Sudan. Studies of this aquifer in Sudan and Egypt indicate that it is largely a fossil groundwater body, with the last major period of recharge occurring in the early to mid-Holocene (Thorweihe, 1990).

The Sahara is characterized by sparse and monotonous vegetation, with many areas (e.g. the *hammada* and the sand sheets of the Western Desert of Egypt) being almost completely devoid of vegetation (Wickens, 1984 and Meadows, this volume). From north to south, there are transitions from the Mediterranean vegetation of the Maghreb region to tropical vegetation types to the south and east (White, 1983). For example, the steppe vegetation of Tunisia, Algeria, and Libya is composed of 80–90 per cent Mediterranean species, whereas that of the central Sahara desert zone with an annual rainfall of 10–25 mm is a fifty-fifty mix of tropical and Mediterranean plants (Le Houerou, 1986). As rainfall decreases away from the Mediterranean coastal zone, vegetation consists of a steppe zone in the area that receives 100–400 mm of rainfall. In some areas the vegetation consists of perennial grasses (*Stipa* and *Stipagrostis*). On silty soils (desert loess), the vegetation consists of low shrubs (*Artemisia herba*). The margins of the chotts have saline soils and a scattered cover of *Salsola* and *Atriplex*. The desert zone receives 50–100 mm of rain and supports a sparse cover of small shrubs with grasses on dune areas; where rainfall decreases to 25–50 mm, vegetation is confined to the main wadis and depressions; and in areas that are hyper-arid (<10 mm of rain) vegetation consists of ephemeral grasses except in groundwater discharge zones of along washes fed by runoff from desert mountains. Mountain areas above 1800–2000 m have a distinctive vegetation that consists of *Artemisia* steppe with sparse evergreen forest which includes shrubs and trees such as *Olea* along the main drainages (Messerli, Winiger *et al.*, 1980). On the southern margins of the Sahara, desert vegetation grades into tropical savannas with a well-developed grass cover and *Acacia* spp. along watercourses.

In all parts of the Sahara, a long history of human occupation has significantly impacted vegetation, leading to destruction of woody plants and expansion of the steppe zone (Le Houérou, 1986). On the southern margin of the desert, pressure on grazing and fuelwood resources and drought have led to reactivation of stabilized dunes in Mauritania, Mali, and the Sudan. Human impacts have also led to the demise of most of the region's once large mammal populations so that oryx and gazelle are very rare or locally extinct.

Deserts in East Africa

Deserts in East Africa are separated from the Sahara by the highlands of Ethiopia, although the Nubian Desert in north-eastern Sudan is essentially an extension of the Sahara. Much of Somalia is semi-arid, whereas the Ogaden, most of which lies in Ethiopia, is arid. The Danakil Desert occupies the northern end of the East African Rift Valley at its junction with the Red Sea Rift. It is an area of active tectonism and volcanic activity. Sedimentation patterns in the arid parts of the rift are strongly controlled by the style of rifting (Frostick and Reid, 1987).

The Chalbi Desert in northern Kenya lies east of Lake Turkana in the arid rainshadow corridor between the Kenya and Ethiopian Highlands. The Chalbi is an enclosed intermontane basin (Nyamweru and Bowman, 1989) with a central *playa* at an elevation of 370 m that is

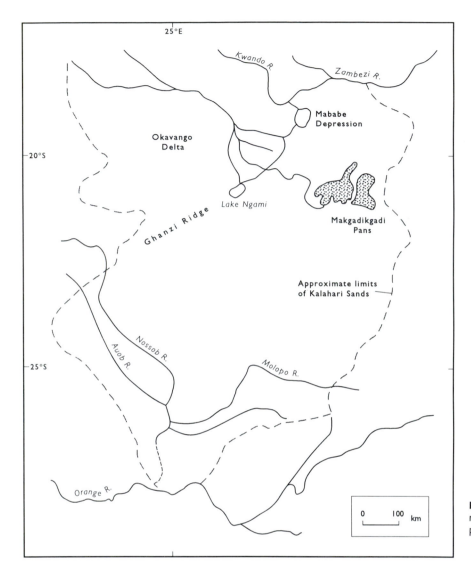

Fig. 13.3. The Kalahari: major drainages and physiographic features.

surrounded by alluvial fans and *bajadas* as well as areas of Tertiary and Quaternary basaltic volcanic rocks. Rainfall is currently 150–200 mm a year, but the basin was the site of a lake or swamp 9.5–11 ka BP (Nyamweru and Bowman, 1989).

The Kalahari

The Kalahari is part of an extensive sand-covered plain at an elevation of 800–1200 m that extends from the Orange River to southern Angola (Figure 13.3). To the east it is bounded by the highlands of eastern Botswana and western Zimbabwe, whilst to the west are the mountains of central Namibia. Most of the drainage in the region is internal and pans (small dry lakes or playas) are common in southern and south-western parts of the Kalahari (Goudie and Thomas, 1985). The Okavango, Kwando, and Cuito rivers rise in the humid highlands of Angola and flow south-east to form large inland deltas and swamps in north-western Botswana and adjacent areas of Namibia and Zambia, where 96 per cent of their discharge is lost by evaporation and transpiration (Wilson and Dincer, 1976). Some water spills over from the Okavango Delta and finds its way to the Makgadikgadi pans. Although often called a desert, the Kalahari is more correctly termed a thirstland or edaphic desert. There is no permanent water outside the swamp zone,

Plate 13.1. The Kalahari in south-western Botswana supports a cover of tree and bush savanna. In its drier portions, as here, dune crests may become reactivated in dry years or under grazing pressure (photo: A. S. Goudie).

but the rainfall of 500–800 mm in northern areas, declining to 150–200 mm in the south-west, supports a cover of tree and bush savanna, largely as a result of the high porosity of the surface sands.

Since the Cretaceous, the Kalahari has acted as a major continental sedimentary basin and is underlain by up to 500 m of sediments, mainly sands and marls with local playa sediments, which appear to have accumulated in mostly semi-arid to arid environments (Partridge, 1993). Many of the sediments have been cemented by calcrete and/or silcrete (Summerfield, 1983). They are capped by the surface Kalahari sands, which are largely aeolian in character (Thomas, 1987) and mantle an area of up to 2.5 million km². The sands are moulded into a variety of largely relict linear and parabolic dune systems. Small pans (ephemeral lakes) are widespread in the southern Kalahari (Lancaster, 1978).

The climate of the Kalahari is semi-arid to arid, with hot summers and winters characterized by warm days and cool nights. Temperatures are however moderated by the high elevation of the region (Thomas and Shaw,

1991). Summer mean daily maximum temperatures reach 32 to 35 °C, whereas winter daily maxima range between 22 and 28 °C. The daily range of temperatures in winter can be as much as 20 °C, with freezing temperatures on many nights. Rainfall occurs mostly during the period October to April and decreases from 500–800 mm in the north-east to as little as 150 mm in the south-west part of the region. Paralleling the rainfall gradient is an increase in its annual variability from 20 to 80 per cent.

The climate of the Kalahari is dominated by the southern African anticyclonic cell (Tyson, 1986). This cell is situated over the eastern highlands of South Africa in winter, resulting in stable, dry descending air masses. In summer, this cell moves offshore, and a low-pressure cell develops over the central African interior. Moist unstable air from the Indian Ocean invades the region, but its moisture content is reduced by its long passage over land to reach the Kalahari. In addition, the ITCZ reaches its southernmost position over the northern Kalahari, enhancing summer convective activity.

The Kalahari forms part of the arid-savanna biome of Southern Africa. Vegetation is dominated by grasses and a variety of shrubs and trees that range in height from 3 to 7 m (Plate 13.1). Large trees (mostly *Acacia* spp.) are found along ephemeral rivers such as the Nossob, Auob, and Molopo. The Kalahari sand plains are characterized by a mosaic of shrub savannas in the drier south-western areas that grade northwards and eastwards into tree savannas (Weare and Yalala, 1971). Grasses are dominantly *Aristida*, *Eragrostis*, and *Stipagrostis* species, whereas trees are mostly *Acacia*, *Commiphora*, *Colophospermum*, and *Terminalia*. The Kalahari supports large populations of herbivores such as the gemsbok, eland, blue wildebeest, and springbok.

The relatively dense vegetation cover and permeable surface sands were believed by some to be indications of significant near-surface groundwater in the Kalahari Group sediments. Subsequent geohydrological investigations showed that this was not the case, but debate has continued on the role of the Kalahari sands in recharge to the deep aquifers in the pre-Kalahari rocks of the region. Earlier workers, e.g. (Boocock and van Straten, 1962) thought that there was no recharge through the sand, but recent isotopic and geochemical studies suggest that recharge does occur where sand cover is thin or there are zones of increased permeability, especially in years of heavy rainfall (Foster *et al.*, 1982; Mazor, 1982). The importance of past periods of increased rainfall to recharge is therefore probably great, as suggested by De Vries (1984) and is supported by geochemical evidence that indicates increased recharge

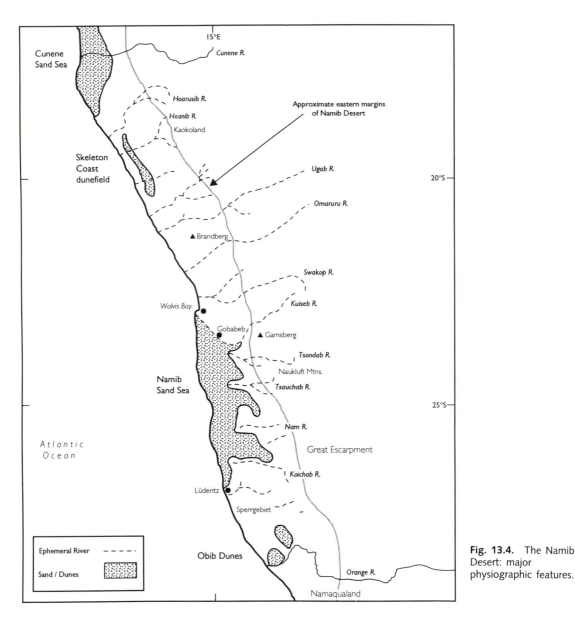

Fig. 13.4. The Namib Desert: major physiographic features.

between 30 and 26 ka and again from 14 to 8 ka BP (Heaton *et al.*, 1983). Groundwater discharge may play a significant role in the formation of the many pans in the region (Lancaster, 1986), as well as some channel systems (Shaw and De Vries, 1988).

The Namib

The Namib Desert extends for over 2000 km along the west coast of Southern Africa from the Olifants River in South Africa (32˚S) to the Carunjamba River at 14˚S in Angola (Figure 13.4). Inland, it is bounded by the Great Escarpment, which lies 120–200 km from the coast and forms the western edge of the interior plateau of Southern Africa.

The Namib can be subdivided into four main areas: the southern or transitional Namib, which includes coastal Namaqualand and the Sperrgebiet (Plate 13.2); the Namib sand sea; the central Namib Plains; and the northern Namib and Skeleton Coast. Much of the southern Namib consists of extensive rocky and sand-covered

Plate 13.2. Kolmanskop, an abandoned diamond mining settlement near Lüderitz in southern Namibia. This is a hyper-arid area with intense wind action (photo: A. S. Goudie).

Plate 13.3. A small granitoid inselberg, Vogelfederberg, in the plains to the north of the Kuiseb River (photo: A. S. Goudie).

plains extending coastwards from the escarpment, with low hills at intervals. The Namib sand sea lies between Lüderitz and the Kuiseb River and has an area of some 34 000 km² (Lancaster, 1989a). North of the sand sea, between the Kuiseb River and the Brandberg is a further extensive rocky plain, cut across by mica schist and syntectonic granitic intrusions, with an average gradient of only 1° between the coast and the 1000 m contour. Rising from the plain are low ranges of hills and isolated inselbergs (Selby, 1977; Ollier, 1978) (Plate

13.3). Extensive pedogenic and groundwater gypsum and calcrete horizons are developed where relief is low (Goudie, 1972; Watson, 1985). The dissected pediplain and sandstone and lava hills of the northern Namib form the area known as the Kaokoveld. The coastal area, which is partly covered by small dunefields, is known as the Skeleton Coast.

The climate of the Namib is arid to hyper-arid but, especially in coastal areas, relatively cool. A major feature is the steep climatic gradient from the cool, foggy hyper-arid coastal zone to the hotter inland areas towards the Great Escarpment which receive scant summer rainfall (Seely, 1978). Mean annual rainfall increases from 15 mm or less on the coast to 80–100 mm near the escarpment (Lancaster, J. *et al.*, 1984). Rainfall occurs mostly during the summer months of January to April, but the southern parts of the desert receive some winter rainfall. Advective fogs, resulting from the cooling of moist oceanic air as it passes over the Benguela Current, are a distinctive feature of the Namib climate and their effects are felt for over 100 km inland. Fog precipitation is a important moisture source for the Namib biota (Seely, 1978) and contributes to rock weathering and mineral breakdown (Goudie, 1972; Sweeting and Lancaster, 1982). The amount of fog precipitation rises from the coast where it averages 34 mm per year to a maximum of 184 mm 35–60 km inland and decreases sharply thereafter to 15 mm in the east of the desert (Lancaster, J. *et al.*, 1984).

Temperatures in the central Namib are moderate by comparison with many other desert regions, reflecting the influence of the cold ocean offshore. Mean annual daily maximum temperatures range from 17 °C at the coast to 28–33 °C inland. Maximum daily temperatures of 38–40 °C occur inland in February–March. Minimum daily temperatures average 13–16 °C throughout the region and range seasonally from 4–8 °C in June–August to 15–18 °C in January–March.

The aridity of the Namib results primarily from the dominant effects of subtropical anticyclonic cells on the regional circulation pattern. In the central Namib, moist air masses can penetrate the area only when this anticyclonic cell is weak. However, their effects are limited, as the moist air is derived from the Indian Ocean and has to cross the subcontinent to reach the Namib. Thus descending, divergent air masses tend to occur throughout the year. The effects of the subsidence-induced stability are reinforced by the presence of the cold Benguela Current offshore, which intensifies the temperature inversion. The southern Namib lies on the equatorial margins of the winter rainfall zone, and is affected only by the strongest frontal systems.

Despite its great aridity, the Namib supports a distinctive biota with a large percentage of endemic species. Vegetation is very sparse except along the larger ephemeral streams (e.g. the Kuiseb, Hoanib) which form linear oases characterized by riparian woodland (dominated by *Acacia, Faidherbia*, and *Tamarix*) and local wetlands from the escarpment zone to the coast. Dune migration into the Kuiseb River valley is strongly influenced by the growth of *Stipagrostis sabilicula* after floods (Ward and von Brunn, 1985) and nara (*Acanthosicyos horridus*)

plants are the nuclei for nebkha (coppice dunes) in the delta region. The Namib plains support a grass cover after rain, whereas the dunes are mostly unvegetated. A unique plant of the central and northern Namib is the Welwitschia. In all areas, there is a diverse and abundant tenebrionid beetle fauna (Seely, 1978).

The Karoo

The Karoo is an arid to semi-arid desert region that extends north from the Cape Fold-Mountain Belt to the Orange River, with an extension into southern Namibia in the area between the Namib and Kalahari. The interior Karoo lies at elevations of 900–1200 m and consists of extensive rocky plains and low hills developed by differential erosion of flat-lying to gently folded Mesozoic sandstone, shale, and tillite (Wellington, 1955) to form a landscape somewhat similar to the Colorado Plateau of the western United States. Mesas, often capped by dolerite sills, occur at intervals and there is a major concentration of pans along the Sak River valley. Towards the western and northern margins of this region, Precambrian rocks crop out and form low ranges of hills. The coastal part of the Karoo below the Great Escarpment consists of sandy plains that form a southern extension of the Namib Desert to 32°S.

The climate of the interior Karoo is more extreme than the other desert regions of Southern Africa. Mean daily maximum temperatures exceeding 30 °C occur 20–25 per cent of the time, and days with minima of 0 °C or less occur on as many as 60 days a year (Tyson, 1986). Precipitation occurs mostly in the summer in the north and during the equinoxes or winter (sometimes as snow) in the south. It ranges between 150 mm per year in the north and west to 400 mm in the north and east. Coastal areas receive 90–290 mm of rain a year, mostly in the winter. The Karoo forms a distinctive biome, with an abundance of succulents along the coast (Lovegrove, 1993). Dwarf shrubs and grasses (mostly *Stipagrostis* spp.) occur in interior regions. Trees are confined to the larger watercourses. Expansion of Karoo shrubs into the grasslands of the South Africa highveld as a result of centuries of overgrazing and desertification was first recognized by Acocks (1953), and is now regarded as a major environmental problem in the subcontinent (Lovegrove, 1993).

Geomorphic Systems
Fluvial Processes and Landforms

Most studies of fluvial systems in the African deserts have concentrated on understanding their response to Quaternary climatic changes (see below). Data on

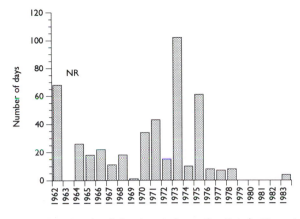

Fig. 13.5.   Length of flood periods on the Kuiseb River at Gobabeb.
(*Source*: data from Ward and von Brunn, 1985.)

modern runoff and sedimentation is sparse except for major exoreic streams such as the Nile, Niger, Senegal, and Orange Rivers which have been intensively studied for their water-supply potential.

Typical of many desert rivers is the Kuiseb River in Namibia, with a catchment area of 14 700 km², 60 per cent of which lies in the highland areas east of the Namib Desert. These areas receive 200–400 mm of rainfall a year and produce all of the runoff water for the river. Median annual runoff at the eastern edge of the desert is 10×10^6 m³ with a maximum of 100 and a minimum of 0.007×10^6 m³. Flow decreases rapidly downstream as water infiltrates into the sandy bed. The river has reached the Atlantic on three occasions in the past century: 1933, 1962–3 (Stengel, 1964), and 1985. Flows as far as the delta region are more frequent. Flood frequency and duration in the lower part of the Kuiseb River have been monitored at Gobabeb since 1960 (Ward and von Brunn, 1985) (Figure 13.5) and indicate that the longest duration floods occur after intense precipitation events in years of higher rainfall in the mountain part of the catchment. Similar patterns have been noted for the Molopo River system in the southern Kalahari (Thomas and Shaw, 1991).

Extreme events accomplish a large proportion of the change in most desert fluvial systems, but they have rarely have been documented in African deserts. One well-documented event took place in January 1976, when exceptional storms near Gobabeb produced a total of 76.5 mm of rain in four days, including 43.5 mm in twelve hours. This produced runoff in a tributary of the Kuiseb for a period of thirteen hours, giving rise to channel incision and removal of 13 200 m³ of material (Marker, 1977).

Alluvial fans are restricted to the more tectonically active parts of the African deserts including Tunisia (White, 1991; Drake *et al.*, 1993) and the East African Rift system (Frostick and Reid, 1987). Many other piedmont areas consist of pediments or *glacis d'érosion* (Barnard, 1975; Selby, 1977; Ollier, 1978). In the central Sahara, such forms appear to be absent (Peel, 1941; Busche and Hagedorn, 1980) and piedmont zones may be relicts of former Tertiary humid climates.

Aeolian Processes and Landforms

Aeolian processes and landforms are a major landscape component in all the African desert regions. The Sahara also plays a dominant role in the global production of wind-blown dust.

Wind-Erosion Processes and Landforms

Erosion of a wide range of materials by wind abrasion and deflation of fines occurs in the African deserts. Quantitative data on rates of erosion are sparse. Williams (1970) estimated that the *bajada* surface in the Biskra region of Algeria had been lowered 1–4 m by deflation in 2000 years, based on radiocarbon-dated deposits. Wilson (1971) provides order of magnitude estimates of erosion rates in the Sahara that range from 10^{-2} to 10^{-4} kg m² yr⁻¹ on desert pavements through 1.0 kg m² yr⁻¹ on *bajadas* to 10^3 to 10^5 kg m² yr⁻¹ for sandy substrates. Measurements of the vertical flux of dust in Mali range between 7.0×10^{-2} and 2.64×10^3 µg m⁻²s⁻¹. The flux increases as a power function of wind shear velocity (Nickling and Gillies, 1993). Because armouring of the surface by coarse material or surface crusts rapidly shuts down wind erosion, deflation rates tend to be greatest in areas of active resupply (e.g. some playas, active floodplains), uniform particle size (dunes, sand sheets), and/or disturbance by humans or animals.

Geomorphic evidence for wind erosion is well-documented from the Sahara and Namib. Wind-eroded landforms can be divided into small-scale (millimetres to centimetres deep and centimentre to metre long) features such as ventifacts (wind-grooved or faceted rock surfaces and individual clasts) and large-scale (metres deep, kilometres long) systems of ridges, and swales and yardangs.

Examples of ventifacts are widespread in the Sahara (e.g. McCauley, 1979; Hagedorn, 1980) and the Namib (Selby, 1977; Sweeting and Lancaster, 1982; Lancaster, 1984). In all areas, grooves parallel sand-transporting winds and facets face into the strongest winds. Grooving, fluting, and faceting are best-developed on fine-grained rocks of intermediate hardness (e.g. basalt, marble, some limestone). They rarely occur on sandstone and granite,

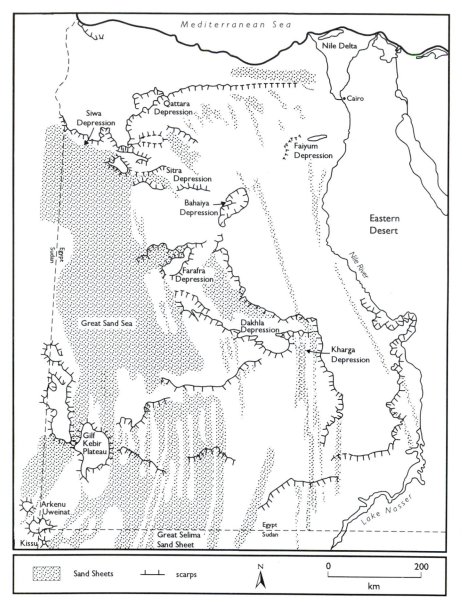

Fig. 13.6. The Western Desert of Egypt showing major sand seas and desert depressions.

(*Source*: El Baz *et al.*, 1979.)

which weather too rapidly to preserve wind-erosion forms.

Large-scale systems of wind-grooved terrain occur in the Borkou region of central Sahara, east of the Tibesti Mountains (Mainguet, 1983) and form a broad NE–SW oriented arc that covers an area of 650 000 km². The ridges are up to 200 m high, 4 km long, and 1 km wide with a spacing of 0.5 to 2 km and are eroded into Palaeozoic and Mesozoic sandstone along a joint pattern that is sub-parallel to the prevailing winds.

The corridors have a variable sand cover and are best-developed where the wind most closely parallels the fracture pattern, but there are also abrupt changes in scale close to escarpments (Mainguet, 1972).

Yardangs are more streamlined residual landforms of wind erosion with a blunt upwind end and a tapering lee side. Width : length ratios of 3 to 1 are common. They can occur in closely spaced fields or as isolated forms. Extensive areas of yardangs occur in the Western Desert of Egypt (El Baz *et al.*, 1979) (Figure 13.6). East

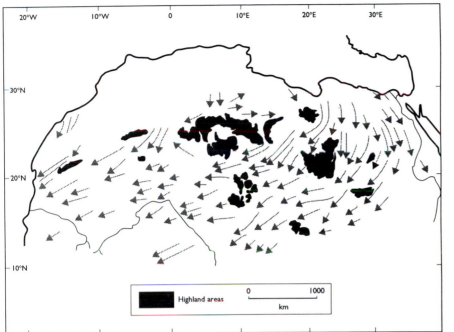

Fig. 13.7. Aeolian sand-transport pathways in the Sahara interpreted from satellite images. (*Source*: Mainguet *et al.*, 1984.)

of the Kharga Depression, they are up to 10 m high and carved into crystalline limestone. Within the depression, yardangs 3–5 m high and tens of metres long are developed in lacustrine sandy clays and Mesozoic sandstone. Yardangs in Namibia are confined to the southern Namib where they are developed in Precambrian dolomite (Kaiser, 1926; Corbett, 1993). Most yardangs also have superimposed flutes and grooves.

Wind erosion plays an important role in the development of desert depressions, ranging in size from those covering thousands of square kilometres such as the Qattara, Dahkla, and Kharga Depressions in Egypt (Figure 13.6) to the tens of square kilometres or less, covered by pans in Southern Africa. The relative importance of deflation, karst processes, salt-weathering promoted by groundwater discharge, and tectonics in the formation of desert depressions has been much debated (e.g. Goudie and Thomas, 1985; Albritton *et al.*, 1990), but it appears that the major process removing material from the basins is wind action as evidenced by the lunette dunes found on the downwind margins of many pans in the Kalahari (Lancaster, 1978, 1986).

Regional-Scale Transport of Aeolian Sand

The Sahara contains a significant proportion of the global inventory of wind-blown sand as well as some of the world's largest sand seas. The major sand accumulations lie each side of the central Saharan uplands, where there is very little sand (Figures 13.2 and 13.7). In the Sahara, sand is moved long distances between source areas and depositional sinks as well as from one sand sea to another. The sand-transport pathways are characterized by sand-choked river valleys, shadow dunes and *barchans*, small climbing and falling dunes, and sand streaks and sheets, and are clearly visible on satellite images (Breed *et al.*, 1979; Mainguet, 1984). Rapid migration of *barchans* along these transport pathways has been documented from Egypt and Mauritania (Table 13.1). The pattern of the sand-transport pathways (Figure 13.7) reflects the dominance of the trade-wind circulation that tends to move sand from the sand-poor eastern and central parts of the desert toward thick sand accumulations in the Sahel, and from the piedmont of the Atlas Mountains and the Mediterranean coast to the northern and western sand seas (Wilson, 1971; Mainguet and Chemin, 1983).

The location of the major Saharan sand seas is controlled both by patterns of winds and by topographic obstacles. The Grand Erg Occidental and the Ubari and Murzuq sand seas of Libya have accumulated against the northern slopes of the central Saharan uplands (Fryberger and Ahlbrandt, 1979). Sand seas in the

Table 13.1 Barchan and crescentic dune migration rates in African sand seas

Source	Advance rate (m yr^{-1})	Dune height (m)
Egypt		
Beadnell (1910)	15.0	12.5
Embabi (1982)	48.4	5.5
Haynes (1989)	7.5	16.6
Namibia		
Kaiser (1926)	48	33
Endrody-Younga (1982)	43	8–10
Barnard (1975)	8.4	11.33
Ward and von Brunn (1985)	0.8–6.4*	10–20
Slattery (1990)	12.4–14.6	8.25

* Crescentic dunes.

Akchar of Mauritania and the Fachi-Bilma region of Tchad lie where winds converge in the lee of escarpments or mountain masses. The thick sand accumulations of the southern Sahara and Sahel have developed because wind energy is reduced on the margins of the trade-wind circulation. Many northern Saharan sand seas (e.g. the Grand Erg Occidental and Oriental) occur in areas of complex or multidirectional wind regimes influenced by both trade-wind and mid-latitude cyclonic circulations, as well as local topographically induced winds (Wilson, 1971).

Similar, but much less extensive, regional-scale sand-transport systems occur in the Namib (Figure 13.8). The southern Namib is characterized by a high-energy unidirectional wind regime that transports sand inland and northwards from beaches, supplied by sand derived from the Orange River by longshore drift, along well-defined sand-transport corridors that extend for distances of up to 120 km (Corbett, 1993). Over time, the position of these transport corridors has changed in response to Quaternary sea-level variations. The Namib sand sea lies in an area in which wind-energy and sand-transport potential decrease from south to north and from west to east (Lancaster, 1989a). There is also a parallel increase in the directional variability of the wind regime. Sand is thus transported into areas of lower wind energy and more variable wind directions where it accumulates in large linear and star dunes.

Dust Transport from the Sahara

The Sahara is a major source of aeolian dust, generating an estimated 300 million tonnes per year, or 60 per cent of the total global production (Junge, 1979). These aerosols are of major importance and may affect the global radiation budget (Carlson and Benjamin, 1980). Regionally, dust loadings affect visibility and air quality in much of arid West Africa and probably lead to health problems. In Mali, the mean ambient dust concentration is 1176 µg m^{-3}, an order of magnitude greater than international health standards (Nickling and Gillies, 1993).

Two major source areas are indicated by observations of dust trajectories and mineralogy: the Chad Basin and the Bodelé Depression of the Faya Largeau region, and an area west of the Hoggar Massif (Kalu, 1979; McTainsh and Walker, 1982). Additional sources occur on the piedmont of the Saharan Atlas and in local areas of alluvial and lacustrine deposits in the southern Sahara and Sahel (Grousset *et al.*, 1992). Dust from the latter area is characterized by kaolinite clays and reddish-stained quartz derived from lateritic crusts (Sarnthein *et al.*, 1981; Rognon *et al.*, 1989).

Three major wind systems transport dust from the Sahara (Figure 13.9). North-easterly trade winds transport dust from the Atlas piedmont and coastal plains in a shallow layer to the vicinity of the Canary and Cape Verde Islands. By far the most important dust-transporting wind system is the harmattan, which transports dust from the southern Sahara and Sahel to the Gulf of Guinea and the subtropical Atlantic Ocean in the period from November to April. Detailed studies of dust emissions and transport processes in the inland delta of the Niger in Mali (Nickling and Gillies, 1993) show that the vertical dust flux is proportional to the fourth power of the wind shear velocity. Dust concentrations in 'dust haze' conditions varied from 26 to 13 735 µg m^{-3}. During dust storm events, the concentration was as much as 100 000 µg m^{-3}.

In summer, dust is mobilized by gusty winds associated with convective activity and E–W travelling squall lines. Once aloft, the dust is incorporated in the 'Saharan Air Layer' associated with the easterly mid-tropospheric jet stream. Outbreaks of Saharan dust over the eastern subtropical Atlantic occur between June and September and are linked to warm, dry plumes of Saharan air circulating around 'easterly waves', which occur with a periodicity of 3–4 days and are associated with the ITCZ (Sarnthein *et al.*, 1981). Most of the dust moves above the trade-wind inversion at a height of approximately 3000 m and circulates around the upper-level high-pressure system situated over the western Sahara. This directs material toward the north (Figure 13.9). The finest material continues westwards toward the Caribbean and eastern South America (Prospero *et al.*, 1981). In winter, intense dust storms with winds of 60 km hr^{-1} for periods of up to forty-eight hours occur in

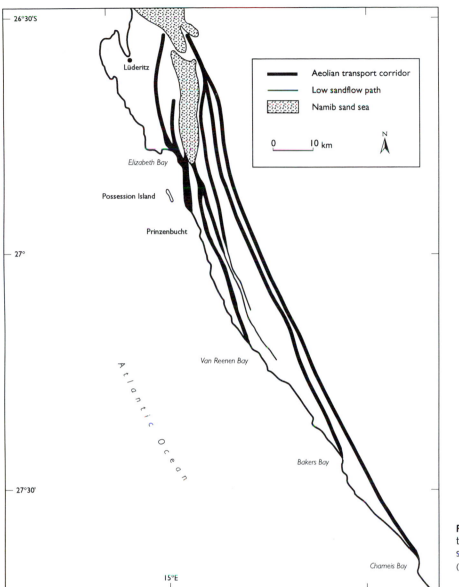

Fig. 13.8. Aeolian sand-transport pathways in the southern Namib. (*Source*: Corbett, 1993.)

the Faya-Largeau and Bilma areas as a result of low-level pressure surges associated with the intensification of anticyclonic cells in the north-west and central Sahara (Kalu, 1979). The dust is derived from lacustrine and alluvial sediments and is carried in a south-westerly direction across northern Nigeria, where deposition rates as high as 22 to 24 tonnes km^{-2} have been recorded (McTainsh and Walker, 1982). Dust from the northern Sahara reaches all parts of the Mediterranean, and is mobilized by winds associated with cold fronts and depressions tracking eastwards along the North African coast (Goudie, 1983).

Sand Seas

Sand seas or ergs represent the depositional sinks for sand transport systems. Within individual sand seas, local changes in wind regimes and sediment supply combine to produce distinct assemblages of dunes of different

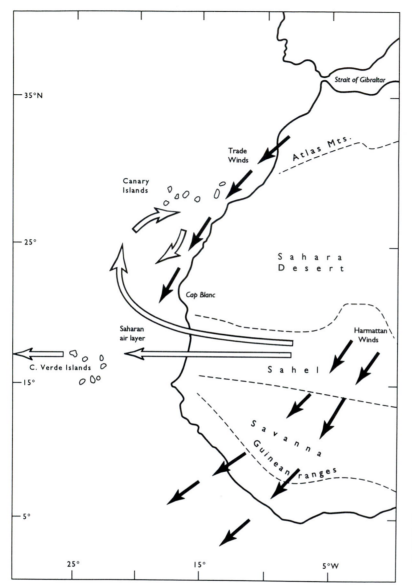

Fig. 13.9. Dust-transporting wind systems in the western Sahara.

(*Source:* Pye, 1987.)

morphological types, the patterns of which are clearly visible on aerial photographs and satellite images (e.g. Mainguet, 1972; Breed *et al.*, 1979; Lancaster, 1983).

Dune patterns in Saharan sand seas have been described by Mainguet (1972, 1984), Mainguet and Callot (1978), and Breed *et al.* (1979) and, are summarized in Tables 13.2 and 13.3. The northern Saharan sand seas (Figure 13.10) are characterized by complex crescentic and star dunes developed in a multidirectional wind regime that results from the interaction of trade-

wind and mid-latitude circulations. By contrast, sand seas in the western and southern Sahara are mainly composed of linear dunes, with small areas of crescentic dunes. The dominance of linear dunes in these areas reflects the persistence of trade-wind circulations. Aeolian accumulations in the eastern Sahara (Figure 13.6) include linear and crescentic dunes (Haynes, 1989), but are dominated by extensive sand sheets and areas of gently undulating dunes with a chevron pattern (Maxwell and Haynes, 1989). The formation of extensive sand sheets

Table 13.2 Percentage of area covered by major dune types in African sand seas

Dune type	Namib	Kalahari	Sahara			
			North	South	North-east	West
Crescentic						
Simple	1.0	0.6	33.4	28.4	14.5	19.2
Compound	12.0					
Complex			26.1	24.3	12.6	18.5
Linear						
Simple and compound	35.0	85.9	5.7	24.1	2.4	35.5
Complex	37.0		13.5		13.5	
Star	9.0		7.9		23.9	
Parabolic		0.6				
Sand sheets and streaks		13.6	36.9	47.5	39.3	45.3
Other/undifferentiated				5.3		

Note: Original data from Landsat MSS images had a spatial resolution of 80 m per pixel. As a result, many dunes classified as simple may be compound forms and some of the sand sheet areas probably represent areas of low dunes.

Sources: Data from Fryberger and Goudie (1981), with additions from Lancaster (1989).

Table 13.3 Dune morphometry in African sand seas (range in parentheses)

	Dune width (m)	Dune spacing (m)
Compound crescentic		
Mali	620	
	(500–800)	
NE Sahara	650	1240
	(300–1500)	(500–2000)
Linear		
Simple Kalahari	290	700
Compound		
Mauritania	940	1930
Niger	1060	1900
Complex		
Algeria	1090	3240
Niger	1280	3280
Namib	880	2200
Star		
Niger	610	1000
	(200–1200)	(150–3000)
Grand Erg	1650	2170
Oriental	(700–3000)	(1500–3100)

Source: Data from Breed *et al.* (1979).

is attributed to the abundance of coarse sand derived from the Mesozoic Nubian Sandstone that underlies much of the area (Breed *et al.*, 1987).

The Namib Sand Sea is characterized by a well-organized dune pattern (Figure 13.11) in which crescentic dunes occur in response to unidirectional (SSW) wind regimes near the coast; compound and complex linear dunes are associated with bi-directional winds (SSW–SW and NE–E) inland (Plate 13.4); and star dunes with multidirectional wind regimes (SW–WSW, NE–E, and N) along the eastern margin of the sand sea (Lancaster, 1983). The general S–N trend of the linear dunes is established by the dominant SSW–SW winds, which blow at an optimal angle for lee side diversion and dune extension, but infrequent strong NE–E winds keep sand on the dune (Livingstone, 1988, 1989). Dune height and spacing are summarized in Table 13.4 and vary together in a systematic way in the Namib sand sea (Figure 13.12), reflecting the regional pattern of sand accumulation in response to declining wind energy and sediment-transport rates toward the north and east. Dunes are highest and most widely spaced in the central and some northern parts of the sand sea, with progressively lower and more closely spaced dunes towards the margins. Northern Namib dune areas consist of small dunefields derived from coastal sources and dominated by crescentic dunes (Lancaster, 1982).

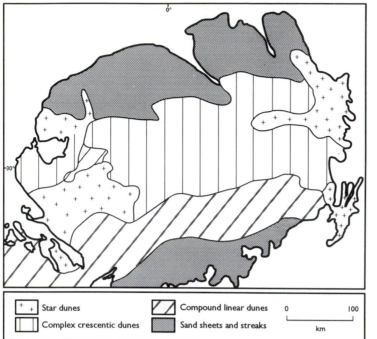

Fig. 13.10. Dune patterns in the Grand Erg Oriental. (Modified from Breed *et al.*, 1979.)

Legend:
- + + Star dunes
- Complex crescentic dunes
- Compound linear dunes
- Sand sheets and streaks
- 0 — 100 km

Impact of Climatic Changes on Landforms and Geomorphic Systems

All parts of the African deserts were affected to some degree by Quaternary climatic changes. The impact of these changes however varied spatially. In part this was the result of the different response of geomorphic systems to climatic change, with aeolian systems tending to accentuate the effects of wet and dry phases (Rognon, 1982). The amplitude of climatic change also varied, with core hyper-arid areas or 'kernwüste' like the central Sahara and Namib remaining relatively unaffected, and desert margin areas (the Sahel and Kalahari) being affected the most.

For example, there is no evidence to suggest that the Namib has experienced any climate wetter than semi-arid at any time during the Quaternary. Pollen in deep-sea cores (Van Zinderen Bakker, 1984; Diester-Haass *et al.*, 1988), as well as the surface survival of calcrete palaeosols (Yaalon and Ward, 1982) and lacustrine carbonates (Selby *et al.*, 1978) indicate continued arid conditions. Ward *et al.* (1983), Lancaster (1984), and Partridge (1993) have suggested that Quaternary climatic fluctuations in the Namib were of low amplitude,

compared to elsewhere in Southern Africa and have been superimposed on a hyper-arid to arid mean.

Relict Dune Systems

Extensive systems of relict linear dunes, mostly fixed by savanna vegetation, occur in the Sahel (e.g. Grove and Warren, 1968; Talbot, 1980; Nichol, 1991) and the Kalahari (Grove, 1969; Lancaster, 1981; Thomas, 1984). These dune systems give rise to distinctive soil and vegetation assemblages (Jaccoberger, 1989) and are very susceptible to remobilization as a result of desertification processes. They have been identified from aerial photographs and satellite images in areas with up to 1000 mm of annual precipitation, many hundreds of kilometres from the margin of areas of active dunes (Figure 13.13). These relict dunes are a major source of evidence for changes in the areas of arid climates during the Quaternary (Sarnthein, 1978).

In the Sahel and southern Sahara, three main dune generations are recognized (Grove and Warren, 1968; Talbot, 1980). The oldest dunes are very degraded and were formed prior to 20 ka BP. A subsequent period of dune formation and/or remobilization occurred during the period 13–20 ka BP and affected a very wide area (Talbot, 1980). Most of the dunes were stabilized by

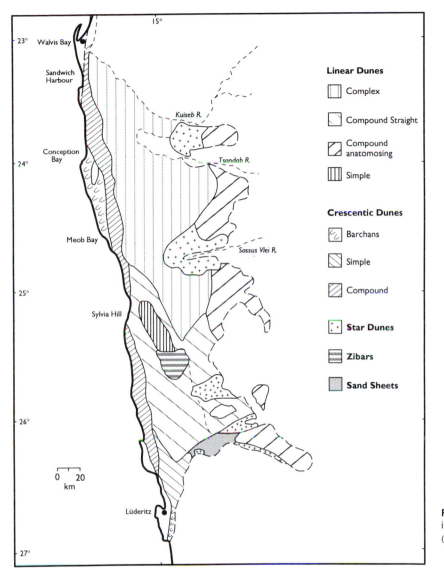

Legend:

Linear Dunes

▦ Complex

◻ Compound Straight

◨ Compound anatomosing

▥ Simple

Crescentic Dunes

⬚ Barchans

◺ Simple

◩ Compound

⊡ **Star Dunes**

▤ **Zibars**

▦ **Sand Sheets**

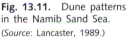

Fig. 13.11. Dune patterns in the Namib Sand Sea. (*Source:* Lancaster, 1989.)

vegetation and soil formation during the early Holocene, 7–11 ka BP (Talbot, 1985; Vökel and Grunert, 1990) during a period of humid conditions throughout the region. The third period of dune reactivation occurred after 5 ka BP. In the past several decades, regional drought and vegetation destruction by human activities have resulted in the remobilization of many dunes and even formation of new dune areas (Nickling, 1993).

Detailed studies of dune and interdune sediments in the Akchar sand sea of Mauritania by Kocurek *et al.* (1991) demonstrate the effects of these changes on dune and sand-sea accumulation. The prominent large complex linear dunes (Figure 13.14) are composite features.

Their core consists of sand deposited during the period 20–13 ka BP. This was stabilized by vegetation during a period of increased rainfall 11–5 ka BP, when pedogenesis altered dune sediments and lakes formed in interdune areas. Further periods of dune formation after 4 ka BP cannibalized existing aeolian deposits on the upwind margin of the sand sea. The currently active 'cap' of crescentic dunes superimposed on the linear dunes dates to the last thirty years.

Dune alignment patterns indicate that most relict dunes in the southern Sahara and Sahel were formed in wind regimes similar to those occurring today (Mainguet and Canon, 1976; Talbot, 1980), with a dominance

Plate 13.4. Large linear dunes in the central Namib near Gobabeb, Namibia (photo: A. S. Goudie).

Table 13.4 Dune morphometry in the Namib sand sea (range in parentheses)

Dune type	Area of sand sea covered (%)	Mean height (m)	Spacing (m)
Crescentic			
Simple	1	8.25 (3–10)	272 100–400
Compound	12	18.6 (10–40)	694 (800–1200)
Linear			
Compound	35	34.25 25–45	1724 (1600–2000)
Complex	37	90.27 (40–170)	2108 (1600–2800)
Star	9	145 (80–300)	1332 (600–2600)

Source: Lancaster, 1989*a*.

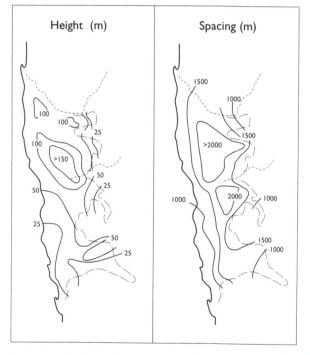

Fig. 13.12. Spatial variations in dune height and spacing in the Namib Sand Sea.
(*Source*: Lancaster, 1989.)

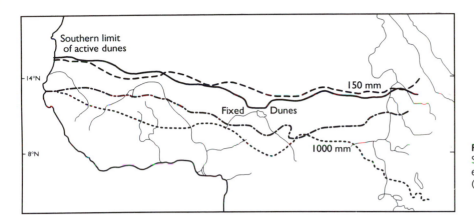

Fig. 13.13. The southern Sahara and Sahel showing extent of active and fixed (vegetation-stabilized) dunes.

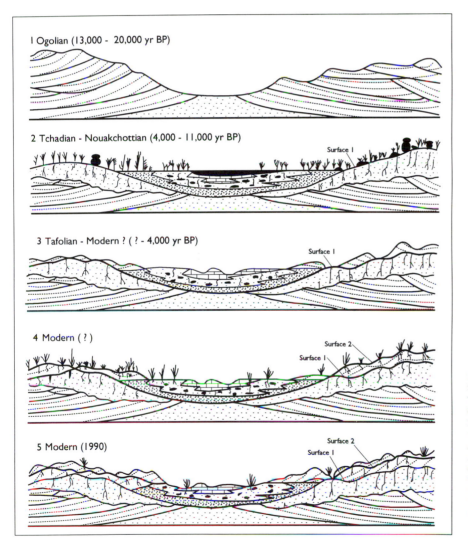

Fig. 13.14. Interpreted sequence of events leading to present complex linear dunes in the Akchar Sand Sea, Mauritania.

1: construction of major linear dunes; 2: stabilization of dunes and formation of lakes in inter-dune areas; 3: desiccation and renewed dune formation; 4: revegetation of dunes; 5: modern reactivation of dunes.

(*Source*: Kocurek *et al.*, 1991.)

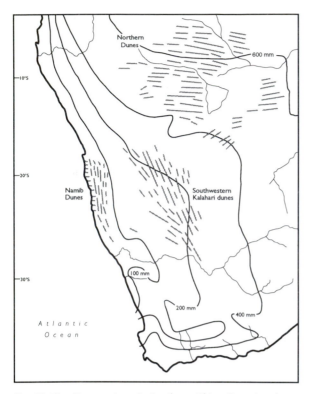

Fig. 13.15.   Dune systems in Southern Africa. Dune trends are schematic.
(*Source*: Lancaster, 1981.)

of easterly to north-easterly winds. Fryberger (1980) however provides evidence from crossing trends of dunes in Mauritania, Mali, and Niger of changes in wind directions between the oldest dunes that were formed by more easterly winds and those formed 13–20 BP ka by winds similar to the north-easterlies occurring today.

In the Sahara, active dunes now occur in areas that receive less than 100–150 mm of rain a year whereas the relict dunes occur in areas that lie 500–700 km south of this limit. Comparison of the precipitation limits of active dunes and relict dunes in the Sahel can therefore provide an indication of the magnitude of changes in the position of climatic belts (Figure 13.13). Although the formation or reactivation of areas of dunes may be the product of shifts in climatic belts and circulation patterns, it is now clear that such models are over-simplified. The importance of interactions between wind velocity and vegetation cover (and hence rainfall) on sand mobility is now being increasingly recognized. Reactivation or formation of relict dunes may therefore be the result of changes in wind velocity and/or the

moisture balance. In the Sahel, Talbot (1984) concluded that, despite an increase in late Pleistocene wind velocities of 50 per cent, dune formation and reactivation in the period 13–20 ka BP would have required a reduction in rainfall to 25–50 per cent of present values.

Extensive systems of dunes (Figure 13.15), mostly of linear form, occur throughout the Kalahari from the Orange River at 28°S to latitude 12°S in Angola and western Zambia. Today, these dunes are stabilized by savanna vegetation and are found in areas where rainfall is up to 800 mm per year (Grove, 1969; Lancaster, 1981; Thomas, 1984). The dunes form a massive semicircular arc with a radius of some 1000 km which corresponds approximately to the pattern of winds outblowing around the southern African anticyclone situated over the northern Transvaal. Within this arc, three distinct subsystems of dunes can be identified. The northern subsystem consists of dunes on E–W or ENE–WSW alignments. North and west of the Okavango Delta, they consist of broad, straight parallel ridges up to 25 m high and 200 km long, with a spacing of 1.5–3 km. The dunes are very degraded and support a dense cover of tree savanna and open savanna woodland. In western Zimbabwe and adjacent areas of Botswana, the second subsystem consists of dunes with a relief of up to 25 m and 1.5–2 km apart (Flint and Bond, 1968; Thomas, 1984). The southern dunes form the third subsystem and are best developed in a 100–200 km-wide belt between the highlands of Namibia and the Orange River near Upington. They consist of NNW–SSE or WNW–ESE trending parallel to subparallel ridges 5–15 m high with a spacing of 0.2–0.4 km, with their steeper slopes facing south-west (Lewis, 1936; Lancaster, 1988).

The ages of the dune systems are presently unknown. The very large dunes of the northern Kalahari and adjacent areas of Zaire, Angola, and Zambia may be of late Pliocene age (Partridge, 1993), whereas linear dunes of the south-western Kalahari may have been formed or reformed in the late Pleistocene (Lancaster, 1989*b*). Thomas and Shaw (1991) and Livingstone and Thomas (1993) have, however, questioned the relict status and therefore the palaeoclimatic significance of these dunes and it appears that many dunes in the south-western Kalahari are episodically active today.

The patterns of winds inferred for all periods of dune formation correspond approximately to those developed around the southern African anticyclone, but with differences that indicate that when the dunes of the south-western Kalahari were formed, circulation patterns were similar to those in modern drought years (Figure 13.16). This suggests that periods of dune formation in the region

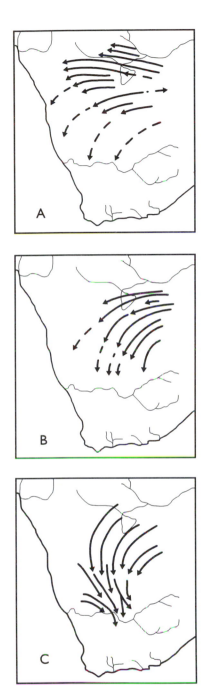

Fig. 13.16. Palaeowinds and circulation patterns inferred from dune alignments in Southern Africa A: oldest relict dunes; B: dunes of north-eastern Botswana and central Kalahari; C: south-western Kalahari dunes.

(*Source*: Lancaster, 1981.)

were associated with strongly anticyclonic circulations and increased regional pressure gradients (Lancaster, 1981).

Present-day active dunes in Southern Africa are restricted to the Namib Desert, and formation of dunes in the central and northern Kalahari would have required a northward shift of the area receiving less than 150 mm of rain a year by as much as 1200 km. Formation of the southern groups of dunes would have required a 250–300 km north-eastward migration of the 150 mm isohyet (Lancaster, 1981). There is little evidence for such major climatic changes in the region, at least during the late Quaternary (Deacon and Lancaster, 1988) and it seems likely that, as in the Sahara, reactivation and/or formation of dunes occurred as a result of changes in wind energy and effective precipitation rather than major shifts in climatic belts. Recent measurements of dune dynamics in the south-western Kalahari suggest that significant sediment transport takes place on these dunes under present climatic conditions. This suggests that relatively minor changes in effective precipitation and vegetation cover could result in major changes in dune activity (Livingstone and Thomas, 1993). These changes can be modelled using the dune mobility index (M) which is the ratio between the amount of time the wind is able to move sand (W) and the effective precipitation P/PE) (Lancaster, 1988):

$$M = W / (P/PE).$$

Threshold values for dune mobility determined from field observations of linear dunes in Southern Africa and crescentic and parabolic dunes in the High Plains of the United States suggest that dunes are fully active when M exceeds 200, and that most dunes are completely stabilized by vegetation when M is less than 50 (Lancaster, 1988; Muhs and Maat, 1993).

In Southern Africa, there is a gradient in sand mobility, as defined by this relationship, from the northern and eastern parts of the south-western Kalahari to the active Namib sand sea (Figure 13.17), that accurately parallels observations of the amounts of sand movement on dunes in the region (Lancaster, 1988).

Fluvial Systems

River systems in the Sahara and Namib exhibit sequences of fluvial deposits that reflect periods of aggradation and incision in a manner similar to those observed in arid regions in Australia and North America. The relative tectonic stability of the African continent however results in a dominance of climatic effects on the behaviour of the fluvial system. In addition to cut-and-fill sequences and river terraces, there is widespread evidence for the

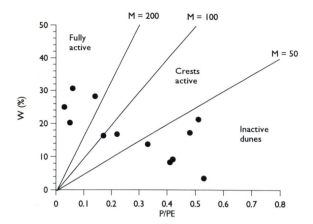

Fig. 13.17. Dune mobility thresholds in Southern Africa.

expansion of integrated drainage systems during periods of humid climates, as well as the formation of large lake systems (Goudie, this volume), and the disintegration of these networks, accompanied by dune blocking of channels, in arid phases (Rognon, 1982). However, not all drainage networks have evolved in this way. In the Kalahari, extensive channel networks (locally called *mekgacha*) extend from the surrounding highlands toward the Makgadikgadi Depression (Figure 13.18). These were thought by earlier workers (e.g. Grove, 1969) to be the products of much wetter periods in the Quaternary, but more recent investigations (Shaw and De Vries, 1988) suggest the importance of groundwater sapping in their formation.

River-terrace sequences along streams draining from the Saharan Atlas record the effects of climatic changes

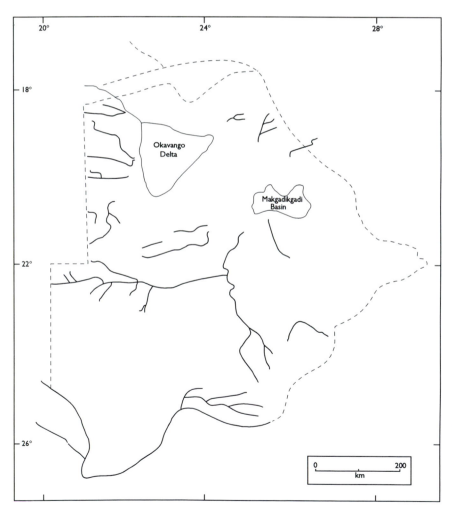

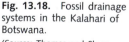

Fig. 13.18. Fossil drainage systems in the Kalahari of Botswana.

(*Source*: Thomas and Shaw 1991.)

in the mountains bordering the Sahara (Rognon and Coudé-Gaussen, 1987). Prior to 40 ka BP, terrace deposits are mostly of gravel, indicating highly competent streams. A sandy river terrace formed in the period 40–20 ka BP in response to more frequent and longer duration floods down the Wadi Saoura, probably as a result of increased and less seasonal runoff in the Saharan Atlas. Elsewhere, silt and sandy silts accumulated in aggrading channels. Reddish soils later formed on these deposits. Increased runoff also fed large lakes in the Chott Djerid region (Rognon and Coudé-Gaussen, 1987). Increasing aridity after c.18 ka BP gave rise to channel incision from 14 to 6 ka BP. The later Holocene is characterized by renewed aggradation of fine-grained alluvial material from 4 to 2 ka BP in a period of increased moisture availability that began as early as 6 ka BP. Interpretation of late Holocene fluvial responses is complicated by human occupation of the region, and all changes after 4 ka BP are to some degree affected by land clearance for agriculture (Rognon and Coudé-Gaussen, 1987).

Fluvial systems in the southern and central Sahara also responded to climatic changes. In the Akator and Tibesti mountains, Messerli et al. (1980) recognize three late Quaternary periods of stream aggradation. The earliest phase is beyond the range of radiocarbon-dating, but contains transported early Stone Age artefacts. Thick sequences of gravels and sands were derived by stripping of adjacent hillslopes in a period of inferred extreme aridity. Following a period that resulted in as much as 20–30 m of incision, there was a further episode of aggradation of silts and clays, with radiocarbon ages of 14–18 ka BP. Cold, humid conditions are indicated by pollen in the fluvial sediments and interstratified frost-shattered debris. A further period of aggradation of fine-grained sediments occurred in a period of increased rainfall in the mid-Holocene (4–6 ka BP). The early Holocene episode of silt accumulation is recognized throughout the region where it marks a period of perennial meandering streams in contrast to the preceding and following periods of coarse sediment-dominated ephemeral braided channels (Talbot, 1980). Small lakes even formed along some streams (Pachur and Kröplein, 1987). This period of stream aggradation can be related to the widespread early-Holocene pluvial period caused by greater penetration of monsoonal moisture from the Gulf of Guinea and a more northerly position of the ITCZ (Petit-Maire et al., 1990).

Perhaps the most dramatic indication of former fluvial regimes in the Sahara is provided by the aptly named 'radar rivers' which have been discovered via radar remote-sensing in the hyper-arid eastern Sahara (Elachi

et al., 1982; McCauley et al., 1982). These channels, which are buried below the surface sand sheets, formed part of an extensive Oligocene to early-Miocene drainage system that extended from the Red Sea Hills to the Chad Basin. Incision of the River Nile around 6 Ma BP beheaded the tributaries and captured the Western Desert sections. The remaining valleys in the eastern Sahara were then infilled with locally derived well-rounded gravels and sands. The region was occupied along these drainage ways by early Stone Age peoples prior to periods of carbonate cementation of these deposits dated by uranium-series techniques to around 141 and 212 ka BP (Szabo et al., 1989).

Large exoreic drainages like the Nile and Senegal Rivers also responded to climatic changes in their headwaters (Michel, 1979; Adamson et al., 1980). The Nile in Egypt was a braided stream, with highly competent discharge from local wadis, from before 32 to around 27 ka BP (Butzer, 1980). During the period of aridity between approximately 24 and 16 ka BP, the Nile in the Sudan was reduced to a braided stream with sharp flood peaks and a highly seasonal discharge (Adamson et al., 1980). In many other areas, dunes were reactivated, leading to blockage of major rivers such as the Senegal, Niger, and White Nile. During the succeeding long period of increased rainfall very high floods on the Nile deposited fine-grained overbank deposits between 12.5 and 5 ka BP (Adamson et al., 1980). The Senegal River cut through the dunes to the sea, and the discharge of the Niger increased significantly, forming a large inland delta 8.5–3.5 ka BP (Petit-Maire and Riser, 1987). After about 4 ka BP, desiccation set in and stream flow in the Nile became more seasonal with higher flood peaks.

Evidence for periods of stream aggradation and incision in response to climatic change is also available from the Namib. In the Namib, the palaeoclimatic significance of early- and middle-Pleistocene fluvial deposits is uncertain. Late-Pleistocene fluvial deposits include a 30 m thickness of silts which accumulated in the Kuiseb Valley between 23 and 19 ka BP in conditions which are interpreted as reduced stream discharge and competence by Vogel (1989) or in embayments as slack-water deposits during major flood events (Ward, 1987). There was a renewed period of aggradation of gravel-sized material in the terminal Pleistocene (Ward, 1987). Shallow lacustrine deposits in interdune areas are interpreted as indicating periods of increased penetration of river systems into the Namib sand sea 25–30 ka BP, with a gradual eastward retreat to around 14 ka BP, as dunes blocked river-courses (Teller et al., 1990).

Conclusions

Deserts in Africa exhibit a wide range of environments from the cool, foggy, hyper-arid Namib to the hot, semi-arid southern Sahara. All areas have been influenced to some degree by Quaternary climatic changes, with the effects apparently being greatest on the desert margins. Much of our knowledge of the geomorphology of these regions, and the effects of climatic changes on landform development are the result of developments in remote-sensing techniques. Use of aerial photographs to map the extent of relict dune systems was pioneered by A. T. Grove in the Sahel and Kalahari. Later workers such as Mainguet, Breed, McCauley, Lancaster, and Maxwell used Landsat and other satellite images extensively for regional syntheses of aeolian and fluvial landforms. These regional studies have provided the background for detailed studies of landforms, geomorphic processes, and deposits. These later studies (e.g. Livingstone and Thomas, 1993) may result in significant changes in the palaeoclimatic interpretations of landforms. The next generation of workers in the African deserts will need to emphasize both detailed investigations of geomorphic processes as well as stratigraphic studies of landforms and deposits to establish both regional and local documentation of the response of landforms to climatic change on time-scales of millennia to decades.

References

Acocks, J. P. H. (1953), 'Veld Types of South Africa', *Botanical Survey of South Africa, Memoirs*, 28: 1–92.

Adamson, D. A., Gasse, F., Street, F. A., and Williams, M. A. J. (1980), 'Late Quaternary History of the Nile', *Nature*, 287: 50–5.

Albritton, C. C., Jr., Brooks, J. E., Issawi, B., and Swedan, A. (1990), 'Origin of the Qattara Depression, Egypt', *Geological Society of America Bulletin*, 102/7: 952–60.

Bagnold, R. A. (1941), *The Physics of Blown Sand and Desert Dunes* (London).

Barnard, W. S. (1975), 'Geomorphologeise processe en die mens: die geval an die Kuisebdelta, Suidwes Afrika', *Acta Geographica*, 2: 20–43.

Beadnell, H. J. L. (1910), 'The Sand Dunes of the Libyan Desert', *Geographical Journal*, 35: 379–95.

Boocock, C. and van Straten, O. J. (1962), 'Notes on the Geology and Hydrology of the Central Kalahari Region, Bechuanaland Protectorate', *Geological Society of South Africa, Transactions*, 65/1: 125–71.

Breed, C. S. *et al.* (1979), 'Regional Studies of Sand Seas using LANDSAT (ERTS) Imagery', in E. D. McKee (ed.), *A Study of Global Sand Seas*, 305–98.

—— and Grow, T. (1979), 'Morphology and Distribution of Dunes in Sand Seas Observed by Remote Sensing', in McKee (ed.), *A Study of Global Sand Seas*, 253–304.

—— McCauley, J. F., and Davis, P. A. (1987), 'Sand Sheets of the Eastern Sahara and Ripple Blankets on Mars', in L. E. Frostick and I. Reid (eds.), *Desert Sediments*, 337–59.

Brookfield, M. E. and Ahlbrandt, T. S. (1983) (eds.), *Eolian Sediments and Processes*, Developments in Sedimentology (Amsterdam).

Busche, D. and Hagedorn, H. (1980), 'Landform Development in Warm Deserts: The Central Saharan Example', *Zeitschrift für Geomorphologie*, Supplementband 36: 123–39.

Butzer, K. W. (1980), 'Pleistocene History of the Nile Valley in Egypt and Lower Nubia', in M. A. J. Williams and H. Faure (eds.), *The Sahara and the Nile*, 253–80.

Carlson, T. N. and Benjamin, S. G. (1980), 'Radiative Heating Rates for Saharan Dust', *Journal of Geophysical Research*, 37: 193–213.

Cloudsley-Thompson, J. L. (1984) (ed.), *Sahara Desert*, Key Environments (Oxford and New York).

Corbett, I. (1993), 'The Modern and Ancient Pattern of Sandflow Through the Southern Namib Deflation Basin', *International Association of Sedimentologists Special Publication*, 16: 45–60.

De Vries, J. J. (1984), 'Holocene Depletion and Active Recharge of the Kalahari Groundwaters: A Review and Indicative Model', *Journal of Hydrology*, 70: 221–32.

Deacon, J. and Lancaster, N. (1988), *Late Quaternary Environments of Southern Africa* (Oxford).

Diester-Haass, L., Heine, K., Rothe, P., and Schrader, H. (1988), 'Late Quaternary History of Continental Climate and the Benguela Current off South West Africa', *Palaeogeography, Palaeoclimatology, Palaeoecology*, 65: 81–91.

Drake, N. A., Heydeman, M. T. and White, K. H. (1993), 'Distribution and Formation of Rock Varnish in Southern Tunisia', *Earth Surface Processes and Landforms*, 18/1: 31–42.

El Baz, F. (1984) (ed.), *Deserts and Arid Lands* (The Hague).

—— and Maxwell, T. A. (1982) (ed.), *Desert Landforms of Egypt: A Basis for Comparison with Mars*. (Washington, DC).

—— Breed, C. S., Grolier, M., and McCauley, J. F. (1979), 'Eolian Features in the Western Desert of Egypt and Some Applications to Mars', *Journal of Geophysical Research*, 84: 8205–21.

Elachi, C. *et al.* (1982), 'Shuttle Imaging Radar Experiment', *Science*, 218/4576: 996–1003.

Embabi, N. S. (1982), 'Barchans of the Kharga Depression', in F. El Baz and T. A. Maxwell (eds.), *Desert Landforms of Egypt*, 141–56.

Endrody-Younga, S. (1982), 'Dispersion and Translocation of Dune Specialist Tenebrionids in the Namib Area', *Cimbebasia (A)*, 5: 257–71.

Evenari, M., Noy-Meir, I. and Goodall, D. W. (1986) (eds.), *Ecosystems of the World*: 12 B, *Hot Deserts and Arid Shrub Lands* (Amsterdam).

Flint, R. F. and Bond, G. (1968), 'Pleistocene Sand Ridges and Pans in Western Rhodesia', *Geological Society of America Bulletin*, 79: 299–314.

Foster, S. S. D., Bath, A., Farr, J. and Lewis, W. (1982), 'The Likelihood of Active Groundwater Recharge in the Botswana Kalahari', *Journal of Hydrology*, 55: 113–36.

Frostick, L. E. and Reid, I. (1987a), 'Tectonic Control of Desert Sediments in Rift Basins Ancient and Modern', in Frostick and Reid (eds.), *Desert Sediments*, 53–68.

—— —— (1987b) (eds.), *Desert Sediments: Ancient and Modern*, Geological Society Special Publication 35 (Oxford).

Fryberger, S. (1980), 'Dune Forms and Wind Regime, Mauritania, West Africa: Implications for Past Climate', *Palaeocology of Africa*, 12: 79–96.

—— and Ahlbrandt, T. S. (1979), 'Mechanisms for the Formation of Aeolian Sand Seas', *Zeitschrift für Geomorphologie*, 23: 440–60.

—— and Goudie, A. S. (1981), 'Arid Geomorphology', *Progress in Physical Geography*, 5/3: 420–8.

Gardner, R. and Scoging, H. (1983) (eds.), *Mega Geomorphology* (Oxford).

Gischler, C. E. (1979), *Water Resources in the Arab Middle East and North Africa* (Cambridge).

Goudie, A. (1972), 'Climate, Weathering, Crust Formation, Dunes, and Fluvial Features of the Central Namib Desert, Near Gobabeb, South West Africa', *Madoqua*, 1/54–62: 15–31.

—— (1983), 'Dust Storms in Space and Time', *Progress in Physical Geography*, 7: 502–29.

—— and Thomas, D. S. G. (1985), 'Pans in Southern Africa with Particular Reference to South Africa and Zimbabwe', *Zeitschrift für Geomorphologie*, 29/1: 1–19.

Grousset, F. E., Rognon, P., Coudé-Gaussen, G., and Pédemay, P. (1992), 'Origins of Peri-Saharan Dust Deposits Traced by Their Nd and Sr Isotopic Composition', *Palaeogeography, Palaeoclimatology, Palaeoecology*, 93: 203–12.

Grove, A. T. (1969), 'Landforms and Climatic Change in the Kalahari and Ngamiland', *Geographical Journal*, 135/2: 190–212.

—— (1980), 'Geomorphic Evolution of the Sahara and the Nile', in Williams and Faure (eds.), *The Sahara and the Nile*, 7–16.

—— and Warren, A. (1968), 'Quaternary Landforms and Climate on the South Side of the Sahara', *Geographical Journal*, 134/2: 189–208.

Hagedorn, H. (1980), 'Geological and Geomorphological Observations on the Northern Slope of the Tibesti Mountains, Central Sahara', in M. J. Salem and M. T. Busrewil (eds.) *Geology of Libya*, 823–35.

Haynes, C. V., Jnr. (1989), 'Bagnold's Barchan: A 57-yr Record of Dune Movement in the Eastern Sahara and Implications for Dune Origin and Palaeoclimate Since Neolithic Times', *Quaternary Research*, 32/2: 153–67.

Heaton, T. H. E., Talma, A. S., and Vogel, J. C. (1983), 'Origin and History of Nitrate in Confined Groundwater in the Western Kalahari', *Journal of Hydrology*, 64: 232–62.

Huntley, B. J. (1985) (ed.), *The Kuiseb Environment: The Development of a Monitoring Baseline* (Pretoria).

Jaccoberger, P. A. (1989), 'Reflectance Characteristics and Surface Processes in Stabilized Dune Environments', *Remote Sensing of Environment*, 28: 287–95.

Junge, C. E. (1979), 'The Importance of Mineral Dust as an Atmospheric Constituent in the Atmosphere', in C. Morales (ed.) *Saharan Dust* (New York), 49–60.

Kaiser, E. (1926), *Die Diamantenwüste Südwestafrikas* (Berlin).

Kalu, A. E. (1979), 'The African Dust Plume: Its Characteristics and Propagation across West Africa in Winter', in Morales (ed.), *Saharan Dust*, 95–118.

Kocurek, G., Havholm, K. G., Deynoux, M., and Blakey, R. C. (1991), 'Amalgamated Accumulations Resulting from Climatic and Eustatic Changes, Akchar Erg, Mauritania', *Sedimentology*, 38/4: 751–72.

Lancaster, J., Lancaster, N., and Seely, M. K. (1984), 'The Climate of the Central Namib Desert', *Madoqua*, 14: 5–61.

Lancaster, N. (1978), 'The Pans of the Southern Kalahari, Botswana', *Geographical Journal*, 144: 81–98.

—— (1981), 'Palaeoenvironmental Implications of Fixed Dune Systems in Southern Africa', *Palaeogeography, Palaeoclimatology, Palaeoecology*, 33: 327–46.

—— (1982), 'Dunes on the Skeleton Coast, SWA/Namibia: Geomorphology and Grain Size Relationships', *Earth Surface Processes and Landforms*, 7: 575–87.

—— (1983), 'Controls of Dune Morphology in the Namib Sand Sea', in M. E. Brookfield and T. S. Ahlbrandt (eds.), *Eolian Sediments and Processes*, 261–89.

—— (1984a), 'Aridity in Southern Africa: Age, Origins and Expression in Landforms and Sediments', in J. C. Vogel (ed.), *Late Cenozoic Palaeoclimates*, 433–44.

—— (1984b), 'Characteristics and Occurrence of Wind Erosion Features in the Namib Desert', *Earth Surface Processes and Landforms*, 9: 469–78.

—— (1986), 'Pans in the Southwestern Kalahari: A Preliminary Report', *Palaeoecology of Africa*, 17: 59–68.

—— (1988), 'Development of Linear Dunes in the Southwestern Kalahari, Southern Africa', *Journal of Arid Environments*, 14: 233–44.

—— (1989a), *The Namib Sand Sea: Dune forms, Processes, and Sediments* (Rotterdam).

—— (1989b), 'Late Quaternary Palaeoenvironments in the Southwestern Kalahari', *Palaeogeography, Palaeoclimatology, Palaeoecology*, 70: 367–76.

Le Houérou, H. N. (1986), 'The Deserts and Arid Zones of Northern Africa', in M. Evenari, I. Noy-Meir, and D. W. Goodall (eds.), *Ecosystems of the World*, 101–48.

Leinen, M. and Sarthein, M. (1989) (ed.), *Paleoclimatology and Palaeometeorology: Modern and Past Patterns of Global Atmospheric Transport* (Dordrecht, Neth. and Boston).

Lewis, A. D. (1936), 'Sand Dunes of the Kalahari Within the Borders of the Union', *South African Geographical Journal*, 14: 22–32.

Livingstone, I. (1988), 'New Models for the Formation of Linear Sand Dunes', *Geography*, 73: 105–15.

—— (1989), 'Monitoring Surface Change on a Namib Linear Dune', *Earth Surface Processes and Landforms*, 14: 317–32.

—— and Thomas, D. S. G. (1993), 'Modes of Linear Dune Activity and Their Significance: An Evaluation With Reference to Southern African Examples', in K. Pye (ed.), *Dynamics and Environmental Context of Aeolian Sedimentary Systems*, 91–102.

Lovegrove, B. (1993), *The Living Deserts of Southern Africa* (Vlaeberg, South Africa).

McCauley, J. F., Breed, C. S., El Baz, F., Whitney, M. I., Grolier, M. J., and Ward, A. W. (1979), 'Pitted and Fluted Rocks in the Western Desert of Egypt: Viking Comparisons', *Journal of Geophysical Research*, 84: 8222–32.

—— et al. (1982), 'Subsurface Valleys and Geoarchaeology of the Eastern Sahara Revealed by Shuttle Radar', *Science*, 218: 1004–20.

McKee, E. D. (1979) (ed.), *A Study of Global Sand Seas*, United States Geological Survey Professional Paper 1052.

McTainsh, G. H. and Walker, P. H. (1982), 'Nature and Distribution of Harmattan Dust', *Zeitschrift für Geomorphologie*, 26: 417–53.

Mainguet, M. (1972a), *Le modelé de grès* (Paris).

—— (1972b), 'Étude d'un erg (Fachi-Bilma) son alimentation, sable-use et sa insertion dans le paysage d'apres les photographies prises par satellites', *Academie des Sciences, Comptes Rendues*, 274: 1633–6.

—— (1983), 'Tentative Mega-Morphological Study of the Sahara', in R. Gardner and H. Scoging (eds.), *Mega Geomorphology*, 113–33.

—— (1984), 'A Classification of Dunes Based on Aeolian Dynamics and the Sand Budget', in El-Baz (ed.), *Deserts and Arid Lands*, 31–58.

—— (1984), 'Space Observations of Saharan Aeolian Dynamics', in El Baz (ed.), *Deserts and Arid Lands*, 59–77.

—— and Callot, Y. (1978), 'L'erg de Fachi-Bilma (Tchad–Niger)', *Mémoires et Documents CNRS*, 18: 178.

—— and Canon, L. (1976), 'Vents et paleovents du Sahara: Tentative d'approache paleoclimatique', *Revue de Géographie Physique et de Géologie Dynamique*, 18/2–3: 241–50.

—— and Chemin, H. C. (1983), 'Sand Seas of the Sahara and Sahel: An Explanation of Their Thickness and Sand Dune Type by the Sand Budget Principle', in Brookfield and Ahlbrandt (eds.), *Eolian Sediments and Processes*, 353–64.

—— Borde, J., and Chemin, M. (1984), 'Sedimentation eolienne au Sahara et sur ses marges: Les images Meteosat et Landsat, outil pour l'analyse des temoignages géodynamiques du transport eolian au sol', *Travaux de l'Institute de Géographie de Reims*, 59–60: 15–27.

Marker, M. E. (1977), 'A Long-return Geomorphic Event in the Namib Desert, South-West Africa', *Area*, 9: 209–13.

Maxwell, T. A. and Haynes, C. V., Jr. (1989), 'Large-scale, Low-Amplitude Bedforms (Chevrons) in the Selima Sand Sheet, Egypt', *Science*, 243: 1179–82.

Mazor, E. (1982), 'Rain Recharge in the Kalahari: A Note of Some Approaches to the Problem', *Journal of Hydrology*, 55: 137–44.

Messerli, B., Winiger, M., and Rognon, P. (1980), 'The Saharan and East African Uplands During the Quaternary', in Williams and Faure (eds.), *The Sahara and the Nile*, 87–132.

Michel, P. (1979), 'The South-West Sahara Margin: Sediments and Climate Changes During the Recent Quarternary', *Palaeoecology of Africa*, 12: 297–306.

Millington, A. C., Jones, A. R., Quarmby, N., and Townshend, J. R. G. (1987), 'Remote Sensing of Sediment Transfer Processes in Playa Basins', in Frostick and Reid, (eds.), *Desert Sediments*, 369–81.

Morales, C. (1979) (ed.), *Saharan Dust* (New York).

Muhs, D. R. and Maat, P. B. (1993), 'The Potential Response of Eolian Sands to Greenhouse Warming and Precipitation Reduction on the Great Plains of the United States', *Journal of Arid Environments*, 251: 351–61.

Nichol, J. E. (1991), 'The Extent of Desert Dunes in Northern Nigeria as Shown by Image Enhancement', *Geographical Journal*, 157/1: 13–24.

Nickling, W. G. and Gillies, J. A. (1993), 'Dust Emission and Transport in Mali, West Africa', *Sedimentology*, 40/5: 859–68.

—— and Wolfe (1994). 'The Morphology and Origin of Nebkhas, Region of Mopti, Mali, West Africa', *Journal of Arid Environments*, 28: 13–30.

Nyamweru, C. K. and Bowman, D. (1989), 'Climatic Changes in the Chalbi Desert, North Kenya', *Journal of Quaternary Science*, 4/2: 131–9.

Ollier, C. D. (1978), 'Inselbergs of the Namib Desert: Processes and History', *Zeitschrift für Geomorphologie N.F.*, Supplementband, 31: 161–76.

Pachur, H. J. and Kröplein, S. (1987), 'Wadi Howar: Paleoclimatic Evidence From an Extinct River System in the South-eastern Sahara', *Science*, 237: 298–300.

Partridge, T. C. (1993), 'The Evidence for Cainozoic Aridification in Southern Africa', *Quaternary International*, 17: 105–10.

Passarge, S. (1904), *Die Kalahari* (Berlin).

Peel, R. F. (1941), 'Denudational Landforms of the Central Libyan Desert', *Journal of Geomorphology*, 4: 3–23.

Petit-Maire, N. (1986), 'Palaeoclimates in the Sahara of Mali: An Interdisciplinary Study', *Episodes*, 9/1: 7–15.

—— (1991), 'Recent Quaternary Climatic Change and Man in the Sahara', *Journal of African Earth Sciences*, 12/1–2: 125–32.

—— and Riser, J. (1987), 'Holocene Palaeohydrography of the Niger', *Palaeoecology of Africa*, 18: 135–42.

—— Commelin, D., Fabre, J., and Fontugne, M. (1990), 'First Evidence for Holocene Rainfall in the Tanzerouft Hyperdesert and its Margins', *Palaeogeography, Palaeoclimatology, Palaeoecology*, 79: 333–9.

Prospero, J. M., Glaccum, R. A., and Nees, R. T. (1981), 'Atmospheric Transport of Soil Dust from Africa to South America', *Nature*, 289: 570–2.

Pye, K. (1987), *Aeolian Dust and Dust Deposits* (London).

—— (1993) (ed.), *Dynamics and Environmental Context of Aeolian Sedimentary Systems*, Geological Society of London, Special Publication 16 (London).

Roberts, C. R. and Mitchell, C. W. (1987), 'Spring Mounds in Southern Tunisia', in Frostick and Reid (eds.), *Desert Sediments* (1987), 321–34.

Rognon, P. (1982), 'Pluvial and Arid Phases in the Sahara: The Role of Non-Climatic Factors', *Palaeoecology of Africa*, 12: 45–62.

—— and Coudé-Gaussen, G. (1987), 'Reconstruction paleoclimatique à partir des sediments du Pleistocene superieur et de l'Holocene du nord de Fuerteventura (Canaries)', *Zeitschrift für Geomorphologie N.F.*, 31/1: 1–19.

—— —— Bergametti, G., and Gomes, L. (1989), 'Relationships Between the Characteristics of Soils, the Wind Energy and Dust Near the Ground in the Western Sand Sea (N. W. Sahara)', in M. Leinen and M. Sarnthein (eds.), *Palaeoclimatology and Palaeometeorology*, 167–84.

Said, R. (1990) (ed.), *The Geology of Egypt* (Rotterdam).

Salem, M. J. and Busrewil iii (1980) (eds.), *The Geology of Libya* (London).

Sarnthein, M. (1978), 'Sand Deserts During Glacial Maximum and Climatic Optima', *Nature*, 272: 43–46.

—— Tetzlaff, G., Koopmann, B., Wolter, K., and Pilaumann, U. (1981), 'Glacial and Interglacial Wind Regimes over the Eastern Subtropical Atlantic and North-West Africa', *Nature*, 293/5829: 193–6.

Seely, M. K. (1978), 'The Namib Dune Desert: An Unusual Ecosystem', *Journal of Arid Environments*, 1: 117–28.

—— (1990) (ed.), *Namib Ecology*, Transvaal Museum Monograph 7 (Pretoria).

Selby, M. J. (1977), 'Bornhardts of the Namib Desert', *Zeitschrift für Geomorphologie N.F.*, 21/1: 1–13.

—— (1977), 'Palaeowind Directions in the Central Namib Desert, as Indicated by Ventifacts', *Madogua*, 10/3: 195–8.

—— Hendy, C. H., and Seely, M. K. (1978), 'A Late Quaternary Lake in the Central Namib Desert, Southern Africa, and Some Implications', *Palaeogeography, Palaeoclimatology, Palaeoecology*, 26: 37–41.

Shaw, P. A. and De Vries, J. J. (1988), 'Duricrust, Groundwater and Valley Development in the Kalahari of Southeast Botswana', *Journal of Arid Environments*, 14: 245–54.

Slattery, M. C. (1990), 'Barchan Migration on the Kuiseb River Delta, Namibia', *South African Geographical Journal*, 72: 5–10.

Smith, G. (1984), 'Climate', in J. L. Cloudsley-Thompson (ed.), *Sahara Desert*, 17–30.

Stengel, H. W. (1964), 'The Rivers of the Namib and Their Discharge into the Atlantic, Part 1: Kuiseb and Swakop', *Scientific Papers of the Namib Desert Research Station*, 22: 1–50.

Summerfield, M. A. (1983), 'Silcrete as a Palaeoclimatic Indicator: Evidence from Southern Africa', *Palaeogeography, Palaeoclimatology, Palaeoecology*, 41: 66–79.

Sweeting, M. M. and Lancaster, N. (1982), 'Solutional and Wind Erosion Forms on Limestone in the Central Namib Desert', *Zeitschrift für Geomorphologie N.F.*, 26: 197–207.

Szabo, B. J., McHugh, W. P., Schaber, G. G., Haynes, C. V., and Breed, C. S. (1989), 'Uranium-series Dated Authigenic Carbonates and Acheulian Sites in Southern Egypt', *Science*, 243: 1053–6.

Talbot, M. R. (1980), 'Environmental Responses to Climatic Change in the West African Sahel Over the Past 20 000 Years', in Williams and Faure (eds.), *The Sahara and the Nile*, 37–62.

Talbot, M. R. (1984), 'Late Pleistocene Dune-Building and Rainfall in the Sahel', *Palaeoecology of Africa*, 16: 203–14.

—— (1985), 'Major Bounding Surfaces in Aeolian Sandstones: A Climatic Model', *Sedimentology*, 32: 257–66.

Teller, J. T., Rutter, N. W., and Lancaster, N. (1990), 'Sedimentology and Palaeohydrology of Late Quaternary Lake Deposits in the Northern Namib Sand Sea, Namibia', *Quaternary Science Reviews*, 9: 343–64.

Thomas, D. S. G. (1984), 'Ancient Ergs of the Former Arid Zones of Zimbabwe, Zambia, and Angola', *Transactions of the Institute of British Geographers*, NS, 9: 75–88.

—— (1987), 'Discrimination of Depositional Environments Using Sedimentary Characteristics in the Mega Kalahari, Central Southern Africa', in Frostick and Reid (eds.), *Desert Sediments*, 293–306.

—— and Shaw, P. A. (1991a), *The Kalahari Environment* (Cambridge and New York).

—— —— (1991b), ' "Relict" Desert Dune Systems: Interpretations and Problems', *Journal of Arid Environments*, 20: 1–14.

Thorweihe, U. (1990), 'Nubian Aquifer System', in R. Said (ed.), *The Geology of Egypt*, 601–14.

Tyson, P. D. (1986), *Climatic Change and Variability in Southern Africa* (Cape Town).

UNESCO (1979), *Map of the World Distribution of Arid Regions* (Paris).

Van Zinderen Bakker, E. M. (1984), 'Aridity Along the Namibian Coast', *Palaeoecology of Africa*, 16: 149–62.

Vogel, J. C. (1984) (ed.), *Late Cenozoic Palaeoclimates of the Southern Hemisphere* (Rotterdam).

—— (1989), 'Evidence of Past Climatic Change in the Namib Desert', *Palaeogeography, Palaeoclimatology, Palaeoecology*, 70: 355–66.

Vökel, J. and Grunert, J. (1990), 'The Problem of Dune Formation and Dune Weathering During the Late Pleistocene and Holocene in the Southern Sahara and Sahel', *Zeitschrift für Geomorphologie* N.F., 34/1: 1–17.

Walther, J. (1924), *Das Gesetz der Wüstenbildung im Gegenwart und Vorzeit* (Berlin).

Ward, J. D. (1987), 'The Cenozoic Succession in the Kuiseb Valley, Central Namib Desert', *Geological Survey of Namibia*, Memoir, 9: 124.

—— and von Brunn, V. (1985), 'Sand Dynamics along the Kuiseb River', in B. J. Huntley (ed.), *The Kuiseb Environment*, 51–72.

—— Seely, M. K., and Lancaster, N. (1983), 'On the Antiquity of the Namib', *South African Journal of Science*, 79: 175–83.

Watson, A. (1985), 'Structure, Chemistry, and Origins of Gypsum Crusts in Southern Tunisia and the Central Namib Desert', *Sedimentology*, 32: 855–75.

Weare, P. R. and Yalala, A. (1971), 'Provisional Vegetation Map', *Botswana Notes and Records*, 3: 131–52.

Wellington, J. H. (1955), *Southern Africa: A Geographical Study*, i, *Physical Geography* (Cambridge).

White, F. (1983), *The Vegetation of Africa* (Paris).

White, K. (1991), 'Geomorphological Analysis of Piedmont Landforms in the Tunisian Southern Atlas Using Ground and Satellite Imagery', *Geographical Journal*, 157/3: 279–94.

Wickens, G. E. (1984), 'Flora', in Cloudsley-Thompson (ed.), *Sahara Desert*, 67–75.

Williams, G. E. (1970), 'Piedmont Sedimentation and late Quaternary Chronology in the Biskra Region of the Northern Sahara', *Zeitschrift für Geomorphologie*, N.F., 10: 40–63.

Williams, M. A. J. (1975), 'Late Pleistocene Tropical Aridity Synchronous in Both Hemispheres?', *Nature*, 253/5493: 617–18.

—— and Faure, H. (1980) (eds.), *The Sahara and the Nile*. (Rotterdam).

Wilson, B. H. and Dincer, T. (1976), 'An Introduction to the Hydrology and Hydrography of the Okavango Delta', in *Proceedings of the Symposium on the Okavango Delta and its Future Utilisation*, Botswana Society (Gaborone), 33–48.

Wilson, I. G. (1971), 'Desert Sandflow Basins and a Model for the Development of Ergs', *Geographical Journal*, 137/2: 180–99.

Yaalon, D. H. and Ward, J. D. (1982), 'Observations on Calcrete and Recent Calcic Horizons in Relation to Landforms in the Central Namib Desert', *Palaeoecology of Africa*, 15: 183–6.

14 Coastal Environments

Antony R. Orme

Introduction

The character of the African coast reflects three inter-active groups of variables: the shape of the continental margin resulting from prolonged tectonic and geomorphic changes; the physical and chemical processes that operate along this margin in response to climatic forcing; and the biological processes which combine with the physical environment to produce distinctive ecological responses. These variables are interdependent because, to a greater or lesser extent, each group influences and is in turn influenced by the others. Thus the shape of the continental shelf affects waves and currents approaching the shore just as the nature of drainage basins influences stream flows and sediment yields reaching the coast. Erosion and sedimentation in turn modify the continental shelf and drainage basins and, in consort with biotic influences, create new coastal environments for subsequent physical, chemical, and biological activity. Underpinning the entire coastal system are the tectonic forces which, though independent of the above variables over the short term, have shifted the African plate to its present location astride the Equator and thus created the framework within which coastal processes and responses function.

The following discussion pursues this interactive theme by viewing the African coast as a series of integrated systems—coastal dunes, mangrove swamps, seacliffs, coral reefs, and the like—produced in various materials by a combination of physical, chemical, and biological processes. After introducing salient aspects of location and shape, the evolution of the African coast since the rupture of Gondwana is examined. Various climatic, oceanic, hydrologic, and other ecological processes are then discussed with emphasis on commonalities.

Finally, to evaluate departures from noted commonalities, the coast is evaluated in terms of its five relatively distinct regional components.

Location and Dimensions

The African continent measures 30×10^6 km^2, about 20 per cent of Earth's land area, and its coastline is about 30 000 km long. In proportion to area this length is less than for any other continent. Asia for example is only 1.5 times larger than Africa but is bounded by 70 000 km of coastline. Over long distances the African coastline is unbroken by sizable inlets and its major river mouths, except the Congo, are either deltaic or blocked by sand bars. This restricts the inland penetration of marine influences while the paucity of protected natural harbours and navigable rivers long hindered the development of the coast and its hinterland. Smoothness in plan is matched by relatively low profile. Despite Africa's high mean elevation of 650 m, lowland coasts with long sandy beaches predominate; rugged mountainous coasts are rare. Excepting Madagascar, there are no large offshore islands to mitigate direct oceanic impacts on the mainland coast.

Offshore Africa's continental shelf covers only 1.28×10^6 km^2, compared with 9.38×10^6 km^2 for Asia and 6.74×10^6 km^2 for North America. The shelf averages only 25 km in width but is wider off southern Tunisia, Guinea, Eritrea, and the Nile, Niger, and Zambezi deltas, and broadens to 240 km on the Agulhas Bank. Elsewhere it is often very narrow, averaging 5 km wide off Somalia, northern Mozambique, and northern Natal. A narrow shelf allows deep ocean waves and surface ocean currents to approach unmodified close to shore where they are unusually influential in sediment transport and coastal ecology.

From Cap Blanc (37° 15′ N) to Cape Agulhas (34° 52′ S), Africa ranges through 72° of latitude. Straddling the Equator in this way ensures the predominance of tropical and subtropical conditions over most of the continent. The coastal zone is no exception, although the Atlantic coast to within 10–15° of the Equator is tempered by the cooling effects of relatively cold ocean currents. In contrast, the Indian Ocean coast is typically tropical but proximity to the Asian land mass imparts strong seasonal (monsoonal) changes to atmospheric and oceanic circulations.

Tectonic Origins and Coastal Evolution

The Rupture of Gondwana

Africa's coastal margins were initially blocked out by the rupture of Gondwana in early Mesozoic times (Figure 14.1) and further modified during progressive opening of the Atlantic and Indian Oceans, gradual closure of the Tethys Sea, and rifting of the Red Sea and Gulf of Aden (see Chapter 1). The massive dissected slope at the outer edge of the continental shelf marks the margins of the African plate, the shelf break normally occurring at depths of 150–200 m.

Geomagnetic lineations, seismic stratigraphy, fission-track analysis, and radiometric data from deep-sea, shelf, and coastal rocks show that tensional stresses began affecting Africa's margins before Mesozoic times but that the subsequent rupture of Gondwana did not affect all parts of the coast simultaneously. The north-west coast was outlined as North America began separating from Africa in mid-Jurassic times around 180 Ma BP. The Guinea coast was initiated by lateral shear as the northern margin of South America shifted along coast-parallel fracture zones from early Cretaceous times (130–100 Ma BP) onward. This in turn caused the west-facing margin farther south to separate rapidly from Brazil while, at Africa's southern tip, the Falkland Plateau sheared off from the Hercynian Cape Fold Belt along a major transform fault. Beyond the continental margin, the Agulhas Plateau was created by ocean volcanism in the fracture zone vacated by the Falkland Plateau.

Following Triassic salt-water incursion, much of Africa's east coast was blocked out during earlier Jurassic times (200–160 Ma BP). In Mozambique, the earliest rift produced large horsts and grabens whose N–S trends are reflected in the coasts south of the Zambezi and Limpopo rivers. In contrast, the NE-trending coasts east of

these rivers reflect right-lateral offsets of the basement, bringing Precambrian rocks close to shore in northern Mozambique. Madagascar is a large continental block that was intermittently uplifted as a complex horst within a system of NNE-trending fractures. Its place in the disruption of Gondwana has been much debated: some think that it occupies its original place or that it formed, with India and the Seychelles, part of a disintegrating Malagasy–Mascarene subcontinent (Kutina, 1975), but others believe it was translated southward to its present location from Tanzania and Kenya, with the reef-capped volcanic Comoro archipelago extruding along resultant fracture zones (Maugé *et al.*, 1982). Farther north, the south-east margins of Somalia and Arabia also formed from the Jurassic rupture of Gondwana.

The Red Sea and Gulf of Aden coasts reflect relatively late intraplate separation of the African–Arabian portion of Gondwana along extensions of the East African Rift system (see Chapter 2). In early Tertiary time, after prolonged stability, the region became the site of continuing intense magmatic and tectonic activity. By late-Eocene times (40 Ma BP) vertical uplift had produced a massive Afro-Arabian dome across which attenuation of weakened sialic crust led to trisection into Somali, Nubian, and Arabian segments. The Somali segment was bounded to the north and west by crustal flexures and tensional faults along which lateral displacement began in the Miocene. Since then, as the Arabian segment has moved north and east away from the Somali and Nubian segments, new oceanic crust some 200 km across has formed in the widening Gulf of Aden and the Red Sea.

Thus, in plate-tectonic terms, the Atlantic and Indian Ocean coasts of Africa occur along passive or divergent margins inasmuch as the African plate has shifted away, relatively, from its former continental neighbours in Gondwana and from the now-intervening oceanic spreading centres. In embryo form, the Red Sea and Gulf of Aden also lie along divergent margins but extensive volcanism and high seismicity belie the term 'passive'.

In contrast, the Mediterranean coast occurs along an active or convergent margin caused by the continuing movement of Africa northwards against southern Europe. Whereas many palaeogeographies of rupturing Gondwana may seem to emphasize the relative stability of Africa compared with rapidly moving India and South America, it should be noted that southern Africa lay beneath continental ice sheets around the South Pole during late Palaeozoic times. The African plate has thus shifted through 60° of latitude over the past 300 Ma and, as its north-west tip foundered against Iberia, has also rotated 15° counter-clockwise and swallowed

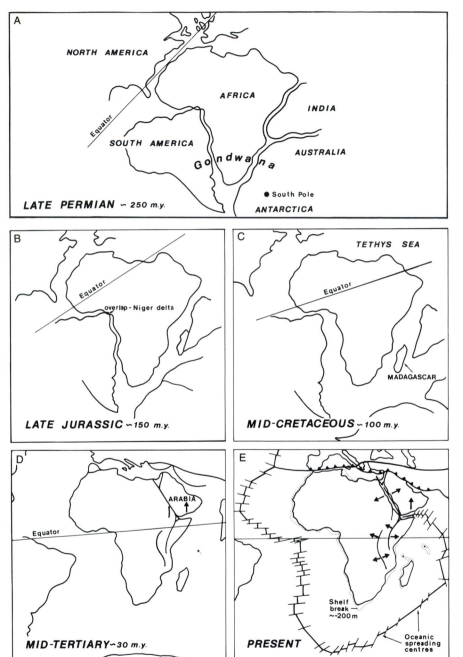

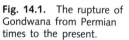

Fig. 14.1. The rupture of Gondwana from Permian times to the present.

Africa's modern coast and orientation are used for recognition relative to the Equator and Gondwana's other components. The best fit between continents occurs around –2000 m deep, down the continental slope.

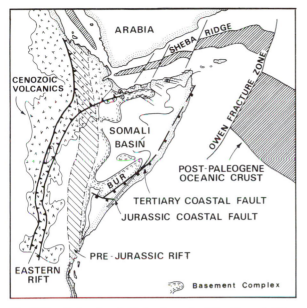

Fig. 14.2. Postpartum tectonic elements of the Horn of Africa.

part of the Tethys Sea. The effects of this northward shift and eventual collision with the soft underbelly of Europe are best seen in the Alpine zone, north of the South Atlas boundary fault between Agadir and Gabès. Farther east, tectonism in the Mediterranean embraces both compressional and extensional interactions between several miniplates, yielding frequent earthquakes and volcanic events, notably where the African plate is subducting at 2.7 cm/yr beneath the Aegean miniplate.

Postpartum Coastal Evolution

The rupture of Gondwana opened Africa's new margins to marine sedimentation, initially in narrow epicontinental seas but later in deep ocean basins. The dozen or so major basins that developed around Africa gradually filled with great thicknesses of Cretaceous and Cenozoic marine sediments. Where these basins extended landward across the present coast, both marine and terrestrial sedimentation occurred, causing variable epeirogenic seaward subsidence of the coastal margin which in turn favoured more sedimentation. The contrast in denudation rates between the evolving coastal margin and its hinterland may have caused significant isostatic uplift in response to differential unloading, reflected for example in 600 m of upwarp along the south-west coast and adjustment of the Orange River below Augrabies Falls (Gilchrist and Summerfield, 1990). Continued warping, faulting and volcanism along these margins have complicated coastal evolution, notably in

extensions of the 7000-km long East African Rift system and the Cameroon volcanic line.

Moving clockwise around Africa from the Gulf of Aden, the first basin encountered is the Somali Basin which merges southward with the Kenya and Dar-es-Salaam Basins (Orme, 1985). When the Somali Basin (Figure 14.2) originated along Gondwana's rifted margin, marine transgressions and mostly carbonate deposition followed. Late-Jurassic dislocation of the Bur basement complex isolated the interior part of this basin but carbonate deposition continued along the subsiding coast, governed by northeasterly faults. In early Tertiary times, these faults were rejuvenated and the coastal basin broke into discrete units where both clastic and carbonate sedimentation occurred. The east coast of Africa from Ras Asir to Tanzania was outlined more precisely at this time. The floor of the Somali Basin now descends from above sea-level in the Haud and Jubaland to −7000 m below sea-level at the coast. Beneath dune sands, it is filled with great thicknesses of postpartum sediments and these outcrop in coastal cliffs in northern Somalia (Jobstraibizer and Cumarshiire, 1977). Farther south, the deep Mozambique Basin was dislocated by Tertiary faulting at the south end of the East African Rift system. Onshore these faults are largely buried by Quaternary fluvial deposits while offshore 6000 m of continental and marine Cretaceous and Cenozoic sediments have accumulated in north–south grabens and the massive Zambezi cone.

The south-west coast of Africa is largely underlain by continental basement blocks downfaulted at the continental margin subparallel with the present shore. These blocks are overlain by a prograding wedge of continental debris eroded from the western Great Escarpment since it first tilted upward in the Cretaceous. Farther north, thick Cretaceous and Cenozoic clastic sediments occur in the narrow onshore portions of the Luanda, Cabinda, and Gabon Basins. Offshore, as the South Atlantic widened, early Cretaceous hydrocarbon-bearing clastic sediments gave way to evaporites and then to late-Cretaceous-Holocene marine carbonates with little terrigenous debris (Reyre, 1984). Postpartum erosion and sedimentation may also be measured by the massive submarine canyon and dissected alluvial cone off the Congo River. The Niger Basin, at the seaward end of the Benue graben, contains up to 10 000 m of Cretaceous and Cenozoic sediments beneath the delta. Farther west, the Ivory Coast Basin is only narrowly present onshore but the broad Senegal Basin plunges to −8000 m at the coast while the Tarfaya Basin, bounded inland by the Zemmour Fault, holds 10 000 m of Mesozoic and Cenozoic sediments at Cape Juby.

Plate 14.1.   Quaternary sea-level change is shown near Durban, South Africa, by horizontal beach deposits, related to Last Interglacial seas about 5 m above the present level, which truncate steeply dipping aeolian-ites deposited during a preceding low sea-level (photo: A. R. Orme).

The early evolution of the Mediterranean coast differs from the other coasts because, as an active margin, its Mesozoic and Cenozoic sediments became involved in later tectonism. Thus from Agadir to Gabès the Alpine orogeny (30–10 Ma BP) is reflected in mountain arcs and deep sedimentary basins that open seawards. Farther east, the effects of plate motion to the north are less visible at the coast. Here, the 5000-m deep Tripoli Basin is separated by NW-trending faults from the extensive Sirte Basin of Cyrenaica. Since the Neogene, the River Nile has discharged sediment beyond its delta onto a deforming abyssal cone.

Relative Sea-Level Change

The elevation of Africa's continental margin relative to sea-level largely explains the comparative importance of terrigenous and marine sedimentation along the coast after Gondwana's rupture. High relative sea-levels lead to extensive marine sedimentation, as in mid-Cretaceous times (90–100 Ma BP) when a shallow marine transgression spread across the Sahara to link the Tethys Sea with the Gulf of Guinea. After Palaeocene regression, Eocene seas again flooded the Sahara and a major gulf persisted into Miocene times in Libya and Egypt. The relative importance of clastic or carbonate sedimentation during such transgressions reflects the regional climate and how much fluvial debris is delivered to the coast. Low relative sea-levels favour terrigenous deposition as rivers extend across the continental shelf and sediments are reworked by winds.

Particular interest in relative sea-level change focuses on the past few million years, not only because

the evidence is more easily read but because recent changes aid prediction of future coastal behaviour. Along the north-west coast, for example, a deformed Pliocene marine surface up to 20-km wide, the Moghrebian rasa, has been raised 100 to 600 m above sea-level, providing a useful measure of Alpine tectonism (Weisrock, 1980). Flights of Quaternary marine terraces are commonly found at lower elevations wherever durable rock reaches the coast, for example along the Mediterranean coast at Monastir, Bizerte, and Algiers where they have been deformed by tectonism, and in Natal and Cape Province in the south (Plate 14.1). Raised coral reefs are found along deformed carbonate coasts, notably in the Red Sea. Oblique Pleistocene beach ridges and shore-parallel Holocene barriers may indicate relative sea-level changes along prograding coasts where sediments are abundant, as along the Guinea coast. Submerged beaches and aeolianites, thick terrigenous deposits on the outer shelf, and certain submarine valleys all testify to low Quaternary sea-levels. The Flandrian transgression that accompanied the melting of the last Pleistocene ice sheets and culminated around 5 ka BP, is reflected in Nouakchottian deposits reaching 2–3 m above present sea-level in Mauritania and Senegal. Whereas both eustatic and epeirogenic forces may be invoked to explain these seemingly high Holocene sea-levels, abandoned estuarine deposits can also be explained by climatically induced changes in hydrology, morphology, and sediment delivery (Ausseil-Badie *et al.*, 1991). Along the Mediterranean, relative sea-level rise during historic times is shown by submerged Phoenician tombs and Roman ports. Overall, compounded by tectonism, isostasy, and eustasy, relative sea-level may have ranged over 500 m during the Quaternary, from 300 m above to 200 m below present.

Relative sea-level change during Quaternary times undoubtedly had a major impact on the ecology and distribution of coastal biota. Eustatic changes have long been invoked to explain the distribution of coral reefs and this theme, in a modern idiom, is also applicable to mangroves and salt-marshes. The drawdown and subsequent rise of coastal waters across the shelf during glacial and interglacial stages respectively, together with related changes in turbidity and circulation, presumably had a dramatic impact on mangroves and salt-marshes, such that their modern distributions essentially postdate the Flandrian transgression. Because sea-surface temperatures fell only 2–3 °C during the Last Glacial Maximum (CLIMAP, 1981), changes in the temperature and salinity of coastal waters were probably less important (Woodroffe and Grindrod, 1991). Rich mangrove pollen counts are recorded from late Quaternary swamp deposits drowned to depths of −60 m by the Flandrian transgression along the Ivory Coast (Fredoux, 1980).

Lithology and Erosional Resistance

The resistance of the coastal zone to wave action and other processes depends in part on the composition and structure of rocks at the shore, which in turn reflect the tectonic history outlined above. Four groups of rocks are involved—Precambrian basement rocks, Palaeozoic tabular and folded cover rocks, consolidated Mesozoic and Cenozoic basinal sediments and volcanics, and mostly unconsolidated Quaternary sediments.

Precambrian rocks outcrop over 57 per cent of Africa, reflecting several major sedimentary cycles whose deposits were intensely folded and fractured, metamorphosed, and often granitized during at least eight orogenic cycles. Gneiss, schist, quartzite, and migmatite are important but post-orogenic molasse deposits and tabular to strongly folded platform covers of sandstone, limestone, tillite, and volcanics also occur. Today, these basement rocks reach the coastal zone in the Anti-Atlas of Morocco, along the Guinea Coast between Monrovia and Accra, at intervals through Angola, Namibia, and western Cape Province, and emerge from beneath later rocks in Mozambique, eastern Madagascar, and the Red Sea. Intense fracturing renders these rocks susceptible to weathering so that they rarely form high cliffs.

Palaeozoic rocks occur mostly as tabular platform covers occupying large basins between swells in the basement complex, but folded covers occur in the Hercynian Cape Fold Belt of South Africa and the Anti-Atlas of Morocco. These mostly continental rocks include the Nubian Supergroup (Cambrian–mid-Cretaceous) near the Red Sea and the 7000-m thick Karoo Supergroup (Carboniferous–Triassic) whose tillites, shales, sandstones, and basalts form bold escarpments inland from the south-east coast and seacliffs at the shore.

The erosional resistance of Mesozoic and Cenozoic rocks depends partly on their structural attitude and partly on how their composition resists weathering. Thus tabular carbonates form vertical seacliffs along arid and semi-arid coasts, for example in Libya and eastern Somalia, but weather to low relief in the humid tropics. Raised Quaternary coral reefs behave in much the same way. Less indurated Quaternary sediments rarely resist erosion.

Coastal Processes

Climatic Factors

Because climate, directly or indirectly, underpins the processes that shape the coast, it is appropriate to outline some salient factors in the present context (see also, Chapter 3). As elsewhere, Africa's coastal climates strongly reflect seasonal changes in Earth–Sun relations, particularly through the impact of these changes on shifts of the intertropical convergence zone (ITCZ), air masses, and the general atmospheric circulation. During the northern winter, the ITCZ shifts southward with the sun, extending from the Guinea coast south-east to Madagascar. Excepting the influx of cool, moist maritime polar air masses from the Atlantic into Morocco and the Mediterranean, most of northern Africa is covered by warm, dry outflowing air associated with continental tropical air masses over the Sahara and south-west Asia, the latter promoting a north-east monsoon along the east coast. This is the *jilaal* season in Somalia when hot, dry dusty surface winds averaging 15–30 km/hr flow along the coast (Orme, 1985). Warm rainy conditions associated with inflowing marine air or local convection cells prevail elsewhere over Africa, except along the fog-shrouded south-west coast where the tempering effect of the cold Benguela Current persists.

During the northern summer, the ITCZ shifts northward to the Sahel. Hot, wet conditions prevail over Central Africa and the Guinea coast, while aridity characterizes the Sahara and Mediterranean and, except for a rainy belt around the Cape, most of Southern Africa. A strong south-west monsoon flows along the northern Indian Ocean coast towards Asia. This is the *hagaa* season in Somalia when south-west winds parallel the east coast with mean velocities over 40 km/hr and a hot,

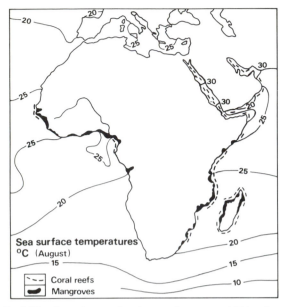

Fig. 14.3. August sea-surface temperatures around Africa showing relation to coral reefs and mangroves.

desiccating offshore wind, the dust-laden *kharif*, blankets the north coast and temperatures in the Gulf of Aden may reach 50 °C.

Inevitably solar radiation and potential evaporation within the coastal zone are highest where outflowing air and cloudless skies are most persistent, for example around the Red Sea and the Horn of Africa and off the western Sahara, and lowest along the Guinea coast. These variables, through their impact on air temperatures and available moisture, affect coastal ecology. For example, mangroves usually flourish where the air temperature of the coldest month does not fall below 20 °C and where the range is around 10 °C, although *Avicennia marina* tolerates winter temperatures as low as 15.5 °C in the northern Red Sea. Further, whereas the amount and seasonality of rainfall strongly influence the nature of terrestrial vegetation reaching the coast, even small amounts of moisture carried by onshore winds are important to coastal plants, notably along the coast of Namibia.

Oceanic Factors

The physical and chemical properties of nearby ocean waters strongly influence the ecology of the African coast (Fig. 14.3). Sea-surface temperature and salinity reflect such variables as solar radiation, air temperature, freshwater inflow, surface wind, and above all the great ocean currents that approach the coast in response to general atmosphere–ocean circulation patterns (Orme,

1982*a,b*). Sea-surface temperatures show significant differences between the east and west coasts. To the east, divergence of the warm South Equatorial Current into the south-flowing Mozambique and Agulhas Currents and the north-flowing East African Coastal Current, extended in the northern summer into the monsoon-induced Somali Current, promotes a powerful mass transfer of warm water along the entire coast. Under the south-west monsoon, the Somali Current may reach surface velocities of 3.7 m/s and transport 60×10^6 m³/s in its upper 200 m. Reversal of the Somali Current by the north-east monsoon in the northern winter restricts the weak East African Coastal Current. Beneath the latter, which is 200 km wide and 100 m deep, more saline waters from the Arabian Sea and even the Red Sea flow southward at depths of up to 700 m (Ochumba, 1988). Mean sea-surface temperatures for August, the coolest month south of the Equator, are everywhere above 20 °C and exceed 30 °C in the Red Sea, affording ideal conditions for coral and mangrove organisms. Indeed, the spread of mangrove species along the east coast is largely attributable to propagules carried on these warm surface currents. The swift Agulhas Current also transports planktonic organisms from tropical waters towards the Cape where the current is diluted by injections of cold water from the West Wind Drift (Figure 14.4).

Along the west coast, warm waters are restricted mainly to the Guinea coast from Cap Vert to the Cuanza River because the equatorward penetration of the cold Canaries and Benguela Currents squeezes warmer waters into a narrower zone than latitude alone would dictate. Indeed, cold waters and lower air temperatures along the coast of Namibia and Cape Province have long formed an effective barrier to the spread of mangrove species and other Indo-West Pacific organisms from the east coast to the west. Even in the warm waters of the Guinea coast, upwelling cold water causes a significant temperature drop and brings increased nutrients to the surface from June to August. This, together with freshwater inflows and turbidity, explains some ecological peculiarities along this coast, including the limited range and restricted biodiversity of coral reefs and coralline algae.

Sea-surface salinity (Figure 14.5) is lowest where solar radiation and potential evaporation are least and rainfall is highest, notably in the innermost Gulf of Guinea where August values below 30 parts per thousand occur. Conversely, where solar radiation and potential evaporation are highest and rainfall least, as in the Red Sea, salinities reach 40 parts per thousand. Seasonal differences occur where rainfall and stream discharge dilute salinities during the rainy season. Such changes particularly affect coastal lagoons, such as Lake St. Lucia,

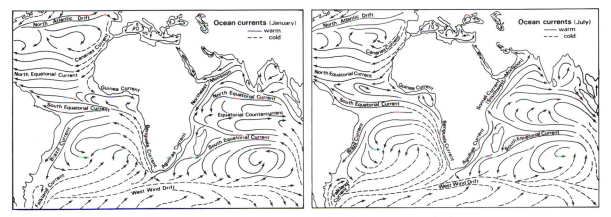

Fig. 14.4. Ocean currents around Africa during January and July.

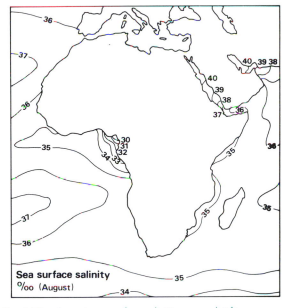

Fig. 14.5. August sea-surface salinities around Africa.

Natal, where salinities fall well below ocean values during floods but may rise to 100 parts per thousand during prolonged droughts, causing dieback of reed-swamp communities (Orme, 1975).

Excepting the Mediterranean and north-west coasts, where north-westerly swells predominate, the wave climate of the African coast is dominated by southerly swells generated by storms in the Southern Ocean (Figure 14.6). From Cape Agulhas to Cap Vert these swells are mainly south-westerly, decreasing in height northwards and in the southern summer. Because of the orientation of Africa's west coast, south-westerly swells generate north-flowing longshore currents north to Cameroon

and east-flowing currents in the Gulf of Guinea. This circulation is reinforced offshore by the Benguela and Guinea Currents. Exceptions to this pattern occur in the lee of major headlands and in currents that diverge from the nose of the Niger Delta. Throughout the Atlantic coast onshore winds are stronger and more frequent during the season when swells are highest—from December to March north of Cap Vert and from June to September farther south. In the southern winter, strong west winds, often exceeding 18 m/s, drive high seas against the Cape.

From Cape Agulhas north to the Horn of Africa southerly swells are refracted to reach the coast from the south-east but, because of the narrow continental shelf, a strong longshore component directed northward remains. This circulation is complicated by the north-east monsoon of the northern winter (Figure 14.7), by westward-moving cyclones and tropical easterlies off Madagascar and East Africa, and by swift ocean currents penetrating close to shore. Thus along the Natal coast, the wave-driven littoral drift of coarse terrigenous sediment is usually northward within the surf zone, but finer sediment that is flushed farther seaward is carried south with the Agulhas Current. Farther north, the Somali Current that sets north-east from April to October reaches surface velocities of 2–4 m/s near the coast, but its role in littoral drift is reduced by the paucity of terrigenous sediments reaching the ocean.

Using mean wave height as a surrogate for coastal energy, the highest energy occurs from Durnford Point to Cape Point. Moderately high energy occurs on both east and west coasts, northward to the Limpopo and Orange Rivers respectively, on the east coast of Madagascar, and the Atlantic coast of Morocco. Relatively low coastal energy occurs in the eastern Mediterranean,

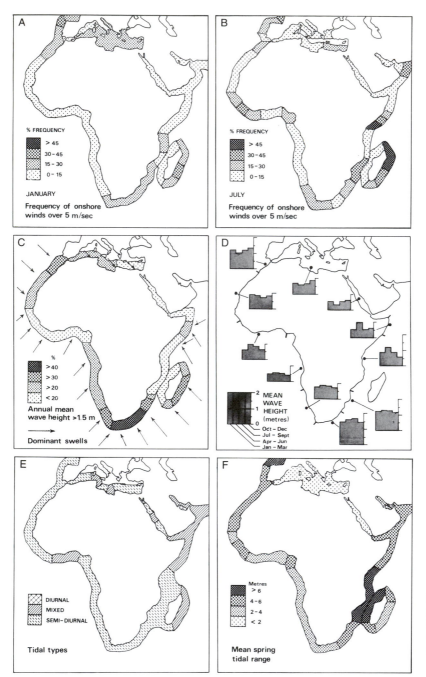

Fig. 14.6. Onshore winds, dominant swells, mean wave heights, and tidal characteristics around Africa. (*Source*: Orme, 1982.)

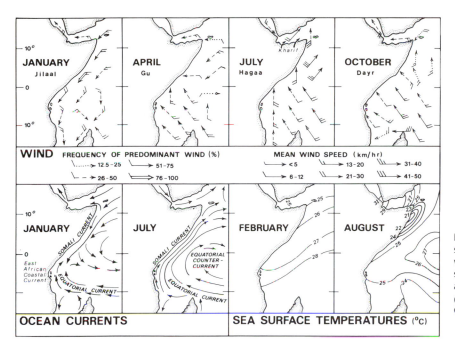

Fig. 14.7. Monsoon-induced seasonal changes in surface winds, ocean currents and sea-surface temperatures around the Horn of Africa. (*Sources: Africa Pilot*, 1967; and Orme, 1985.)

the Red Sea, the Mozambique Channel, and along the Guinea coast.

Semi-diurnal tides predominate around Africa, with mixed regimes confined to the central Mediterranean and Red seas, and the Horn, eastern Madagascar, and Guinea coasts. Mean spring tidal range is much less than 2 m throughout the Red Sea and the Mediterranean, mostly 2–6 m elsewhere, but exceeds 6 m in the Bight of Sofala off Beira where the shallow continental shelf widens to 140 km.

Hydrologic Factors

The net balance between precipitation inputs and evapotranspiration losses experienced over drainage basins inland is eventually reflected by the volume of river runoff and sediment reaching the coast, which in turn affect coastal water salinities and sediment budgets. Because of its mainly tropical location compared with other continents, Africa as a whole loses about 80 per cent of its precipitation to evapotranspiration, leaving only 20 per cent as runoff. Most of Africa's great rivers, such as the Nile, Niger, and Orange which have their sources in humid uplands, lose water downstream through seepage and evaporation, lose sediment loads through floodplain storage, and thus reach the sea much atrophied. Also, about one-third of the African continent is characterized by internal drainage whose runoff

is of no consequence to the open coast. Nevertheless, the spatial and temporal variability that typifies African hydrology provides some interesting contrasts at the coast, notably between desert coasts that rarely see river runoff and fluvial sediments, and those coasts that receive seasonal floods and abundant sediment from high magnitude storms.

In terms of how the hydrologic regime affects the coast through freshwater inputs and sediment delivery, there are basically three environments: arid, seasonally wet, and humid. Along the arid coasts of Mauritania, Namibia, and north-east Africa, ocean waters are rarely diluted with fresh river waters, surface salinities often exceed average ocean values and, lacking fluvial inputs, carbonate deposition predominates. Noteworthy exceptions include the Nile, Senegal, and Orange Rivers which, though losing 64, 53, and 44 per cent of their respective potential basin runoff to evaporation, still reach desert coasts as significant rivers. Others are less effective. In the Saloum and Casamance estuaries of Senegal, tidal flows far exceed river flows and net discharge flows upstream owing to massive evapotranspiration from wetlands. Because salinity increases inland, ponded hypersalinities form a salt wedge directed downstream towards the ocean, reversing normal geomorphic and sedimentary patterns (Barusseau *et al.*, 1985). In Somalia, Quaternary coastal dunes have long blocked the direct seaward path of the Shebele River.

Though encompassing several climatic types, the seasonally wet environments share one feature in common, namely episodic floods which can deliver large quantities of sediment to the coast and beyond. Thus, in Algeria and Morocco, tectonically active mountains and high magnitude winter storms combine to promote slope failures and floods which quickly deliver large quantities of sediment downstream. Heavy convectional summer rains in Mozambique and Natal entrain erodible soils on steep slopes below the Great Escarpment, materials important to beach and delta construction at the coast. The Tugela, which drains 28 000 km^2 of Natal, varies in discharge from 74 m^3/sec in the dry winter season to 480 m^3/sec during the summer rainy season. Its suspended sediment yield represents mechanical denudation of its drainage basin of about 375 t/(km^2 yr), much greater than the value of 75 t/(km^2 yr) for the Zambezi Basin (Orme, 1973).

Rivers draining perpetually humid environments suffer comparatively low net losses in basin runoff and their discharge at the coast commonly lowers salinities. The Congo, Africa's largest river in terms of discharge, loses only 3 per cent of its potential basin runoff in reaching the coast. On the other hand, such rivers usually drain areas of luxuriant vegetation and riparian forest, which diminish basin sediment yields and bank erosion respectively. Even where soil erosion is intense, the coincidence of humid regimes with low gradient rivers allows entrained sediments to be redeposited downstream in swamps and floodplains. Thus, suspended sediment yields from the vast Congo Basin only amount to 14 t/(km^2 yr) (Nkounkou and Probst, 1987; Walling, this volume).

Despite high sediment yields from seasonally wet environments, the vast internal drainages and often arid low gradient external basins combine to restrict Africa's total sediment yields to the coast. Thus, compared with other continents, direct wave attack on pre-existing clastic and carbonate rocks is a more important source of sediment. Further, the wide extent of arid, semi-arid and sub-humid regimes means that, for all or much of the year, ocean processes predominate over fluvial processes around river mouths. Thus the rivers of Angola and the Guinea coast are commonly deflected by barrier beaches formed by the *kalema*, a powerful combination of high surf and littoral drift produced by south-westerly swells. During drought, the estuaries of most African rivers are closed by littoral drift. This significantly alters estuarine ecology although, depending on drought duration, such changes are reversible when barriers are breached by subsequent floods.

Other Ecological Factors

The character of the African coast also reflects several other factors. Some, such as the vertical and lateral zonation of plant and animal communities, are directly related to the processes noted above. Others, such as competition and predation among species, are often less readily explained. Human activity is an increasingly pervasive influence.

The vertical zonation of plant and animal life is expressed by the distinction between marine communities that live permanently submerged in sea water; littoral communities in the intertidal zone; beach, dune, and cliff communities living just beyond the direct reach of the sea; and terrestrial communities that penetrate towards the shore. The width of these zones and the quality of life therein reflects such controls as tidal range, slope, exposure to waves and currents, and the effect of river discharge on sediment delivery, water quality, and nutrient availability. Steeply shelving coasts compress the above zones, whereas broad intertidal shallows favour mangrove or salt-marsh communities, although neither develops on exposed coasts where wave action inhibits the establishment of seedlings. Zonation is frequently dynamic in time and space, as shown by seasonal changes in the extent of salt, brackish, and freshwater habitats.

Edaphic factors such as soil texture, water chemistry, and nutrient availability play a major role in coastal ecology. Sandy substrates predominate around Africa, often over long distances in north-west Africa, Namibia, and Somalia where extensive dunes occur. Muddy substrates are important along low-energy lowland coasts where fine river sediments settle or carbonate muds accumulate. Where mangroves are well developed, for example along sheltered parts of the Guinea and Indian Ocean coasts, the roots of these plants encourage further deposition. Conversely, the absence of mud along rocky and arid coasts edaphically limits mangroves. Rocky shores are locally significant, notably in West Africa and the Cape where intertidal ecology has been much studied, but cliffed coasts need further research.

Competition and predation among plants and animals play an important role in African coastal ecology and, although the linkages may be obscure, these forces are often triggered by changes in physical factors. For example, the prolonged closure of estuaries by littoral drift inhibits the exchange of water and biota with the sea, affecting estuarine salinities and interrupting the feeding and breeding patterns of plankton, prawns, fish, and organisms even farther up the food chain. Trapped inside a closed estuary, sharks must compete with other carnivorous fish and crocodiles for available food resources (Orme, 1975).

Although the African coast has suffered less from human activity than that of Europe or North America, some historic impacts are noteworthy and the pace of change has recently accelerated. The mangal vegetation of the Indian Ocean coast has suffered many centuries of exploitation, mangrove poles or *boritis* being shipped from such ports as Lamu, Kenya, to treeless regions of south-west Asia for building purposes. Although this trade has now declined, mangrove harvesting continues and where 25–50 m trees once stood, 2–5 m shrubs now stand. During and since the colonial interlude, plantations of coconuts and other commercial crops have often replaced the natural vegetation. Further, land-use practices inland have often led to extensive soil erosion and increased sediment yields. At the coast, sediment plumes have been dispersed by waves and currents, blanketing existing marine communities while creating fresh habitats among extensive mudflats and new beach-dune ridges. In northern Natal, for example, present sediment delivery rates are two to three times the average for the past 5000 years, owing in part to overgrazing and accelerated erosion of coastal watersheds (Orme, 1975). Along Somalia's east coast, Pleistocene coastal dunes have been destabilized by overgrazing of vegetation by cattle, goats, and camels, aggravated more recently by tillage practised by an increasingly desperate local population swelled by refugees. Destabilization has led to severe gullying and dune reactivation, causing the dune front to invade valuable farmlands in the Shebele valley (Orme, 1985). Elsewhere, dunes have been disrupted by mining for heavy minerals while beaches and estuaries near major ports have been progressively consumed by harbour and urban development. Finally, the construction of major dams inland, such as the Aswan High Dam (completed 1964, 162 × 10^6 m^3 capacity) on the Nile, the Kariba (160 × 10^6 m^3) and Cabora Bassa (63 × 10^6 m^3) dams on the Zambezi, and the Akosombo Dam on the Volta, has had significant coastal impact by reducing river discharge and sediment yields (see Walling, Chapter 6, this volume).

The Mediterranean Coast

Africa's 5000-km long north coast (Figure 14.8) is physically divisible into two contrasting parts: a 3300-km eastern segment in Egypt, Libya, and southern Tunisia underlain by little disturbed Cenozoic sediments that reach the shore in low cliffs or beneath barrier beaches, lagoons, and dunes; and west of Cap Bon a 1700-km western segment in northern Tunisia, Algeria, and eastern Morocco where Alpine tectonism has caused cliffed mountainous stretches to alternate with small coastal basins (Figure 14.2). Throughout this coast, prevailing north-west winds generate predominant north-westerly swells and wind waves, strongest in winter, which in turn promote net east-flowing littoral drift. Both wave and current energy diminish eastward. North-easterly swells are locally important in summer. With a mean tidal range of 0.2–0.5 m, tidal forces are insignificant, although in the shallow Gulf of Gabès the range rises to 1.8 m and swift tidal currents flow around Djerba. Contrasts between the eastern and western Mediterranean coasts are reinforced by the precipitation regime: annual rainfall over the mountainous west reaches 500–1000 mm, mostly in winter when floods deliver much sediment to the coast. As rainfall diminishes eastward, arid or semi-arid conditions prevail at the coast and, excepting the Nile, perennial drainage is lacking.

Geomorphology

The 12 500 km^2 Nile Delta dominates the Egyptian coast, extending over 280 km and contributing its dark sediments, enriched in Fe and Al from the Ethiopian Highlands, to beaches eastward into Israel. Athough the delta forms the outlet for a vast 2.7 × 10^6 km^2 drainage basin with an average precipitation of 870 mm/yr, much potential basin runoff is removed by evaporation and seepage, and by irrigation and reservoir construction such as the Delta Barrages (1881), the Aswan Low Dam (1902), and the Aswan High Dam (1964). Thus discharge through the delta now averages only 1500 m^3/sec and with sediment discharge, which averaged 80 × 10^6 t/yr before completion of the Aswan High Dam, much reduced, the delta coast suffers locally severe erosion. In early historic times, the Nile delta contained seven distributaries but, owing to reduced flows, sedimentation and human activity, only the Damietta and Rosetta branches now function and even their prominent outlets are now eroding at 60 m/yr for want of sediment (El-Ashry, 1985). High density magnetite, ilmenite, zircon, and garnet tend to remain as black-sand placers on the eroding beaches while lighter quartz, feldspar, and hornblende move away (Frihy and Komar, 1991). Elsewhere, a distinctive 40-km wide littoral zone, the *Barari* (barren) fronts the delta, comprising salt-marshes, salt-flats, and shallow lagoons behind discontinuous barrier beaches. The largest lagoons are Manzala (1450 km^2), Burullus (560 km^2), Idku (140 km^2), and Maryut (200 km^2). Over the past 8000 years, delta subsidence has ranged from 1mm/yr near Alexandria to 5 mm/yr near

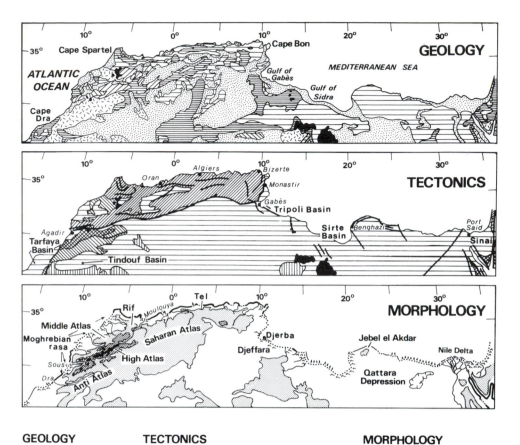

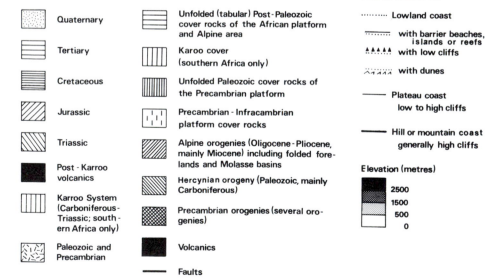

Fig. 14.8. The Mediterranean coast: geology, tectonics, and morphology.

Port Said (Stanley, 1990). During historic times, while delta subsidence was offset by sedimentation, coastline location did not change much but, more recently, reduced sediment delivery has caused several areas to subside below sea-level, expanding lagoons and aggravating erosion.

Eastward from the delta, the north coast of Sinai is a broad sandy plain with prominent dunes and a 500-m-wide barrier fronting Bardawil Lagoon along whose shallow hypersaline margins gypsum and halite readily precipitate. Westward, dark Nile sediments decrease in importance while carbonate grains increase. Several beach-dune ridges separated by elongate sabkhas parallel the coast, reflecting episodic barrier formation during later Quaternary times. In Cyrenaica, the Jebel el Akdar limestone plateau reaches the coast in rocky cliffs and narrow terraces, but from Benghazi to Cap Bon the low-lying coast is again dominated by barriers, sabkhas, sand dunes, and desert. The sheltered Gulf of Gabès is a low energy environment where north-easterly waves dissipate across the broad shallow continental shelf and inshore across submarine meadows of *Cymodocea nodosa* and *Posidonia oceanica*. The 150-km^2 Sabkha el Melah in southern Tunisia affords an excellent example of Holocene evaporitic lagoonal sedimentation in which marine detrital carbonates, euxinic beds, magnesian carbonates, gypsum, polyhalite, and halite have successively accumulated over the past 6000 years (Busson and Perthuisot, 1977).

West of Cap Bon, coastal character changes dramatically as Alpine structures form prominent cliffs and subsiding basins, and as the seasonally wetter, stormier climate mobilizes fluvial and coastal sediments. In northern Tunisia, 400 m of Quaternary littoral sediments have accumulated beneath the subsiding Medjerda delta whose outlet has changed frequently in historic times in response to flood activity, as in March 1973 when flows reached 3500 m^3/sec. (Paskoff, 1978). Farther west, along the Algerian Tell and Moroccan Rif, structural basins floored with Quaternary sediments reach the coast through swampy plains, such as the Fetzara, Soummam, and La Macta lowlands which open respectively onto Skikda, Bejaia, and Arzew Bays. Wadis may discharge large sediment loads to the coast in winter, to be partly worked back onshore in aeolian dunes. The Isser and Soummam yield 4.8 and 4.0×10^6 t/yr respectively, while sediments transported by the Moulouya from the High Atlas contribute to impressive shingle beaches east of Nador, Morocco (Mahrour and Dagorne, 1985). Between the basins, Alpine structures form massifs up to 1000 m above sea-level which reach the shore in active seacliffs and relict bevels 100–300 m high and

then, beyond the narrow continental shelf, plunge more than 2000 m down the Habibas and related escarpments. These bold picturesque cliffs are lithologically varied, with crystalline basement, Mesozoic limestones and sandstones, Cenozoic molasse deposits, and volcanics all represented.

Ecology

Barrier beaches, abandoned beach ridges, aeolian dunes, salt-marshes, barren salt-flats, lagoons, and seacliffs provide distinctive habitats throughout the Mediterranean but there is little ecological variety along the coast, largely because limited latitudinal range restricts fluctuations in solar radiation, air and sea temperatures, evaporation and salinity. Seacliffs are an exception because increased storminess and precipitation toward the west affect the size of the splash zone and the density of shrubs and trees on higher slopes.

Dry coastal formations are exemplified by sand dunes and abandoned beach ridges along the coast west of the Nile Delta. Floristically, this is the richest part of Egypt, comprising about 1000 species of flowering plants or 50 per cent of the Egyptian flora (Ayyad, 1973). The dunes are fashioned from coarse carbonate grains by predominant north-west winds and then cemented by secondary calcite. Initially these dunes are unstable and only a few plant species are able to grow, specifically *Ammophila arenaria* and *Euphorbia paralias* that resist burial and bind the sand. With increasing stability, shelter and soil development, the number of species rises, with *Crucianella maritima*, *Launaea resedifolia*, *Agropyron junceum* and *Echinops spinosissimus* eventually achieving co-dominance with *Ammophila arenaria*. Destruction by animals and people has, however, blurred the relationship between species distribution and soil character. Behind the dunes, abandoned beach ridges, 20–40 m above sea-level, composed of 40–60 per cent oolitic carbonate grains, form pale brown loamy soils. *Thymelaea hirsuta* and *Gymnocarpos decandrum* are dominant on shallower soils, *Plantago albicans* and *Asphodelus microcarpus* on deeper soils.

Wet coastal formations of the Mediterranean shore are typified by salt-marshes that continue southward to Mauritania and the Red Sea. It is often difficult to determine where salt-marsh ends and salt desert begins. Along the coast of Tunisia and Algeria, *Halocnemon strobilaceum* is a widespread pioneer, trapping sediment that is then colonized by *Cutandra memphitica*, *Bassia muricata*, and *Traganum nudatum*. In Tunisia the Halocnemetum zone is succeeded by a *Limoniastrum quyonianum* community in which *Zygophyllum album*, *Nitraria retusa*, and *Suaeda vermiculata* occur. Around Oran, Algeria, the pioneer community is dominated by *Salicornia arabica* and *S. herbacea*, succeeded by a floristically rich Limonietum dominated by the pan-Mediterranean *L. sinuatum* and localized *L. sebkarum* and *L. spathularum*. On solonchak soils, the primary colonist is either the shrubby perennial *Arthrocnemum glaucum* or *A. macrostachyum*, succeeded landward by *Salicornia fruticosa, Centaurium spicatum* (with submergence), or *Monerma cylindrica* (salt desert). *Suaeda fruticosa* forms an important landward community in Morocco. *Spartina maritima* dominates river mouths in Algeria, associated with *Puccinellia palustris, P. distans*, and *Crypsis aculeata* (Chapman, 1977). This Spartinetum is essentially an eastern Atlantic community that has spread into the Mediterranean.

The Atlantic Coast

The Atlantic coast of Africa may be divided into three broad sectors: the semi-arid to arid north-west coast from Cap Spartel to Cap Vert dominated by the cool Canaries Current, north-westerly swells and south-flowing littoral drift; the humid tropical coast from Cap Vert to beyond the Congo River, dominated by the Guinea Current, southerly swells and east-flowing littoral drift; and the sub-humid to arid south-west coast from central Angola south to the Cape characterized by the cool Benguela Current, south-westerly swells and strong north-flowing littoral drift (Figure 14.9).

Geomorphology

Despite its underlying Hercynian and Alpine framework, the Moroccan coast from Cap Spartel to Cap Dra has a relatively smooth outline. This is because orogenic structures were largely planed off by late Cenozoic marine erosion, forming the Moghrebian rasa, since deformed to between 100 and 600 m above sea-level, and several less deformed Pleistocene terraces that reach the sea in low cliffs (Weisrock, 1980). Winter floods transport large sediment loads through the Sebou, Rharb, and Sous basins to feed local beaches and dunes up to 80 m high which, shaped by north-west winds in winter and north-east trade winds in summer, enclose many lagoons, marshes, and sabkhas.

From Cap Dra to Cap Vert, the coast overlies the Tarfaya and Senegal structural basins whose thick postpartum sediments, planed by Mohgrebian and later transgressions, form low coastal cliffs. Elsewhere, Saharan sands blanket the coast along a broad front. Lacking both rainfall and a mountain backdrop, water and fluvial sediment discharge are negligible except during rare storms. Aeolian processes have long dominated this coast, driven by the *alisio*, a persistent northerly onshore wind, and the *irifi*, a hot easterly wind from the Sahara. One must distinguish, therefore, between the vast amount of Saharan sand being worked toward the coast from inland sources and the more restricted coastal dunes being reworked by onshore winds. Of the former, generations of late Cenozoic dunes occur, ranging from fixed shore-parallel ridges at Afrafir and Aguerguer which block coastal wadis in Saraoui, to linear dunes (*draas*) and superimposed barchans trending obliquely across the Mauritanian coast between Cap Blanc and the Senegal River, notably in the reactivated Azefal and Akchar ergs (Kocurek *et al.*, 1991). Saharan sands now shed 5 to 13×10^6 m^3 of quartz sand to the continental shelf annually, but during low Pleistocene sea-levels sand discharge into the Atlantic was 5–10 times greater (Sarnthein and Walger, 1974). Quaternary aeolianites lie below −50 m off Cap Vert and aeolian dust is found far beyond the shelf break. This sediment source, together with local carbonate and biogenic products, creates a distinctive sedimentary environment off the Sahara, in contrast to areas north of the Sous and south of the Senegal River where fluvial sediments are more important. Coastal dunes of reworked inland and littoral sands, 200–2000 m wide and 30–40 m high, trap sabkhas (*niayes*) seaward of Saharan draas in southern Mauritania and Senegal.

The 4250 km^2 Senegal Delta forms the outlet for a 196×10^3 km^2 drainage basin whose mostly summer rains yield an average rainfall of 1381 mm/yr. Although much upland discharge is lost downstream through evaporation, annual discharge through the delta averages 870 m^3/s but ranges from negligible during prolonged drought to 3700 m^3/s during floods (Guilcher, 1985). The delta formed in a bay created by the Flandrian (Nouakchottian) transgression but, as the delta plain accreted, high wave energy and south-flowing littoral drift along the open coast deflected the Senegal river mouth 120 km southward behind a massive barrier spit. The distal end of this barrier, the unstable 30-km-long Langue de Barbarie, has been breached twenty-five times by storm waves or river floods since 1850. Although spring tidal range is only 1.2 m, the low gradient of the Senegal River allows tidal forces to penetrate over 400 km upstream, and salinities within the delta 60 km from the sea may reach 30 parts per thousand. Southward littoral drift along the open coast ranges from 0.5 to 1.0×10^6 t/yr, part of which is lost down the Kayor submarine canyon which heads close inshore north of Cap Vert.

From Cap Vert to the Niger Delta, cliffed coasts formed in hard Precambrian and Palaeozoic rocks and later basinal sediments alternate with lowlands typified by

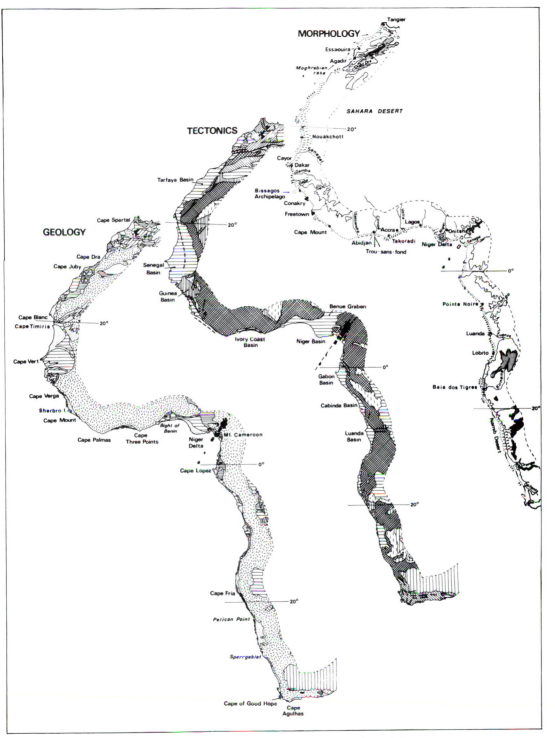

Fig. 14.9. The Atlantic coast: geology, tectonics and morphology.
(See Figure 14.8 for key.)

mangrove swamps and prominent beach ridges. Morphology is much influenced by diverse lithologies and shore-parallel faults: Palaeozoic rocks emerge along the south coast of Guinea-Bissau and at Cap Verga; basic intrusives form the Kaloum and Freetown peninsulas; Precambrian basement rocks underlie the coasts of Liberia and western Ivory Coast and reappear with Palaeozoic rocks along the low-cliffed coast of Ghana as far east as the Volta Delta.

The coast between Cap Vert and Sherbro Island is transitional as the more northerly forces and arid conditions in the north weaken in the face of more southerly forces drawn into an increasingly humid coast. Tropical rainfall of 2000–3000 mm/yr, enhanced by south-west winds drawn onshore during the northern summer, is reflected in high river discharge and sediment delivery. Rocky headlands apart, the coast is dominated by broad, often deltaic estuaries, such as those of the Casamance, Geba, Konkoure, and Scarcies Rivers, flanked by intricate networks of tidal creeks, extensive mangrove swamps and supratidal mudflats or *tannes*. The continental shelf is 200 km wide off the Bissagos Archipelago, near which the 60-km-wide Great Geba Flat is less than 10 m deep. This shelf contains many drowned river channels and submerged deltas, and old barriers and seacliffs indicating low Pleistocene sea-levels. Coral 20 750 years old has been dredged from the shelf-edge. The subsequent Flandrian transgression may have risen 2–3 m above the present, cliffing older estuarine deposits and creating the *tannes* (Debenay and Bellion, 1983). Alternatively, relict Holocene estuarine features may have formed as a result of climatically induced changes in hydrology and sediment delivery (Ausseil-Badie *et al.*, 1991). Today the surface Canaries Current, strong tidal currents, and north-westerly swells transport fine terrigenous sediment south-eastward across this shelf, while vigorous bottom counter-currents and southerly swells sweep material northwards, thereby maintaining high turbidity in shelf waters. Farther south, powerful onshore swells and seasonally reversing littoral drift have combined with terrigenous sediment to form a series of Holocene beach ridges backed by lagoons and mangrove swamps, notably on Sherbro Island and Turner Peninsula, Sierra Leone, and across many Liberian estuaries.

From Cape Palmas eastward, the E–W alignment of the Guinea coast causes south-south-westerly swells to promote east-flowing littoral drift whose effect is conditioned by arcuate coastal orientation and available sediments. Thus, although swells approach the coast at 45° between Cape Palmas and Fresco, littoral drift amounts to only 0.2×10^6 m³/yr because Precambrian basement rocks reach the coast and sediment yields are small. Along the central Ivory Coast, the presence of erodible Tertiary rocks and larger fluvial inflows increases littoral drift to 0.8×10^6 m³/yr, forming massive barrier beaches in front of elongate lagoons. Drift then decreases towards Cape Three Points as waves approach perpendicular to the shore and strong rip currents develop (Hinschberger, 1985). There the sequence begins again: there is little transport at Takoradi, more at Sekondi, and much from Accra to Lagos where barrier beaches again block many river mouths (Martin, 1971). Where littoral drift is interrupted, notably by upcoast jetties protecting the Vridi Canal and Lagos Harbour, downdrift beach erosion is significant. Completion of the Akosombo Dam in 1961 has controlled 99 per cent of the Volta drainage basin, further aggravating erosion downdrift from its delta. Offshore the continental shelf is relatively narrow (10–70 km), smooth, and featureless. Exceptions include the Trou-sans-Fond, a submarine canyon that heads at −10 m off the Vridi Canal but descends to −2000 m and traps 0.4×10^6 m³/yr of littoral drift; the steep deforming front of the Volta delta; and a series of submerged beach-aeolianite ridges that indicate a late Quaternary regression to a shelf break at −120 m.

The Niger Delta is the outlet for a drainage basin covering 1.11×10^6 km² whose main artery, the 4460-km Niger River, rises in the Guinea mountains only 250 km from the Atlantic. After separation of the South American plate, the delta developed in a Tertiary basin whose shoreline lay near Onitsha, 250 km inland from the modern coast. This basin acquired its present form during a later Quaternary pluvial stage when the NE-flowing Soudanese Niger overflowed near Bourem, Mali, into the SE-flowing Nigerian Niger. Subsequent accretion, subsidence, and eustatic oscillations culminated in the Flandrian transgression which drowned the outer 50 km of the delta's late Pleistocene distributary system. The delta now covers 28 827 km², of which 19 135 km² lie above sea-level. About 500 000 km³ of Eocene–Holocene sediment fill the delta to a maximum thickness of 8 km (Hospers, 1971). These sediments are broken by arcuate gravity faults downthrown toward the gulf. Present yearly discharge of freshwater into the delta is about 200×10^9 m³ and of sediment about 18×10^6 m³, mostly as silt and clay. Water and sediment are then discharged radially but unequally through a dozen major distributaries into the gulf. Depositional environments and corresponding sedimentary facies are distributed concentrically within the delta (Allen, 1970). The delta plain comprises upper and lower floodplains, meandering tidal creeks, mangrove swamps, and barrier beaches. The submerged delta comprises river-mouth bars, delta-front ramps, prodelta slope, and relict Pleistocene littoral deposits. Prevailing south-westerly swells strike the delta nose symmetrically, causing divergent littoral drift, east toward the Cross River and north-west toward the Benin River. Submarine canyons channel 1 million m^3 of sediment a year from these drifts into fans beyond the delta foot. Mangrove clearance and natural subsidence augmented by subsurface oil and gas withdrawal have increased erosion to 30–40 m/yr along the exposed muddy coast north-west of the Benin River (Ebisemiju, 1987).

Southward from the cliffed volcanics of Mount Cameroun (4070 m), the west coast of African comprises several narrow postpartum basins separated by basement rocks and veneered with Quaternary sediments. Powerful south-westerly swells promote mostly northward littoral drift, reflected in massive barrier spits fronting elongate lagoons and swampy estuaries. The Congo River is a notable exception because its large discharge jets up to 50×10^6 m^3 / yr of sediment seaward into a submarine canyon, denying most of this load to the littoral drift. The steep V-shaped canyon descends to –2700 m, beyond which a fan valley with leveed distributaries slopes to –4900 m. Rocky coasts are rare north of the Congo estuary, but farther south postpartum carbonate and clastic sediments form seacliffs 50–150 m high fronting parts of the Luanda Basin. At major river mouths, however, northward-oriented spits (*restingas*) may reach exceptional lengths—Luanda spit is 12 km long, Palmeirinhas spit 34 km, Lobito spit 8 km, and before breaching in 1962 the 60-m-thick spit enclosing Baia dos Tigres was 37 km long (Guilcher *et al.*, 1974). The Kunene River is a mirror image of the Senegal River, losing perennial discharge from humid uplands *en route* to an arid coast where it is deflected northward by strong swells. Trade winds drawn onshore form the SSW shape barchanoid dunes along the coast.

Although Precambrian basement rocks and Cretaceous basalts underlie much of the coast of Namibia and may form rugged cliffs, they are usually masked by relict alluvial fans (van Zyl and Scheepers, 1992) and by active and relict Quaternary coastal dunes, notably along the 400-km seaward edge of the Namib sand sea and again for 150 km north from the Orange River. This river, though blocked by a coastal barrier during the dry season, has long been the source of the highly mobile beach and dune sand that moves north with the dominant south-westerly swells and winds, prograding the coast and deflecting the mouths of the Swakop and Kuiseb rivers. These and other underfit streams now blocked by drifting sand supplied more sediment to the coast during Pleistocene pluvial stages. Between the Orange and Olifants Rivers, short ephemeral streams now supply little sand, exposing cliffs of Precambrian basement rocks to wave erosion. Notable among the Quaternary shorelines of this coast are the diamantiferous terrace deposits that range from 90 m above to –60 m below sea-level along the Sperrgebiet and trace their provenance to switching Cenozoic outlets of the Orange River (Dingle and Hendey, 1984). The shelf varies from less than 30 km wide off rocky coasts to more than 160 km wide off the Orange River where terrigenous sediment has formed a large submarine delta. Off the Kunene sand sea, terrigenous sediment from the Kunene River is also common. Off the intervening Namib sand sea, however, terrigenous sand is limited and Holocene sediments are mostly represented by diatomaceous muds and foraminiferal sands. Two shelves separated by a 50-m-high submarine cliff occur south of Cape Fria, reflecting differential erosion and accretion of shelf sediments.

Ecology

With the cool Canaries Current offshore, the north-west coast is characterized by frequent advection fogs, high inversion mists, and cool temperatures. Wet formations are confined to small salt-marshes similar in flora to Mediterranean marshes. Dry formations on the strand and coastal dunes merge inland with typical Saharan associations of chenopods and acacias. Cap Vert is the southern limit of cold-water organisms and cool coastal floras. The Mauritanian coast between Cap Blanc and Cap Timiris, part of the Atlantic flyway for birds migrating between Greenland, Europe, and tropical Africa, became a National Park in 1976.

From Senegal to central Angola annual average air and soil temperatures increase, temperature range decreases, relative humidity and precipitation rise, and fluvial sediments reach coasts bathed in warm ocean currents. Two ecological features are noteworthy: the flora and fauna of open rocky shores and sandy beaches, and the mangrove vegetation of muddy estuaries and lagoons. On open rocky shores the supratidal zone is usually bare except for halophytes such as *Sesuvium portulacastrum* which grows in soil-filled cracks. Where more soil occurs, evergreen coastal shrubs are found. The supralittoral fringe is characterized by the snails *Littorina punctata* and *Tectarius granosus*, and such algae as *Entophysalis crustacea* (Lawson, 1966). The main littoral zone may be divided into an upper belt of barnacles (*Chthamalus* spp.) together with *Siphonaria pectinata*, *Nerita senegalensis*, *Ostrea cucullata*, and such algae as *Enteromorpha* and *Ulva*, and a lower belt of pink encrusting sheets of lithothamnia algae and such animals as

Mytilus perna, Patella safiana, Fissulrella nubecula, and *Thais* spp. The infralittoral fringe is typified by algae such as *Sargassum* and *Dictyopteris deliculata,* and the sea urchin *Echinometra lucunter.*

Open sandy beaches are generally bare of plant life, although intertidal sand is often coloured green by *Melosira* diatoms while sea-grasses such as *Cymodocea* may be exposed at very low water. The backshore vegetation comprises specialized flowering plants. Steep coarse sandy beaches are almost devoid of animal life except for the ghost crab *Ocypoda cursor.* Flatter, fine sandy beaches have a much richer fauna including *Ocypoda cursor* and *O. africana,* the mole crab *Hippa cubensis,* the polychaete *Nerine cirratulus,* the isopod *Excirolana latipes,* and the small lamellibranch *Donax pulchellus* (Lawson, 1966).

Six mangrove species occur along the West African coast between Mauritania and central Angola, namely *Rhizophora racemosa,* the common stilt-root pioneer of low intertidal muds which unless coppiced may grow to 40 m in height, *R. harrisonii* and *R. mangle* which grow to 5–7 m on wetter and drier more saline substrates respectively, *Avicennia africana* (similar to the American *A. germinans*) in the higher intertidal zone, and *Laguncularia racemosa* and *Conocarpus erectus.* These species differ from those found along the east coast of Africa, showing affinity with mangroves of western Atlantic shores.

From Cap Timiris to the Gambia River, the long dry season and low winter temperatures (10–15 °C) restrict *Rhizophora* so that *Avicennia* dominates and barren *tannes* lie inland. Southward from the Gambia River, however, co-dominant *Rhizophora-Avicennia* swamps line most estuaries, behind which herbaceous *tannes* are characterized by the halophytes *Paspalum vaginatum, Sesuvium portulacastrum,* and *Philoxerus vermicularis.* In Sierra Leone, large areas of former mangrove swamp have been poldered for rice cultivation. Along the Guinea coast, *Rhizophora racemosa* and *R. harrisonii* typify lagoons with continuous open access to the sea, while shrubby *Avicennia* occurs around seasonally closed lagoons. Vertical zonation is reflected by a lower zone of the red algae *Bostrychia* and *Calaglossa,* a middle zone of the oyster *Ostrea tulipa* in *Rhizophora* stilt roots, and an upper zone of barnacles *Chthamalus rhizophorae* and the snail *Littorina angulifera.* Mangroves reach their richest development between the Niger and Ogooue Rivers; all species are present and riparian mangroves extend 100 km up the Niger. The transition to fresh water is revealed where *Rhizophora mangle* is invaded by the screw pine *Pandanus candelabrum* and the palms *Phoenix spinosa* and *Raphia.* Where protected from cutting, *Rhizophora racemosa* may reach

50 m and *Avicennia africana* 30 m. The rich animal life includes two crocodiles, the Mona monkey, mudskipper fish, pelicans, fish eagles, sandpipers, terns, and crocodile birds. This broad mangrove belt narrows south of the Ogooue River and in northern Angola forms only a sporadic fringe along estuaries. As coastal waters cool, *Rhizophora* species are replaced as principal colonists by *Avicennia* (on mud) and *Laguncularia* (on sand). Mangroves disappear south of Lobito.

From Lobito to the Olifants River, negligible rainfall combines with the cold Benguela Current to promote semi-arid to arid conditions under which desert biota reach the shore. Plants of this region include *Mesembryanthemum salicornioides, Hydrodea bossiana, Drosanthemum paxianum,* and dense cushions of *Aizoon dinteri* and *Zygophyllum simplex.* Advection fogs that form over the Benguela Current may be driven 100 km inland by onshore winds, only to dissipate and curl back toward the coast during the day. At Pelican Point on the central Namib coast 200 fog-days may occur annually but this declines inland as orographic enhancement raises the moist marine air to low stratus. Seasonal and diurnal variations in the fog climate increase inland while fog density decreases until only 5 fog-days occur annually 100 km from the coast (Olivier, 1993). These fogs are ecologically important. With 200 foggy days, fog precipitation (dew) amounts to 50 mm/yr. On open ground this soon evaporates as the sun rises and is of little use to plants. Where driven against cliffs, however, fog precipitation may seep into crevices and be more useful to plants. Seacliffs thus tend to have a richer flora than open strand and dune. Despite the fog, the coastal Namib is essentially an environment of low habitat diversity, low primary productivity, and therefore few strongly localized primary consumers (herbivores). However, a surprising array of secondary consumers (carnivores) such as black-backed jackal (*Canis mesomelas*), brown hyaena (*Hyaena brunnea*), and less common cheetah and lion walk the beach, subsisting largely on seals and seabirds (Nel, 1992).

The Southern Cape

Africa's southern tip is in many respects transitional between Atlantic and Indian Ocean influences but it contains sufficient peculiarities, for example in the Cape Fold Belt and the distinctive phytogeography of the Cape flora, to be treated separately.

Geomorphology

Between the Olifants and Tugela rivers, strong southerly swells and winter storms have fashioned a rugged, often cliffed coast from Precambrian and Palaeozoic rocks of the Cape Fold Belt and thick formations of the Karoo Supergroup, interspersed with sandy bays of varying size. Thus massive sandstone-on-granite cliffs of the Cape Peninsula are separated from fold mountains farther east by the extensive, dune-mantled Cape Flats. Between Cape Agulhas and Algoa Bay, the coast cuts obliquely across the easterly strike of the Cape Fold Belt, causing cliffed anticlinal ridges to alternate with synclinal corridors fronted by barrier beaches and extensive active and relict dune systems. In the Alexandria Basin, a 2–3-km wide transverse dunefield is actively transgressing inland along a 50-km broad front driven before dominant WSW winds off Algoa Bay. Sand enters this dunefield at a rate of 375 000 m^3/yr and leaves at 45 000 m^3/yr, for a net gain of 330 000 m^3/yr (Illenberger and Rust, 1988). Farther north-east, gently tilted rocks of the Karoo Supergroup form majestic cliffs. Perennial and intermittent rivers deliver much sediment to local beaches, from winter (May–August) rains in the Cape and summer (December–March) thunderstorms in Natal. Rivers flowing swiftly from the Drakensberg enter the sea through Pleistocene channels excavated to 50–100 m below present sea-level, now partly filled with Holocene estuarine sediments (Orme, 1974, 1976; Cooper, 1993). Over the past 200 years, sediment yields have been accelerated by soil erosion and diminished by dam construction.

Quaternary marine terraces occur locally to over 60 m but are often poorly preserved; a high Holocene shoreline may occur around Langebaan Lagoon (Fleming, 1977). Beach deposits and aeolianites related to low Quaternary sea-levels are preserved offshore as submerged ridges (Plate 14.1; Orme, 1973). Relict beachrock, ranging in age from 25 ka to 1 ka BP, forms a broken pavement along much of the shore. It typically consists of quartz grains and skeletal fragments cemented by micrite or aragonite overlain by a laminated calcrete. Off Cape Agulhas, the continental shelf forms the 240-km-wide Agulhas Bank, relatively featureless except for a shelf edge at −110 to −380 m which is cut by several canyons. The bank comprises over 6000 m of postpartum sediments lying on basement rocks and during low Quaternary sea-levels was exposed over 22 500 km^2.

Ecology

The southern tip of Africa constitutes a distinct ecological entity, separating the biota of the east and west coasts with a region of temperate forms. Evergreen sclerophyllous shrubs of the fire-prone coastal *fynbos*, dominated by Ericaceae, Proteaceae, and Restionaceae, favour slopes with acidic sandy soils and low nutrient

status (Cowling, 1992; Allen, this volume). Locally, luxuriant coastal forests also occur, dominated by yellowwood (*Podocarpus* spp.) at Knysna, euphorbias and aloes at Alexandria, and white milkwood (*Sideroxylon inerme*) near Cape Agulhas. The *fynbos* apart, this region's boundaries are less well defined. For example, whereas mangroves extend only as far south as Lobito and the Kei River on the west and east coasts respectively, impoverished tropical marine faunas extend down the west coast to Walvis Bay and down the east coast to the Cape Peninsula. The latitudinal imbalance in both wetland vegetation and intertidal biota is largely explained by the contrasting effect of the cool Benguela and warm Agulhas Currents along their respective coasts. These elements will suffice to illustrate some aspects of the Cape's coastal ecology.

Because temperatures are too cool for mangroves, wet formations around the southern Cape are represented by salt-marsh. Above an infralittoral fringe of eel grass *Zostera capensis*, the low marsh is dominated by *Spartina maritima*, associated with *Bostrychia* red algae, the mud prawn *Upogebia africana*, the crabs *Sesarma catenata* and *Cyclograpsus punctatus*, and the barnacles *Balanus elizabethae* and *B. amphitrite* (Macnae, 1963). The plant *Arthrocnemum perenne* and the snail *Cerithidea decollata* occur somewhat higher. The high marsh is dominated by a Limonietum (*L. linifolium*), with *Chenolea diffusa*, *Crassula maritima*, *Suaeda fruticosa*, and *S. maritima* among the flora. Above the Limonietum, muddy areas are dominated by *Arthrocnemum africanum* and *A. pillansii*, whereas *Sporobolus virginicus* typifies sandy areas. This sequence, well seen in the Zwartkops estuary near Port Elizabeth, is similar to Mediterranean salt-marshes.

Intertidal biota comprise three distinct but overlapping populations: the cold-water Namaqua element of the west coast, forty-two species of which (fourteen animals and twenty-eight algae) extend some distance along the south coast; the temperate Cape element of the south coast which extends some distance north and east; and the tropical Indo-West Pacific element of the south-east coast which penetrates west towards Cape Point but rarely beyond (Stephenson and Stephenson, 1972).

The Namaqua supralittoral fringe is dominated by *Littorina knysnaensis* and by *Porphyra capensis* algae. The upper littoral zone is either bare or supports *Patella granularis* and a few barnacles, or in more sheltered sites a rich algal succession of *Porphyra capensis*, through *Chaetangium* spp. and *Ulva lactuca*, to *Aeodes orbitosa*, *Iridophycus capensis*, and kelp. The lower littoral zone supports sheets of lithothamnia, the worm *Gunnarea capensis*, various algae, and descends into a limpet belt of *Patella argenvillei*, *P. cochlear*, and the mussels *Mytilus meridionalis* and *M. crenatus*. The infralittoral fringe is transitional between the limpet belt and the giant kelp of the subtidal zone among which *Ecklonia buccinalis*, *Laminaria pallida*, and *Macrocystis pyrifera* are noteworthy.

The Cape fauna is typically temperate but includes both tropical and cold-water forms. The supralittoral fringe is dominated by *Littorina knysnaensis* with *L. africana* east of Cape Agulhas. The upper littoral zone is dominated by tropical barnacles and limpets and by the local periwinkle *Oxystele variegata*. The lower littoral zone has broad stretches of the gregarious tube-building polychaet *Pomatoleiso crosslandi* and the algae *Gelidium pristoides* and *Colpomenia capensis*, passing down into a limpet belt dominated by *Patella cochlear* and lithothamnia or, where exposed, into a belt of mussels, barnacles, and algae. The *Patella cochlear* belt is replaced at the infralittoral fringe by the leathery ascidian *Pyura stolonifera* and by algae such as *Gelidium cartaligineum* and *Sargassum* spp. In contrast to Natal, no reef corals or xeniids, and only one zoanthid occur on the Cape coast.

To the north-east, as the effects of the Agulhas Current intensify, intertidal biota become typically tropical. The supralittoral fringe is dominated by *Littorina africana*, *L. obesa*, and *Tectarius natalensis*. The upper littoral zone contains barnacles such as *Chthamalus dentatus*, *Tetraclita serrata*, and *Octomeris angulosa*, limpets such as *Patella granularis*, *P. variabilis*, *Cellana capensis*, and *Siphonaria* spp., the oyster *Crassostrea cucullata*, and the snail *Oxystele tabularis*. The lower littoral zone is dominated by algae such as *Gelidium reptans*, *Caulacanthus ustulatus*, *Gigartina minima*, and *Centroceras clavulatum*, replaced toward the infralittoral fringe by the bright green *Hypnea spicifera* and various zoanthids, and then by swards of red and purple algae such as *Hypnea rosea*, *Phodymenia natalensis*, and *Sargassum* spp. Healthy reef-building corals such as *Anamastrea*, *Favia*, *Goniastrea*, *Pocillipora*, *Psammocora*, and *Stylophora* may form large colonies, notably in rock pools.

The Indian Ocean Coast

From the Tugela estuary to Ras Asir, Africa's east coast is essentially tropical and low-lying, seasonally humid in the south, humid in Tanzania, to semi-arid farther north (Figure 14.10). Relentless south-easterly swells characterize the entire coast, enhanced by tropical easterlies and periodic cyclones in the south and by reversing monsoonal winds farther north. The warm waters of the south-flowing Mozambique and Agulhas currents south of 10°S, of the Equatorial Current off Tanzania, and of the reversing Somali Current farther north extend tropical conditions beyond comparable latitudes on the west coast. Sediment yields reflect climate and vegetation cover or land use, diminishing from south to north.

Geomorphology

The coastal plain overlying the Mozambique Basin is 300 km wide in central Mozambique but pinches out northward at Mocambo Bay and southward near the Tugela estuary. Its 2100-km-long coastline is formed by long barrier beaches topped by parabolic and lobate dunes behind which lie numerous lagoons, swamps, old barriers, and aeolian cover sands. In Kwa-Zulu the massive coastal barrier rises to 180 m and, although veneered with Holocene sands, is superimposed on a major Pleistocene barrier-lagoon complex, the Port Durnford Formation, whose facies relate to a high interglacial sea-level (Hobday and Orme, 1975). Late Pleistocene drawdown is shown by bedrock channels cut to −60 m beneath present estuaries and by beachrock and aeolianites found down to −100 m on the continental shelf (Orme, 1974, 1976). The Pleistocene coastal barrier was later breached by the Flandrian transgression forming, for example, Lake St. Lucia and Delagoa Bay (Figure 14.11). Shorezone dynamics are conditioned by strong south-easterly swells in the south which, although muted in the lee of Madagascar, are augmented by cyclones penetrating the Mozambique Channel. Sediment plumes flushed seasonally seawards onto the shelf are readily entrained by powerful longshore currents while finer sediments drift south at up to 6 km/h with the Mozambique–Agulhas Current or are reworked inshore by counter-currents. The Tugela River discharges 10.5×10^6 t/yr of sediment and its plume may be traced far out into the Agulhas Current (Plate 14.2). The Zambezi delta covers 7150 km² at the mouth of a 1.33×10^6 km² drainage basin and a 3540-km-long mainstream. Discharge through the delta averages 7000 m³/sec and sediment not trapped within coastal lagoons or entrained as littoral drift is flushed onto a massive cone beyond the shelf edge. Abundant sediment and a 6-m tidal range favour extensive intertidal flats in the Sofala Bight.

From Mocambo Bay northward to Kisware Bay, faulted Cretaceous and Tertiary rocks underlie an embayed, often cliffed coast fronted by raised coral bluffs and fringing reefs. Off northern Mozambique, the continental shelf is often only a few hundred metres wide, plunging to over −2500 m within 30 km of the shore. Prolonged N–S faulting and subsidence have also outlined the Tanzanian coast and its offshore islands where warped Pleistocene marine terraces and relict coral reefs testify to complex eustatic and tectonic forces (Alexander, 1968). As the Dar es Salaam Basin widens northward,

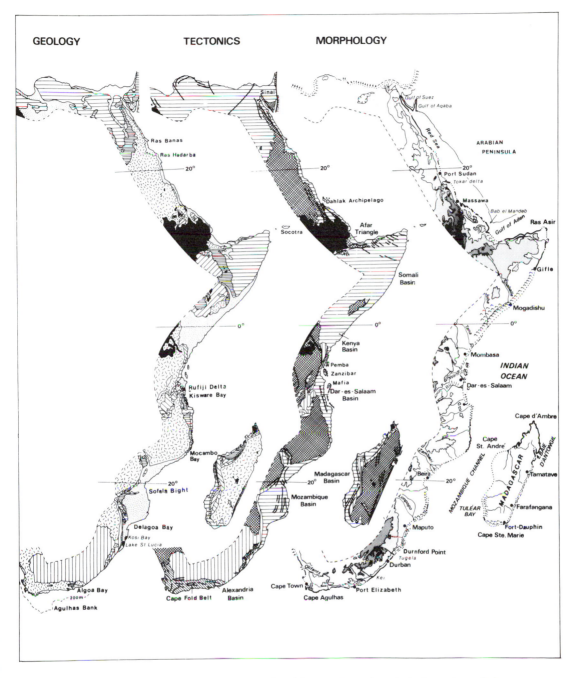

Fig. 14.10. The Indian Ocean and Red Sea coasts: geology, tectonics and morphology.
(See Figure 14.8 for key.)

Plate 14.2. The Mtunzini estuary, Natal, South Africa, showing double breaker zone induced by massive offshore bars and the predominant northward littoral drift of Tugela River sediments (photo: A. R. Orme).

the coastal plain has prograded through fluvial sedimentation, notably in the Rufiji and Wami deltas. Net littoral drift is northward though nearshore circulation is complicated by monsoon effects. The continental shelf widens to contain extensive coral reefs and the islands of Mafia and Zanzibar (1660 km²), but Pemba (984 km²) is separated from the mainland by a 700-m deep graben and all three islands are defined eastwards by massive faults. These islands are underlain by Cretaceous–Miocene marine sediments veneered with later corals and associated marine terraces. Coral barrier islands and fringing reefs characterize much of the mainland shore from the Pangani River northward along the Kenya coast and similar formations are associated with marine terraces rising inland to 140 m (Hori, 1970). From the Tana River for 1600 km northward to Gifle in Somalia (Plate 14.3), although fringing reefs persist, the coast is dominated by massive aeolian deposition promoted by the semi-arid conditions, strong monsoon winds, and episodic supplies of fluvial sediment, especially during past pluvial stages. The Tana delta is blocked seaward by a series of Quaternary dunes while dune systems up to 200-m high have deflected the Shebele River for 400 km parallel with the coast allowing it to reach the sea after heavy rains through the Juba estuary (Orme, 1985). As noted earlier, palaeodunes along the Somali coast have been widely destabilized and gullied as a result of overgrazing and poor tillage practices, notably where

the local population has been swelled recently by refugees. Dune stabilization efforts have so far met with limited success. From Gifle to Ras Asir, Tertiary carbonate rocks form rocky marine-terrace terrain and seacliffs 100–300 m high. The 1100 km-long *guban* (sunburnt plain) coast along the Gulf of Aden is partly cliffed, partly backed by alluvial plains, *sabkhas*, and marine terraces. Thick Plio-Pleistocene marine sequences and coral reefs raised 180–280 m above sea-level around Berbera emphasize the relatively recent tectonic emergence of this coastal zone (Orme, 1985).

The coast of Madagascar reflects the structural and topographic asymmetry of this large island (587 000 km²). The linear, fault-controlled east coast is broken by only one major embayment, the Baie d'Antongil graben. Basement rocks plunge from 2000-m highlands onto a narrow coastal plain and shelf underlain by a ribbon of Cretaceous sediments. Persistent south-east trade winds, storms in the Southern Ocean, and westward-moving cyclones from December to March combine to promote strong wave action. In response, beach barriers and shallow linear lagoons separated by low rises or *pangalanes* have formed along 800 km of coast from Fenerive to Fort-Dauphin. In contrast, the west coast fronts a broad alluvial lowland underlain by thick postpartum sediments and the shelf widens to 150 km off Cap St. André and on the Leven Bank. Along the north-west coast between rugged Cap d'Ambre and Cap

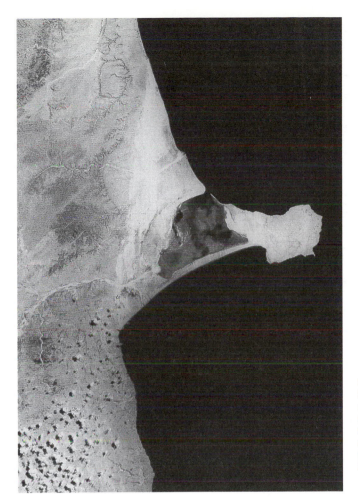

Plate 14.3. Ras Hafun, Somalia, is tied to the mainland by a 25-km-long sand tombolo which also shelters mangroves and evaporite deposits in the nearby lagoon. (photo: NASA)

St. André, the Flandrian transgression breached a barrier system to produce a complex of rias, partly drowned cuestas, deltas, mangrove swamps, and islands. The semi-arid south-west coast is dominated by active parabolic dunes and oxidized palaeodunes shaped by prevailing southerly winds. The shelf is narrower here and in Tuléar Bay is notched by massive submarine canyons, that off the Onilahy River descending to −2600 m. Reef formations of many ages and types occur: barrier reefs, fringing reefs, coral banks, and islets. As in Pleistocene times, Holocene reefs are best developed between major estuaries whose muddy detritus inhibits reef growth (Weydert, 1974). Relative sea-level change since the last interglacial is expressed in the barrier that lies 8–12 m above sea-level behind the east-coast *pangalanes*, in uplifted coral reefs at the northern tip, and in shorelines deformed elsewhere to below sea-level (Battistini, 1978).

Ecology

Because the Indian Ocean coast is bathed in warm air and warm waters throughout the year, mangroves and reef-building corals thrive under suitable conditions, and both sandy habitats and rocky shores exhibit a rich biota (Figure 14.11). More than half the plants found along the coast have wide Indo-West Pacific ranges or are pan-tropical, reflecting adaptation for long-range dispersal by ocean currents. Systematic studies are limited but a sampling of dry habitats and wetland communities along the coast will indicate the general ecology.

In the dry plant communities of northern Natal and southern Mozambique, important strand pioneers such as *Scaevola thunbergii*, *Ipomoea biloba*, and the grass *Digitaria erianthia* flourish in shifting sand and promote foredune accumulation (Orme, 1973). *Mesembryanthemum*

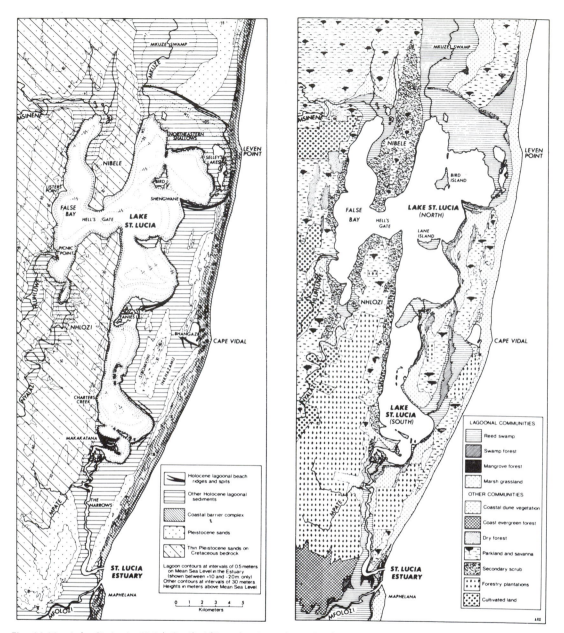

Fig. 14.11. Lake St. Lucia, Natal, South Africa, showing relationship between morphology, bathymetry, and ecology.

edule, Tephrosia canescens, Gazania uniflora, and *Osteospermum moniliferum* follow and are in turn invaded by *Othonna sarnosa, Gloriosa virescens, Passerina rigida,* and *Anthospermum littoreum.* This plant mat secures the outer dunes and enables a dense dune scrub to grow to about 2 m. Beyond the extreme effects of wind, shifting sand, and salt spray, dune scrub matures into luxuriant coastal

dune forests, composed of *Mimusops caffra, Diospyros rotundifolia, Brachylaena discolor, Euclea natalensis* and other trees 10–25 m tall, which afford a rich habitat for animals (Figure 14.11).

The sandy backshores of northern Kenya and Somalia are colonized by the sprawling beach vine *Ipomoae pescaprae* associated inland with *Scaevola taccada* and the

Plate 14.4. Coastal strand at Marka, Somalia, showing pioneer colonization of the backshore by the trailing beach vine *Ipomoea pes-caprae* (photo: A. R. Orme).

Fig. 14.12. Transect through a typical East African mangrove association.

grasses *Lepturus repens* and *Sporobolus virginicus* (Plate 14.4). Beyond the highest tides, thickets of *Canavalia maritima*, *Colubrina asiatica*, and *Cordia somaliensis* occur. The physiognomy of this seaward zone is one of low, succulent plants and small, thick-leaved bushes. Farther back, 1–2 m-high dune scrub contains *Grewia glandulosa*, *Cadaba farinosa*, *Cleome strigosa*, *Pemphis acidula*, and *Sideroxylon diospyroides*. In Kenya, coral cliffs in the spray zone carry a dense evergreen scrub dominated by the spiny caper bush, *Capparis cartilaginea*. Doum palms (*Hyphaene coriacea*) occur on thin sands but where dunes overlie coral, low trees like *Afzelia cuanzensis*, *Ficus tremula*, and *Vitex amboniensis* rise to 5 m above dense undercover. Relict stands of evergreen high forest may occur on coral soils but disturbed shrubland and secondary savanna are more common. Human disturbance has played a major role here, notably in two centuries of clearance for coconut and *Casuarina* plantations along the Kenya coast and in the introduction of *Opuntia* cactus, *Euphorbia* spp., and *Commiphora* shelterbelts for dune stabilization in Somalia (Orme, 1985).

Along the central Indian Ocean coast, between the Tana and Zambezi deltas, and along the relatively sheltered west coast of Madagascar, climatic, edaphic, and geomorphic conditions provide for luxuriant mangrove forests with some trees reaching 15–25 m high, although long exploitation for timber and tanbark means that most are shorter. Five mangrove species occur: *Sonneratia alba* as a pioneer of exposed habitats; *Avicennia marina* as the primary seaward colonist on firmer sandy soils; *Rhizophora mucronata* on muddy soils of the creeks; *Ceriops tagal* in drier areas; and *Bruguiera gymnorrhiza* in wetter areas. Although mangrove zonation is crudely related to exposure and edaphic factors, there is much overlap and role reversal such that the succession concept is difficult to sustain.

Thus in the Zambezi delta, *Bruguiera* forms the main mangrove zone between seaward *Rhizophora* and the landward margin. Elsewhere, *Avicennia* may be found landward of other species (Figure 14.12). Fiddler crabs (*Uca* spp.) are typically found among the more seaward mangroves, whiles *Sesarma* crabs and various snails live in the dense shade of the *Ceriops-Bruguiera* community. Where mangroves are not crowded, a ground cover of *Sesuvium portulacastrum*, *Arthrocnemum* spp., *Suaeda monoica*, the grass *Sporobolus virginicus*, and, in the south, a summer growth of *Salicornia pachystachya* occur. The transition to brackish water is marked by the appearance of *Lumnitzera racemosa*, *Xylocarpus granatum*, *X. moluccensis*, the fern *Acrostichum aureum*, the freshwater mangrove *Hibiscus tiliaceus*, and the tree *Heritiera littoralis*. The transition to fresh water is through a reedswamp of *Phragmites australis*, *Cyperus papyrus*, *Juncus*, and *Scirpus*. *Sonneratia* is not found south of Delagoa Bay, *Ceriops* dies out at Kosi Bay, but the other three mangroves extend as impoverished patches south to the Kei River. Northward, the geomorphic and edaphic nature of the Somali coast is not well suited to mangroves. *Avicennia* is dominant, indeed the only species on Socotra offshore. Throughout the coast, these wetlands afford excellent habitat for crocodiles (*Crocodylus niloticus*), monitor lizards (*Varanus niloticus*), hippopotamus (*Hippopotamus amphibius*), and many birds, fishes, and insects, as at Lake St. Lucia in northern Natal (Figure 14.11; Orme, 1975).

The Red Sea Coast

Geomorphology

The Red Sea extends 2000 km NNW from the Bab el Mandeb narrows, ranges from 150 to 300 km wide, and deepens to −2300 m along its central axis. This embryo ocean occupies a complex rift system between the African and Arabian plates whose separation began with left-lateral movement in Cretaceous–Eocene times but accelerated with the 7° counter-clockwise rotation of the Arabian block during the Neogene. The Red Sea opened to the Indian Ocean in Miocene times but this link has not been continuous; it remained open to the Mediterranean until the early Quaternary. Then, closure of the Mediterranean link and eustatic and tectonic changes at Bab el Mandeb, isolated the Red Sea from time to time during the later Quaternary. With evaporation around 2000 mm/yr and rainfall ranging from 3 mm/yr in the north to 150 mm/yr in the south, isolation of the Red Sea from the Indian Ocean has a dramatic effect on water volume and thus on coastal environments.

Owing to the recency of plate separation, faulted basement rocks lie close to the coast and dissected mountains plunge toward a narrow ribbon of coastal plain underlain by Neogene and Quaternary deposits. The northern Red Sea reflects mainly normal tension faulting of continental margins, as do its extensions: the 10–25-km-wide Gulf of Aqaba which descends to −1800 m, and the shallower Gulf of Suez where Oligocene graben faulting allowed thick Miocene evaporites to accumulate (Hassan and El Dashlouty, 1970). South of Ras Banas, the Red Sea is structurally more complex

and falls stepwise into a narrow central trough where ascending volcanics have engulfed remnants of continental crust.

The Red Sea coast is characterized by a faulted staircase of Quaternary coral reefs that form wedgelike plates up to 10 m thick, several kilometres wide, and range from 50 m above to −100 m below present sea-level. Onshore these reefs are often covered by aeolian sand, lagoonal evaporites, estuarine gravels, or by alluvial deposits. The coastal plain widens only where major valleys reach the coast, as in the Tokar delta into which the swift Baraka River floods between June and September, or where structural troughs occur, such as the Dallol Basin, a saline depression up to −116 m below sea-level where the Red Sea rift turns inland south of Massawa. Offshore, faulted palaeoreefs form the Dahlak Archipelago at the northern end of the Danakil horst. The paucity of terrigenous sediments allows modern fringing reefs to thrive in warm transparent waters. Reef margins are dominated to windward by *Acropora* associations, to leeward by *Porites* associations (Braithwaite, 1982). Lateral reef growth of several cm/yr even occurs in front of now dry early Holocene river deltas and abandoned *marsas* (*sharms*), corridors cut by wadis draining to low Pleistocene sea-levels. Barrier reefs and atolls are less common but have developed over subsiding fault blocks from Port Sudan to Ras Hadarba (Mergner and Schuhmacher, 1985).

Ecology

The Red Sea coast is bordered by hot desert. Mean temperatures for the hottest months are 30–35 °C, humidity is high but rainfall is everywhere deficient. Further, the shallow submarine sill at Bab el Mandeb, only −125 m below sea-level, restricts the exchange of water with the Indian Ocean. Accordingly, the Red Sea has acquired some distinctive ecological features: warm water up to 30 °C in August and, with high evaporation rates and negligible stream inflows, mean salinities of 37–43 parts per thousand. Thus conditions for life are rather hostile, barren *sabkhas* common, and most interest attaches to the wetlands.

Coastal wetland communities range from mangrove and reed-swamp to salt-marsh, and reflect a change from truly tropical vegetation types in the south to plants with closer Mediterranean affinities in the north. Of the mangroves, *Avicennia*, *Rhizophora*, *Ceriops*, and *Bruguiera* all penetrate the southern Red Sea coast, typically growing in black sandy muds rich in decaying organic matter. Farther north, however, the mangrove community is dominated by *Avicennia marina*, pure stands of which vary from small patches to continuous dense growth over several kilometres (Zahran, 1977). Mangroves disappear north of Myos Hormos Bay, Egypt, and Wadi-kid (28° 10′ N), Sinai. Reed-swamps typified by *Phragmites australis* and *Typha domingensis* occur in brackish water near the mouths of large wadis such as Wadi el Ghweibba. Twenty community types of salt-marsh occur, variations in the ecological amplitudes of dominant species being reflected in differences in distribution, density, stratification, zonation, and floristic composition. The southern coast is dominated by the *Arthrocnemum glaucum* community type with *Halopeplis perfoliata*, *Limonium axillare*, and *Suaeda monoica*. The *Zygophyllum album* type occurs along the Sudanese and Egyptian coast farther north, while the *Halocnemon strobilaceum* type with *Limonium pruinosum* and *Nitraria retusa* dominates the Gulf of Suez.

Conclusion

Early scientific interest in the African coast arose from the needs of navigators and traders for information on winds, currents, safe anchorages, and navigable rivers. The Nile Delta intrigued classical scholars such as Herodotus and Ptolemy, and knowledge of the Mediterranean coast was disseminated by Greek, Phoenician, Roman, Byzantine, and Arab traders long before the remaining African coast was known other than locally. The rapid expansion of Islam after AD 632 saw Arab navigators extend south to the Senegal River, and to Sofala near the Zambezi Delta linking with Asian traders who also availed of the monsoon circulations of the Indian Ocean. Sustained European interest in the African coast developed from the fifteenth century as the Portuguese established west-coast trading posts, for example at Elmina in 1482 and Luanda in 1575. After Vasco da Gama aided by Arab pilots had navigated a sea-route to India in 1498, traders from many nations shared east-coast ports such as Mombasa. Lacking navigable rivers to the interior, however, the ocean-orientation of these ports restricted scientific enquiry. In the nineteenth century, as private European trading companies gave way to colonial authorities, coastal and hydrographic surveys were initiated for commercial, military, and administrative purposes. The colonial interlude, which began in Algeria in 1830 and largely ended in Southern Africa in the 1990s, saw many scientific studies of the coast, initially as resource inventories but later as detailed process and integrated environmental studies. Since independence, most nation states have continued research programmes, often in concert with institutions in former colonial nations, with the USA, the former USSR, and China, and with such

international agencies as UNO, UNESCO, and the World Bank.

Until recently, the African coast had probably suffered less human interference than any other continental coast beyond Antarctica. Much early change was exploitive, for example in mangrove cutting and sand mining, and this continues. In recent times, however, the scale of interference has increased significantly, for example in dam construction, oil and gas recovery, port development, dredging, land reclamation, and pollution. The impacts of the High Aswan and Akosombo Dams on sediment delivery to the coast, of increased sediment yields reflecting soil erosion caused by unwise farming practices, of accelerated subsidence in the Niger Delta caused by oil and gas withdrawal, of new ports like Richards Bay, and of mangrove clearance for paddy-rice cultivation illustrate the increased role that people are now playing in the evolution of the African coast.

References

Alexander, C. S. (1968), 'The Marine Terraces of the Northeast Coast of Tanganyika', *Zeitschrift für Geomorphologie*, 7: 133–54.

Ausseil-Badie, J., Barusseau, J. P., Descamps, C., Diop, E. H. S., Giresse, P., and Pazdur, M. (1991), 'Holocene Deltaic Sequence in the Saloum Estuary, Senegal', *Quaternary Research*, 36: 178–94.

Ayyad, M. A. (1973), 'Vegetation and Environment of the Western Mediterranean Coastal Land of Egypt: The Habitat of Sand Dunes', *Journal of Ecology*, 61: 509–23.

Barusseau, J. P., Diop, E. H. S., and Saos, J. L. (1985), 'Evidence of Dynamic Reversal in Tropical Estuaries: Geomorphological and Sedimentological Consequences (Saloum and Casamance Rivers, Senegal)', *Sedimentology*, 32: 543–52.

Battistini, R. (1978), 'Observations sur les cordons littoraux pleistocènes et holocènes de la côte est de Madagascar', *Madagascar Revue de Géographie*, 35: 9–37.

Braithwaite, C. J. R. (1982), 'Patterns of Accretion of Reefs in the Sudanese Red Sea', *Marine Geology*, 46: 297–325.

Busson, G., and Perthuisot, J.-P. (1977), 'Intérêt de la Sebkha el Melah (sudtunisien) pour l'interprétation des séries evaporitiques anciennes', *Sedimentary Geology*, 19: 139–64.

Chapman, V. J. (1977), *Wet Coastal Ecosystems* (New York).

Cooper, J. A. G. (1993), 'Sedimentation in a River-Dominated Estuary', *Sedimentology*, 40: 979–1017.

CLIMAP (1981), 'Seasonal Reconstructions of the Earth's Surface at the Last Glacial Maximum', *Geological Society of America, Map and Chart 36* (Boulder Colo.).

Cowling, R. M. (1992) (ed.), *The Ecology of Fynbos* (Cape Town).

Debeney, J. P., and Bellion, Y. (1983), 'Le Quaternaire récent des microfalaises de Mbodiène (Sénégal): Stratigraphie, variation de niveau marin', *Bulletin de l'Association Sénégalaise d'Etude du Quaternaire Africain*, 70/71: 73–81.

Dingle, R. V., and Hendey, Q. B. (1984), 'Late Mesozoic and Tertiary Sediment Supply to the Eastern Cape Basin (SE Atlantic) and Palaeodrainage Systems in Southernmost Africa', *Marine Geology*, 56: 13–26.

Ebisemiju, F. S. (1987), 'An Evaluation of Factors Controlling Present Rates of Shoreline Retrogradation in the Western Niger Delta, Nigeria', *Catena*, 14: 1–12.

El Ashry, M. (1985), 'Egypt', in E. C. Bird and M. L. Schwartz (eds.), *The World's Coastline* (New York), 513–17.

Fleming, B. W. (1977), 'Langebaan Lagoon: A Mixed Carbonate-Siliclastic Tidal Environment in a Semiarid Climate', *Sedimentary Geology*, 18: 61–95.

Fredoux, A. (1980), 'Evolution de la mangrove près d'Abidjan (Côte d'Ivoire) au cours des quarante derniers millénaires', in *Les Rivages Tropicaux: Mangroves d'Afrique et d'Asie* (Bordeaux), 49–88.

Frihy, O. E., and Komar, P. D. (1991), 'Patterns of Beach-Sand Sorting and Shoreline Erosion of the Nile Delta', *Journal of Sedimentary Petrology*, 61: 544–50.

Gilchrist, A. R., and Summerfield, M. A. (1990), 'Differential Denudation and Flexural Isostasy in Formation of Rifted-Margin Upwarps', *Nature*, 346: 739–42.

Guilcher, A. (1985), 'Senegal and Gambia', in Bird and Schwartz (eds.), *The World's Coastline*: 555–60.

—— Medeiros, C. A., Matos, J. E., and Oliveira, J. T. (1974), 'Les restingas (flèches littorales) d'Angola', *Finisterra*, 9: 171–211.

Hassan, F., and El Dashlouty, S. (1970), 'Miocene Evaporites of the Gulf of Suez Region and their Significance', *American Association of Petroleum Geologists Bulletin*, 54: 1686–96.

Hinschberger, F. (1985), 'Ivory Coast', in Bird and Schwartz (eds.), *The World's Coastline*: 585–9.

Hobday, D. K., and Orme, A. R. (1975), 'The Port Durnford Formation: A Major Pleistocene Barrier-Lagoon Complex Along the Zululand Coast', *Transactions of the Geological Society of South Africa*, 77: 141–9.

Hospers, J. (1971), 'The Geology of the Niger Delta Area', in F. M. Delaney (ed.), *The Geology of the East Atlantic Continental Margin* (London).

Hori, N. (1970), 'Raised Coral Reefs along the Southeastern Coast of Kenya, East Africa', *Geographical Reports, Tokyo Metropolitan University*, 5: 25–47.

Jobstraibizer, G., and Cumarshiire, J. (1977), 'Il Quaternario della Somalia', *Quad. Geologia Somalia*, 1: 51–9.

Illenberger, W. K., and Rust, I. C. (1988), 'A Sand Budget for the Alexandria Coastal Dunefield, South Africa', *Sedimentology*, 35: 513–21.

Kocurek, G., Havholm, K. G., Deynoux, M., and Blakey, R. C. (1991), 'Amalgamated Accumulations Resulting from Climatic and Eustatic Changes, Akchar Erg, Mauritania', *Sedimentology*, 38: 751–72.

Kutina, J. (1975), 'Tectonic Development and Metallogeny of Madagascar with References to the Fracture Pattern of the Indian Ocean', *Geological Society of America Bulletin*, 86: 582–92.

Lawson, G. W. (1966), 'The Littoral Ecology of West Africa', *Oceanography and Marine Biology Annual Reviews*, 4: 405–48.

Mahrour, M., and Dagorne, A. (1985), 'Algeria', in Bird and Schwartz (eds.), *The World's Coastline*: 531–6.

Macnae, W., (1963), 'Mangrove Swamps in South Africa', *Journal of Ecology*, 51: 1–25.

Martin, L. (1971), 'The Continental Margin from Cape Palmas to Lagos: Bottom Sediments and Submarine Morphology', in Delaney (ed.,), *The Geology of the East Atlantic Continental Margin*, 83–95.

Maugé, L. A., Ségoufin, J., Vernier, E., and Froget, C. (1982), 'Géomorphologie et origine des bancs du nord-est du Canal de Mozambique–Ocean Indien Occidental', *Marine Geology*, 47: 37–55.

Mergner, H., and Schuhmacher, H. (1985), 'Quantitative Analyse von Korallengmeinschafter des Danganeb Atolls, mittleres Rotes Meer', *Helgoländer Meeresuntersuchungen*, 39: 375–417.

Nel, J. A. J. (1993), 'The Coastal Namib Desert as Habitat for Mammals', *South African Geographer*, 19: 127–35.

Nkounkou, R. R., and Probst, J.-L. (1987), 'Hydrology and Geochemistry of the Congo River System', in E. Degens, S. Kempe, and G. Weibin (eds.), *Transport of Carbon and Minerals in Major World Rivers* (Hamburg).

Ochumba, P. B. O. (1988), 'The Distribution of Skates and Rays Along the Kenya Coast', *Journal of the East African Natural History Society and National Museum*, 78: 25–45.

Olivier, J. (1993), 'Some Spatial and Temporal Aspects of Fog in the Namib', *South African Geographer*, 19: 106–26.

Orme, A. R. (1973), 'Barrier and Lagoon Systems Along the Zululand Coast, South Africa', in D. R. Coates (ed.), *Coastal Geomorphology* (New York), 181–217.

—— (1974), *Estuarine Sedimentation along the Natal Coast* (Washington, DC).

—— (1975), 'Ecological Stress in a Subtropical Coastal Lagoon: Lake St. Lucia, Zululand', *Geoscience and Man*, 12: 9–22.

—— (1976), 'Late Pleistocene Channels and Flandrian Sediments Beneath Natal Estuaries', *Annals of the South African Museum*, 71: 78–85.

—— (1982a), 'Africa, Coastal Ecology', in M. L. Schwartz (ed.), *The Encyclopedia of Beaches and Coastal Environments* (Stroudsburg, Penn.), 3–16.

—— (1982b), 'Africa, Coastal Morphology', in M. L. Schwartz (ed.), *The Encyclopedia of Beaches and Coastal Environments*: 17–32.

—— (1985) 'Somalia', in E. C. Bird and M. L. Schwartz (eds.), *The World's Coastline*: 703–11.

Paskoff, R. P. (1978), 'Evolution de l'embouchure de la Medjerda (Tunisie)', *Photo-Interprétation*, 5: 1–23.

Reyre, D. (1984), 'Remarques sur l'origine et l'evolution des bassins sédimentaires de la côte Atlantique', *Bulletin de la Société Géologique de France*, 6: 1041–59.

Sarnthein, M., and Walger, E. (1974), 'Der äolische Sandstrom aus der W-Sahara zur Atlantikkuste', *Geologische Rundschau*, 63: 1065–87.

Stanley, D. J. (1990), 'Recent Subsidence and Northeast Tilting of the Nile Delta, Egypt', *Marine Geology*, 94: 147–54.

Stephenson, T. A., and Stephenson, A. (1972), *Life Between Tidemarks on Rocky Shores* (San Francisco).

Weisrock, A (1980), *Géomorphologie et Paléo-environments de l'Atlas atlantique* (Service Géologique du Maroc, Rabat).

Weydert, P. (1974), 'Sur l'existence d'une topographie anté-récifale dans la région de Tuléar (Côte sud-ouest de Madagascar)', *Marine Geology*, 16: 39–46.

Wilcox, B. H. R. (1985), 'Angiosperm Flora of the Mangrove Ecosystem of the Niger Delta', in B. H. R. Wilcox and C. B. Powell (eds.), *The Mangrove Ecosystem of the Niger Delta* (Port Harcourt), 33–44.

Woodroffe, C. D. and Grindrod, J. (1991), 'Mangrove Biogeography: The Role of Quaternary Environmental and Sea-level Change', *Journal of Biogeography*, 18: 479–92.

Zahran, M. A. (1977), 'Africa: Wet Formations of the African Red Sea Coast', in V. J. Chapman (ed.), *Wet Coastal Ecosystems* (New York), 215–31.

15 Wetlands

Francine M. R. Hughes

Introduction

Wetlands are by definition lands with water-tables at or near the surface, either seasonally or permanently. On the African continent, which is semi-arid or arid over much of its area, wetlands have long been of great importance to its people and wildlife, providing all important seasonal or year-round water, fodder, and food (Scoones, 1991; Adams, 1992). It is only relatively recently that scholars and development practitioners have begun to appreciate and write about the importance of African wetlands.

The aim of this chapter is to consider the functioning of Africa's interior wetlands, concentrating particularly on hydrological linkages within wetland areas and between wetlands and their surroundings. No attempt is made to list African wetlands in any comprehensive way but rather to highlight, through the use of examples, the importance of understanding these linkages if 'wise use' principles of wetland use are eventually to be implemented. Useful descriptions of many African wetlands can be found in Hughes and Hughes (1992), Howard-Williams and Gaudet (1985), Thompson and Hamilton (1983), Denny (1985, 1991, 1993), Ingram (1991), and Welcomme (1979). Coastal wetlands are considered in Chapter 14.

My thanks are due to Dick Grove who first encouraged me to work in Africa and who shared his great store of knowledge on the waterways of that continent. I would also like to thank the following for commenting on various drafts of this chapter or for providing valuable information for it: W. M. Adams, A. S. Goudie, T. S. McCarthy, A. R. Orme, T. Scudder, and The World Conservation Monitoring Centre, Cambridge.

Wetland Classification

Africa's major interior wetlands are concentrated south of the Sahara Desert and north of the Tropic of Capricorn and the Kalahari Desert, in a broad area straddling the Equator (Figure 15.1). Their distribution reflects patterns of available moisture, which are in turn a complex function of rainfall, temperature, evaporation, and evapotranspiration. In the equatorial belt there are two rainy seasons per year and it is here that many of the largest wetlands occur. Beyond the equatorial belt, wetlands may be sustained in areas of lower rainfall because the timing of flooding and local rainfall may be out of synchrony, giving an extended flood period. In the case of the Okavango Delta, for example, the timing of floods derived from headwaters in the highlands of Angola and Namibia is out of phase with local rainfall by two to three months.

Although in total area Africa is the driest continent, having around 21 million km^2 (72.5 per cent) classified as drylands (Hopkins and Jones, 1983) at least 345 000 km^2 (1 per cent) are occupied by wetlands (Denny, 1991). The area of wetland will vary depending on two quite separate factors, namely the wetland classification used and whether the wetland is defined in terms of its maximum seasonal extent or some lesser value. Both factors highlight problems involved in describing wetlands.

Wetlands were defined at the Ramsar Convention (1971) as: 'areas of marsh, fen, peatland or water, whether natural or artificial, permanent or temporary, with water that is static or flowing, fresh, brackish or salt, including areas of marine water the depth of which at low tide does not exceed six metres'. This definition is only one of very many separate definitions in use

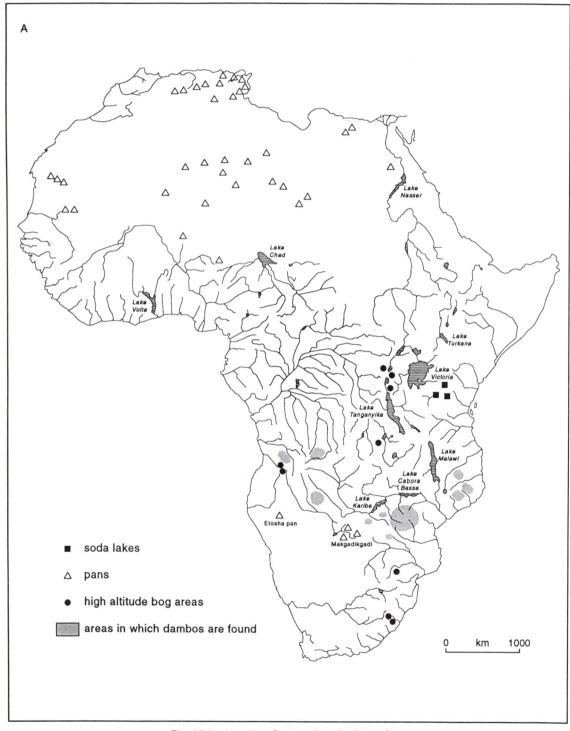

Fig. 15.1. Location of principal wetlands in Africa

(A) Pans, soda lakes, high altitude bogs, and areas with dambos; (B) Swamp forests, fringing floodplains, freshwater marshes, and herbaceous swamps.

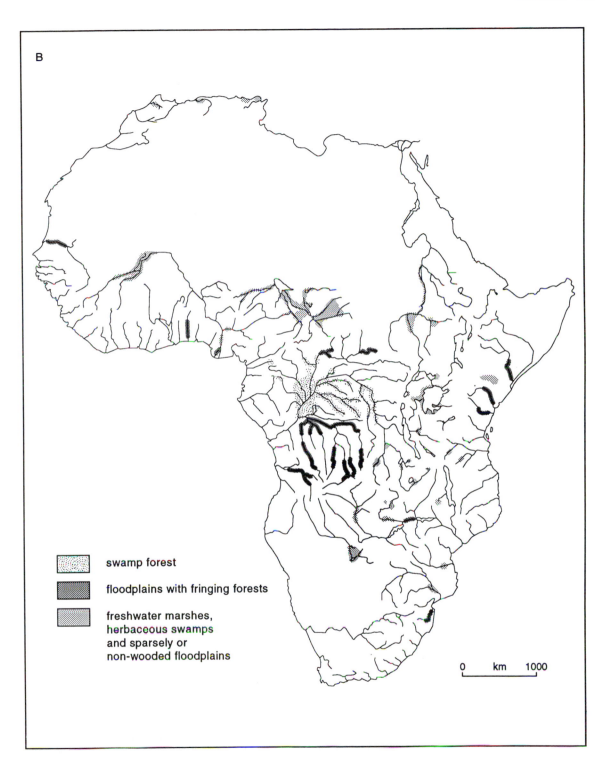

B

swamp forest

floodplains with fringing forests

freshwater marshes,
herbaceous swamps
and sparsely or
non-wooded floodplains

0   km   1000

Plate 15.1. An oblique aerial photograph of the Lower Tana River, Kenya, downstream of the Tana River National Primate Reserve. Ox-bow lakes and extensive areas of floodplain are shown inundated during the November floods. While sections of evergreen forest can be seen, this photograph also shows areas of floodplain grassland and agriculture, used by Pokomo people (photo W. M. Adams).

today and one of the broadest (Dugan, 1990). Narrower definitions have been used to suit particular projects and research purposes. Many different wetland classifications have been described and compared, some of which are very complex, e.g. Dugan (1990); Mitsch and Gosselink (1993); Scoones (1991); Muthuri (1992). The Ramsar Convention wetland classification has thirty wetland categories based on their basic biological and physical characteristics.

Whatever classification system is used, it soon becomes clear that many wetlands fall between categories, change from one category to another at different seasons, or are hydrologically linked complexes of various wetland types. Furthermore, it is difficult to define the spatial extent of many wetlands because their water levels fluctuate seasonally. The classification used by Dugan (1990) is one of the most straightforward and includes seven basic landscape units including: (1) estuaries, (2) open coasts, (3) floodplains, (4) freshwater marshes, (5) lakes, (6) peatlands, and (7) swamp forests. Of these, the last five are relevant to wetlands discussed in this chapter. Each unit can include a wide range of wetland types and several wetlands can occur in more than one unit since the category titles are a mixture of geomorphological and vegetation descriptions. Dugan (1990) also emphasizes that wetlands are very heterogeneous and frequently linked with one another and with surrounding dryland or coastal areas. The following discussion outlines some basic characteristics of these major wetland types.

In *river floodplains*, wetlands commonly occur between an active river channel and valley sides or river terraces in areas which are periodically inundated by rising river water. Floodplain wetlands can occur far inland or near the coast, often in pre-delta locations. They can be narrow strips as along the Tana River floodplain (Plate 15.1) in Kenya, or vast complexes as in the lower Zaire Basin where a number of rivers and permanent lakes are included. In some cases, such as in the Niger floodplain, they have been called inland deltas, but they behave as complex floodplains sometimes with a vast system of anastamosing channels. The floodplain area, however broad or narrow, usually has quite distinctive geomorphological features supporting different wetland vegetation types ranging from those of ox-bow lakes to herbaceous marshes or swamp forests. Some tree species such as *Diospyros mespiliformis* and *Tamarindus indica* have a pan-Africa distribution on floodplains, though they may also be found away from floodplains in areas of higher rainfall. *Trichilia emetica* and many species of *Ficus* are strictly riparian in distribution.

Freshwater marshes can be part of large floodplain complexes or may border shallow lakes and are frequently referred to as swamps in the literature. They have a widespread distribution, and many different geomorphological origins based on their source of water. In Africa there are well-known marshes dominated by papyrus (*Cyperus papyrus*) and cattail (*Typha* spp.), one of the largest being the Sudd in the Sudan. Some authors use the term 'swamp' to refer to wetlands dominated by herbaceous species (e.g. Howard-Williams and Gaudet, 1985). Others also include swamp forest in this category (e.g. Thompson and Hamilton, 1983). It is impossible to give a precise definition of terms such as 'marsh' or 'swamp' or to state categorically that certain types tend to have dominantly peaty or mainly mineral substrates since there are many gradations between the two.

Swamp forests frequently develop around lake margins or in still-water areas on floodplains, often around oxbow lakes. They are well represented in the Zaire Basin, and were once very widespread around the margins of Lake Victoria prior to recent lake-level rises that introduced unfavourable ecological conditions. The vast forested swamp and river floodplain complex of the Zaire River has been poorly studied, but is the only large remaining swamp forest in Africa. Below Kisangani, the course of the Zaire broadens and so does the floodable area. This culminates in the Bangweulu Swamps and an extensive complex of floodlands at the confluence of the Zaire, Ubangui, and Sangha Rivers. Thompson and Hamilton (1983) suggest that there are over 8000 km² of permanent or seasonal swamp within the forested lands of the Zaire Basin, dominated largely by *Cyrtosperma senegalense* and *Raphia farinifera*. *Ficus* spp. and borassus palms, though typical of these wetland types, also occur on levees on many floodplains, usually only on the highest areas receiving least flooding, so that their occurrence doesn't necessarily signify swamp forest. In general they have lower nutrient levels than freshwater marshes and also lower pH values.

Lakes and ponds have many different origins including volcanic activity, wind action in arid areas, or various fluvial processes. Their margins can be shallow or steep with many different substrates, and this will greatly affect the type of wetland community that develops. They often intercept overland runoff, the quality of which affects their nutrient content, and these in turn affect the water quality of the adjoining lake and the range of aquatic, semi-aquatic and emergent vegetation species. For example, *Potomageton pectinatus* and *Najas marina* can live in lakes with salinities of up to 30 per cent of seawater but prefer freshwater lakes (Denny, 1991). Soda lakes are usually dominated by blue-green algae such as *Spirulina* spp. Lacustrine environments are discussed further by Adams (this volume).

Peatlands generally develop in oxygen-deficient environments. Their origins and distribution are very variable. There are relatively few ombrogenous peatlands in Africa, and most of these are located in highland areas receiving abundant rainfall, notably Lesotho and East Africa. Peats also form under many swamp forests and in some marsh areas. The chemistry of their waters and associated soils varies from acidic to alkaline depending on peatland origins and location. Peat in lowland swamp forests has an organic content of around 30 per cent, whereas at high altitudes *Syzgium/Erica* peat may exceed 95 per cent organic carbon (Thompson and Hamilton, 1983). Woody species found on bogs in Africa include species of *Senecio* and *Helichrysum* and some *Erica* and *Vaccinium* (Thompson and Hamilton, 1983).

These brief descriptions of broad wetland types demonstrate the difficulty in separating them based on species presence, substrate types, or origins and highlights the importance of looking at how wetlands function through the hydrological linkages within wetlands and between wetlands and their surrounding areas.

Hydrological Linkages in Wetlands

The hydrology of any wetland creates unique physical and chemical conditions that distinguish it from surrounding terrestrial or aquatic environments. Many hydrological pathways contribute water to wetland development, including precipitation, slope runoff, groundwater, and flooding from rivers. Water losses occur through evapotranspiration, seepage, and runoff. Whichever pathway is dominant in a particular wetland at any time, it is the avenue for the import and export of water, nutrients, and to some extent energy to and from the wetland. Along with a wetland's hydro-period, these factors affect the biochemistry of wetland soils and therefore the primary and secondary producers that will ultimately inhabit and use them. As emphasized by Mitsch and Gosselink (1993) 'the hydrology of a wetland is probably the single most important determinant for the establishment and maintenance of specific types of wetlands and wetland processes' (p. 68).

In addition, as hydrological inputs change over time due to either allogenic changes caused by climate-forcing or autogenic changes such as peat growth, wetlands evolve as a set of responses each of which will have different lag times. For example, soil chemistry may respond rapidly to inundation while herbaceous species may respond over a period of several years and swamp-forest trees respond over many decades. On islands in the Nile near Khartoum, changes from *Tamarix nilotica* on new sandbars to *Salix* spp., *Acacia arabica*, and *Ziziphus* spp. reflected increased height of substrate and decreased flood frequency, finer soil composition, and increased water retention capacity. There was a shift from allogenic to autogenic influences over a period of >65 years (Halwagy, 1963).

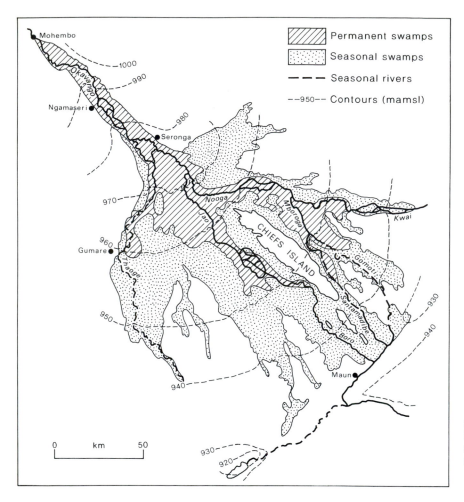

Fig. 15.2. The Okavango Delta showing permanent and seasonal swamps. (Redrawn from McCarthy, 1992.)

Fish and animal users can respond very quickly if physical habitat or food sources are altered rapidly. Following construction of the Kainji Dam in 1968, fish-catch sizes and fish-species composition changed on the Niger River downstream (Lowe-McConnell, 1985). The living components of the wetland will respond to a complexity of changes in, for example, quantities of available water, availability of regeneration sites, and chemical environments, and will be shown in changes in mortality patterns, species mixes and richness, the timing and quantity of seed production, regeneration strategies (e.g. seeding or suckering), and ecosystem productivity. Similarly, when hydrologic pathways remain similar over a period of time, the structure and functional integrity of the wetland persist. The internal mosaic of vegetation patterns may fluctuate but the same mix of species can persist for a long time. This emphasizes the dependence of wetland ecosystems on hydrological inputs and

outputs. Further, beyond the local links between hydrological processes and biotic responses in a wetland, longitudinal links exist between components of larger hydrological systems of which a particular wetland might physically be only a small part. This aspect is particularly well illustrated by the Okavango and Sudd wetlands in Botswana and the Sudan respectively.

The Okavango River system, which probably originated in pre-Cretaceous times (Thomas and Shaw, 1988; Summerfield, this volume) has been subjected to numerous movements of the Earth's crust (Du Toit, 1926; McCarthy, 1992). Most of its catchment is directly underlain by Kalahari sands which have been discharged into a partially depressed fault block by the Okavango River. Sedimentologically, therefore, the so-called Okavango Delta is in fact a very large alluvial fan (McCarthy, 1992) (see Figure 15.2). Above the alluvial fan, the Okavango River valley is more confined and this has significant effects on the hydrology of the delta. The source of water for the Okavango lies in a series of headwater rivers in the highlands of Angola and Namibia. At Mohembo, in the confined part of the panhandle above the main delta, the seasonal flood peaks between February and April, while at Maun, downstream of the delta, the peak is between June and August (UNDP, 1977). Annual inflow is 11×10^9 m³ to which 5×10^9 m³ is added from rainfall onto the delta although this falls mainly between November and March (McCarthy, 1992; Hughes and Hughes, 1992). Wilson and Dincer (1976) suggest that only 0.3×10^9 m³ of water leaves the delta through the Boteti River, peaking between June and August, and a further estimated 0.3×10^9 m³ is lost to groundwater flows. The remaining 15.4×10^9 m³ must be lost by evapotranspiration (Table 15.1). There are about 6000 km² of permanent swamp in the Okavango, an area which more than doubles during the flooding period and varies considerably in depth from one location to another (McCarthy, 1992). The whole delta slows down the floods since a large percentage of floodwater is rapidly shed by distributary channels into extensive peaty swamp areas between them. The downstream movement of the flood is slow and takes about four months to flow from Mohembo to Maun. Each area of swamp is dependent on the channel system that serves it for its maintenance as a wetland ecosystem. Downstream areas are also dependent on the particular timing and delivery patterns which are created by the swamp and which spread delivery of water over a longer period of time.

Within the delta there are numerous wetlands characterized by different origins, substrates, and vegetation. The patterns of sedimentation in the delta reflect the hydrological pathways. The bedload consists of Kalahari sand eroded from the catchment and has an annual input of 170 000 t, 90 per cent of which is deposited in the panhandle (McCarthy et al., 1991). Suspended sediment has an annual input of 30 000 t and consists mainly of clays, organic matter, and siliceous bodies of plant origin called phytoliths (McCarthy, 1992). The annual input of total dissolved solids is large at 456 000 t (McCarthy and Metcalfe, 1990) although its concentration is low (35 ppm in the Okavango river, increasing to 95 ppm in the outflow: Wilson and Dincer, 1976; McCarthy and Ellery, 1994). It has a high silica content because much of its catchment is dominated by siliceous sands and granites. In places where channels aggrade and begin to carry less water, vegetation blockages occur, frequently of papyrus (Ellery et al., 1989) and flow is further reduced. Adjacent swamps become desiccated and the peat catches fire and burns off to give a thin deposit of fertile, clay-rich material on which new

grasslands develop, suitable for large herbivores such as hippopotamuses (Ellery et al., 1989). In this way, patterns of wetland vegetation change over space and time as a direct and indirect result of changing hydrological pathways.

Species such as *Cyperus papyrus* and *Phragmites australis* line channel margins and form permeable barriers which allow widespread lateral movement of water but filter sediments. Animals that use the delta can also exert a powerful influence on the direction of the hydrological changes. Hippopotamuses create trails which can develop into new channels (McCarthy, 1992).

It is clear that the Okavango Delta has a very important attenuation effect on downstream floods in the Boteti River system, acting as a sponge which gradually releases water following floods. It therefore reduces the flashy nature of the hydrological regime in areas downstream.

The links between upstream and downstream areas are again apparent in the Sudd of the southern Sudan. This, the most extensive wetland system in Africa is located at the confluence of the Nile and Bahr-el-Ghazal (Figure 15.3) and is fed annually by the floods of the Nile as they move northwards. Welcomme (1979) notes that the 10 000 km² permanent complex of papyrus swamps and open water swells to over ten times this area when the Nile floods arrive from the south (Table 15.1). Inflow to the Sudd is primarily a combination of outflows from Lake Victoria (80 per cent), Lake Kioga, and Lake Albert. Unlike the Okavango, the wet season (between April/May and October/November) is the result of simultaneous river floods and rainfall. Local rainfall, however, is far less important than rainfall over upstream areas of the catchment (Johnson, 1988). The resulting downstream floods are filtered during a four-month period through the Sudd's permanent swamps and lakes which act as a temporary storage, attenuating the effects of the floods (see Figures 15.3 and 15.4). In the Sudd, permanent and temporary swamps are replenished with water during this period, providing areas of livelihood for local pastoralists.

In the Kafue River Basin of Zambia, a number of floodplain and swamp areas have an important effect on the hydrograph. For example, the Lufupa River flows into the Busanga swamp where a large quantity of water is lost through evaporation. The outflow from the swamp is relatively slow, implying that the swamp functions as a natural temporary storage basin (Ellenbroeck, 1987).

In the mega-wetlands discussed above, there is an enormous dependence on surface flows, particularly in the form of overbank flooding. This is also the case in many smaller wetland areas of which a great number are linked to and in some way dependent on river

Table 15.1 Some hydrological characteristics of selected African wetlands

River system and/or wetland name	Wet season area (km²)	Dry season area (km²)	Mean annual or seasonal inflow (million m³)	Mean annual or seasonal outflow (million m³)	Evaporation (million m³ or %)	Sources
Nile River:						
Sudd (total)	92 000	10 000				Rzoska (1974)
Bahr-el-Jebel (below Bor)	29 800	16 200 (Permanent, swamp)	26 800 (1905–60) 33 000 (1905–80) 50 300 (1961–80)	14 200 16 100 21 400		Howell, Lock, and Cobb (1988)
Zaire River:						
Lualaba R.						
Kamolondo depression	14 400 11 840	7040				Hughes and Hughes (1992) Welcomme (1979)
Luapula R.						
floodplain on south shore of L. Mweru	3000					
Bangweulu swamp	7000					
Bangweulu Lake	2733					
Kwilu R.	1550					Hughes and Hughes (1992)
Lualua R.	4500					
Oubangui R.	30 000					
Middle Zaire	50 000 40 500					Gaudet (1992)
Niger River:						
Central Delta (wet year)	32 000 20 000	16 000 3877	50 000 (Niger R.) 20 000 (Bani R.) 250–500 (Yame R.)	35 000 (below Timbuktu)	50% of inflow	Hughes and Hughes (1992) Welcomme (1979); Hughes and Hughes (1992); Raimondo (1975)
fringing plains						
In Niger	907					
In Benin	274					FAO/UN (1970, 1971)
In Nigeria	4800					
Chari and Logone Rivers:						
Bahr Aouk and Bahr Salamat floodplains,	49 950					Hughes and Hughes (1992)
Yaéres	7000					
Lower Chari and Logone Rivers. (whole system)	63 000	6300	26 500 (Chari above Logone confluence) 12 600 (Logone above Chari confluence)	2400 (inflow to Lake Chad)	10 550	Welcomme (1979); Blache (1964) Welcomme (1979); Hughes and Hughes (1992)
Chari	29 150					Hughes and Hughes (1992)
Logone and El Beid	33 000					
Okavango River:						
Okavango Delta	18 000	6000	11 000 (Okavango river inflow) 5000 (rainfall)	300 (outflow) 300 (groundwater flow)	15 400 (96%) of inflow	McCarthy (1992); McCarthy et al. (1988); McCarthy and Metcalfe (1990); Wilson and Dincer (1976)
Zambezi River:						
Barotse floodplain	10 752 7700	537				Welcomme (1979) Hughes and Hughes (1992)
Kafue R.						
Kafue Flats	5666		10 716	8876	1839	FAO (1968); Coopconsult (1982); Hughes and Hughes (1992);
Lukanga swamps	4340 2100	1456	7381 (capacity between high and low water)			Welcomme (1979) Hughes and Hughes (1992)
Shire R.						
all swamps	1030	480				Welcomme (1979)
Elephant and	665	200				
Ndinde marshes	770					Hughes and Hughes (1992)
Volta River:						
fringing plain in Ghana	8532	1022				Welcomme (1979)
Rufiji River:						
Kilombero	6650 6625					Gaudet (1992) Hughes and Hughes (1992)
Ruaha R.						
Usangu Flats	1500					Hughes and Hughes (1992)
Senegal River:						
fringing floodplain	5000	500	22 809 (1903–87) 12 137 (1972 onwards)			Hollis (1990); Welcomme (1979)
Benue River:						
fringing plain	3100	1290				Welcomme (1979)
Komadugu Yobe River:						
Hadejia-Nguru floodplain	2000 (1964–71) 1–2000 (1972–82) 900 (post-1982)		4391 (river inflow) 567 (rain)	1055 (river outflow) 1465 (groundwater recharge, 24%)	2827.4 (64%)	Hollis, Adams and Aminu-Kano (1993); Schultz (1986); Adams and Hollis (1988)
Tana River:						
fringing floodplain	1500		6105 (pre-Masinga Dam 1944–78) 3935 (post-Masinga Dam 1982–7)			Hughes (1990); Hughes and Hughes (1992)
Pongola River:						
fringing floodplain	130		1082 (pre-Pongolapoort dam) 862 (post-dam)			Hughes and Hughes (1992)
	100	26				Furness and Breen (1980); Heeg et al. (1980)

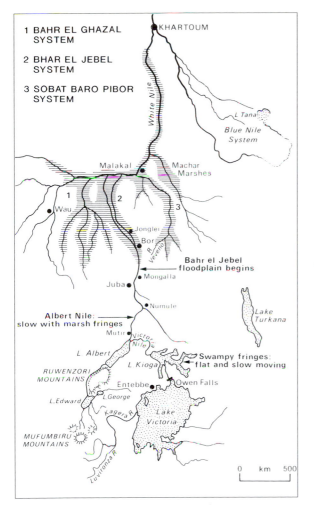

Fig. 15.3. The location of the Sudd within the hydrological pathways of the Nile Basin.

(Redrawn from Howell, Lock, and Cobb, 1988.)

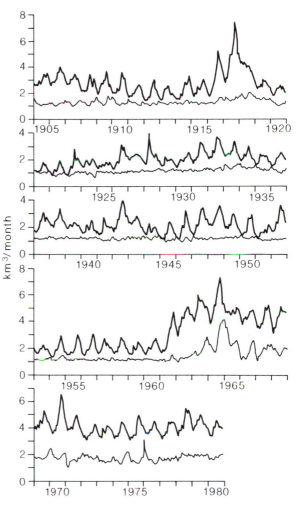

Fig. 15.4. Inflows to (–) and outflows from (–) the Sudd, 1905–80, in km³ per month.

Inflows measured at Mongalla; outflows derived by subtracting River Sobat flows at Hillet Doleib from White Nile flows at Malakal. The attenuating effect of the Sudd on the flows is evident here. (Redrawn from Sutcliffe and Parkes, 1987.)

systems. Most African rivers have well-developed fringing floodplains, and like the large wetland complexes also have large seasonal extents of flood water. For example, in the Senegal River floodplain, during the dry season, there are about 500 km² of water bodies in small lakes, 150 km² in the Lac de Guiers, and 281 km² in the main channel. At peak floods the river covers 5000 km² (Welcomme, 1979) (Table 15.1). Prior to construction of the Kainji Dam, the Niger River in the dry season covered about 35 per cent of its wet-season flooded extent. In a comparative study of four major African wetlands, the Senegal, Niger, Sudd, and Okavango, Sutcliffe and Parks (1989) show through the use of

water-balance models that the proportional areas of permanent and seasonal flooding vary between different wetlands and that these differences are reflected in the dominant vegetation species. Where seasonal fluctuations are great, seasonally flooded areas are more extensive than permanently flooded areas (6 : 1 ratio for the Niger and Senegal). Where seasonal fluctuations are less, perennial flooding creates extensive areas of permanent swamps (0.6 : 1 in the Sudd and less for the Okavango). Table 15.1 lists a range of hydrological characteristics for a selection of Africa's major wetlands.

Hydrologically Independent Wetlands

Relatively few African wetlands are hydrologically independent of river systems. This reflects the aridity of the continent and the paucity of cool, wet places in which ombrogenous peaty wetlands can develop. The few there are reflect the distribution of high land and precipitation. Peat develops in high-altitude bogs, medium-altitude swamps, and in brackish coastal swamps (Thompson and Hamilton, 1983) but the best conditions for peat development are low temperatures and consistently high water-tables resulting from high rainfall and/or considerable groundwater flows.

Although Mt. Cameroun may receive 9000 mm/yr of orographic rainfall (Thompson and Hamilton, 1983), it is small in area and the terrain is not suitable for development of peat bogs. In the Fouta Djallon and Nimba Mountains of Guinea, rainfall averages 2500–3000 mm/yr. Many peaks exceed 1000 m above sea-level, the highest, Mount Nimba lying at 1752 m. However, little is known about the flora or types of upland wetlands which occur here although one highland marsh near Gueckedou has an area approaching 5000 ha (Hughes and Hughes, 1992). Most African bogs lie in the equatorial zone of East Africa where the combination of high mountains and two rainy seasons favour their development. There is also a small area in the Drakensberg of South Africa and Lesotho. Most African upland bogs are dependent to some extent on groundwater movement. Several different types are distinguished by Thompson and Hamilton (1983) and Denny (1993), based on characteristic species although their distribution is related to hydrological factors which are in turn related to altitude. For example, *Sphagnum*-dominated bogs are rare and only occur in the highest rainfall areas of the Ruwenzori and in the Rukiga highlands of south-west Uganda. *Carex runssoroensis* mires are usually found between 3300 and 4500 m on all but the driest mountains, conditions at lower altitudes being too dry. A good example is Bigo Bog at 3500 m in the Ruwenzori of western Uganda (Thompson and Hamilton, 1983). In Lesotho, which has the highest mean altitude in Africa, bogs and mires occur in the subalpine and alpine belts, above about 2300 m, on basaltic rocks which allow ombrogenous bog development. For example, the summit of Thabana Ntlenyana, the highest mountain in Southern Africa, is partly encircled by bogs and they are also very extensive on the high plateau west of the Drakensberg escarpment (Hughes and Hughes, 1992). Most of the bogs in the Drakensberg have surface-water supplies with pH values of 7–8.5 making them unusu-

ally species-rich since most upland wetlands are normally subjected to strong leaching.

A number of ombrogenous mires have an effect on downstream hydrological pathways through their attenuation effects. At lower altitudes but still in zones of high precipitation, are the valley swamps in the volcanic mountains of the Ruwenzori where volcanic activity has dammed valleys allowing the formation of lakes and swamps. Some of these have developed extensive peat deposits. Many are found between 2000 and 3000 m above sea-level at the Equator but can occur at lower altitudes at higher latitudes. Good examples are to be found in Uganda's upland areas in the Ruwenzori and in the south-west corner within the Virunga volcanoes adjacent to Rwanda. For example, Lake Bunyoni in south-west Uganda was formed when a steep-sided dendritic valley system was blocked by volcanic activity around 18 000 BP (Denny, 1972). It has extensive papyrus swamps and drains to the Ruvuma and Mutanda Swamps. At a higher altitude, Muchoya Swamp, at 2256 m above sea-level just west of Lake Bunyoni, is dominated by large tussocks of the sedge *Pycreus nigricans* which is the main peat-former of this fen-like swamp. The pollen record preserved in these swamps provides an excellent record of their evolution. Morrison (1968) studied the pollen profile of a 1.2-metre sediment core from the centre of Muchoya Swamp. If the hydrophytes from the pollen diagram are considered, fine lake muds with *Potamogeton spp.* can be seen at the bottom of the profile dated around 24 000 BP, overlain by coarser sediments. Around 6000 BP, these mostly inorganic sediments give way to swamp peats containing macrofossils and pollen, especially of *Myrica kantdiana*, a swamp forest species (Figure 15.5) (Denny, 1985). This swamp forest finally gave way to the *Pycreus* swamp of today which is only 0.2 metres deep. While these changes in sediment type and vegetation reflect progressive changes in environmental factors such as sediment and nutrient availability they could also reflect changes in climate (see Goudie, this volume). Both autogenic and allogenic changes are well demonstrated in a core from the Kamiranzovu Swamp in Rwanda where the present-day cover of *Syzygium* swamp forest is the result of catchment downwash, dust accumulation, and recycling which have increased the nutrient status of the bog since it ceased to accumulate peat around 11 000 BP (Thompson and Hamilton, 1983).

Dambos are also wetland types with a strong measure of hydrological independence. Dambos are seasonally waterlogged, usually grass-covered, shallow, linear depressions, frequently without a marked stream channel (Mäckel, 1974; Acres *et al.*, 1985; Boast, 1990).

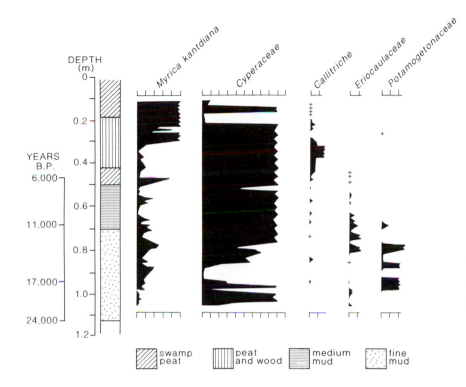

Fig. 15.5. Pollen diagram to show succession from submerged plants (e.g. Potamogetonacea) to emergent plants such as the swamp forest tree *Myrica kantdiana* in Muchoya Swamp, Uganda.

The increase in swamp forest pollen around 4000 BP, may reflect progressive sedimentation and succession but could also reflect a generally drier climate in tropical Africa at that time. (Redrawn from Denny, 1985; and Morrison, 1968.)

Slopes are mainly gentle and smooth and dambo boundaries are sharp and clear, marked by changes in vegetation that reflect changes in hydrology and soil (Acres *et al.*, 1985). While some subcategories include a river channel, most do not. They are usually found in the annual rainfall belt of 600–1500 mm at altitudes of up to 2700 m and their characteristic seasonal waterlogging is primarily related to the climatic regime although local geology, soils, and vegetation play a major role in local variations. Their hydrology is complex. Water movement within them reflects rainfall, the water-table rising with seepage from surrounding uplands, giving rise temporarily or in some cases permanently to waterlogged soil (Adams and Carter, 1987). Detailed water budgets of four dambos in Zambia were studied by Balek and Perry (1973). Surface runoff was the most important element in total runoff, with slope soil characteristics and vegetation cover affecting the runoff hydrograph, and evapotranspiration rather than baseflow being the key factor in groundwater storage depletion.

Hydroperiods and Wetland Ecology

Hydrological pathways and linkages are the prime determinant of the ecology of a wetland. The seasonal pattern of water-level changes in a wetland, the hydroperiod, is defined by Lugo *et al.* (1989) as the depth, duration and timing of flooding. The exact hydroperiod of a wetland is described by Mitsch and Gosselink (1993) as its 'hydrologic signature' since it determines the exact nature of the wetland. Wetland flooding regimes are usually strongly seasonal. In equatorial Africa many wetlands have bi-annual floods reflecting the movement of the Intertropical Convergence Zone and movement of westerly winds approaching from the Atlantic, which bring rain in two separate rainy seasons. The seasonal floods experienced by a wetland reflect rainfall over the wetland itself and, in the case of wetlands connected to a river system, rainfall patterns in remoter catchment areas. Seasonal floods bring new water and sediments to a wetland ecosystem and often remove accumulated organic debris. This is the case, not only in wetlands experiencing a very marked flood each year, as in many floodplains but also in those experiencing more stable hydrological regimes such as in wetlands on the shore of Lake Victoria where much organic material is exported from the lake-edge swamps each year, decreasing the likelihood of peat formation.

Much of tropical Africa is drained by a small number of large rivers such as the Zaire, Niger, Nile, and Zambezi. They and many smaller basins are characterized by marked flood periods which result from runoff during

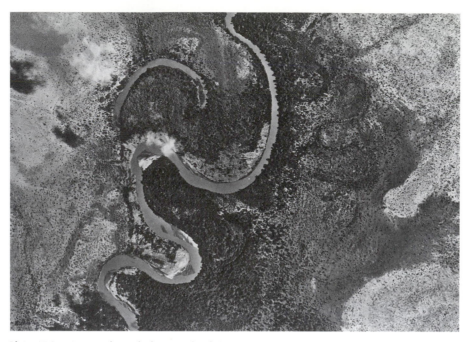

Plate 15.2. A vertical aerial photograph of the Lower Tana River, Kenya, upstream of Nanighi. The abrupt transition between the floodplain zone and the surrounding *Acacia-Commiphora* bushland is very clear on the right-hand side of the photograph. Extensive patches of evergreen forest, ox-bow lakes, and scars are clearly visible as well as areas of new sedimentation on the tips of point bars (photo: F. M. R. Hughes).

and following the rainy seasons. Many rivers have a mean maximum discharge up to ten times that of mean minimum discharge (Scudder, 1991). As shown, this marked flooding regime greatly alters the seasonal area of many wetlands such as the Okavango and the Sudd. However, the elevation range and duration of floods also have a marked effect on vegetation distribution in floodplain areas. In the Pongola floodplain of northern Natal, Furness and Breen (1980) emphasize the dependence of riparian vegetation composition on particular flooding conditions. They recognize six main communities characterized by relative durations of exposure and inundation, with forest types dominating in the higher parts of the floodplain. Hughes (1990) recognizes seven different floodplain vegetation types in the Tana River floodplain of Kenya (Plate 15.2), also strongly related to duration of flooding. Here, forest trees can only grow above elevations receiving floods of a critical maximum frequency and duration (around eleven days of continuous flooding), emphasizing that tolerance to flooding is a crucial determinant of species distribution.

Micro-elevational variations across a floodplain, related to geomorphological features, are important in determining the frequency and duration of flooding received by individual areas of the floodplain and therefore the species that will grow there. On the Bahr-el-Ghazal floodplain in the Sudan, Morison *et al.* (1948) noted that grassland vegetation dominated areas subject to frequent overbank flooding while evergreen trees grew most densely on higher ground, often on termite mounds. Michelmore (1939) pointed out that floodplains liable to frequent flooding are nearly treeless and dominated by grasses but that in areas of infrequent flooding, members of tree and bush genera such as *Hyphaene*, *Borassus*, *Acacia*, *Ficus*, and *Kigelia* may grow. These relationships are repeated in the floodplains of Lengwe National Park in Malawi (Hall-Martin, 1975) and in the Chobe river floodplain of Botswana (Simpson, 1975). A very clear vegetation zonation is also present on the Kafue Flats of Zambia (Ellenbroek, 1987), with some *Acacia albida* and *Borassus aethiopium* and tall grasses on levees, various different grassland communities at different elevations in the seasonally inundated floodplain proper, *Acacia* trees and *Hyphaene* palms on the highest termite mounds, and palm savannas in areas with higher water-tables near to the floodplain. The location of each vegetation type is determined by a combination of substrate type and water-table or flood regime, highlighting the importance of both flood duration and flood depth across the microtopography created by geomorphological processes.

Apart from separating vegetation types out along an elevational gradient, the flood regime has a less direct effect on vegetation types through its influence on plant regeneration. Thus, particularly in semi-arid areas, many species depend on floods for water-table recharge and for their annual water supplies while others rely on very particular flooding patterns for successful establishment of seedlings. Timing and magnitude of floods have direct impacts on the regeneration of floodplain tree species (Hughes, 1994). These impacts are effected through the geomorphological changes brought about by successive floods, the ratio of erosion to deposition in a floodplain and the availability of suitable sites for seedlings to flourish. Phenological characteristics of individual species are important in this regard, especially the timing of seeding with respect to flood timing. Studies of tree seedling regeneration and stand structures in floodplains of semi-arid parts of North America indicate a marked age structure reflecting a relatively infrequent occurrence of ideal conditions for successful riparian tree seedling establishment (Bradley and Smith, 1986; Hughes, 1994). It is likely that this situation is also common in many African floodplains. For example, in the Turkwel River floodplain of northern Kenya, *Hyphaene compressa* woodlands germinate and establish well during periods of unusually high flooding (Oba, 1991). Infrequent regeneration of floodplain tree species in association with high flood events is also suggested for the Tana River in Kenya (Hughes, 1984, 1990; Medley, 1992).

The timing of rainfall with regard to peak floods is also critical in many wetlands. Thus some floodplain areas such as the Bahr Aouk and Bahr Salamat floodplains in south-east Chad and north-east Central African Republic are flooded each year with water from the Sudan and the Central African Republic but also depend on sheet flooding from local rainfall and runoff (Welcomme, 1979).

Primary Production in Wetlands

Rivers transport both dissolved and solid sediment loads into wetlands. These loads and their associated hydrological conditions will in turn affect biogeochemical cycling and nutrient status within wetlands, and these will determine primary productivity. Chemical cycling can be divided into transformation processes which involve cycling within the wetland area and exchange of chemicals between a wetland and its surrounding area (Mitsch and Gosselink, 1993). The biogeochemical cycles obviously vary greatly between different wetland types.

Many African soils tend to be low in nutrients due to leaching and flood waters tend to be oligotrophic with a low buffering capacity (Walling, this volume). However, species present in many African wetlands have adapted their nutrient conservation mechanisms so that although their net primary production may be low in a nutrient-stressed environment, their standing biomass may not. Thompson *et al.* (1979) list productivities of dry matter as 7.1 t/(ha yr) for a nutrient-stressed papyrus swamp compared to over 20.3 t/(ha yr) for one which is more nutrient-rich. The actual biomass of these papyrus communities, however, is very similar and does not appear to be affected by nutrient shortage although it may take longer to accumulate and turn over. Wetland vegetation stands out for its high productivity and standing biomass in semi-arid zones. In the Tana river floodplain, levee forests had 499 m^3/ha of standing biomass compared with 32 m^3/ha in nearby *Acacia/Commiphora* bush (Hughes, 1990). Denny (1991) gives values of 17 t/ha for *Paspalidium* water meadows and 40 t/ha for *Vosia/Echinochloa scabra* grasslands in the Kafue Flats and of 8–10 t/ha for Sudd grasslands. Pools along the Nile and Zaire Rivers can be covered by *Eichornia crassipes* which is capable of achieving a productivity of 11–33 t/(ha yr) (Westlake, 1975), and Gaudet (1992) estimates that overall, most African floodplains should be capable of a sustained net production approaching 20 t/(ha yr). In more humid zones, wetlands can be less productive than surrounding areas because of water stress. An important input to floodplains can be dung from cattle and wild animals grazing during the dry season. Shepherd (1976) estimates that 3000 cattle visiting the 1000 km^2 Bangula lagoon in Malawi drop 500 kg of dung/yr in every 100 m stretch around the shore, making significant contributions to nutrient concentrations.

The availability of nutrients to a wetland ecosystem is closely linked to its flooding regime. Thus, seasonal floods on floodplains, in swamps and on lake margins will usually bring water with a high silt and dissolved nutrient load to the ecosystem and will usually also export organic matter from the ecosystem. Rainfall is an important factor in water movement in swamps. Following heavy rainfall there is a gradual increase in throughflow and a seasonal flushing of swamps. This seasonal rainfall flushing is particularly common in African lake-side swamps (Howard-Williams and Howard-Williams, 1978). It can result in considerable movement of bottom organic and inorganic material.

Despite year-round warm temperatures and high

radiation in tropical wetlands, productivity in bottom-rooted tropical swamps is no higher than in temperate swamps. Howard-Williams and Gaudet (1985) compare temperate prairie marshes with a productivity of 15–84 t/(ha yr) (Van der Valk and Davies, 1978) and sedge meadows with a productivity of 9–17 t/(ha yr) (Bernard and Gorham, 1978) with a *Typha* swamp in Africa which has a productivity of 16–30 t/(ha yr). (Howard-Williams and Lenton, 1975). There are two main reasons for this difference. First, high tropical temperatures increase respiration rates giving higher gross but not net production. Secondly, in temperate regions longer days during the growing season cancel the apparent advantage of continuously high radiation and little change in daylength in the tropics.

Wetland soils are the medium in which many chemical transformations take place and a major storage component of available chemicals for wetland plants. Within most wetland complexes, the variety of substrates will vary greatly. They range from the highly organic or peaty soils of swamp forests in the Zaire Basin, where low nutrient levels and high levels of humic acid give pH values sometimes as low as 3.8–5.0 to herbaceous swamps which can have organic substrates with an average pH of 6.7 (Thompson and Hamilton, 1983). Many fringing floodplains in semi-arid areas are dominated by mineral soils. Here, high organic contents are rare but mineral soils can range from sandy to clayey over short distances in response to the variety of sedimentation processes. Turner (1986) also comments on the variability of soils in dambos. The hydroperiod of a wetland is important in this regard because where there are marked seasonal flooding regimes, water movement and its import and export functions tend to inhibit accumulation of organic material. Only in valley swamps can humic acids accumulate sufficiently to raise the organic content of the substrate to higher levels (Thompson and Hamilton, 1983). Flood periods in the Zaire Basin have a bimodal pattern and can last for several months compared with the Tana floodplain where floods occur for only eight days on levees, twenty-eight days on point-bars, and forty-nine days in lowest parts of ox bows (Hughes, 1990), and with the Pongola floodplain where floods last for 8–60 days depending on elevation (Furness and Breen, 1980). The wide difference in the duration of anoxic conditions on different floodplains will predispose some to be dominated by peat and others by mineral soils. Ingram (1991) concludes that periodicity of saturation is the most important determinant of African wetland soil processes.

The quality of river water that inundates African floodplains, and therefore provides nutrients to these eco-systems, varies greatly depending on precipitation and the rocks and sediments over which the river flows. In semiarid areas, dissolved salts tend to become more concentrated because of the evaporation of river water (Welcomme, 1979). In the tropical rain-forest region of Zaire, rivers tend to rise on poor leached podzolic soils and are characterized by high levels of humic acids, low pH, and low dissolved nutrient concentrations (Matthes, 1964). Sulphates are common in rivers flowing from volcanic regions. Temperature variations and levels of electrical conductivity in different water bodies within wetlands reflect water mobility, vegetation cover, and catchment characteristics. Most river water will be well mixed and not show any thermal stratification but in floodplain lakes and other kinds of lakes, stratification can be pronounced and greatly affect the availability of nutrients for plants. It varies seasonally and sometimes diurnally.

Vegetation on a floodplain and in other wetland types plays a major role in trapping sediment and with it, nutrients, during floods. On lake shores, vegetation provides a diverse habitat for animals and plants, and acts as a filter and trap for allochthonous and autochthonous materials which in turn serve as nutrients for the plant communities themselves and secondary producers that use them (Howard-Williams and Lenton, 1978). The nutrient pump effect of the vegetation also reputedly increases the concentration of elements in littoral areas and contributes to autotrophic production in that, as vegetation decays, it forms a rich detritus which is utilized as food by many organisms.

Most of the primary productivity of floodplain wetlands is concentrated in the higher vegetation, frequently as on the Kafue Flats in perennial grasses. Plants act as nutrient sinks by taking up nutrients during flooding and returning them to the soil in the wet season. In this respect, it is clear from studies in various swamp types that root and rhizome development are very important. Howard-Williams (1973) found 50 per cent of the maximum biomass underground while Thompson (1975) found 30–40 per cent of papyrus biomass was root and rhizome material to which carbon is translocated when the aerial parts die off. If this 'recycling' is taken into account for various chemicals, net primary productivity is probably greater than once thought and possibly in the region of 48–143 t/(ha yr) (Thompson, 1979). Interestingly, crop plants can produce similar productivities but only when considerable levels of fertilizer are applied whereas natural swamps are achieving these levels of productivity in a low-nutrient environment (Howard-Williams and Gaudet, 1985).

In the Sudd, it appears that water flowing out has

Plate 15.3. Ox-bow lakes along the Lower Tana floodplain, Kenya provide fishing for local people and wildlife. Dense evergreen forest growth on the former cutbank includes many species such as *Hyphaene coriacea, Ficus sycomorus, Tamarindus indica, Trichilia emetica*, and several species of *Acacia. Terminalia brevipes* now colonizes the new lake edge (photo: F. M. R. Hughes).

lower oxygen, lower pH, lower levels of nitrates, phosphates and sulphates, and increased carbon dioxide than water flowing in (Gaudet, 1992). This implies that wetlands do act as nutrient traps, although it is hard to quantify how many nutrients are taken directly out of the water and how much from soils as they flood. Nutrient flushes with the onset of floods are obviously of great importance for wetland productivities and are recorded by Beadle (1974) in the Yaéres floodplain of the Chad Basin, by Reavell (1977) in the Okavango Delta following a drought in 1973, by Reizer (1974) on the Senegal floodplain, and by Ellenbroek (1987) on the Kafue Flats.

Secondary Production in Wetlands

In any wetland ecosystem, the consumers occupy various trophic levels ranging from certain bacteria and insects which feed directly on vegetation or decomposing organic matter, to top herbivores and carnivores such as lechwe, crocodiles, birds, fish, and humans. Seasonal use of floodplain resources by wildlife and humans has always been vital to both groups and are closely linked.

In the Pongola floodplain of South Africa fifty species of fish have been recorded (Heeg *et al.*, 1980). This large number of separate species coexists through opportunistic feeding and spawning habits, showing some specialization. Fish spawning on floodplains, and therefore fisheries on floodplains, are directly dependent on flooding (Welcomme, 1979) (Plate 15.3). Fish-tagging experiments along the Pongola show a high degree of endemism in separate floodplain waterbodies or pans, although during floods the river provides routes between pans to supplement breeding stocks (Heeg *et al.*, 1980). As flood waters cover floodplain substrates, nutrients and organic matter are released into them and they can support a rich invertebrate population on which juvenile fish feed. Furthermore, shallow flooded areas afford protection to young fish from aquatic predators.

The proportion of the primary productivity of different floodplains used by herbivores and detrivores varies greatly. Frequently, a relatively small percentage of the aquatic production is actually used by grazers. For example 5 per cent of *Potamogeton crispus*, one of two major aquatic communities in the Pongola floodplain is used by ducks, the main grazers; the rest enters into the detritus system and is consumed by snails. Usually, grazers use the seasonally flooded rather than the aquatic parts of the floodplain ecosystem. Examples of levels of use of seasonally inundated floodplains by grazers include 55 per cent of *Cynodon dactylon* (grass) removed by grazing cattle and hippopotamuses on the Pongola floodplain (Rogers, 1980); and 80 per cent of *Paspalidium* water-meadows and 40 per cent of *Echinochloa* grasslands removed by grazing lechwe in the Kafue Flats (Ellenbroek, 1987). Despite the high productivities of many wetlands, relatively few herbivores actually use permanently wet areas since only a few large animals such as the hippopotamus and lechwe are adapted to very wet environments or unstable substrates.

In contrast, the margins of many permanent water bodies and swamps and many seasonally flooded areas are very rich in both bird and animal life that makes use of water and food sources. The main reasons why many *Papyrus* and *Typha* swamps are not used is their inaccessibility, their unpalatability, and the possibility that grazing management such as controlled burning and occasional heavy grazing reduce productivity and therefore further grazing. Moss (1979) suggests that some wetland ecosystems such as papyrus swamps may have evolved to eliminate any species which can change the swamp structure. Floodplain grasslands however, are clearly very productive, highly palatable, and heavily grazed by many herbivores. They are more productive when grazed and tend to be replaced by unpalatable grasses when they become overgrazed (Howard-Williams and Gaudet, 1985).

Forested floodplain areas are notable in several places for their primate populations. In the Tana River Primate Reserve in Kenya, two species of endemic monkeys have been intensively studied (Homewood, 1978; Marsh, 1978, 1985; Kinnaird, 1992). Here flowering and fruiting phenologies of important food items vary through the year giving year-round food availability (Kinnaird, 1992). Dam construction in the Tana headwaters may alter not only germination patterns and potential in these floodplain forests but also flowering phenologies and hence availability of fruit and seeds and in turn pollinators and seed dispersers (Kinnaird, 1992). Here, changes in the hydrological pathways of the wetland ecosystem may have a major and catastrophic effect on a very limited and rare primate population through a number of linked routes. Wooded termitaria on the Kafue Flats are primarily used as shelter for many birds and for animals such as the aardvark and honey badger which otherwise use open grasslands (Ellenbroek, 1987). The presence of some animal species can improve germination potential in some areas. For example, the passage of seeds through ruminants such as impala, gazelle, and elephants can improve germination success (Lamprey *et al.*, 1974) and on the Turkwel River floodplain in

northern Kenya, domestic herbivores such as sheep, goats, and cattle play an important role in improving germination of *Acacia tortilis* seedlings (Oba, 1991; Adams, 1990). Successful germination of *Acacia tortilis* can however be adversely affected by bruchid beetle larvae, while the seedlings can suffer from predation by herbivores or from desiccation. Butter Barbel fish (*Eutropius depressirostris*) eat the fruit of *Ficus sycomorus* in the Pongola floodplain and play a role in its dispersal. It is not possible to give a full account of animal use of wetlands here, but it is clear that it is closely related both directly and indirectly to flooding regimes and to the maintained integrity of a wetland's hydrological cycle.

Human Use and Management of Wetlands

Floodplains both large and small are the most important wetland areas for both wildlife and human use. Their location within the hydrological system ensures that flood waters and sediments maintain their productivity. Fishing, grazing by domestic cattle, hunting, and the collection of plant products (food, fuel, and fibre) form important elements within the indigenous economy of large African floodplain wetlands (Adams, 1992; Scoones, 1991). Floodplain fisheries support many rural economies (Welcomme, 1979, 1985). In the Chari-Logone floodplain in Cameroon and the Niger Inland Delta in Mali as many as 90 000 tons of fish have been landed in a single year (Konare, 1977). Most of the floodplains of West Africa and many of those of East and Central Africa support indigenous agricultural systems dependent on flood-related cropping. The indigenous rice cultivation of small wetlands of Sierra Leone or the Niger Inland Delta, or the sorghum cultivation of the Logone-Chari floodplain or Lake Chad are particularly notable (Adams and Carter, 1987). In some West African wetlands, intensive cultivation has led to a long tradition of irrigation based on the *shadoof* (Adams, 1992).

The availability of grazing resources in the dry season in floodplain wetlands often allows pastoralists to exploit very large areas of surrounding land where forage is only available for limited parts of the year, and the economic and ecological integration of wetland and dryland is often close (Scoones, 1991). Grazing, fishing, and agriculture are also often integrated closely within or between human communities, the floodplain of the Senegal River being a good example (Adams, 1992; Scudder, 1991). There is growing awareness that these human activities in African wetlands are of very real economic value (Dugan, 1990; Adams, 1993). The importance of direct economic products (agricultural crops, fish, fuelwood, grazing and browse resources, building materials) is now being recognized and measured, for example at Lac d'Ichkeul in Tunisia (Thomas et al., 1991) and in the Hadejia-Jama'are floodplain in Nigeria (Barbier et al., 1991; Adams and Kimmage, 1992). There are also important indirect economic benefits of wetlands, including their capacity to recharge groundwater in aquifers that supply extensive areas.

Conventional economic appraisal of wetlands in Africa has usually failed to recognize the scale and importance of their existing economy, and has concentrated instead on the development of new resources through investment in development projects, particularly irrigation. The economic performance of irrigation schemes in Africa has been poor, and the costs of development high. A reassessment of the relative merits of the transformation of wetland ecosystems in this way has now begun, leading to the more sensitive development of existing resources and economic activities (Adams, 1992). This is matched by a renewed interest in the conservation of wetland resources in Africa. Their importance is based on their resident populations of wildlife, their importance for migratory species (particularly birds), and the extent to which large human populations depend on the natural cycle of flood and desiccation (Dugan, 1993). Knowledge of the extent and nature of wetlands is slowly improving (e.g. Hughes and Hughes, 1992; Crafter et al., 1992). The Wetlands Programme of the World Conservation Union (IUCN), for example, is supporting national programmes aimed at the conservation and sustainable use of wetlands.

Dam construction has had a significant impact on many wetlands in Africa through the alteration of the hydroperiods of rivers. Not only have dams frequently reduced overall flows owing to evaporation in the reservoir area, but flood peaks are usually reduced and the timing of floods has often been greatly altered. Naturally, floodplains evolve over time in response to changing climate and runoff patterns, changing patterns of sediment delivery and availability within the catchment, and land-use change within the catchment. The species composition on floodplains has altered in response to these changes and individual species have evolved adaptations to cope with the disturbances caused by floods, in some instances becoming dependent on floods for their regeneration. However, dam construction and other schemes designed to manipulate water, cause a sudden change in hydrological pathways with major consequences for downstream wetlands. These consequences may cause immediate mortality followed by new

successional directions or causes slower, more subtle changes such as reducing regeneration potential through loss of suitable regeneration sites. On the Kafue Flats, construction of dams both upstream at Itezhitezhi and downstream at Kafue Gorge has had radical effects on the natural flooding cycle, leading to a decrease of the temporarily flooded area, an increase in the permanently flooded area, a delay in the start of the river floods, and a reduction in flood velocities. Comparison of flooded areas in the flats before and after dam construction illustrates this (Figure 15.6). Tall floodplain grasslands are important for lechwe feeding and require a short period of deep flooding (50 cm or more) to develop properly. Decreased flooding of these areas since dam construction has allowed their continuous grazing by lechwe with overall loss of productivity and carrying capacity (Chabwela and Ellenbroek, 1990).

In the Tana River of Kenya, dam construction has altered the timing and magnitude of floods and relatively low regeneration levels of some floodplain species have been noted (Hughes, 1990; Medley, 1992). In the Pongola floodplain of South Africa, about 8000 ha lie between high flood level and the level at which floodplain pans and the main river lose contact with one another. This area is normally flooded for very short periods (Breen et al., 1978). Construction of the Pongolapoort dam extended the flooding period but the dam was then managed to give two annual releases instead of the single pre-dam flood (Scudder, 1991). Changed timing of floods has had a significant and adverse effect on various ecological communities (Heeg et al., 1980). Reduced flooding can have a significant effect on soil chemistry. In the Okavango Delta, precipitation of salts is a major depositional process, particularly in permanently flooded areas. In seasonal swamps, flood waters leach salts downwards (McCarthy and Metcalfe, 1990). Reduced flooding could increase the incidence of shallow calcic and sodic horizons if dams and channelization proposed in the Southern Okavango Integrated Water Development Project are implemented (IUCN, 1992).

Clearly the economic value of wetlands in Africa is significant, and so too are the threats to those values. The sustainable management of wetlands and their natural resources is a major challenge that is only now starting to be addressed. It will have to happen through better environmental appraisal, better consideration of costs and benefits, and of who bears the costs and enjoys the benefits (Adams, 1993). Controlled releases of water from dams in order to satisfy some needs of both human agricultural systems and natural ecosystems located downstream are being tried on the Pongola and Senegal

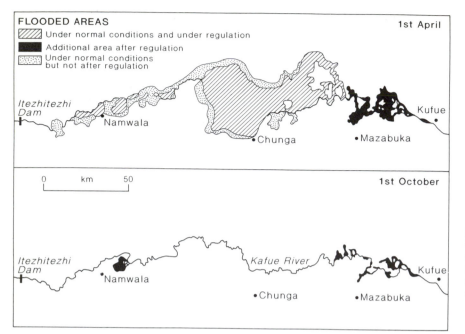

Fig. 15.6. Flooded areas of the Kafue Flats before and after regulation in dry years on 1 April and on 1 October. (Redrawn from Ellenbroeck, 1987; and SWECO, 1971.)

Rivers with some encouraging results (Scudder, 1991). It seems possible that such compromise solutions to restoring hydrological pathways could contribute to the sustainable development of wetlands.

References

Acres, B.D., Blair Rains, A., King, R. B. Lawton, R. M., Mitchell, A. J. B., and Rackham, L. J. (1985), 'African Dambos: Their Distribution, Characteristics and Use', *Zeitschrift für Geomorphologie*, 52: 63–87.

Adams, W. M. (1990), 'Dam Construction and the Degradation of Floodplain Forest on the Turkwel River, Kenya', *Land Degradation and Rehabilitation*, 1: 189–98.

—— (1992), *Wasting the Rain: Rivers, People and Planning in Africa* (London).

—— (1993), 'Indigenous Use of Wetlands and Sustainable Development in West Africa', *Geographical Journal*, 159/2: 209–18.

—— and Carter, R. C. (1987), 'Small-scale Irrigation in Sub-Saharan Africa', *Progress in Physical Geography*, 11: 1–27.

—— and Hollis, G. E. (1988), *Hydrology and Sustainable Resource Development of a Sahelian Floodplain Wetland*, Report for the Hadejia-Nguru Wetland Conservation Project to RSPB (Sandy, England and Gland, Switz.).

—— and Kimmage, K. (1992), 'Wetland Agricultural Production and River Basin Development in the Hadejia-Jama'are Valley, Nigeria', *Geographical Journal*, 158: 1–12.

Amoros, C. Roux, A. L. and Reygrobellet, J. L. (1987), 'A Method for Applied Ecological Studies of Fluvial Hydrosystems', *Regulated Rivers*, 1: 17–36.

Balek, J. and Perry, J. E. (1973), 'Hydrology of Seasonally Inundated African Headwater Swamps', *Journal of Hydrology*, 19: 227–49.

Barbier, E. B., Adams, W. M. and Kimmage, K. (1991), *Economic Valuation of Wetland Benefits: The Hadejia-Jama'are Floodplain, Nigeria*, IIED-LEEC Paper DP 91–02 (London).

Beadle, L. C. (1974), *The Inland Waters of Tropical Africa* (London).

Blache, J. (1964), *Les poissons du Bassin du Tchad et du bassin adjacent du Mayo Kebbi*, Mem. ORSTOM 4.

Bernard, I. M. and Gorham, E. (1978), 'Life History Aspects of Primary Production in Sedge Wetlands', in R. E. Good, D. F. Whigham, and R. L. Simpson (eds.), *Freshwater Wetlands* (New York), 39–51.

Boast, R. (1990), 'Dambos: A Review', *Progress in Physical Geography*, 14: 153–77.

Bradley, C. E. and Smith, D. G. (1986), 'Plains Cottonwood Recruitment and Survival on a Prairie Meandering River Floodplain, Milk River, Southern Alberta and Northern Montana', *Canadian Journal of Botany*, 64: 1433–42.

Breen, C. M., Furness, H. D., Heeg, J., and Kok, H., 'Bathymetric Studies on the Pongolo River Floodplain', *Journal of Limnological Society of South Africa*, 4: 95–100.

Chabwela, H. N. and Ellenbroek, G. A. (1990), 'The Impacts of Hydroelectric Developments on the Lechwe and its Feeding grounds at Kafue Flats, Zambia', in D. F. Whigham, R. E. Good, and Kvet (eds.) *Wetland Ecology and Management: Case Studies*, Tasks for Vegetation Science 25 (Dordrecht, Neth.), 95–103.

Coopconsult (1982), *Zambia Inland Fisheries Development Project* (Rome).

Crafter, S. A., Njuguna, S. A., and Howard, G. W. (1992), *Wetlands of Kenya*, IUCN Wetlands Programme (Gland, Switz.).

Denny, P. (1985), 'Wetland Vegetation and Associated Plant Life-Forms', in P. Denny (ed.), *The Ecology and Management of African Wetland Vegetation* (Dordrecht, Neth.), 1–18.

—— (1991), 'Africa', in M. Finlayson and M. Moser (eds.), *Wetlands* (London).

—— (1992), 'Lakes of South-Western Uganda, I: Physical and Chemical Studies on Lake Bunyoni', *Freshwater Biology*, 2: 143–58.

—— (1993), 'Africa', in D. Wigham, et al., Wetlands of the World, 1. Inventory, Ecology and Management, Handbook of Vegetation Science 15/2 (Dordrect, Neth.), 1–45.

Ellenbroek, G. A. (1987), 'Ecology and Productivity of an African Wetland System' (The Hague).

Ellery, W. N., Ellery, K., McCarthy, T. S., Cairncross, B., and Oelofse, R. (1989), 'A Peat Fire in the Okavango Delta, Botswana and its Importance as an Ecosystem Process', African Journal of Ecology, 27: 7–21.

FAO (1968), Multipurpose Survey of the Kafue River Basin, Zambia, 7 vols. (Rome).

FAO-UN (1970), Report to the Government of Nigeria on Fishery Investigation on the Niger and Benue Rivers in the Northern Region and Development of a Programme of Riverine Fishery Management and Training, Based on the Work of M. P. Motwani (Rome).

—— (1971), Rapport au gouvernement du Niger sur le development et la rationalization de la peche sur le fleuve Niger, etablis sur la base des travaux de N. Bacalbasa-Dobrovici, technologist des peches (Rome).

Furness, H. D, and Breen, H. C. (1980), 'The Vegetation of Seasonally Flooded Areas of the Pongolo River Floodplain', Bothalia, 13: 217–30.

Gaudet, J. J. (1992), 'Structure and Function of African Floodplains', Journal of East African Natural History Society, 82/199: 1–32.

Hall-Martin, A. J. (1975), 'Classification and Ordination of Forest Thicket Vegetation of the Lengwe National Park, Malawi', Kirkia, 10: 131–84.

Halwagy, R. (1963), 'Studies on the Succession of Vegetation on Some Islands and Sandbanks in the Nile near Khartoum, Sudan', Vegetatio, 11: 217–34.

Heeg, J. Breen, C. M., and Rogers, K. H. (1980), 'The Pongolo Floodplain: A Unique Ecosystem Threatened', in M. N. Bruton and K. H. Cooper (eds.), Studies on the Ecology of Maputaland (Capetown), 374–81.

Hollis, G. E. (1990), Senegal River Basin Monitoring Activity: Hydrological Issues: Part II (Binghampton, New York).

—— Adams, W. M., and Aminu-Kano, M. (1993), The Hadejia-Nguru Wetlands: Environment, Economy and Sustainable Development of a Sahelian Floodplain Wetland (Gland, Switz.).

Hopkins, S. T., and Jones, D. E. (1983), Research Guide to the Arid Lands of the World (Phoenix, Ariz.).

Howard-Williams, C. and Gaudet, J. J. (1985), 'The Structure and Functioning of African Swamps', in P. Denny (ed.), The Ecology and Management of African Wetland Vegetation (Dordrecht, Neth.), 153–75.

—— and Howard-Williams, W. A. (1978), 'Nutrient Leaching From the Swamp Vegetation of Lake Chilwa, a Shallow African Lake', Aquatic Botany, 4: 257–67.

—— and Lenton, G. M. (1975), 'The Role of the Littoral Zone in the Functioning of a Shallow Tropical Lake Ecosystem', Freshwater Biology, 5: 445–59.

Howell, P., Lock, M., and Cobb, S. (1988) (eds.), The Jonglei Canal: Impact and Opportunity (Cambridge).

Hughes, F. M. R. (1984), 'A Comment on the Impact of Development Schemes on the Floodplain Forests of the Lower Tana River in Kenya', Geographical Journal, 150: 230–45.

—— (1990), 'The Influence of Flooding Regimes on Forest Distribution and Composition in the Tana River Floodplain, Kenya', Journal of Applied Ecology, 27: 475–91.

—— (1994), 'Environmental Change, Disturbance and Regeneration in Semi-arid Floodplain Forests', in A. C. Millington and K. Pye (eds.), Environmental Change in Drylands (Chichester), 321–45.

Hughes, R. H. and Hughes, J. S. (1992), A Directory of African Wetlands, IUCN (Gland, Switzerland and Cambridge; UK/UNEP) (Nairobi, Kenya); WCMC (Cambridge), 820 pp.

Ingram, J. (1991), Soil and Water Processes, Pt 2: Wetlands in Drylands: The Agroecology of Savanna Systems in Africa (London).

IUCN (1992), The IUCN Review of the Southern Okavango Integrated Water Development Project, Final Report ed. T. Scudder et al. (Gland, Switz.).

Johnson, D. H. (1988), 'Adaptations to Floods in the Jonglei Canal Area of the Sudan: An Historical Analysis', in D. H. Johnson and D. M. Anderson (eds.), The Ecology of Survival: Case Studies from N. E. African History (London and Boulder, Colo.), 173–93.

Kinnaird, M. F. (1992), 'Phenology of Flowering and Fruiting of an East African Riverine Forest Ecosystem', Biotropica, 24(2a): 157–94.

Konare, A. (1977), Collecte, traitment et commercialisation du poisson en plaines inondables, I CIFA Working Party on River and Floodplain Fisheries, mimeo, 211, 215, 32–45.

Moss, B. (1979), 'The Lake Chilwa Ecosystem: A Limnological Overview', in M. Kalk, A. J. McLachlan, and C. Howard-Williams (eds.), Lake Chilwa (The Hague), 401–15.

Lamprey, H. F., Halevy, G., and Makacha, S. (1974), 'Interactions Between Acacia Seed Bruchid Beetles and Large Herbivores', East African Wildlife Journal, 12: 81–5.

Lowe-McConnell, R. H. (1985), 'The Biology of the River Systems with Particular Reference to the Fishes', in A. T. Grove (ed.), The Niger and its Neighbours: Environment, History, Hydrobiology, Human Use and Health Hazards of the Major West African Rivers (Rotterdam).

Lugo, A. E., Brown, S., and Brinson, M. M. (1989), 'Concepts in Wetland Ecology', in A. E. Lugo, M. M. Brinson, and S. Brown (eds.), Forested Wetlands (New York), 53–85.

McCarthy, T. S. (1992), 'Physical and Biological Processes Controlling the Okavango Delta: A Review of Recent Research', Botswana Notes and Records, 24: 57–86.

—— and Metcalfe, J. (1990), 'Chemical Sedimentation in the Semi-arid Environment of the Okavango Delta, Botswana', Chemical Geology, 89: 157–78.

—— and Ellery, W. N. (1994), 'The Effect of Vegetation on Soil and Groundwater Chemistry and Hydrology of Islands in the Seasonal Swamps of the Okavango Fan, Botswana', Journal of Hydrology, 54: 169–93.

—— Stanistreet, I. G., and Cairncross, B. (1991), 'The Sedimentary Dynamics of Active Fluvial Channels on the Okavango Fan, Botswana', Sedimentology, 38: 471–87.

—— —— Ellery, W. N., Ellery, K., Oelofse, R., and Grobicki, T. S. A. (1988), 'Incremental Aggradation on the Okavango Delta-Fan, Botswana', Geomorphology, 1: 267–78.

Mäckel, R. (1974), 'Dambos: A Study in Morphodynamic Activity on the Plateau Regions of Zambia', Catena, 1: 327–65.

Marsh, C. (1978), 'Ecology and Social Organisation of the Tana River Red Colobus, Colobus badius rufomitratus' (Ph.D. thesis, University of Bristol).

Matthes, H. (1964), 'Les poissons du Lac Tumba et de la region d'Ikela: Etude systematique et ecologique', Annales Musée Royal de l'Afrique Centrale, 126, 204 pp.

Medley, K. E. (1992), 'Patterns of Forest Diversity along the Tana River in Kenya', Journal of Tropical Ecology, 8/4: 353–73.

Michelmore, A. P. G. (1939), 'Observations on Tropical African Grasslands', Journal of Ecology, 27: 282–312.

Mitsch, W. J. and Gosselink, J. G. (1993), Wetlands (New York).

Morison, C. G. T., Hoyle, A. C., and Hope-Simpson, J. F. (1948), 'Tropical Soil-Vegetation Catenas and Mosaics', Journal of Ecology, 38: 1–84.

Morrison, M. E. S. (1968), 'Vegetation and Climate in the Uplands of South-western Uganda during the Late Pleistocene Period', Journal of Ecology, 56: 363–84.

Moss, B. (1979), 'The Lake Chilwa Ecosystem: A Limnological

Overview', in M. Kalk, A. J. McLachlan, and C. Howard-Williams (eds.), *Lake Chilwa* (The Hague), 401–15.

Muthuri, F. (1992), 'Classification and Vegetation of Freshwater Wetlands', in S. A. Crafter, S. G. Njuguna, and G. W., Howard, *Wetlands of Kenya*, IUCN Wetlands Programme (Gland, Switz.), 79–85.

Oba, G. (1991), *The Ecology of the Floodplain Woodlands of the Turkwel River, Turkana, Kenya*, TREMU Technical Report No. D-2 (Nairobi).

Raimondo, P. (1975), *Monograph on Operation Fisheries, Mopti' in Consultation on Fisheries Problems in the Sahelian Zone*, CIFA Occasional Paper 4 (Bamako, Mali), 294–311.

Reavell, P. E. (1977), 'A Discussion of Factors Limiting Plant Plankton Growth in the Water of the Okavango Delta', *Botswana Notes and Records*, 9: 129–37.

Reizer, C. (1974), 'Définition d'une politique d'amenagement des resources halieutiques d'un ecosysteme aquatique complexe par l'étude de son environment abiotique, biotique et anthropique. Le fleuve Sénégal Moyen et Inferieur, D. Sc. thesis, Fondation Universitaire Luxembourgeoise, Arlon.

Rogers, K. H. (1980), 'The Vegetation of the Pongolo Floodplain: Distribution and Utilization', in M. N. Bruton and K. H. Cooper (eds.), *Studies on the Ecology of Maputaland* (Capetown), 69–77.

Rzoska, J. (1974), 'The Upper Nile Swamps: A Tropical Wetland Study', *Freshwater Biology*, 4: 1–30.

Schultz International Limited (1976), *Hadejia River Basin Study*, CIDA, 8 vols in an interim and final report.

Scoones, I. (1991), *Overview: Ecological, Economic and Social Issues. Part 1: Wetlands in Drylands: The Agroecology of Savanna Systems in Africa*, IIED Drylands Programme.

Scudder, T. (1991), 'The Need and Justification for Maintaining Transboundary Flood Regimes: The Africa Case', *Natural Resources Journal*, 31: 75–107.

Shepherd, C. J. (1976) (ed.), *Investigation into Fish Productivity in a Shallow Freshwater Lagoon in Malawi* (London).

Simpson, C. D. (1975), 'A Detailed Vegetation Study on the Chobe River in N. E. Botswana', *Kirkia*, 10: 185–227.

Sutcliffe, J. V. and Parks, Y. P. (1989), 'Comparative Water Balances of Selected African Wetlands', *Hydrological Sciences Journal*, 34/1: 49–62.

SWECO (1971), *Kafue River Hydroelectric Power Development, Stage 2. Part 1: Main report* (Lusaka).

Thomas, D. H. L., Ayache, F., and Hollis, G. E. (1991), 'Use and Non-use Values in the Conservation of Ichkeul National Park, Tunisia', *Environmental Conservation*, 18: 119–30.

Thomas, D. S. G. and Shaw, P. A. (1988), 'Late Cainozoic Drainage Evolution in the Zambesi Basin: Geomorphological Evidence from the Kalahari Rim', *Journal of African Earth Science*, 7: 611–18.

Thompson, K. (1985), 'Emergent Plants of Permanent and Seasonally-Flooded Wetlands', in Denny (ed.), *The Ecology and Management of African Wetland Vegetation*.

—— and Hamilton, A. C. (1983), 'Peatlands and Swamps of the African Continent', in A. J. P. Gore (ed.), *Mires: Swamp, Bog, Fen and Moor*, Ecosystems of the World 4B (Amsterdam), 331–73.

—— Shewry, P. R., and Woolhouse, H. W. (1979), 'Papyrus Swamp Development in the Upemba Basin, Zaire: Studies of Population Structure in *Cyperus papyrus* Stands', *Botanical Journal of Linnean Society*, 78/4: 299–316.

Turner, B. (1986), 'The Importance of Dambos in African Agriculture', *Land Use Policy*, 3: 343–7.

UNDP (1977), *Investigation of the Okavango Delta as a Primary Source of Water for Botswana*, UN/FAO Dev. Proj. DP/BOT/71/506, Technical Report (Gaborone, Botswana).

van der Valk, A. G. and Davis, C. B. (1978), 'Primary Production of Prairie Glacial Marshes', in Good, Whigham, and Simpson (eds.), *Freshwater Wetlands*, 21–37.

Welcomme, R. L. (1979), *Fisheries Ecology of Floodplain Rivers* (New York).

—— (1985), *River Fisheries*, FAO Fisheries Technical Paper 262 (Rome).

Westlake, D. F. (1975), 'Primary Production of Freshwater Macrophytes', in J. P. Cooper (ed.), *Photosynthesis and Productivity in Different Environments* (Cambridge), 189–206.

Wilson, B. H. and Dincer, T. (1976), 'An Introduction to the Hydrology and Hydrography of the Okavango Delta', in *Proceedings of the Symposium on the Okavango Delta* (Gaborone, Botswana), 33–48.

Worthington, E. B. (1976), 'The Conservation of Wetlands in Africa', in *Proceedings of the Symposium on the Okavango Delta*, 61–6.

16 Mountains

David Taylor

Introduction

Mountains are characterized by high altitude, steep slopes, and sharp climatic and ecological gradients. Such a definition precludes a majority of the highland areas in Africa, such as the extensive plateaux of Eastern and Southern Africa, because these tend to exhibit much more gradual topographic, climatic, and ecological variation. Instead it includes a relatively small number of widely distributed, generally isolated features, with the largest concentration found in association with the great rift valleys in Central and Eastern Africa (Figure 16.1).

Mountain systems are strongly influenced by levels of temperature and available moisture, both of which can vary greatly over relatively small distances and produce steep altitudinal gradients in climatic conditions and in the composition and structure of ecological communities. Indeed mountains may be more sensitive to shifts in climate than landforms with a more subdued topography, with relatively small climatic fluctuations producing major environmental responses. This has prompted fears that mountains may be particularly sensitive to global warming. Mountains certainly appear more prone to degradation, as slopes denuded of their vegetation cover become unstable and susceptible to erosion and collapse. This sensitivity has been put to

Thanks are due to Keith Scurr and John Garner of the School of Geography of Earth Resources, University of Hull (SGER) for assistance with the diagrams, and to Robert Marchant (SGER) for the use of his unpublished data and for critically reading the manuscript. Particular thanks are owed to Rob Wilde, DTC Uganda, and Jonathan Baranga, Institute for Tropical Forest Conservation in Uganda, for support during numerous recent visits to the mountains of south-west Uganda.

good use through the use of mountain systems as barometers of environmental change. Many studies into the Quaternary climatic and vegetation history of the African continent have utilized mountain-based evidence, as have studies into environmental change over much shorter time-scales. For example, fluctuations in the volume of glacier ice on Mount Kenya have been used to track variations in climate for the past hundred years or so. Both these research areas have contributed to our understanding of the past. They have also helped in understanding present environments and in reducing the uncertainty in predicting the future.

Eastern and Southern Africa are thought to have been important centres for hominid evolution (Leakey and Lewin, 1992). The mountains there may have played an important role in this by providing a varied environment in which natural resources were abundant, but where there was a lower incidence of debilitating diseases such as malaria. Subsequently climatic conditions more suited to rain-fed agriculture than those at lower altitudes would have encouraged the settlement of early farming communities, with immigration and natural increase resulting in the high rural population levels that are today a common feature of many mountainous regions. In south-west Uganda, for example, which includes the Rukiga Highlands (Plate 16.1) and part of the Virunga range of volcanoes, a long period of agricultural settlement, relative political stability, and an equable environment have all contributed to some of the highest rural population densities on the continent. Densities greater than 300 persons per km^2 have been recently recorded at District level (Statistics Department, 1992), with some parishes containing more than 600 persons per km^2 (Cary Farley, personal communication).

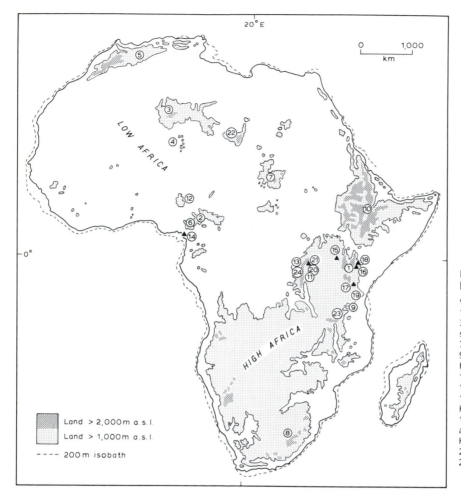

Fig. 16.1. The principal highlands and mountains of Africa mentioned in the text.

1 = Aberdares; 2 = Adamawa; 3 = Ahaggar; 4 = Air; 5 = Atlas system; 6 = Bamboutos; 7 = Dafur; 8 = Drakensberg; 9 = Eastern Arc; 10 = Ethiopian Highlands; 11 = Interlacustrine; 12 = Jos Plateau; 13 = Mitumbe; 14 = Mount Cameroun; 15 = Mount Elgon; 16 = Mount Kenya; 17 = Mount Kilimanjaro; 18 = Mount Meru; 19 = Pare and Usambara; 20 = Rukiga Highlands; 21 = Ruwenzori; 22 = Tibesti; 23 = Uzungwa; 24 = Virunga.

The Origin of African Mountains

Constructional Processes

The majority of Africa's high mountains originated in tectonic dislocations that began in the Mesozoic, but have been most active since mid-Tertiary times (see also Summerfield, Chapter 1; this volume). There were earlier periods of mountain building on the continent, such as during the Precambrian, although the supporting evidence has mainly been obscured or destroyed through erosion and further tectonic activity. Exceptions are the cratons which presently form the Congo and Kalahari Basins and which were uplifted more than 1500 million years ago. Within these 'older' cratons are found most of Africa's deposits of gold, diamonds, asbestos, and iron. Between them lie the 'younger orogens' which were formed during the last 1200 million years and which contain the majority of the continent's copper, lead, and zinc ores.

The Atlas system of North Africa (including the High Atlas, Atlas Saharien, and Rif Mountains) originated from the folding of basement rock along the line of an ancient rift, during the Alpine orogenic events that began in the late Cretaceous period and led also to the formation of the European Alps (Stets and Wurster, 1981). Previous phases of orogenesis in North Africa are indicated by Palaeozoic structures underlying the Alpine imprint. Rifting in the western part of the continent, which formed the Bight of Benin rifts including the Benue Trough, is of late Jurassic to early Cretaceous age and is associated with the opening of the South Atlantic (Wright, 1978, 1981; Fairhead and Binks, 1991; Wilson and Guiraud, 1992). Grove (1986) suggests that the Adamawa (or Cameroon) Highlands and the Jos Plateau

Plate 16.1. The Rukiga Highlands in south-west Uganda, showing intensive cultivation on the steep hillsides. Bunding is carried out extensively to minimize the risk from soil erosion and slope collapse. Lake Bunyonyi and an area of papyrus swamp are in the centre of the field of view (photo: D. Taylor).

may be the remnants of Cretaceous updoming across which the Benue Trough and the associated volcanics of the Cameroon Volcanic Line (CVL) developed. Volcanoes along the CVL have been active throughout the Tertiary and Quaternary periods and some are still active as three moderate earthquakes were felt around Mount Cameroun in 1989 alone (Ateba *et al.*, 1992). To the north and north-east a number of uplifted domes of basement rock occur. These are mantled with volcanic deposits, forming the Ahaggar, Air, Tibesti, and Darfur highlands, which date to the Tertiary and Quaternary periods (Vincent, 1970; Bermingham *et al.*, 1983; Pouclet and Karche, 1988; Dautria and Lesquer, 1989).

The eastern part of the continent is dominated by the Eastern and Western Rifts (see Chapter 2). These are, in geological terms at least, relatively young features and probably date from the late Miocene (Grove, 1983, 1986). The Eastern or Great Rift Valley actually comprises three rifts, centred upon Djibouti, presently occupied by the Red Sea, the Gulf of Aden, and the East African rift valleys see Nyamweru Chapter 2, this volume. They meet in an uplifted dome of approximately 1000 km in diameter which has been split into two; the smaller north-eastern segment now forms the highlands of southern Yemen whilst the remainder forms the Ethiopian Highlands. The Ethiopian Highlands are further divided into northern and southern massifs by the northern reaches of the Great Rift Valley, the opening up of which during the Tertiary also led to the formation of volcanoes along its margins. Recently active volcanoes are also associated with the Great Rift Valley in Kenya and northern Tanzania, the most impressive of which are the Aberdares, Mount Kenya, Mount Elgon, Mount Meru, and Mount Kilimanjaro. The highest of these, Kibo peak on Kilimanjaro (5895 m), is also the highest mountain in Africa and was last active during the late Quaternary, some 36 000 years ago (Kingdon, 1990).

The Western Rift system extends in an arc from southern Malawi to north-east Zaire. Numerous large mountains are also found in association with this feature and many, although not all, are volcanic in origin. The Virunga volcanoes in eastern Zaire, northern Rwanda, and western Uganda are of Quaternary, age and two, Nyamlagira and Nyiragongo in Zaire, have been active during the past two decades. In contrast, Ruwenzori (5109 m) is not volcanic but is formed from a block of Precambrian crystalline rock, that has been pushed up between two branches of the rift (Livingstone, 1967; Gerrard, 1990). Ancient, non-volcanic mountains are also found to the east of the Western Rift in the form of the Eastern Arc Mountains, a line of isolated blocks of crystalline Precambrian (Usagaran) gneisses, granulites, and amphibolites which sweep down through the eastern part of Tanzania. They include the Pare and Usambara Mountains in the north and extend south to the Uzungwa Mountains. Their progenitors are thought to have originated more than 100 million years ago from the fault-directed uplift of basement rock (Hamilton, 1989*a*; Lovett, 1993), although the present-day Eastern Arcs probably date only to mid-Tertiary tectonic activity (Burke and Wilson, 1972). The latter was associated with a general uplift of the Central African plateau and subsidence of the Congo Basin which ultimately led to altered drainage patterns and accentuated the east–west continental climatic and biogeographic divide.

The relief of the southernmost part of the continent is also a product of uplift, volcanism, warping, and rifting (King, 1978*a*; Grove, 1986). During the Precambrian, Southern Africa formed the central part of the supercontinent of Gondwanaland, and the previously mentioned cratons owe their origin to outpourings of basalt during this period. Subsequent subsidence during the late Palaeozoic to mid-Mesozoic allowed thick deposits of freshwater and deltaic sandstones and shales to accumulate in lower-lying areas forming the Karoo system. Similar deposits are found on all of the southern continents and provide compelling evidence for the former unity of Gondwanaland. During the late Mesozoic, subsidence was followed by massive emissions of basalt, as Gondwanaland fragmented and the southern continents drifted apart (King, 1978*b*). At this point Africa's continental margins were strongly monoclinal. Subsequent differential uplift, flexing, warping, and denudation has, through the resurrection of ancient mountains such as the Cape Fold Belt and the creation of new ranges such as the Drakensberg, resulted in the present topography of Southern Africa.

Denudational Processes

Denudational processes lead eventually to an overall levelling of mountain topography and comprise weathering, mass movement, and erosion by water. The activity of ice is important at the highest altitudes and would have been more so during Pleistocene ice ages (see Chapter 3). Both physical and chemical weathering occur, with their relative importance at any one location being a function of temperature, moisture availability, and the extent of diurnal and seasonal variations in climate. Thus at high altitudes on mountains in East Africa, freeze–thaw is thought to be an important agent in rock weathering (Hedberg, 1964; Hastenrath, 1984; Mahaney, 1990), as is exfoliation brought about by the frequent expansion and contraction of rocks and their minerals. At lower altitudes, particularly in warm and moist environments, chemical weathering becomes relatively more important. During wet periods water can be an important agent of erosion and downslope movement of material, particularly when rates of sheet and channel flow are enhanced by shallow soils, high relative relief, and low vegetation cover.

The greater part of the African continent has remained above sea-level since the fragmentation of Gondwanaland, and has thus been subjected to a long, uninterrupted period of denudation. Denudation, particularly during the late Cretaceous to mid-Tertiary periods, produced the great 'African' planation in which the immense savanna-dominated plateaux of Eastern and Southern Africa was fashioned (King, 1978*b*). Many of the products of this planation were either shed from the continent and into the adjacent oceans, or were deposited in the Kalahari and Congo Basins rather than accumulating *in situ*.

Contemporary Conditions on African Mountains

Mountains in Africa are distributed across some seventy degrees of latitude, from the Atlas system in the north-west to the Drakensberg in the south-east. The broad range of environmental conditions they incorporate is accentuated by variations in size and structure of the mountains themselves, as conditions can vary enormously on individual mountains, owing to differences in continentality, altitude, topography, and aspect.

Latitude

Latitude exerts its influence in a number of ways. In the absence of cloud cover, solar radiation received at the Earth's surface, and therefore mean annual temperatures, is greatest in the tropics and declines towards the poles. As a result, the tree-line is 300–400 m higher on equatorial mountains, such as Mount Kenya, than it is on mountains at mid-latitudes, such as the High Atlas (Messerli and Winiger, 1992; Figure 16.2). Latitude also influences the relative importance of seasonal and diurnal variations in climate, with diurnal variations tending to predominate on equatorial mountains, the direction of the prevailing surface winds, and the nature and timing of rainfall. Easterlies generally give way to westerlies at higher latitudes, whilst in low latitudes the precipitation system is primarily convective and localized, and associated with the passage overhead of the Intertropical Convergence Zone (ITCZ). The ITCZ is an ill-defined area of low pressure, towards which the trade winds of the two hemispheres converge, that follows the sun in its annual migration between the tropics of Cancer and Capricorn (see also Chapter 3). The ITCZ passes over the Equator twice a year and equatorial mountains therefore experience two wet seasons. The importance of the ITCZ declines towards the poles and cool air in the descending arm of the Hadley circulation predominates,

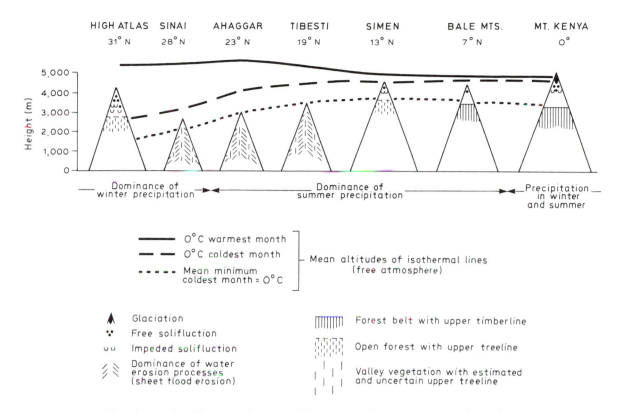

Fig. 16.2. Variation in altitudinal belts on selelcted mountains from the Equator to the Mediterranean.
(*Source:* Messerli and Winiger, 1992.)

resulting in low levels of precipitation. This is most pronounced in a large slice of Africa between 10 and 30° north of the Equator. At higher latitudes at the north-western and south-western tips of the continent, most rainfall is related to cyclonic activity during the winter months.

Continentality

Continentality exerts its influence through differences in heat capacity between land and water. As a result, the gap between maximum and minimum temperatures, on both a yearly and daily basis, is greater over continents than over oceans. Similarly, because land gains and loses heat much faster than water, there is a much shorter time-lag between maximum levels of insolation and maximum temperatures over land. Thus, on mountains that lie relatively close to substantial bodies of water, smaller annual and diurnal temperature ranges and longer lags between temperature and insolation maxima are to be expected when compared to those farther inland. A greater incidence of clouds and mist is also found on mountains situated near the coast

and these cause overall reductions in temperature and allow montane plant taxa to descend to abnormally low altitudes.

The degree of continentality is also determined by the extent of a mountainous area. Extensive upland areas, such as the high plateaux of Eastern and Southern Africa, can act as 'continents within continents' thereby enhancing the continentality effect and further influencing atmospheric circulation. The impact of large mountain systems on climates is more commonly known as the 'Massenerhebung' or 'mountain-mass effect'. Large mountain ranges can also affect atmospheric circulation by acting as a barrier to the flow of air. For example, the Mitumbe Mountains in Central Africa form a barrier to moisture-bearing airmasses moving eastwards from the Atlantic and thus limit the penetration of rainfall into Eastern Africa.

Altitude

Altitude, through its effect on levels of temperature and precipitation over relatively short distances, is responsible for the sharp climatic and ecological gradients found

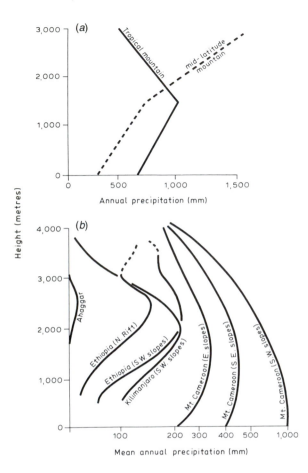

Fig. 16.3. Precipitation / altitude profiles; (a) comparison between tropical and mid-altitude mountains
(*Source*: Lauscher, 1976);

(b) comparison between African mountains.
(*Source*: modified from Lauer, 1975).

on mountains. In general, daytime temperatures fall with increasing altitude as a result of the 'adiabatic lapse rate'. The absolute rate at which this can occur is the rate at which an unsaturated air parcel cools when displaced upwards (the dry adiabatic lapse rate, DALR) and is 9.8 °C per 1000 m. Under normal conditions however the air parcel will be partly saturated and, owing to the release of latent heat from condensation, the actual or environmental lapse rate (ELR) is thus usually less and closer to 6 °C per 1000 m.

Spatial and temporal variations in atmospheric humidity mean that ELRs are never permanently fixed for a particular mountainous region. Low night temperatures on upper slopes and summits can lead to the formation

of katabatic winds and the accumulation at lower altitudes of cold, dense air, and eventually a temperature inversion, in which ambient temperatures actually increase with increasing altitude, rather than decrease. Fluctuations also occur on a seasonal basis. For example, Yacono (1968) states that ELRs on the Ahaggar Mountains differ by as much as 3 °C per km between January and July. ELRs are also likely to have been different during Quaternary cold stages, as a result of major variations in atmospheric moisture levels. Levels of precipitation are also influenced by altitude. According to Lauscher (1976) tropical mountains exhibit precipitation maxima at 1500 m, above which levels can rapidly decline (Figure 16.3a). This can be seen on Mount Kenya and Mount Cameroun, where precipitation levels above 3000 m are only 10 to 30 per cent of the maximum. Mount Kilimanjaro also shows a unimodal precipitation–altitude profile, with the peak in precipitation falling around 2000 m (Lauer, 1975; and Figure 16.3b). By comparison, the Ethiopian Highlands, with maxima at 2000–2500 m and again at 3000–3500 m, have a bimodal distribution of rainfall. At higher latitudes the trend is somewhat different, with precipitation levels tending to increase progressively with altitude up to 3000 m.

The altitude-driven differences in climatic conditions described above have significant impacts on the composition of mountain biotas, and on the morphology of the component organisms. They also determine the predominant weathering and mass movement processes, and therefore the nature and stability of hill-slopes (Gerrard, 1990).

Topography and Aspect

Topography and aspect are important determinants of micro-variations in mountain environments, as they influence levels of insolation received, rates of evaporation, soil chemistry and stability, and the availability of moisture. Slopes hidden from the sun will experience lower than ambient temperatures, with the effect being most pronounced in valleys within which cool air has accumulated at night. Aspect also affects levels of precipitation (Figure 16.3b). On Mount Cameroun, for example, the highest levels of precipitation occur on south-west-facing slopes as these lie directly in the path of the summer monsoon. Mountains close to the Equator on the eastern side of the continent also show differences in rainfall levels according to aspect. For example, the precipitation maxima for Mount Elgon and Mount Kenya occur on their western and south-eastern flanks (Winiger, 1981; Hamilton, 1987).

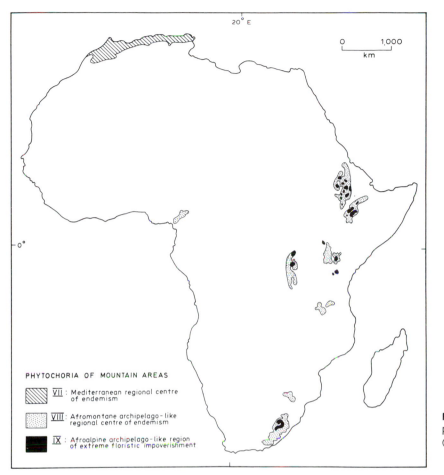

PHYTOCHORIA OF MOUNTAIN AREAS

VII : Mediterranean regional centre
of endemism

VIII : Afromontane archipelago-like
regional centre of endemism

IX : Afroalpine archipelago-like region
of extreme floristic impoverishment

Fig. 16.4. Mountain
phytochoria in Africa.
(*Source*: White, 1978, 1983.)

Africa's Mountain Biota

White (1978, 1983) has classified the flora of continental Africa into a total of eighteen phytochoria. These are regional divisions of vegetation, based upon floristic composition, with three largely being restricted to mountains (Figure 16.4). The 'Afromontane archipelago-like centre of endemism' is the most extensive mountain-based phytochoria and is generally found at altitudes greater than 2000 m, below which it grades into lowland vegetation types. On the highest mountains it passes into the 'Afroalpine achipelago-like region of extreme floristic impoverishment'. This has an extremely limited extent, as it tends only to be found on mountains higher than 3600 m and is thus restricted to Ruwenzori, the Virunga volcanoes, Mount Elgon, the Aberdares, Mount Kenya, Kilimanjaro, Mount Meru, the Ethiopian Highlands, and the Drakensberg. The third of the mountain-based phytochoria is the 'Mediterranean regional centre of endemism', comprising the vegetation of the Atlas system.

According to Dowsett (1986) it is possible to recognize a montane avifauna which comprises some 222 species of birds confined to high altitude during their breeding season in at least part of their range. These species have been separated into seven main groupings (Cameroon, Angola, South-eastern, Tanganyika–Nyasa, East Congo, Kenya, and Ethiopia), or montane 'islands', each of which has a characteristic avifauna (Moreau, 1966; Dowsett, 1971). Some sharing of species is apparent, however, and the relative level of this between the seven grouping has been used to shed light on the possible origins of the montane biota. For example, East Congo and Tanganyika–Nyasa share a high proportion

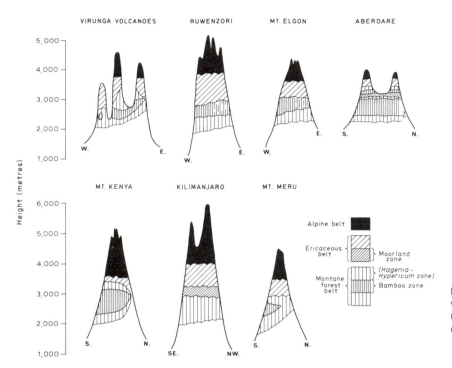

Fig. 16.5. Zonation of vegetation on East Africa's mountains.
(*Source:* Hedberg 1951.)

of their avifauna with western and eastern groupings, possibly suggesting that the mountains of Central Africa provided a conduit for dispersal across the continent.

White sets the boundary between lowland rain forests (i.e. the Guineo-Congolian phytochoria) and those of the Afromontane at around 2000 m. However, very little forest presently exists between the altitudes of 1500 and 2100 m, because these are the altitudes most favoured by farmers. Field-based studies have shown that those that do remain share many taxa of plants with lowland and montane types, particularly at the genus level, and species of fauna (Carcasson, 1964; Coe 1989). It therefore seems likely that today's remnants of lowland and montane forests are extremes of the same phytochoria, with human activity accentuating any differences now apparent between the two through clearing transitional forms. Such an explanation receives support from research in the few cases where a complete altitudinal range of forests still remains. In Uganda, for example, surveys of altitudinal changes in forest composition do not indicate the presence of an abrupt boundary, or 'critical altitude', between montane and lowland forest types, where large turnovers of species occur over relatively short distances. Instead they show a gradual gradation with some taxa more common in one formation also occurring well within the confines of the other (Hamilton, 1975).

Altitudinal Zonation

Altitudinal gradients also exist in the physiognomy of the dominant vegetation. They have been most fully described for tropical Africa by Hedberg (1951) who classified the vegetation into a series of *zones*, or locally based altitudinal bands, and *belts*, which tend to be broader and more widely visible on a regional scale (Figure 16.5). Hedberg recognized three belts, which replace one another with increasing altitude; the Montane Forest Belt below 3000 m, the Ericaceae Belt between 3000 m and 3600 m, and the Alpine Belt (subsequently renamed Afroalpine after a term coined by Hauman (1933)), above 3600 m. The latter correlates with White's 'Afroalpine achipelago-like region of extreme floristic impoverishment' mentioned above, whilst the Montane Forest and Ericaceae belts correlate with his 'Afromontane archipelago-like centre of endemism'.

Hedberg further divided the Montane Forest Belt into three zones; the Montane Rain-forest Zone of broad-leaved hardwood trees such as *Prunus africana*, the Bamboo Zone dominated by mountain bamboo (*Synarundinaria* spp.), and the *Hagenia-Hypericum* Zone, an upper montane forest type in which *Hagenia abyssinica* and *Hypericum revolutum* are frequently the most important trees. The lower boundary of the Montane Forest Belt is usually found between 1500 m and 2100

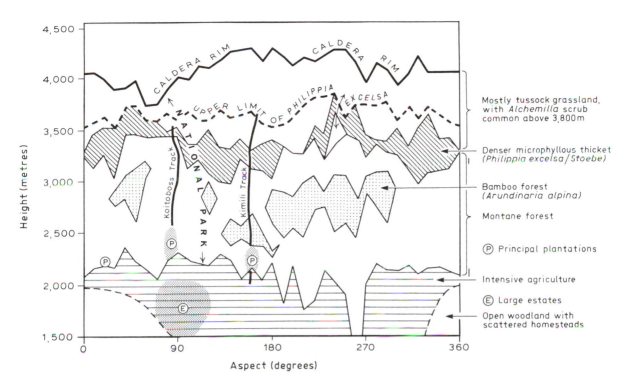

Fig. 16.6. Zonation of vegetation on Mount Elgon.
(*Source*: Hamilton and Perrott, 1981.)

m. On drier slopes it normally passes into dry wood-land/savanna, whilst in wetter areas it grades into low-land forest. Structural differences exist between lowland and montane forests, with montane forest tending to have more gaps, and fewer synusia (or strata), and a lower canopy.

Above the Montane Forest Belt is the Ericaceae Belt, which is usually dominated by ericaceous shrubs such as *Phillipia* spp. and *Erica arborea*. It may also include some broad-leaved trees, such as *Hagenia*, and in some cases, such as on the Aberdares, can be dominated by tussock-forming grasses. Where the mountains attain sufficient altitude the Ericaceae Belt passes into the Afroalpine Belt. The latter is commonly characterized by Giant Lobelias and Giant Senecios, although these are only common on the wetter mountains. On drier mountains, such as Mount Meru, extensive grasslands replace the characteristic Afroalpine flora.

Although relatively poor in species, the Afroalpine Belt has long held a special fascination for European scientists. J. W. Gregory was possibly the first to collect plant specimens from this zone during his expedition to the East African mountains in 1893, although these were subsequently lost on the return journey; he also

noted that the Afroalpine flora contained a number of north-temperate genera, and hence was probably the first to speculate over its origin, a subject which will be returned to later. The most complete study of Afroalpine ecology has been made by Hedberg (1964). He stressed the importance of the rarefied ambient atmosphere, diurnal fluctuations in temperature, and variations in microclimate in shaping morphological features adopted by plants, such as densely pubescent leaves, tussock form, giant leaf rosettes, and shortened stems.

Since the publication of Hedberg's work, several alternative models of altitudinal zonation have been published. Some of these are specific to a particular mountain, others to a group of mountains or region. An example of the former was produced for Mount Elgon by Hamilton and Perrott (1981) (Figure 16.6). The data were collected in part to assist in the climatic interpretation of sequences of plant fossils obtained from Mount Elgon, although the overriding influence of human activity on the vegetation to altitudes greater than 3000 m has limited this application. Differences in the intensity of human activity have caused both the upper and lower forest boundaries on Mount Elgon to fluctuate over a wide altitudinal range, with the most extensive stands of

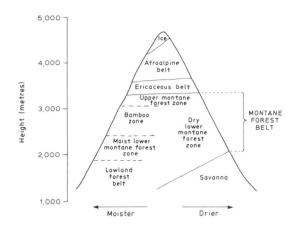

Fig. 16.7. Zonation of vegetation in East Africa.
(*Source*: Hamilton, 1982.)

montane forests and woodlands now located on the west-facing slopes which, in addition to being the wettest, are also the least attractive for farming.

Other models are intended for wider and more regional applications. One such example is the scheme put forward by Langdale-Brown *et al.* (1964) for Uganda (including Mount Elgon). They recognized four mountain forest zones: *Pygeum* (= *Prunus*) Moist Mountain Forest and *Juniperus-Podocarpus* Dry Mountain Forest between 1500 m and 2750 m on wetter and drier slopes respectively, *Hagenia-Rapanea* Moist Mountain Forest on all aspects at high altitudes (between 2600 m and 3200 m), and Mountain Bamboo Forest on wetter slopes above the *Prunus* Moist Mountain Forest. On the taller mountains in Uganda the *Hagenia-Rapanea* forest grades into a High Mountain Health community (the equivalent of Hedberg's Ericaceae Belt). At higher altitudes (3200–3800 m) this passes into High Mountain Woodland (the equivalent of Hedberg's Afroalpine Belt).

A later model of altitudinal zonation for East Africa is described by Hamilton (1982) (Figure 16.7). The Montane Forest Belt is divided into upper and lower altitudinal types, with the adjectives 'Moist' and 'Dry' being applied where appropriate. The Upper Montane Forest Zone occurs on all mountains of sufficient height and exhibits little floristic variation between mountains, indicating in part the uniformity in climate which exists at high altitudes in East Africa. More floristic variation is found in the Lower Montane Forest Zone, a moist form of which occurs on the climatically wet mountains and mountain slopes, where it is separated from the Upper Montane Forest Zone by the Bamboo Forest Zone. Common trees in the Moist Lower Montane Forest Zone on Mount Elgon are *Prunus africana*, *Polyscias* spp., and

Entandrophragma excelsum. In drier areas, a dry form of lower montane forest is found. This Dry Lower Montane Forest Zone passes directly into upper montane forest, and savanna grassland and scrub, at higher and lower altitudes, respectively. It is found on the drier north- and east-facing slopes of Mount Elgon, where it includes trees such as *Celtis africana*, *Juniperus procera*, *Olea africana*, and *Podocarpus gracilior* (Hamilton, 1982).

Vegetation also varies with altitude on mountains at higher latitudes in Africa. Although the ordering of the zones shows some similarities in terms of structure, they are not directly compatible on floristic grounds. Nevertheless Messerli and Winiger (1992) have attempted to correlate vegetation belts along a transect from Mount Kenya to the High Atlas. Significantly, montane forest has an upper limit at 3200 m on Mount Kenya, whilst its equivalent on the High Atlas, the evergreen oak forest, dominated by *Quercus ilex*, grades into woodland and scrub at 2800 m.

An example of altitudinal zonation of vegetation in Southern Africa is described by Hilliard and Burtt (1987) for the escarpment of the southern Natal Drakensberg, where forests dominated by the conifer *Podocarpus* extend to 2000 m. Above these, grassland and scrub predominates, with summit areas between 2900 and 3050 m dominated in places by a low heath community. The authors also describe the stark contrasts in floras that exist within the Drakensberg, with forests more extensive and plants able to extend their ranges to higher altitudes in northern compared to southern parts, and between the Drakensberg and other parts of the Afromontane phytochoria.

The Origin of the Mountain Biota

From a biogeographer's viewpoint, mountains in Africa can be seen as an archipelago of oceanic islands in which lowland vegetation types have replaced wide expanses of ocean as the major isolating factor. As a result less mobile members of the mountain biota, such as plants dependent on heavy seeds for dispersal, show relatively high levels of endemism (i.e. species with a distribution restricted to one or a few mountains) and vicariance (i.e. similar niches on different mountains are occupied by different species). High levels of endemism mean that a large proportion of the available gene pool is unique to that site, and mountains therefore have an important role to play in the maintenance of genetic diversity. They can have an even greater importance biologically when compared to surrounding lowland areas, because of the comparatively high level of habitat diversity, and therefore biological richness. Supporting evidence for this comes from the results of a study made by Lauer and Frankenberg (1979) in North Africa, in which the number of plant species per unit area on the Atlas, Ahaggar, Tibesti, and Air Mountains is shown to be upwards of twenty times greater than in the surrounding lowlands of the western Sahara.

The origin of this rich biota has long been debated. Similarities between present-day remnants of lowland and montane rain forest indicate that they share broadly similar histories. They thus probably date to the period during and since the later part of the Mesozoic, and the origination, diversification and spread of an angiosperm-dominated tropical flora. In addition, similarities between presently isolated blocks of montane forest in the western, central, and eastern parts of the continent suggest that they may have been more closely connected in the past. Alternatively, and more likely, the distributional patterns represent pathways of long-distance dispersal.

Unlike the more extensive Afromontane communities, approximately 80 per cent of the Afroalpine plant taxa on Equatorial mountains are endemic (Coe, 1989; Hedberg, 1961, 1970). Levels of vicariance are also high (Hedberg, 1970), which suggests that individual mountains have been mutually isolated for a considerable length of time. Some biological affinities exist between Afroalpine communities of individual mountains, and Hedberg (1961) uses these to divide the Afroalpine region into two subregions; eastern (along the Eastern Rift) and western (along the Western Rift). Of 278 Afroalpine plant taxa, 145 (52 per cent) are confined to the eastern and 34 (12 per cent) to the western groups. Only 66 species (24 per cent of the total) are found in both subregions, which is a testament to the efficiency of the largely hot and semi-arid African plateau as a barrier to dispersal. A number of Afroalpine species are also found at altitudes below the Afroalpine Belt (Hedberg, 1961; Mabberley, 1974), although more have their closest relatives on mountains in temperate latitudes, with the northern- and southern-temperate elements comprising 21 per cent and 8 per cent of the flora's composition respectively (Smith and Cleef, 1988). This suggests that a major route by which newly forming Equatorial mountains were colonized was from high to low latitudes in both hemispheres. As there are no interconnecting mountain ranges, many Afroalpine taxa may have established themselves during Pleistocene ice ages or earlier, when cold regions were more extensive and distances between alpine environments consequently less.

The extensive montane grasslands of Southern Africa, in which tussocks of *Themeda triandra* are almost ubiquitous, are commonly assumed to have had an anthropogenic origin (e.g. Chapman and White, 1970) and therefore similar to that of the high-altitude paramo vegetation of South America (Jon Fjeldsa, personal communication). This is partly based on the presence of forest remnants within areas of grassland, which has been taken to indicate the past dominance of trees before the relatively recent appearance of large human populations. Meadows and Linder (1993) have, however, recently proposed an alternative theory. Whilst accepting that the present landscape is maintained though human activity, they argue that a great deal of phytogeographical and geoscience data points to a substantial age for the Afromontane grasslands and to their maintenance over a long period of time by local soil conditions and the occurrence of natural fires.

Environmental Change on Africa's Mountains

The glaciated and snow-covered mountains of Africa were discovered only relatively recently. John Ludwig Krapf reported seeing a snow-capped Mount Kenya on 3 December 1849 (Krapf, 1860), whilst Rebman (1849) described seeing snow on the summit of Kilimanjaro. These reports were initially met with scepticism, although the presence of snow-capped mountains on the Equator in South America had been known for more than 300 years. The sceptics were not finally silenced until subsequent expeditions recorded similar sightings, and also described snow on Ruwenzori (Stuhlmann, 1894). Towards the end of the nineteenth century, Hans Meyer led a series of expeditions to Kilimanjaro, reaching the summit of Kibo in 1889, and made the first systematic observations of both the present and past extents of ice in East Africa (Meyer, 1890). Old moraines were also recognized on Mount Kenya by Gregory in 1893 and by Ross and Hutchins (Hutchins, 1909), and it soon became apparent that parts of Equatorial Africa had experienced major changes in environment in the past, the most immediately obvious effect of which was an extension of glacial features to altitudes 1000 m below those of the present. Ross and Hutchins also managed to photograph the Cesar and Tyndall glaciers, and their photographs have since proved extremely useful as a basis for determining recent rates of glacier retreat.

The first attempt to compare the evidence for past glaciations in East Africa was made by Eric Nilsson between 1927 and 1932 (Nilsson, 1931, 1935). He mapped and correlated tills and moraines on Mount Elgon, Mount Kenya, and Kilimanjaro. When the exercise is extended to Ruwenzori (Osmaston, 1965, 1975) and radio-carbon age determinations are included, two stages of glacier advance stand out: a Main Glaciation, the onset of which is still not well dated, and a Recent Glaciation which commenced during the last two millennia.

Studies on glacier ice fluctuations during the Quaternary are best advanced for Africa on Mount Kenya and are described in detail by Mahaney (1982, 1984, 1987, 1988) and Mahaney et al. (1991). Five glacial–interglacial cycles are presently recognized, beginning prior to the Bruhes–Matuyama boundary around 0.75 million years ago. The most recent of these are the Liki (Main) and Neoglacial (Recent) advances, which began 100 000 and 1000 years ago respectively. A second Liki advance (Liki II) may have taken place around 25 000 yr BP, whilst retreat of Liki glaciers commenced between 12 500 yr BP and 15 000 yr BP. The Neoglacial advances comprise the Tyndall and the Lewis stades. These commenced around 1000 yr BP and towards the end of the eighteenth century respectively.

Evidence for expanded glaciation during the late Pleistocene is also known from the Ethiopian Highlands (Hastenrath, 1977; Messerli et al., 1980; Potter, 1976; Hurni, 1989) and the High Atlas (Messerli and Winiger, 1992). Messerli et al. (1980) suggest that the Ahaggar Mountains may also have been glaciated, whilst evidence of nivational and periglacial activity in the past, common on mountains in East Africa, has been recorded on the High Atlas (Mensching, 1953), the Ahaggar and Tibesti Mountains (Rognon, 1967; Messerli, 1972; Messerli et al., 1980) and in Southern Africa (Harper, 1969; Hagedorn, 1984).

Causes of Environmental Change

A number of explanations have been put forward for greater extents of glacial and periglacial activity in the past. For example, Gregory (1894) suggested that the mountains themselves formerly reached to higher altitudes, whilst Nilsson (1931, 1935) proposed a pluvial theory, namely that the old tills and moraines represented cooler and wetter conditions during past Pleistocene ice ages. Pluvial theory was also used to explain formerly high lake-levels in East Africa, and dominated ideas about past climates in Africa, and in the tropics in general, until well into the 1960s. However, recent palaeoclimatic research has rejected the pluvial theory, at least for tropical Africa, in favour of the belief that conditions were significantly cooler and at times more arid than the present during many Pleistocene ice ages, including the most recent.

Climate has fluctuated widely during the last ice age, from 70 000 to 10 000 yr BP (Hamilton and Taylor, 1991). For example, the interval between 30 000 and 21 000 yr BP may have been a cool and relatively humid period when glaciers reached their greatest extent in the late Pleistocene. Maximum coldness and aridity was experienced throughout much of Africa around 18 000 yr BP, or the last glacial maximum (LGM), probably as a consequence of much reduced levels of insolation. The resultant lower ambient temperatures would have reduced rates of evaporation (therefore resulting in drier atmospheric conditions) and monsoonal activity, causing precipitation levels to fall throughout tropical Africa. In middle latitudes the LGM may have been marked by relatively moist conditions owing to the adoption of a more equatorward track by winter cyclones (Butzer, 1984; Messerli and Winiger, 1992).

Deglaciation commenced on Ruwenzori and on Mount Kenya 15 000 yr BP (Livingstone, 1967; Mahaney, 1987). Warmer temperatures during late glacial times led eventually to the development of wetter conditions throughout much of Africa, especially after monsoonal activity increased around 12 500 yr BP (Kutzbach and Street-Perrott, 1985; Kutzbach and Guetter, 1986). Warm, humid conditions in tropical Africa are believed to have reached their post-glacial maximum during the early Holocene (Hamilton and Taylor, 1991).

Although it has been possible to interpret the evidence for past climates obtained from mountains in broad terms, such as cooler and drier or warmer and wetter conditions, quantification of temperature and rainfall has proven to be much more difficult. Part of the problem lies in determining by how much the temperature – precipitation – altitude relationship has differed in the past as this has never been done to any satisfactory degree. Consequently the calculation of the levels of temperature and rainfall necessary for the past extents of glacier ice has tended to be based on modern environmental lapse rates (ELRs), even though they must have varied widely through time. Thus the descent of glaciers and associated features to 1000 m below their present levels on some East African mountains has been taken to imply continental temperature reductions of around 6 °C (Osmaston, 1965; Livingstone, 1975). Subsequently, Livingstone (1980) adjusted his estimates to 7.5 °C less than the present for the LGM, after incorporating the effect on ELRs of a 29 per cent reduction in precipitation, determined from changes in lake volume by Harvey (1976). Significant temperature reductions on the African continent do not, however, fit with results obtained from ocean sediments, which indicate that sea-surface temperatures in the tropics during the LGM remained much the same as today (CLIMAPP, 1981). In order to explain this apparent discrepancy, temperature–altitude gradients would need to have been much steeper than now. Some evidence in support of this is provided in the following section.

A more recent quantification of the terrestrial evidence has been attempted by Bonnefille *et al.* (1990), using Transfer Functions to translate fossil-pollen data directly into past levels of temperature and precipitation. Using this method, levels of rainfall in tropical Africa are shown to have been around 30 per cent less than the present during the LGM. Estimated temperatures are also less, by between 2 and 6 °C, which come closer to the CLIMAPP estimates for sea-surface temperature changes. However, these estimates should be treated with caution, because of doubts over the sensitivity of certain pollen types selected as proxy indicators of climate. Many of these, such as Gramineae, Ericaceae, *Alchemilla*, and *Myrica*, potentially include pollen from a broad range of habitats and altitudes, and hence cannot be used with any great precision.

Climatic change has continued to affect mountains in Africa to the present time. On Mount Kenya for example, the Lewis glacier has formed the basis for glaciological studies aimed at quantifying the glacier–climate relationship over the last 100 years or so, and at providing a means of monitoring contemporary variations in temperature and precipitation. The full range of studies includes investigations of ice-surface topography and velocity, the nature of bed-rock, ice thickness and stratigraphy, and glacier net balance (Hastenrath, 1984), and has highlighted the complexity of the relationship between climate and ice volume. This complexity did not, however, prevent Hastenrath (1984) from converting figures for ice volume into levels of solid precipitation and air temperature for the period 1889 to 1978, which suggest that the climate has got somewhat warmer and drier since 1889. Although the reductions in precipitation are not borne out by rainfall measurements made in the surrounding highlands during the present century, they are supported to some extent by variations in lake-levels in Kenya (Butzer, 1971). Hastenrath suggests that the differences between actual and estimated precipitation may be due to feedback processes initiated by variations in albedo, evaporation, and cloudiness during the period of study. He goes on to describe a 'most plausible' climate-change scenario for the last 100 years drawn from measured variations in ice volume. This comprises reductions in precipitation (by about 150 mm), cloudiness (by about 10 per cent), and albedo (by about 2 per cent) during the latter part of the nineteenth century, and warming by a few tenths of a °C and fall in albedo of around 1 per cent between the turn of the century and the 1950s.

The Biotic Impact of Late Quaternary Climatic Change

It has been possible to investigate the impact of substantial temperature and precipitation fluctuations during the last glaciation, and in particular the last 30 000 years, through the use of plant fossils which are abundantly preserved in mires and lakes in highland areas. Thus the late Quaternary vegetation history has been reconstructed, largely on the basis of fossil pollen and spores, for the mountains of North Africa (Lamb *et al.*, 1989), Equatorial Africa (e.g. Coetzee, 1967; Hamilton, 1982, 1987; Bonnefille and Riollet, 1988; Taylor, 1990, 1992, 1993; Jolly and Bonnefille, 1991), and Southern Africa (e.g. Coetzee, 1967; van Zinderen Bakker and Coetzee, 1988).

Reconstructions for the period during and since the LGM show substantial alterations in the relative extent and distribution of plant communities. In tropical Africa the extent of montane forest was much reduced 18 000 yr BP, even when compared to today's human-modified environment. It was replaced by high-altitude vegetation types, through a movement to lower altitudes of taxa presently associated with the Ericaceae and Afroalpine belts. Changes in pollen spectra in sediments deposited during the terminal stages of the last ice age indicate that rising temperatures and precipitation led to the expansion of montane forest at the expense of the scrub-grassland communities. Forest may have spread from small refuge areas, located on the wettest mountains. This spread appears from the pollen records to have continued, possibly in an episodic fashion rather than a smooth advance, until the maximum Holocene extent was reached 8000 years ago (Hamilton and Taylor, 1991).

Models of whole vegetation communities moving down (and up) the sides of mountains have tended to dominate theories of the history of vegetation on mountains. However, palaeoecological data from three sites in the Rukiga Highlands (Taylor, 1990, 1992, 1993; Marchant, unpublished data), provide good evidence that the biotic response was much more complex. The three sites (Ahakagyezi, 1830 m; Mubwindi, 2100 m; and Muchoya, 2230 m) are all peat-accumulating systems and are located within the current altitudinal range of the lower montane forest zone. They are particularly well suited to the role of monitoring past changes in the composition of the surrounding vegetation because their sediments are well dated (forty-nine radio-carbon ages), incorporate the period during and since the LGM, and lie along a 400 m altitudinal gradient. The sites also have small, well-defined catchments, with the bulk of

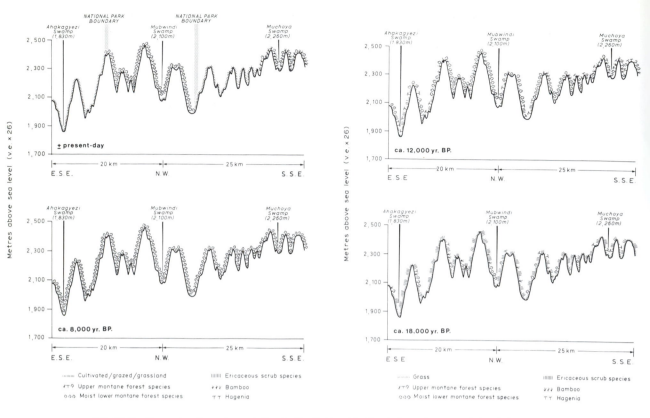

Fig. 16.8. Reconstruction of vegetation in the Rukiga Highlands from the present to 18 000 yr BP.

the pollen deposited originating between the altitudes of 1700 m and 2500 m above sea-level. Using sedimentary evidence from the three mires it has been possible to reconstruct not only the compositional changes in vegetation at a point, which tends to be the norm for palaeoecological investigations, but also to develop a more regionally based model. This is illustrated through four time 'windows'; the present, 8000 yr BP, 12 000 yr BP, and 18 000 yr BP (Figure 16.8).

1. Present

Degraded scrub and agricultural land occupies hillsides adjacent to Ahakagyezi Swamp, the lowest site. As moist lower montane forest is today found at similar altitudes in other parts of the Rukiga Highlands and as human activity is intense around Ahakagyezi, it can be assumed that the scrub is derived. Moist lower montane forest presently surrounds Mubwindi Swamp, midway along the altitudinal transect, as Impenetrable (or Bwindi) Forest. By contrast Muchoya Swamp, the highest altitude site considered here, is surrounded by mountain bamboo forest. As the latter is at an unusually low

altitude and contains numerous secondary forest trees, it probably represents a seral stage in the regeneration of lower montane forest.

2. 8000 yr BP

All of the land along the transect supported lower moist montane forest, which according to the fossil-pollen data seems to have been a wetter type than that presently found in Impenetrable Forest.

3. 12 000 yr BP

Vegetation was predominantly grassland and ericaceous scrub at the lowermost altitudes around Ahakagyezi Swamp; moist lower montane forest, with secondary forest taxa more common than the present, at mid-altitudes around Mubwindi Swamp; and upper montane forest (possibly including bamboo) at the highest altitudes around Muchoya Swamp.

4. 18 000 yr BP

Ericaceous scrub and grassland were present around Ahakagyezi Swamp. Some montane forest was present around Mubwindi Swamp, as were some taxa presently associated in Eastern Africa with high-altitude vegetation types. Possibly small fragments of forest were present at middle altitudes in the Rukiga Highlands, maintained by a combination of orographic precipitation and high soil moisture, whilst high-altitude vegetation descended to lower altitudes on the driest hillslopes. This high-altitude vegetation, presently associated with the Ericaceae Belt, was predominant around Muchoya Swamp.

It seems therefore that precipitation levels at the lowermost altitudes in the Rukiga Highlands, particularly during the LGM, were generally too low to support large expanses of montane forest. Instead, small patches may have survived on the wettest hillsides, whilst dry scrub predominated elsewhere. The pollen therefore more closely reflect conditions found today on the driest mountains in East Africa, where upper montane communities pass directly into semi-arid scrubland at lower altitudes, rather than a simple displacement, by about 1000 m, of entire vegetation belts. The pollen data also imply that small patches of montane forest were present at the same time as taxa presently associated with the Ericaceae Belt. As the latter are presently absent from the Rukiga Highlands, the data thus appear to indicate that either temperature–altitude gradients were steeper during the LGM, or that differences in vegetation cover according to aspect were greater than today.

In comparison to Equatorial Africa, the evidence from climatically sensitive mountains in northern and southern parts of the continent is more fragmentary. Lamb *et al.* (1989) describe expansions of evergreen oak forest on the High Atlas during the latest Pleistocene and Holocene times. The biotic history of the Ethiopian Highlands during the late Quaternary has been relatively poorly studied, despite their great biological importance, in terms of numbers of endemics (Kingdon, 1990); the distributional and climate modelling evidence to indicate that they may have provided a refuge for forests during the LGM (COHMAP, 1988; Hamilton, 1976); and the presence of suitable study sites. Hamilton (1982) describes pollen data from cores of sediment collected by A. T. Grove and F. A. Street-Perrott from two high-altitude mires (3830 m, Danka Valley, Bale Mountains, and 4040 m, Mt. Badda, Arussi Mountains) in glacier-cut valleys. The core from Mt. Badda provides the longer record of the two, but even this does not extend back beyond 11 500 yr BP, and both sites appear to contain incomplete records for the Holocene. The pollen spectra from Mt. Badda indicate the presence of montane forest on the mountain during the period immediately before 10 000 yr BP. They also appear to indicate a more

extensive cover of dry vegetation types at low altitudes, and possibly the colonization of exposed rift valley lake sediments by halophytic vegetation. Clearly much more remains to be learnt about the climatic and biological history of this important mountainous region. South of the Equator, Scott (1982) describes the displacement of vegetation in the southern Cape to 1000 m below its present altitude during the LGM, whilst *Olea*-dominated woodland replaced the previously open vegetation in the coastal region of South Africa between 14 200 and 12 000 yr BP (Deacon *et al.*, 1984).

The Impact of Humans on Africa's Mountains

Human impact on mountain systems is widespread throughout Africa (Grosjean and Messerli, 1988). This is not a recent phenomenon as human-induced forest clearance associated with soil erosion began some 2000 years ago in the Atlas Mountains (Lamb *et al.*, 1991) and the interlacustrine highlands of Central Africa (Jolly and Bonnefille, 1991; Taylor, 1990, 1992). Environmental degradation as a result of human activity is also thought to have been partly responsible for the demise of the civilization of Axum on the Ethiopian plateau during the first millennium AD (Butzer, 1981). In the Rukiga Highlands, forest clearance and soil erosion appear to have taken place in what are presently forested catchments (Taylor, 1990, 1992; Marchant, unpublished data), indicating that not only have changes taken place in the pattern of settlement in the area, but also that degraded mountain environments can recover, provided sufficient time is available between phases of disturbance.

The prevalence and antiquity of environmental degradation in mountain systems is often taken as evidence of poor techniques of traditional land management and, *ipso facto*, justification for the introduction of 'improved' agricultural techniques. Such a response ignores the fact that it is often not poor land husbandry that has caused land degradation, but other factors, such as the enforced fragmentation of holdings and land abandonment. Indeed the latter are probably the most common causes of highland degradation in Africa, as in other mountainous regions such as the Himalaya (Ives and Messerli, 1989). For the Rukiga Highlands, at least, it seems that environmental degradation, in the form of soil erosion, long preceded the introduction of agriculture. For example, 6 m of inorganic sediments accumulated in Muchoya Swamp in little over 100 years 11 000 yr BP, most

Plate 16.2. Part of the boundary between Impenetrable Forest National Park and the surrounding, heavily cultivated land. The boundary is marked by introduced cypress trees (photo: D. Taylor).

likely from hillslopes in the catchment rendered unstable by changes in climate and vegetation. Furthermore, aerial photographs and an analysis of swamp sediments indicate that a huge land slip more than 30 000 years ago, possibly owing to tectonic activity, could have impeded drainage and led to the development of mire conditions at Mubwindi.

The Rukiga Highlands provide a good example of how successful traditional agriculture can be. The region is dominated by parallel-running, steeply sloping ridges that are blanketed by infertile altisols. Between these are poorly drained valleys within which peaty soils have accumulated. Agriculture under such conditions is difficult, more so during periods of drought, yet the area has long supported high population densities (Kagambirwe, 1972) without the input of western technology. Largely this has been achieved through farming methods developed by the local Bachiga people, who have an appreciation of the dangers of soil erosion and exhaustion that is believed to pre-date the first contact with colonial administrators (Cary Farley, personal

communication). Fortunately, the value of this indigenous knowledge is increasingly being realized and accommodated within development plans for the region. An example of this is a Development Through Conservation (DTC) project that is currently in operation in and around Impenetrable Forest and Mgahinga National Parks. These two parks, though relatively small (total area about 350 km^2), include areas of montane forest with immense biological and socio-economic value (for example, they are home to the majority of the world's surviving population of mountain gorilla, *Gorilla gorilla berengei*, and provide both timber and non-timber resources). Unfortunately, as forest products are in short supply in the area, they are under intense pressure from the surrounding population (Plate 16.2). In an effort to integrate better the demands of interested parties in management strategies, heavy reliance has been placed upon the accumulated knowledge of indigenous people, including Bachiga and Batwa (pygmy) communities. Particular emphasis has been placed on those groups which have enjoyed a long-standing involvement in the use of the forests' natural resources. As part of the project, Cunningham (1992) has identified practices and levels of exploitation that are unlikely to threaten survival of the forests, whilst providing a useful source of revenue for local people, and thus worthy of support in the future. Bee-keeping is one such example. At the same time Cunningham has also highlighted resources, such as saplings and bamboo stems, used extensively in construction, which are threatened by harvesting and which therefore need to be provided, or produced, outside the boundaries of the forests.

In addition to human activity, today's mountain environments are being affected by climatic change. As mentioned previously, a tendency towards warmer and drier conditions this century has been revealed by glaciological studies on Mount Kenya. In addition, a trend towards decreased annual rainfall reliability since the 1960s is recorded in meteorological records from the Usambara Mountains in Tanzania (Hamilton and Macfadyen, 1989), whilst long-time residents there note a reduced incidence of mist, less predictable rainfall, and warmer temperatures during the last ten to fifteen years (Hamilton, 1989*b*). The cause of these changes is debated, although widespread environmental degradation, particularly the clearance of forests and drainage of wetlands, and increased levels of atmospheric CO_2 are probable contributing factors.

Predictions from General Circulation Models (GCMs) suggest that conditions in much of Africa are likely to continue to get warmer and drier during the coming 100 years or so (IPCC, 1992). The combination of higher temperatures with lower effective precipitation may be unique for the late Quaternary, when warmer climates are thought to have led to more rainfall through increased monsoonal activity. The possibility that future climatic conditions might be unlike any experienced in the relatively recent geological past has made predictions of their impact on the biota difficult. However, we can expect further increases in temperature and decreases in humidity to alter denudation rates, the distribution and composition of biotas, land-use pressures, and the incidence of disease. In fact many of these impacts are already apparent. For example, over the past few decades the composition of montane forest on the Usambara Mountains has been altered by the deaths, without replacement, of several species of mature forest trees, which may have been induced by climatic change (Hall, 1985; Binggeli, 1989). On the same mountains a recent upslope migration of the cultivation of fruit crops (Hamilton, 1989*b*) and an increased incidence of malaria above 1000 m (Matola *et al.*, 1987) could also have been assisted by climate change.

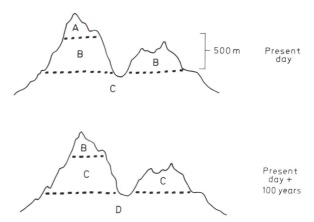

Fig. 16.9. Altitudinal shifts in species distributions as a result of global warming.
(*Source*: Peters, 1992.)

The Future of Mountain Environments in Africa

Global warming is widely seen as a threat to the world's alpine regions, including those in Africa. According to Peters (1992) a 3 °C temperature increase over the coming 100 years, predicted for the tropics by some GCMs (IPCC, 1992), is equivalent to a 500 m upward shift in altitudinal zones (Figure 16.9). Although this is an oversimplification, because temperature is but one of a complex of factors that determine the distributions of plants and animals, it is not improbable that climatic warming in the future will be instrumental in forcing the movement of some taxa to higher altitudes. This will effectively reduce the extent of the Afroalpine zone relative to that of today, thereby further increasing its isolation. In the more immediate future, montane habitats will be placed under increased pressure by changes in land use, as the altitudinal restrictions on certain crops are lifted, and by increased rural population densities. Without active regulation, an intensification of degradation will result, as new areas of montane forest are cleared and shorter periods of fallow introduced. The effects will not be restricted to high or even middle altitudes as the degradation of mountains systems will have a far-reaching impact. For example Egypt and the Sudan, two predominantly low-lying countries, possess agricultural economies that are heavily dependent on what happens in relatively remote highland areas. In both cases irrigation-fed agriculture predominates, which is largely based on water supplied via the Nile and its tributaries from the densely populated highlands of

Central and Eastern Africa. Massive environmental degradation in these headwaters could eventually lead to the collapse of agriculture along the Nile's lower reaches by interrupting water supplies and causing siltation of reservoirs and irrigation channels.

Because of their critical importance to the future of Africa, there is an urgent need for an improved understanding of the continent's mountain systems, of their functioning, and of their likely responses to environmental change in the coming years. Human populations have long had an impact in Africa and for at least the last two millennia this has included farming in highland areas. During this time the communities involved have accumulated a great deal of knowledge on the dynamism and productivity of mountain environments, which must be respected in any future development plans. In order to provide the time for an improved understanding of mountain systems, conservation in mountainous areas needs to be prioritized. This is already being done in some countries, notably in the highlands of Cameroon (Alpert 1993), Ethiopia, and Uganda, but more involvement is needed in order to accommodate the full range of mountain environments in Africa.

References

Alpert, P. (1993), 'Conserving Biodiversity in Cameroon', *Ambio*, 22/1: 44–9.

Ateba, B., Ntepe, N., Ekodeck, G. E. Soba, D., and Fairhead, J. D. (1992), 'The Recent Earthquakes of South Cameroon and Their Possible Relationship with Main Geological Features of Central Africa', *Journal of African Earth Sciences*, 14/3: 365–9.

Bermingham, P. M., Fairhead, J. D. and Stuart, G. W. (1983), 'Gravity Model of the Central African Rift System: A Model of Continental Disruption. 2: The Darfur Domal Uplift and Associated Cainozoic Volcanism', *Tectonophysics*, 4: 205–22.

Bingelli, P. (1989), 'The Ecology of Maesopsis Invasion and Dynamics of the Evergreen Forest of the East Usambaras, and Their Implications for Forest Conservation and Forestry Practices', in A. C. Hamilton and R. Bensted-Smith (eds.), *Forest Conservation in the East Usambara Mountains, Tanzania* (Gland, Switz.), 269–300.

Bonnefille, R. and Riollet, G. (1988), 'The Kashiru Pollen Sequence (Burundi) Palaeoclimatic Implications for the Last 40 000 yr BP in Tropical Africa', *Quaternary Research*, 30: 19–35.

Bonnefille, R., Roeland, J. C. and Guiot, J. (1990), 'Temperature and Rainfall Estimates for the Past 40 000 Years in Equatorial Africa', *Nature*, 346: 347–9.

Burke, K. and Wilson, J. T. (1972), 'Is the African Plate Stationary?', *Nature*, 239: 287–9.

Butzer, K. W. (1971), 'Recent History of an Ethiopian Delta', Department of Geography Research Paper no. 136 (Chicago University).

—— (1981), 'Rise and Fall of Axum, Ethiopia: A Geo-archaeological Interpretation', *American Antiquity*, 46/3: 471–95.

—— (1984), 'Late Quaternary Environments in South Africa', in J. C. Vogel (ed.), *Late Cenozoic Palaeoclimates of the Southern Hemisphere* (Rotterdam), 235–65.

Carcasson, R. H. (1964), 'A Preliminary Survey of the Zoogeography of African Butterflies', *East African Wildlife Journal*, 2: 122–57.

Chapman, J. D. and White, F. (1970), *The Evergreen Forests of Malawi* (Oxford).

CLIMAPP Project Members (1981), 'Seasonal Reconstruction of the Earth's Surface at the Last Glacial Maximum', *Geological Society of America, Map and Chart Series MC-36*.

Coe, M. (1989), 'Biogeographic Affinities of the High mountains of Tropical Africa', in W. C. Mahaney (ed.), *Quaternary and Environmental Research on East African Mountains* (Rotterdam), 257–75.

Coetzee, J. A. (1967), 'Pollen Analytical Studies in East and Southern Africa', *Palaeoecology of Africa*, 3: 1–145.

COHMAP Members (1988), 'Climatic Changes of the Last 18 000 Years: Observations and Model Simulations', *Science*, 241: 1043–52.

Cunningham, T. (1992), 'People, Park and Plant Use Research and Recommendations for Multiple-use Zones and Development Alternatives Around Bwindi-Impenetrable National Park, Uganda', Report prepared for CARE-International (Kampala, Uganda).

Dautria, J. M. and Lesquer, A. (1989), 'An Example of the Relationship Between Rift and Dome: Recent Geodynamic Evolution of the Joggar Swell and of its Nearby Regions (Central Sahara, Southern Algeria and Eastern Niger)', *Tectonophysics*, 163: 45–61.

Deacon, H. J., Deacon, J., Scholtz, A., Thackeray, J. F., Brink, J. S., and Vogel, J. C. (1984), 'Correlation of Palaeo-environmental Data From the Late Pleistocene and Holocene Deposits at Boomplaas Cave, Southern Cape', in Vogel (ed.), *Late Cainozoic Palaeoclimates of the Southern Hemisphere*: 339–51.

Dowsett, R. J. (1971), 'The Avifauna of the Makutu Plateau, Zambia', *Review of Zoology and Botany of Africa*, 84: 312–33.

—— (1986), 'Origins of the High-altitude Avifaunas of Tropical Africa', in F. Vuilleumier and M. Monasterio (eds.), *High Altitude Tropical Biogeography* (Oxford), 557–85.

Fairhead, J. D. and Binks, R. (1991), 'Differential Opening of the Central and South Atlantic Oceans and the Opening of the West African Rift System', *Tectonophysics*, 187: 191–203.

Gerrard, A. J. (1990), *Mountain Environments* (London).

Grove, A. T. (1978), *Africa*, 3rd edn. (Oxford).

—— (1983), 'Evolution of the Physical Geography of the East African Rift Valley Region', in R. W. Sims and J. H. Price (eds.), *Evolution, Time and Space: The Emergence of the Biosphere* (London), 115–55.

—— (1986), 'Geomorphology of the African Rift System', in L. E. Frostick, R. W. Renant, I. Reid, and J. J. Tiercelin (eds.), *Sedimentation in the African Rifts*. Geological Society Special Publication, No. 25 (Oxford), 9–16.

Gregory, J. W. (1894), 'Contributions to the Physical Geography of British East Africa', *Geographical Journal*, 4: 289–315, 408–25, 505–24.

Grosjean, M. and Messerli, B. (1988), 'African Mountains and Highlands: Potential and Constraints', *Mountain Research and Development*, 8/2–3: 111–22.

Hagedorn, J. (1984), 'Pleistozane Periglazial-Formen in Gebirgen des sudlichen Kaplandes (Sudafrika) und ihre Bedeutung als Palaoklima=Indikatoren', *Palaeoecology of Africa*, 16: 405–10.

Hall, J. B. (1985), *Mazumbai Forest: Report on Large Tree Survey 1981–84*, Department of Forestry and Wood Science, University College North Wales, mimeo.

Hamilton, A. C. (1975), 'A Quantitative Analysis of Altitudinal Zonation in Ugandan forests', *Vegetatio*, 30: 99–106.

—— (1976), 'The Significance of Patterns of Distribution Shown by Forest Plants and Animals in Tropical Africa for the Reconstruction of Upper Pleistocene Palaeo-environments: A Review', *Palaeoecology of Africa*, 9: 63–97.

—— (1982), *Environmental History of East Africa* (London).

—— (1987), 'Vegetation and Climate of Mt. Elgon During the Late Pleistocene and Holocene', *Palaeoecology of Africa*, 18: 283–304.

—— (1989a), 'The Place and the Problem', in Hamilton and Bensted-

Smith (eds.), *Forest Conservation in the East Usambara Mountains, Tanzania*: 29–33.

—— (1989b), 'Climate Change on the East Usambaras: Statement on Climatic and Environmental change', in Hamilton and Bensted-Smith (eds.), *Forest Conservation in the East Usambara Mountains, Tanzania*: 115–16.

—— and Perrott, R. A. (1981), 'A Study of Altitudinal Zonation in the Montane Forest Belt of Mount Elgon', *Vegetatio*, 45: 107–25.

—— and Macfadyen, A. (1989), 'Climatic Change on the East Usambaras: Evidence from Records from Meteorological Stations', in Hamilton and Bensted-Smith (eds.), *Forest Conservation in the East Usambara Mountains, Tanzania*: 103–7.

—— and Taylor, D. (1991), 'History of Climate and Forests in Tropical Africa During the Last 8 million years', *Climatic Change*, 19: 65–78.

Harper, G. (1969), 'Periglacial Evidence in Southern Africa During the Pleistocene Epoch', *Palaeoecology of Africa*, 4: 71–101.

Harvey, T. J. (1976), '*The Palaeolimnology of Lake Mobuto Sese Seko, Uganda–Zaire: The Last 28 000 Years*', Doctoral diss., Duke University, Oh.

Hastenrath, S. (1977), 'Pleistocene Mountain Glaciation in Ethiopia', *Journal of Glaciology*, 18: 309–13.

—— (1984), *Glaciers of Equatorial East Africa* (Dordrecht, Neth.).

Hauman, L. (1933), 'Esquisse de la végétation des hautes altitudes sur le Ruwenzori', *Bulletin Academie Royale de Belgique, Classe des Sciences*, 5/19: 602–16, 702–17, 900–17.

Hedberg, O. (1951), 'Vegetation Belts of the East African Mountains', *Svensk Botanisk Tidskrift*, 45/1: 140–202.

—— (1961), 'The Phytogeographical Position of the Afroalpine Flora', *Recent Advances in Botany*, 1: 914–19.

—— (1964), 'Features of Afroalpine Plant Ecology', *Acta Phytogeographica Suecia*, 49: 1–144.

—— (1970), 'Evolution of the Afroalpine Flora', *Biotropica*, 2/1: 16–23.

Hilliard, O. M. and Burtt, B. L. (1987), *The Botany of the Southern Natal Drakensberg* (Cape Town).

Hurni, H. (1989), 'Late Quaternary of Simen and Other Mountains in Ethiopia', in W. C. Mahaney (ed.), *Quaternary and Environmental Research on East African Mountains* (Rotterdam), 105–20.

Hutchins, D. E. (1909), *Report on the Forests of British East Africa* (London).

IPCC (1992), *Climate Change 1992*, Supplementary Report to the IPCC Scientific Assessment (Cambridge).

Ives, J. D. and Messerli, B. (1989), *The Himalayan Dilemma: Reconciling Development and Conservation* (London).

Jolly, D. and Bonnefille, R. (1991), 'Diagramme pollinique d'un sondage Holocène de la Kuruyange (Burundi Afrique Centrale)', *Palaeoecology of Africa*, 22: 265–74.

Kagambirwe, E. R. (1972), *Causes and Consequences of Land Shortages in Kigezi*, Department of Geography, Makerere University, Occasional Paper no. 22 (Kampala, Uganda).

King, L. C. (1978a), 'A Comparison between the Older (Karoo) Rifts and the Younger (Cenozoic) Rifts of Eastern Africa', in I. B. Ramberg and E. R. Neumann (eds.), *Tectonics and Geophysics of Continental Rifts* (Dordrecht, Neth.), 347–50.

—— (1978b), The Geomorphology of Central and Southern Africa, in M. J. A. Werger (ed.), *Biogeography and Ecology of Southern Africa* (The Hague), 3–17.

Kingdon, J. (1990), *Island Africa The Evolution of Africa's Rare Animals and Plants* (London).

Krapf, J. L. (1860), *Travels, Researches and Missionary Labours in East Africa* (London).

Kutzbach, J. E. and Street-Perrott, F. A. (1985), 'Milankovitch Forcing of Fluctuations in the Level of Tropical Lakes From 18 to 0 Kyr BP', *Nature*, 317: 130–4.

—— and Guetter, P. J. (1986), 'The Influence of Changing Orbital Parameters and Surface Boundary Conditions on Climate Simulations for the Past 18 000 Years', *Journal of Atmospheric Sciences*, 43: 1726–59.

Lamb, H. F., Eicher, U., and Switsur, V. R. (1989), 'An 18 000 Year Record of Vegetation, Lake-level and Climatic History from the Middle Atlas, Morroco', *Journal of Biogeography*, 16: 65–74.

—— Damblon, F., and Maxted, R. W. (1991), 'Human Impact on the Vegetation of the Middle Atlas, Morocco, during the Last 5000 Years', *Journal of Biogeography*, 18: 519–32.

Langdale-Brown, I., Osmaston, H. A., and Wilson, J. G. (1964), *The Vegetation of Uganda and its Bearing on Land-use* (Entebbe, Uganda).

Lauer, W. (1975), 'Klimatische Grundzuga der Hohenstufung tropischer Gebirge', *Deutscher Geographentag Verhandlungen*, 40: 76–90.

—— and Frankenberg, P. (1979), 'Zur Klima- und Vegetationsgeschichte der westlichen Sahara', *Akademie der Wissenschaften und der Literatur in Mainz, Abhandlungen der* Math ematisch-Naturwissenschaften, 1: 61.

Lauscher, F. (1976), 'Weltweite Typan der Hohenabhanm gigkeit des Niederschlags', *Wetter und Leben*, 28: 80–90.

Leakey, R. and Lewin, R. (1992), *Origins Reconsidered in Search of What Makes us Human* (London).

Livingstone, D. A. (1967), 'Postglacial Vegetation of the Ruwenzori Mountains in Equatorial Africa', *Ecological Monographs*, 37: 25–52.

—— (1975), 'Late Quaternary Climatic Change in Africa', *Annual Review of Ecology and Systematics*, 6: 249–80.

—— (1980), 'Environmental Changes in the Nile Headwaters', in M. A. J. Williams and H. Faure (eds.), *The Sahara and the Nile* (Rotterdam), 336–60.

Lovett, J. C. (1993), 'Climate History and Forest Distribution in Eastern Africa', in J. C. Lovett and S. K. Wasser (eds.), *Biogeography and Ecology of the Rain Forests of Eastern Africa* (Cambridge), 23–9.

Mabberley, D. J. (1974), 'Branching in Pachycaul Senecios: The Durian Theory and the Evolution of Angiospermous Trees and Shrubs', *New Phytologist*, 73: 967–75.

Mahaney, W. C. (1982), 'Chronology of Glacial Deposits on Mount Kenya, East Africa', *Palaeoecology of Africa*, 14: 25–43.

—— (1984), 'Late Glacial and Postglacial Chronology of Mount Kenya, East Africa', *Palaeoecology of Africa*, 16: 327–41.

—— (1987), 'Dating of a Moraine on Mount Kenya: Discussion', *Geografiska Annaler*, 69A/2: 359–63.

—— (1988), 'Holocene Glaciations and Palaeoclimate of Mount Kenya and other East African Mountains', *Quaternary Science Review*, 7: 211–25.

—— (1990), *Ice on the Equator* (Sister Bay, Wis.).

—— Harmsen, R., and Spence, J. R. (1991), 'Glacial-interglacial Cycles and Development of the Afroalpine Ecosystem on East African Mountains. I: Glacial and Postglacial Geological Record and Paleoclimate of Mount Kenya', *Journal of African Earth Sciences*, 12/3: 505–12.

Matola, V. G., White, G. B., and Magayuka, S. A. (1987), 'The Changed Pattern of Malaria Endemicity and Transmission at Amani in the Eastern Usambara mountains, North-eastern Tanzania', *Journal of Tropical Medicine and Hygiene*, 90: 127–34.

Meadows, M. E. and Linder, H. P. (1993), 'A Palaeoecological Perspective on the Origin of Afromontane Grasslands', *Journal of Biogeography*, 20: 345–55.

Mensching, H. (1953), 'Morphologische Studien im Hohen Atlas von Marokko', *Wuerzburger Geographische Arbeiten*, H. 1: 104 S.

Messerli, B. (1972), 'Formen und Formungsporzesse in den Hochgebirgen des Tibesti', *Hochgebirgsforschung*, H. 2: 23–86.

—— Winiger, M., and Rognon, P. (1980), 'The Saharan and East African Uplands during the Quaternary', in Williams and Faure (eds.), *The Sahara and the Nile*: 87–118.

Messerli, B. and Winiger, M. (1992), 'Climate, Environmental Change, and Resources of the African Mountains From the Mediterranean to the Equator', *Mountain Research and Development*, 12/4: 315–36.

Meyer, H. (1890), *Ostafrikanische Gletscherfahrten* (Leipzig).

Moreau, R. E. (1966), *The Bird Faunas of Africa and its Islands* (New York).

Nilsson, E. (1931), 'Quaternary Glaciation and Pluvial Lakes in British East Africa and Abyssinia', *Geografiska Annaler*, 13: 241–8.

—— (1935), 'Traces of Ancient Changes of Climate in East Africa', *Geografiska Annaler.*, 1–2: 1–21.

Osmaston, H. A. (1965), 'The Past and Present Climate and Vegetation of Ruwenzori and its Neighbourhood', D.Phil. thesis. Oxford.

—— (1975), 'Models for the Estimation of Firnlines of Present and Pleistocene Glaciers', in R. F. Peel, M. Chisholm, and P. Haggett (eds.), *Processes in Physical and Human Geography* (London), 218–45.

Peters, R. L. (1992), 'Conservation of Biological Diversity in the Face of Climatic Change', in R. L. Peters and T. E. Lovejoy (eds.), *Global Warming and Biological Diversity* (New Haven), 15–30.

Potter, E. C. (1976), 'Pleistocene Glaciation in Ethiopia: New Evidence', *Journal of Glaciology*, 17: 148–50.

Pouclet, A. and Karche, J. P. (1988), 'Petrology and Geochemistry of the Cenozoic Volcanism in Eastern Niger (West Africa)', in UNESCO-UISG 227, *Mesozoic to Present-day Magmatism of the African Plate and its Structural Setting, Abstracts* (Paris), 194–5.

Rebmann, J. (1849), 'Narrative of a Journey to Jagga, the Snow Country of Eastern Africa', *Church Missionary Review*, 2: 12–23.

Rognon, P. (1967), *Le massif de l'Atakor et ses bordures*, Études Géomorphique (Paris).

Scott, L. (1982), 'A Late Quaternary Pollen Diagram From the Transvaal, Bushveld, South Africa', *Quaternary Research*, 17/3: 339–70.

Smith, J. M. B. and Cleef, A. M. (1988), 'Composition and Origins of the World's Tropicalpine Floras', *Journal of Biogeography*, 15: 631–45.

Statistics Department, Ministry of Finance and Economic Planning (1992), *Final Results of the 1991 Population and Housing Census* (pre-release) (Entebbe, Uganda).

Stets, V. J. and Wurster, P. (1981), 'Zur strukturgeschichte des Hohen Atlas in Marokko', *Geologische Rundschau*, 70/3, 801–41.

Stuhlmann, F. (1894), *Mit Emin Pascha ins Herz von Afrika* (Berlin).

Taylor, D. (1990), 'Late Quaternary Pollen Records From Two Ugandan Mires: Evidence for Environmental Change in the Rukiga Highlands of South-west Uganda', *Palaeogeography, Palaeoclimatology, Palaeoecology*, 80: 283–300.

—— (1992), 'Pollen Evidence from Muchoya Swamp, Rukiga Highlands (Uganda), for Abrupt Changes in Vegetation During the Last 21 000 Years', *Bulletin Société Géologique de France*, 163/1: 77–82.

—— (1993), 'Environmental Change in Montane South-west Uganda: A Pollen Record for the Holocene from Ahakagyezi Swamp', *The Holocene*, 3/4, 324–32.

van Zinderen Bakker, E. M. and Coetzee, J. A. (1988), 'A Review of Late Quaternary Pollen Studies in East, Central, and Southern Africa', *Review of Palaeobotany and Palynology*, 55: 155–74.

Vincent, P. M. (1970), 'The Evolution of the Tibesti Volcanic Province', in T. Clifford and I. A. Gass (eds.), *African Magmatism and Tectonics* (Edinburgh), 285–303.

White, F. (1978), 'The Afromontane region', in Werner (ed.), *Biogeography and Ecology of Southern Africa*: 463–513.

—— (1983), *The Vegetation of Africa* (Paris).

Wilson, M. and Guiraud, R. (1992), 'Magmatism and Rifting in Western and Central Africa, from Late Jurassic to Recent Times', *Tectonophysics*, 213: 203–25.

Winiger, M. (1981), 'Zur thermisch-hygrischen Gliederung des Mt. Kenya', *Erdkunde*, 35: 248–63.

Wright, J. B. (1978), 'Origins of the Benue Trough', in C. A. Kogbe (ed.), *Geology of Nigeria* (Ile-Ife Nigeria).

—— (1981), 'Review of the Origin and Evolution of the Benue Trough in Nigeria', *Earth Evolutionary Science*, 2: 98–103.

Yacono, D. (1968), *L'Ahaggar: Essai sur le Climat de Montagne au Sahara*, Travaux de l'Institut de Récherches Sahariennes, 27 (Algiers).

17 Mediterranean Environments

Harriet D. Allen

Introduction

Regions with a mediterranean climate can be broadly characterized as having a concentration of precipitation in winter months, with sufficient rainfall to support a continuous vegetation cover on all but the rockiest of sites. Winter months are those with an average temperature of less than 15 °C; frosts can occur and can locally be widespread but in general are rare. In contrast, summer is a period of drought resulting from high temperatures and little or no rainfall (Aschmann, 1973). Such climates are transitional between dry tropical and temperate climates, and in Africa two regions of mediterranean environment can be identified, the *Maghreb in North Africa* and the *south-western tip of Africa in Cape Province*.

Both areas have vegetation communities that can be recognized as typically mediterranean, with a predominance of evergreen sclerophyllous shrubs and trees. However, many of these communities, especially in the Maghreb, are considerably degraded following centuries of human activity. In addition, both regions have experienced Quaternary climatic fluctuations. Thus the present vegetation is a reflection of contemporary climate, human activity, and past environmental change, as well as other physical factors such as the nature of the terrain, geology, and soils. The latter are especially important at the local scale where the floristic and structural composition of vegetation communities alters over short distances; Cape Province is a good example.

Whilst it is possible to make these broad generalizations it is important to recognize that, as in all regions with a mediterranean-type climate, climate can vary considerably over short distances. This is particularly so in the mountainous regions of the Maghreb and the

Cape. Thus it is difficult, on the basis of climate alone, to define accurately the extent of both regions, and small-scale maps will exaggerate the problem. Maps showing areas of mediterranean-type climate will also differ from those of mediterranean-type vegetation despite the relatively close relationship between climate and vegetation at the macroscale. For example, in Cape Province, the *fynbos* biome of mediterranean-type climate extends into areas where there is a more even distribution of precipitation through the year. The UNESCO-FAO *Bioclimatic Map of the Mediterranean Zone* (1963), the UNESCO-FAO *Carte de la végétation de la région mediterranéenne* (1970), and White's *UNESCO Vegetation Map of Africa* (White, 1983) depict these areas of mediterranean-type climate and vegetation. It is the last which provides the basis for the delineation of the two regions considered here (Figure 17.1).

The aim of this chapter is to review and compare the biogeography and changing environments of these two regions, particularly in response to climatic change and human activities, to natural and human-induced fire, and to biogeographical invasions. Of necessity the approach is multidisciplinary and draws on the work of a variety of researchers from different fields.

Much of the research on the functioning of mediterranean ecosystems (both in Southern Africa and the Maghreb, as well as in south–western Australia, Chile, California, and the rest of the Mediterranean Basin) has been carried out within the framework of the International Biological Programme. In 1971 there was a symposium on biogeographical convergence (di Castri and Mooney, 1973), another in 1977 on the role of fire in resource management in mediterranean ecosystems (Mooney and Conrad, 1977), and a further one in 1980 on the role of nutrients in species and ecosystem convergence (Kruger

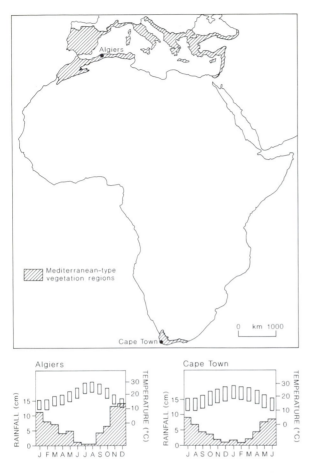

Fig. 17.1. Mediterranean-type vegetation regions in Africa.

et al., 1983). The biogeography of mediterranean invasions has been reviewed by di Castri *et al.* (1990) and Groves and di Castri (1991).

Multidisciplinary research is needed to determine and evaluate the environmental history of these areas. North Africa falls within the domain of the INQUA subcommission on circum-Mediterranean Holocene history. There is also a long tradition of research by French scientists in North Africa and so a review of the French literature on the environmental history of the Maghreb is included. For the Cape *fynbos* biome, Cowling (1992) has produced a synthesis of research undertaken by the Fynbos Biome Project, a co-operative programme launched in 1978 to stimulate research on the functioning of *fynbos* ecosystems.

Inevitably in a review of limited length, as here, there are omissions. In particular little consideration is given to the zoogeography of the mediterranean regions of

Africa. This is partly a reflection of the greater extent of our knowledge of the origins, functioning, and response of vegetation communities to environmental factors and change.

The Maghreb
Physical Background
The mediterranean region of North Africa, the Maghreb, extends from Morocco to Tunisia. It is an area of fold mountains with a narrow lowland along the Mediterranean coastline (Figure 17.2). The region is dominated by the Atlas Mountains which extend for over 3000 km, trending WSW–ENE approximately parallel to the mediterranean coastline. The range can be divided into the Rif Atlas, High Atlas, and Middle Atlas, with the Anti-Atlas Mountains separating the mediterranean region from the Saharan region to the south. Although earlier orogenies are expressed in the Anti-Atlas, the present mountains were mostly formed during the Alpine orogeny of the Tertiary Period when profound folding and faulting of sediments of the Tethys Sea occurred. The oldest fold system is the Rif Atlas, a coastal range running south-eastwards from Tangier in northern Morocco and continuing into Algeria as the Tell Atlas. The Oued (river) Molouya cuts through the Rif to the Mediterranean Sea east of Melilla. The High Atlas extends north-eastwards from the Atlantic coast between Essaouira and Agadir into Algeria where it becomes the Saharan Atlas. The highest elevation is Mount Toubkal (4165 m) in Morocco. There are a number of other peaks over 3500 m in Morocco. In Algeria, the Saharan Atlas rarely exceeds 2000 m, with the highest elevation at Jebel Chelia (2328 m) in eastern Algeria where the Saharan and Tell Atlas approach each other. The Middle Atlas is a plateau region in Morocco bounded by the high peaks of the High Atlas to the south. The Anti-Atlas marks the edge of the African shield and is generally above 1500 m. At its eastern extreme it converges with the High Atlas at Jebel Siroua, a volcanic outcrop. In the west, the Sous valley divides the Anti-Atlas from the High Atlas. In Tunisia, the Atlas ranges are known as the Northern, High, and Low Tell.

The latitude of the Maghreb, between 30 and 37°N, places it on the boundary between arctic or polar air masses and those of the mid-latitude, subtropical high-pressure systems. In winter months rainfall and lower temperatures result from the southward migration of the mid-latitude westerly winds and associated cyclonic disturbances. In summer, the area is dominated by the expansion of the North Atlantic subtropical high-pressure system which penetrates the Mediterranean

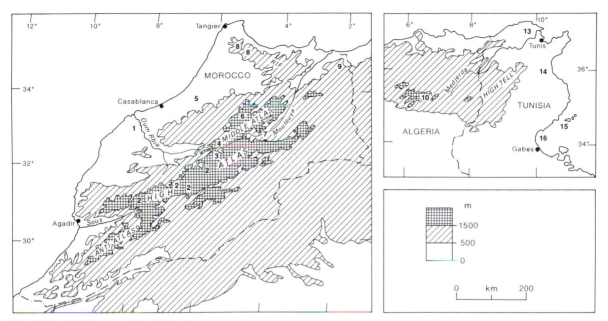

Fig. 17.2. Relief map of the Maghreb.

Numbers refer to sites used in palaeoenvironmental studies: 1 = Oualida; 2 = the High Atlas; 3 = Lake Agziza; 4 = Tigalmamine; 5 = Dayat-er-Roumi; 6 = Middle Atlas; 7 = Middle Moulouya; 8 = Central/western Rif; 9 = Oujda Mountains; 10 = Capeletti Cave–Aures; 11 = Tebessa; 12 = Kroumirie; 13 = Garaet el Ichkeul; 14 = Sebkhet Kelbia; 15 = Kneiss and Kerkennah; 16 = Gulf of Gabes.

region. The Saharan high-pressure system also migrates north, blocking all precipitation except that associated with convectional or orographic processes. However, regional climates are complicated by considerable local variability as different mountain ranges are separated from one another by often deep basins and plateaux. Distance from the sea, variations in relief, and exposure or shelter provided by the mountains result in considerable differences in climate over short distances. In general, however, annual rainfall ranges between 250 and 1000 mm.

Palaeoclimatic History

The Maghreb is an important region in the reconstruction of climatic history during the last 30 000 years because of its transitional position between the temperate regions of Europe, especially the Mediterranean region of southern Europe, and the subtropical regions of Saharan Africa. In addition, the terrain imparts an altitudinal climatic gradient which overlies the latitudinal gradient. Researchers have been interested in the extent to which climatic changes in these regions have paralleled those of northern Europe (Rognon, 1987; Brun, 1989; Lamb *et al.*, 1989). Sites in the Maghreb are also sensitive indicators of contemporary environmental change, both natural and as a result of human activity (Hollis

and Kallel, 1986; Flower *et al.*, 1989; Till and Guiot, 1990; Stevenson and Battarbee, 1991; Thomas *et al.*, 1991; Flower and Foster, 1992).

Much of the work on palaeoenvironmental reconstruction in North Africa has been undertaken by French researchers. For example Rognon (1987) and Brun (1989) have considered events during the last 30 000 years based on analyses of pollen records and sedimentary deposits throughout the Maghreb. Much of the evidence is fragmentary and there are too few reliable radio-carbon ages (Figure 17.3). Between about 40 000 and 20 000 years BP, evidence from palaeosols, palynology, and alluvial and aeolian deposits indicates a wetter period than at present, which extended into the northern Sahara, with deciduous oak and pine pollen predominant. This was a consequence of the southerly displacement of westerly winds from more northerly latitudes and their contribution of cyclonic rain, in response to glacial conditions in northern Europe. It is suggested that the rain would have been better distributed through the year with decreased temperatures. However, it is impossible to determine whether this meant an end to summer aridity. From about 30 000 years BP, there was a progressive environmental deterioration with an increase in aridity. In Tunisia, two particularly wet phases are dated to 28 000–27 000 years BP and 22 000–17 000

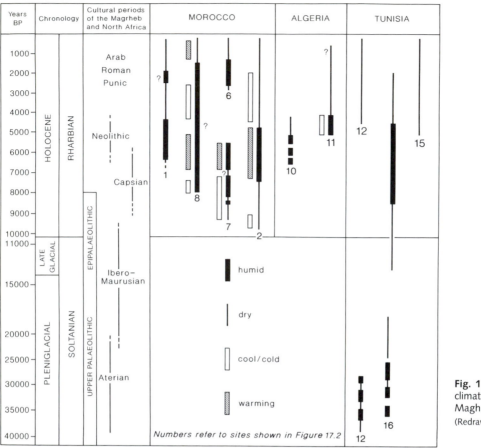

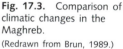

Fig. 17.3. Comparison of climatic changes in the Maghreb.
(Redrawn from Brun, 1989.)

years BP. Pollen evidence from the Gulf of Gabès indicates a diverse forest and a hot, humid climate at around 27 000 years BP with a pine forest predominant at around 21 000 years BP (Rognon, 1987).

From 20 000 years ago there is pollen evidence for an increasingly arid period characterized by steppe vegetation dominated by *Artemisia* and Chenopodiaceae from sediments of the Gulf of Gabès and from the Kerkennah Plateau, both in Tunisia (Brun, 1991, 1992). A sedimentary record from Tigalmamine, a series of adjacent solution lakes in the Middle Atlas of Morocco, indicates an arid phase when moisture levels would have been too low to support the growth of trees. Steppe-like, herb-rich grasslands were present, dominated from 18 000 years BP to 8500 years BP by Gramineae-Chenopodiaceae-*Artemisia*, but with a change between 14 000 and 12 000 years BP when the pollen record shows scattered evergreen oaks in the locality (Lamb *et al.*, 1989). This period is synchronous with the deglacial periods in northern

Europe and may be climatically significant. However, the sediment record from the Oujda Mountains, to the east of Tigalmamine, has been interpreted as an indication of a more arid climate (Wengler and Vernet, 1992).

For the Holocene (or Rhabian period in North Africa) more palynological results are available for the reconstruction of palaeoenvironments. These are summarized by Brun (1989) and shown in Figure 17.3. It is concluded that by about 7000 years BP the climate was warmer and more humid than previously. Analyses of sediments from the Gulf of Gabès and from Tigalmamine date the beginning of this amelioration to about 8500 years BP (Lamb *et al.*, 1989; Brun, 1991, 1992). The pollen record of this period shows a marked increase in both evergreen and deciduous oak forest at Tigalmamine where a wetter climate is evidenced by an increase in lake-level, perhaps by as much as 9 m. The coincidence of the lake-level rise and the forest development implies a rapid response of vegetation to climatic change probably attributable to a close source of seeds for the re-establishment of the forest. Thus there was no migrational lag between climate and vegetation response.

From about 5000 years BP it is difficult to distinguish the climatic signals in the palaeoecological evidence from those which indicate human activity and their determination of the present vegetation communities. However, there does appear to be a difference between the western and eastern Maghreb in the climatic interpretations for this period. These regional differences may be explained by changes in the pattern of North African atmospheric circulation at about 4500 to 4000 years BP (Lamb *et al.*, 1989). It has been suggested that there has been an increase in aridity in Tunisia from about 5000 years ago. This is based on analysis of sediment cores from the Gulf of Gabès and the Kerkennah Plateau. In contrast, the record for northern Algeria and the coastal and mountainous regions of Morocco suggests that the aridity was accompanied by a period of cooler temperatures (Brun, 1989, 1991, 1992). At Tigalma-mine, the pollen record reveals a regeneration of cedar, possibly because of a fall in temperature (Lamb *et al.*, 1991). The δO^{18} records indicate an increase in temperatures which may be due to a change in the source and amount of the precipitation, possibly indicating the start of the present-day circulation of westerly storm tracks which bring winter rain to the region.

Contemporary Flora and Vegetation

The Mediterranean Basin can be described as a 'biogeographical cross-roads', an area of successive invasions by plant species migrating in response to the climatic changes of the Quaternary. This makes it difficult to define clearly what is meant by the term 'mediterranean species'. Four meanings for a mediterranean species have been given by di Castri (1991). The first is a species which has evolved *in situ* within an area of mediterranean-type climate. The second is a species which evolved *in situ*, but in a climatic regime that existed prior to the establishment of a mediterranean climate and which managed to adapt to the new climate. The third is a species which evolved outside the region but was able to colonize the Mediterranean region as it migrated in response to climatic changes. The fourth is an invasive species which arrived 'recently', but which has become so established that it is accepted as an indicator of a mediterranean climate. There are many good examples of this last group. For instance, there is *Opuntia ficus-indica* or prickly pear, a native of tropical South America, reputedly introduced to the Mediterranean Basin by Christopher Columbus, or the cultivated olive, *Olea*

europea, which is frequently described as a true Mediterranean species, but which is probably not native and is, anyway, a cultivated plant (Polunin and Huxley, 1987).

The examples of *Opuntia ficus-indica* and *Olea europea* also highlight the importance of species whose existence in a region is due to human occupation and activities. The flora of the Mediterranean Basin are therefore a mosaic of species from different biogeographical source areas. Several species should be considered as naturalized old invaders with species migrating either naturally or together with people, both within the region from east to west and from outside the region (di Castri, 1991). This is not surprising given the long and rich history of human activity in the Mediterranean Basin.

Compared with other regions of the world which have a mediterranean-type climate (south-western Australia, the Cape Province, central Chile, and California), the Mediterranean Basin is the least isolated in terms of evolution and biogeography. Species diversity and rates of endemism are thus lower, although Mediterranean Europe has a greater diversity of plant species than any other European area. It is on the basis of endemism that White (1983) published his vegetation map of Africa. A regional centre of endemism is a phytochorion which has both more than 50 per cent of species restricted to it and more than 1000 endemic species. In the Mediterranean centre of endemism of North Africa there are about 4000 species of plants, of which about 73 per cent are endemic to the Mediterranean Basin although only 20 per cent of these are confined to North Africa. The non-endemic species show links with a variety of other source areas reflecting the region's position at the biogeographical cross-roads: 2.2 per cent are Irano-Turanian link species, 3 per cent are Saharan link species, and 20 per cent are Boreal link species. This last figure contrasts with faunal exchange between Europe and Africa which is very uncommon given the independent evolution of mammals on both continents during the Quaternary and the presence of the Mediterranean Sea as a barrier to migration (Cheylan, 1991).

Palynological studies have shown that before large-scale clearances occurred, much of the Maghreb was covered with forest. Few undisturbed stands of forest remain and details of severe degradation during the later Holocene are given below. However, these undisturbed, or relict, stands contain at least sixty species (White, 1983). Figure 17.4 shows White's three vegetation mapping units for the Mediterranean centre of endemism: Mediterranean sclerophyllous forest; Mediterranean montane forest and altimontane shrubland; and anthropic landscapes. The designation 'anthropic landscape' is not meant to imply that the other units have not been

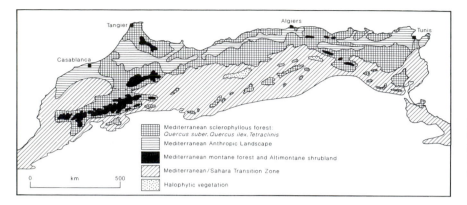

Fig. 17.4. The Mediterranean regional centre of endemism of North Africa. (Redrawn from White, 1983.)

subjected to human activity but are those dominated by cultivation. Within the sclerophyllous and montane forests White identified three different forest types: evergreen sclerophyllous forest, coniferous forest, and deciduous forest.

Evergreen sclerophyllous forest includes *Quercus ilex* (Holm oak) forest, *Q. suber* (cork oak) forest, and *Q. coccifera* (Kermes or Holy oak) forest. These are generally mutually exclusive, but taken together comprise about half the total contemporary forest area of the Maghreb. Of these *Q. ilex* is the most extensive and is a mountainous species forming forest above 2400 m in the Atlas, although it is also found at elevations as low as 400 m. It can tolerate a range of precipitation regimes from semi-arid (500–700 mm/yr) through subhumid (700–1000 mm/yr) to humid (more than 1000 mm/yr). Within these precipitation belts it can tolerate a range of minimum monthly temperatures of between −5 °C and 7 °C. It has survived severe forest degradation because it is fire-resistant, suckers from stumps, and can survive beneath the shade of cedar woods. When a gap in the canopy of a cedar forest opens, previously suppressed individuals of *Q. ilex* grow rapidly until overshadowed again by regenerating cedars (White, 1983).

Q. suber is a valuable tree exploited for its cork. It will tolerate a variety of precipitation regimes similar to those of *Q. ilex* but grows at lower elevations. It is also adapted to fire which allows it to be more dominant than deciduous oaks where they grow as mixed stands. *Q. coccifera* is the least extensive of the three evergreen oak forests, being more important in Algeria and Tunisia than in Morocco where it is restricted to the Rif Atlas. It tolerates a range of annual precipitation regimes, from 450 mm/yr to more than 1000 mm/yr. Its growing habit varies from multibranched, dense shrub thickets to trees; occasionally it attains heights of 12 m. It is

frequently cut for charcoal production but suckers freely from coppiced stools.

Coniferous forest is almost as important in area as evergreen sclerophyllous forest (White, 1983). There are ten main dominant species: *Abies numidica*, *A. pinsapo* subsp. *marocana*, *Cedrus atlantica*, *Cupressus atlantica*, *C. sepervirens*, *Juniperus phoenicea*, *J. thurifera*, *Pinus halepensis*, *P. pinaster*, and *Tetraclinis articulata*. Four other conifers are also important: *Taxus baccata*, *Pinus nigra*, *Juniperus oxycedrus*, and *J. communis*.

There are three species which comprise the deciduous oak forests: *Quercus faginea*, *Q. pyrenaica*, and *Q. afarensis* (White, 1983). *Q. faginea* is widespread in the Maghreb but is found only in widely scattered, small stands. *Q. pyrenaica* is found in Morocco and *Q. afarensis* is restricted to Algeria. All grow in a humid precipitation regime of more than 1000 mm/yr.

Olea-Pistacia communities are common throughout the Maghreb, and it possible that in some semi-arid areas (500–700 mm/yr) the climax vegetation may not have been forest but rather bushland, thicket, or scrub forest with *Olea europea* dominant. In the original scrub forest it is likely that *Q. europea* grew alongside *Pistacia lentiscus*, *P. atlantica* and *Chamaerops humilis* (the dwarf Maghreb palm). However, these communities have been severely degraded and to a large extent resemble maquis and garrigue, termed secondary shrubland. As elsewhere in the Mediterranean Basin, a sequence of degradation from forest to scrub forest to maquis can be identified. The difference between maquis and garrigue can be one of height and vegetation density, but there is considerable inconsistency and confusion in the way in which these terms are used. The plant communities represent a spatially and temporally varying balance between exploitation and abandonment, and between degradation and regeneration. Commonly, maquis communities are dense and expected to be more than 2 m tall. Maquis may be derived from degenerating forest or scrub forest, or from abandoned and regenerating garrigue—where the term garrigue is used for low, bushy communities often dominated by *Q coccifera*. The great variety of these secondary shrublands in the Maghreb, as elsewhere in the Mediterranean region, means that there is little merit in perpetuating the use of the terms maquis and garrigue, especially given that their species composition may be a function of soil as much as floristic origin (White, 1983).

Altimontane shrubland of dense cushion-shaped spiny shrubs (*xérophytes épineux en coussinets* in French) occurs at high elevations above the tree-line. These grow where soils are poor and climate is extreme, with snow in winter and extreme aridity in summer. However, it is difficult to determine the lower climatic limits of altimontane shrubland because of widespread clearance of forests; undoubtedly, some areas presently occupied by these vegetation communities were once forested.

Anthropogenic Activity

The pattern of human activity in the Maghreb during the last 5000 years has been one of deforestation. Until about 2000 years ago it is likely that the pastoral economy in the forests of the Middle Atlas was of local and temporary significance (Lamb *et al.*, 1991). At Dayat Iffir in the Middle Atlas, pine was dominant or codominant with evergreen oak until it declined between 2300 and 2000 years BP. This decline is ascribed to anthropogenic clearance rather than natural causes such as forest fire. From 2300 BP there was an increase in Gramineae pollen and of arable weeds after 1000 BP. These are interpreted as evidence of first pasture and then cultivation in the area. From 1500 BP there are also sedimentary signs of soil erosion from the catchment into the lake. At the nearby lake, Dayat Affougah (or Afourgagh), forest clearance began, on the basis of the pollen record, about 1700 years ago, at approximately the same time as a major episode of soil erosion. There may have been earlier clearances of pine about 2500 years BP. At Tigalmamine, further south in the Middle Atlas, a relatively diverse *Cedrus* forest was important from 4000 years BP which included *Quercus, Fraxinus, Pistacia, Phillyrea*, and *Vitis*. The *Fraxinus* decline from 2250 years BP is interpreted as the first indicator of human activity. For the last 2000 to 1500 years the three sites reveal indications of arable farming and forest clearance, often coincident with topsoil erosion.

At the Col du Zad in the Middle Atlas, a decline of cedar at about 2860 years BP is interpreted as corresponding to the Phoenician, Carthaginian, and Roman occupations (Reille, 1976). By reference to dated pollen diagrams from the Rif, the migration of the Berber people into the mountains following the Arab invasions of the eighth century AD can also be recognized. Lamb *et al.* (1991) discuss the pollen record of *Cedrus* at sites in the Moroccan Atlas and note that despite the contemporary degradation of cedar forest as a result of overgrazing, pollen records do not show any major decline in abundance. In fact, at Tigalmamine it appears to have regenerated. Modern pollen rain of cedar is high, probably because there are few other trees present. Thus interpretation of cedar pollen may not be revealing in terms of forest clearance.

Other techniques are now being applied to elucidate the anthropogenic history of an area. These include mineral magnetic stratigraphy, diatom analysis, and sediment geochemistry. These palaeolimnological techniques have proved useful in attempting to reconstruct the impact of human activity in more recent, historical periods, the last 500 or so years. This is especially important as there are few documentary records available to interpret this period of human impact in the Maghreb.

At two sites in the Middle Atlas (Dayat Affougah and Lake Agziza) and one on the coastal plain near Rabat (Dayat-er-Roumi), palaeolimnological techniques have been used to reconstruct sediment accumulation rates and catchment disturbance (Flower *et al.*, 1989). The cores were radiometrically dated using ^{210}Pb and ^{137}Cs. Dayat-er-Roumi and Dayat Affougah show indicators of vegetation disturbance, woodland clearance, and soil erosion, and are today intensively agricultural; Lake Agziza is still dominated by natural *Cedrus* woodlands. These land-use patterns are reflected in the sediment accumulation rates of the basins. In general, Dayat-er-Roumi and Dayat Affougah have had much higher sedimentation rates in the recent past than Lake Agziza. For example, for the period prior to 1954 there was accelerated erosion of catchment topsoils with sedimentation rates greater than 1.0 cm y^{-1}. Between 1963 and 1984 there were accumulations of 30 cm at Dayat-er-Roumi and 25 cm at Dayat Affougah. These figures are also lower than for earlier periods when rates of more than 1.5 cm y^{-1} were measured. The most recent past has shown some regeneration of woodland. The greatest sedimentation rates are associated with woodland clearance and wetland drainage during the 1940s and 1950s. In contrast, the reconstructed sedimentation rates for Lake Agziza are lower. It is a less disturbed catchment with accumulation of 6 cm y^{-1} between 1963 and 1984. However, sedimentary analyses show that there was a major catchment disturbance prior to 1630.

Stevenson and Battarbee (1991) have applied similar palaeolimnological techniques at Garaet el Ichkeul in Tunisia. They have shown that soil erosion resulted from the increase in agricultural activity by French settlers in the late 1880s. This is associated with the replacement of *Pistacia* by farmland, as indicated in the pollen record. Sedimentation rates continued to increase in the twentieth century as a result of hydrological engineering in the catchment, with a sixfold increase in sediment accumulation rates associated with these schemes.

Garaet el Ichkeul illustrates the recent historical impact of anthropogenic activity. It is a National Park and Ramsar site, and an important overwintering place for wildfowl which graze the extensive beds of *Potamogeton pectinatus*. *Potamogeton* spread at the end of the last century as a result of the construction of the nearby Bizerte ship-canal and exchanges of water between Garaet el Ichkeul and Lac Bizerte. Although a National Park, Garaet el Ichkeul is threatened by proposed river channelization, dam construction, and agricultural improvements. There are similar threats to another internationally important wetland farther south in Tunisia, Sebkhet Kelbia (Hollis and Kallel, 1986).

The Fynbos Biome of the South-western Cape

Physical Background

In Southern Africa, the area of mediterranean-type climate is confined to a narrow coastal belt at the southwest tip of Cape Province. Offshore to the east is the warm Agulhas Current, while to the west there is the cold Benguela Current, across which prevailing westerly winds blow in winter. The position of the South Atlantic High Pressure cell determines, in part, the seasonal rainfall pattern (Deacon *et al.*, 1992). In winter it is located near to 32°S and the prevailing westerlies bring orographic rain to the west-facing slopes of the Cape. In addition, frontal systems are important and precipitation is also associated with the resultant fogs. In summer, the anticyclone is centred near 37°S and dry easterly winds prevail. Rainfall totals are generally greater than 250 mm/yr, mostly in the range of 300–2500 mm, although reaching 5000 mm in the mountains.

The location of the coastal mountain belt means that there is a marked windward–leeward wind pattern with föhn-like winds and consequent rainshadows. However, the west to east precipitation pattern is also influenced by southerly winds which bring orographic rain to the south-facing coastal mountains, especially in spring and autumn, to areas south of 33°S and east of 20°E. As a result, stations to the east of this show a bimodal rainfall pattern with winter, spring, and autumn rain; thus, the climate in this region is not strictly mediterranean as compared with the western area which receives 60–80 per cent of its rain in the winter (Deacon *et al.*, 1992; White, 1983).

The strong oceanic influence results in lower annual temperatures and smaller seasonal and daily temperature ranges in the coastal regions. Summers are cool, winters are mild, and frosts are rare. Meteorological stations in mountain areas have a more continental climate, although at higher altitudes orographic rain ameliorates the summer drought. In the sheltered valleys, temperatures are higher in summer and there is a greater danger of frost in winter.

The relief of the *fynbos* area is dominated by the southern Cape Fold Mountain Belt; a complex structure with parallel ridges separated by long valleys. Rocks of the Table Mountain Group comprise highly resistant sandstones and are associated with relatively base-poor substrates. These contrast with the weaker shales of the lowlands, which tend to be associated with relatively base-rich substrates. The average elevation of these mountain ranges is between 1000 and 1500 m, with individual peaks higher than 2000 m. The topography of the coastal forelands is irregular. To the west and north of Table Mountain (1087 m) is the Malmesbury Plain of slates and shales. Along the coast, older rocks are mantled by deposits of Tertiary and Quaternary age. During the Quaternary, extensive coastal dunefields accumulated across the Cape Flats east of Cape Town. During interglacial stages, sea-levels were only a few metres higher than those of today. Lower sea-levels during the glacials resulted in extension of the coastal foreland.

Palaeoclimatic History

A mediterranean-type climate first developed in the Cape region at the end of the Pliocene, around 3 million years BP, when the South Atlantic High Pressure cell became fixed relative to Southern Africa. There it has remained, even during the last glacial maximum at 18 000 years BP, the coldest and driest stage of the last 125 000 years (CLIMAP, 1976; Deacon and Lancaster, 1988). At that time the vegetation was grassier and more open than at present. Glaciers did not develop in the Cape Mountains because they were below the permanent snowline. However, block fields and block streams did form (Deacon *et al.*, 1992). It is estimated that mean annual temperatures were about 5 °C lower. Summer temperatures at these times would have been equivalent to present-day winter temperatures and, as the interannual temperature range was no different from the present, winter temperature would have been accordingly lower.

The transition to the warmer conditions of the Holocene occurred before 14 000 years BP, indicating that the postglacial warming began about 2000 years earlier in the southern hemisphere than in the northern hemisphere. The early Holocene was wetter than both the previous and later stages. The contemporary climate patterns were in place from about 5000 years BP, including those of the eastern Cape, where today there is a bimodal seasonal precipitation regime. Prior to 5000 years BP, this region would have been one of winter-dominated rainfall (Deacon *et al.*, 1992).

There have been very few vegetation reconstructions for the Cape for the late-glacial and Holocene. One of the most recent is that of Meadows and Sugden (1991). They analysed the pollen stratigraphy from two vleis on Cederberg Mountain, 250 km north of Cape Town, where there are contemporary *fynbos* communities. The record covers the last 14 000 years and indicates that there has been relatively little environmental change in the Cederberg during that time. Subtle changes in the vegetation community have occurred, but no marked climatic change occurred at the Pleistocene–Holocene transition. This is explained in terms of the geology of the region, the nutrient-poor substrates of the Table Mountain Group being more influential than climate. The significance of this finding is that it gives a longer time-frame for the development of mountain *fynbos*.

Flora and Vegetation

The natural vegetation of the mediterranean-type climatic zone of Southern Africa is commonly known as *fynbos*. The term was first used by Bews (1916) when he described the fine-leaved bushy habit of the South African shrublands. The *fynbos* communities have been the subject of considerable study including that of Acocks (1953), Taylor (1978), and Goldblatt (1978), and summarized by White (1983). More recently the Fynbos Biome Project has had the objective 'to provide sound scientific knowledge of the structure and functioning of the constituent ecosystems as a basis for the conservation and management of the fynbos biome' (Anon., 1984). Major reviews of the work of the Fynbos Biome Project include Day (1983); Deacon *et al.* (1983); Kruger *et al.* (1983); Booysen and Tainton (1984); Huntley (1987); Cowling (1992); and van Wilgen *et al.* (1992).

The *fynbos* region is a distinct phytogeographic unit that is commonly recognized as a floral kingdom of its own: the Cape Floristic Kingdom, or Capensis (Taylor, 1978). It covers 90 000 km² and contains 152 families, 986 genera, and 8504 species; 68 per cent of the species are endemic (Bond and Goldblatt, 1984). Taylor (1978) gives figures of 282 genera with their centres of origin in the Cape Floristic Kingdom, 212 of which are confined to the Kingdom. These figures may represent the highest rate of generic endemism in the world (Good, 1974). Seven families are entirely confined to the Kingdom: Bruniaceae, Geissolomataceae, Grubbiaceae, Penaeaceae, Retziaceae, Roridulaceae, and Stilbaceae.

Definitions of Fynbos

The *fynbos* biome has been defined in a number of different ways. There is a general consensus that it is a fine-leaved evergreen sclerophyllous shrub vegetation (Moll and Jarman, 1984a), but it can be defined further based on climatic characteristics, floristic composition, the structure and function of the biome, and edaphic characteristics. Thus it is not always clear which vegetation types should be included within the biome.

Floristically, *fynbos* has one or two important features, notably 'lack of single species dominance, and/or the conspicuous presence of members of the family Restionaceae'. Physiognomically, 'fynbos is characterized by three elements, restioid, ericoid, and proteoid'. These elements comprise plants that resemble typical members of the Restionaceae, Ericaceae, and Proteaceae respectively in growth form but do not necessarily belong to these families (Taylor, 1978). The restioid growth form is of tufted or rhyzomatous plants, 0.2 m to 2 m tall, with green tubular or wiry non-woody stems with reduced, non-photosynthetic leaf scales. These persist for more than one year but usually less than four years. Ericoid leaves are small, narrow, and often rolled, typical of the Ericaceae family, but also found in other families such as Bruniaceae, Polygalaceae, Rutaceae, Thymelaeaceae and in species in genera like *Cliffortia* (Rosaceae), *Phylica* (Rhamnaceae) and *Metalasia* and *Stoebe*, both Asteraceae. The proteoid element is associated with less-branched, usually taller bushy plants with moderate-sized hard leaves with a dull surface, typified by members of the the family Proteaceae, such as *Leucadendron, Leucospermum, Mimetes,* and often *Protea* itself. At drier, higher altitudes the proteoid element may be absent, while the restioids and ericoids are usually always present in fynbos. Some workers have questioned the almost universal presence of the restioid element in Taylor's definition, but it is, in general, the most acceptable (Moll and Jarman, 1984a).

Typical *fynbos* vegetation can be found outside the south-western Cape region and indeed small patches are found at low altitudes in the Transkei, in southern Natal, and further north in the Chimanimani Mountains of Zimbabwe. A detailed phytogeographical study is

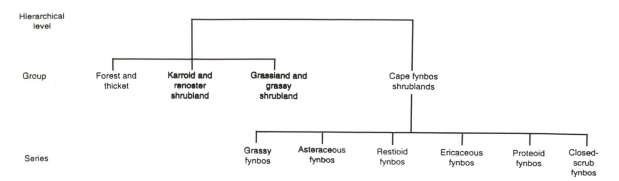

Fig. 17.5. Hierarchical arrangements of *fynbos* communities.
(Modified and reproduced with permission from Cowling and Holmes, 1992, in Cowling, 1992: 36.)

needed to understand these related *fynbos* types (Moll and Jarman, 1984*a*). Thus *fynbos* is not confined to those areas with a mediterranean-type climate. Indeed, even though it is considered as one of the five mediterranean-type ecosystems of the world, its extent in Cape Province stretches into the area of bimodal seasonal precipitation to the east of about 20°E. Several terms are used to distinguish between different *fynbos* ecosystems. Among these are *fynbos* itself, strandveld, macchia, and coastal renosterveld. These may all be considered as ecosystems within the Cape Floristic Kingdom, the area of mediterranean-type climate, or within the *fynbos* biome. There is some confusion in that the terms are defined differently by different authors. According to Moll and Jarman (1984*a*), climatically the *fynbos* biome comprises three veld types defined by Acocks (1953) as strandveld, coastal renosterveld, and coastal macchia. These are mediterranean *fynbos*. Two further Acocks' veld types are often included floristically within the *fynbos* biome, macchia, and false macchia (or collectively mountain *fynbos*), but are not strictly mediterranean-type shrublands.

Distinctions are made between coastal *fynbos* in the lowlands (coastal macchia) and mountain *fynbos* (macchia and false macchia) (Acocks, 1953; Taylor, 1978). These are largely based on floristic characteristics. However, there is a high geographic turnover in the *fynbos* biome, and therefore mountain landscapes widely separated may have flora which show less similarity than between adjacent mountain and coastal sites. Thus floristic divisions into coastal and mountain *fynbos* can be misleading (Cowling and Holmes, 1992).

Structural classification of *fynbos* could be more effective in revealing ecological gradients than a purely floristic classification (Campbell, 1985, 1986*a*, 1986*b*). In a floristically complex region such as the Cape Floristic Kingdom there are considerable practical problems in

identifying the hundreds of species present, especially given the rapid geographic turnover of species. This results in communities with similar environmental and structural characteristics having very different floristic communities. There are, however, problems with the structural approach in the fire-susceptible ecosystems of the *fynbos* (Cowling and Holmes, 1992). Structural classifications are based on mature vegetation but the effects of fire mean that many vegetation communities may be immature. However, it is unlikely that a floristically based scheme for delineating different *fynbos* biome vegetation communities will be produced before the end of the century.

Attempts at assessment (including maps) of the extent of the natural vegetation of the *fynbos* biome have been made partly on a floristic basis. Specific spectral signatures from Landsat imagery have been used to delineate various *fynbos* vegetation types, based on the characteristic reflectance values (Moll and Bossi, 1984). Eight different vegetation units have been identified: strandveld, coastal renosterveld and coastal macchia; macchia and false macchia, which together comprise mountain *fynbos*; and mountain renosterveld and Knysna Forest. The resulting map is the most detailed yet of the *fynbos* communities, but the bases of the delineation are too weak to test in the field. Thus Cowling and Holmes (1992) advocate Campbell's vegetation classification which recognizes six series in the Cape Fynbos shrubland group (Figure 17.5). Figure 17.6 delineates the general area of the *fynbos* biome, but no attempt is made to subdivide the biome into its different ecosystem types.

The first of Campbell's series is the grassy *fynbos* with a high-grass cover and a relatively high cover of non-proteoid nanophyllous vegetation and forbs. It is found mostly in the east but also in the southern interior and southern coastal areas. It is the easternmost *fynbos* where there is a high proportion of summer rains on finer

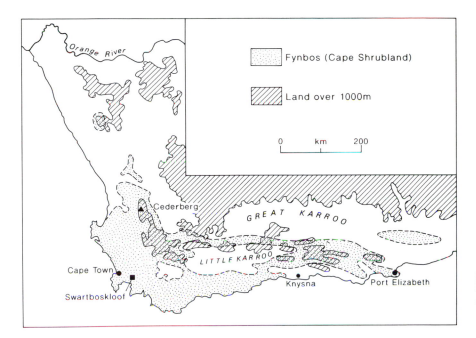

Fig. 17.6. The *fynbos* biome of Cape Province (As shown in White, 1983.)

textured and more fertile soils. Asteraceous *fynbos* has a low total cover, often with high grasses and *Elytropappus* cover and a high cover of non-ericaceous ericoids. It is found throughout the biome in the eastern mountains on the driest sites and on a range of substrata. It links *fynbos* to the karroid and renoster shrublands. It is widespread on coastal forelands, on calcareous dunes, and on shales and sil-ferricretes where precipitation is less than 550 mm/yr (although in the mountains precipitation ranges from 450 mm to 950 mm/yr). Restioid *fynbos* has a high restioid cover of more than 60 per cent and a low shrub cover. It has the lowest constancy of tall (i.e. more than 1.5 m) shrubs of all the *fynbos* series. It is found throughout the biome, except in the mountains in the east and is rare in the southern coastal regions. It occurs where excessive waterlogging or drainage lead to limiting conditions for shrub growth. It is found on sites which are more mesic than asteraceous *fynbos*; soils may be deep, shallow or rocky.

Ericaceous *fynbos* is a leptophyllous shrubland like asteraceous *fynbos*, but has a high cover of restioids, and the shrubs are mainly ericaceous ericoids. Total cover and shrub cover are also higher than for asteraceous *fynbos*. It is found mainly in south-western and southern coastal areas and is predominantly confined to south-facing and wet slopes of the coastal mountains (where precipitation averages more than 1500 mm/yr). Soils have a high carbon content, low pH and high fine-particle fraction. Proteoid *fynbos* has more than 10 per cent cover of mid-high to tall non-sprouting proteoid shrubs. In some coastal foreland communities canopy proteoids are low. It is found throughout the biome, especially in southern coastal areas, but is generally rare in north-western and eastern areas. Like grassy *fynbos*, it occurs at lower altitudes than do the other *fynbos* series, and it is found on a wide range of substrata including non-aeolian sand and limestone on the coastal forelands. Its altitudinal range is from sea-level to 950 mm with precipitation from 400 mm/yr at the coast to 1100 mm/yr at higher elevations.

Finally, closed-scrub *fynbos* is found throughout the biome except in eastern and interior areas. It is associated with well-drained riparian habitats in the mountains and is similar to forest and thicket cover in its relatively high percentage cover of mesophyllous, non-proteoid plants. It has a higher percentage cover of restioids, with frequent presence of Ericaceae, than forest of thicket.

These six *fynbos* series all fall within mountain and coastal *fynbos* and strandveld communities as defined by Taylor (1978). However, Campbell's structural classification does not include coastal renosterveld (*sensu* Taylor) as a member of the *fynbos* shrubland group, although renosterveld falls within the mediterranean-type climatic region of the Cape and is found throughout the *fynbos* biome on lower mountains, interior valleys and coastal foreland sites on the ecotone between *fynbos* and succulent shrubland. It is a low to mid-high, open to mid-dense shrubland, often with an open, grassy understorey and dominated by *Elytropappus*. Coastal renosterveld can be regarded as secondary Cape shrubland which readily encroaches on agricultural land (White, 1983).

Edaphic factors are also important in the determination of *fynbos*. Soil nutrients are an important control on *fynbos* community structure. *Fynbos* soils tend to be nutrient-poor, but need not be. Reviews of these relationships are given by Groves (1983), Cowling and Holmes (1992), and Le Maitre and Midgley (1992).

Fynbos may be recognized as heathland despite the much better nutrient status of some of the soils in the *fynbos* biome (Moll and Jarman, 1984b). For the southern hemisphere, Specht's (1979) definition of heathlands is redefined as fire-maintained, evergreen sclerophyllous communities dominated by Ericaceae and occurring on oligotrophic podsolized soils, to encompass what might be described as non-typical heath soils.

Fire and Invasive Species in the Fynbos

Fynbos is a fire-type community with species which are adapted and pre-adapted to survive frequent fires. There has been considerable research on the effects of fire on *fynbos* communities: it has focused in particular on effects on structure, species composition, and species response to fire; on fire frequency, both natural and as a result of deliberate and accidental combustion; on fire as a management tool; and on the relationship between fire and invasive, alien plant species in *fynbos* communities. Reviews of *fynbos* and fire can be found in Taylor (1978), Kruger (1979), Kruger and Bigalke (1984), van Wilgen, Bond and Richardson (1992), van Wilgen *et al.* (1992), and Rebelo (1992).

Fires in *fynbos* may occur naturally from lightning strikes or from the sparking of rolling or falling rocks as a result of erosional processes involved in scarp retreat or after earth tremors. Examples of the latter include fires in the Cederberg Range in September 1969 (Taylor, 1978). Fires may also be started deliberately, often as a management strategy. The use of fire in Southern Africa is first associated with the Chelles-Acheulian people and possibly also with the Fauresmith Industries over 53 000 years ago (Taylor, 1978). Evidence of Acheulian settlements is found in the intermontane valleys and coastal forelands (Deacon, 1991). Late Stone Age populations would have used fire to manage food resources and to maintain the productivity of the *fynbos*. Today, most fires occur near to centres of population, as in the Cape Peninsula, and, although it is not always easy to establish the causes, these would include arson, cigarettes, honey-hunters, locomotives, and controlled burns (Kruger, 1979; Kruger and Bigalke, 1984). Natural fires are still important in remote areas such as the Cederberg.

While it is difficult to distinguish clearly between adaptive and pre-adaptive strategies for surviving fires, *fynbos* species show a range of typical strategies. Plants have been categorized into fire life forms comparable to Raunkiaer's life forms (van der Merwe, 1966). Four of these cover the main strategies of *fynbos* plants: fire geophytes, fire hemicryptophytes, fire chamaephytes and nanophanerophytes, and fire therophytes. Fire geophytes such as Liliaceae and Iridaceae grow from dormant buds on underground storage organs. Some of these will flower immediately after fire and rarely at other times, for example the fire lily (*Cyrtanthus contractus*). Others will flower in greater abundance shortly after a fire and less frequently later. For example, 5 per cent or less of all *Watsonia pyramidata* ramets flower annually, but there is abundant flowering (about 50 per cent) following an autumn burn. There is no response to burning in spring. Thus fires have to occur at the 'right' season for a species like *W. pyramidata* to respond (Kruger and Bigalke, 1984).

Fire hemicryptophytes show a rapid response to recovery after fire. These are plants which sprout from rhyzomes at or near the soil surface, including most graminoid and restioid components. These are among the quickest to respond to fire and, temporarily, may become dominant in the period up to 4 or 5 years after a fire, the youthful phase (Kruger and Bigalke, 1984). At Swartboskloof, 65 per cent of the vegetation reproduces vegetatively, and thus is very important in the successional stages of *fynbos*. In particular, these plants help to stabilize soils and prevent erosion because their basal parts are not destroyed by fire and also bind the soil. In addition, their rapid growth immediately after a fire provides new ground cover.

Species which regenerate from seed following fires are mostly trees and shrubs. Some species have seeds which remain on the plant, only being released when the plant dies. For example, forty-nine out of the eighty-two species of *Leucadendron* have seeds which remain in cone-like fruiting heads for up to eight years (Kruger and Bigalke, 1984). Other seeds stay dormant in the seedbank and only germinate after a fire. Heat may be needed for scarification before germination. Seeds can remain dormant for long periods. *Elytropappus glandulosus* produces vast quantities of seed which are viable for up to seven years. It is important to consider fire frequency with respect to those species which regenerate from seed. Maturity can be reached between one and ten years depending on species. The most vulnerable species are

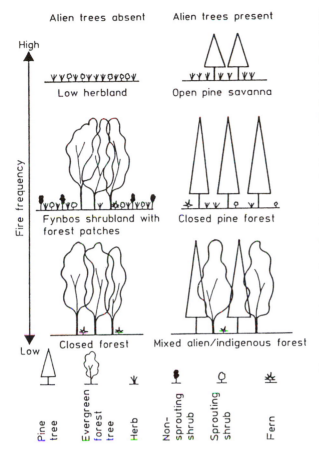

Fig. 17.7. Hierarchical stable vegetation formations under different fire frequencies and without alien trees.

(Reproduced with permission of Springer-Verlag, from van Wilgen and Richardson, 1992, in van Wilgen *et al.*, 1992.)

height and flowering ability (for example, longer-lived Proteaceae). Low-seeding shrubs begin to die and litter accumulates more rapidly. This is followed by senescence of the older seed-regenerating shrubs. Where the canopy opens there may be some seed regeneration but litter continues to accumulate. The senescent stage can only occur if fire has not spread again through the community. The accumulation of litter, dead shoots and dead standing wood, all highly inflammable, promotes the likelihood of fire. Usually at least four years regrowth following a previous burn is needed to sustain a fire (Kruger and Bigalke, 1984).

Experiments in the Swartboskloof have resulted in an hypothesis which examines the effects of variations in fire frequency on the structure and composition of *fynbos* communities (see Figure 17.7). Where fire frequency is low, indigenous forest elements could become established, but this is not practically attainable and would probably reduce species diversity in *fynbos*. Moderate fire frequencies maintain a mosaic of shrub and forest. Where fire frequencies are high, the forest areas and re-seeding shrubs are replaced by low-herb land (Richardson and van Wilgen, 1992).

The hypothesis also examines the role of fire with respect to alien invading plants. *Fynbos* communities are often characterized by a large element of non-native (i.e. introduced and naturalized) plant species. Many of these are fire-adapted and the regular occurrence of fires promotes invasion and population growth (for example, *Hakea* and *Pinus* spp.). Sixteen of the twenty-three alien weeds in the *fynbos* biome are trees and shrubs which regenerate from seed (Hall and Boucher, 1977). Although their initial colonization of a site may result from seed dispersal by birds (including introduced species such as the European starling) or animals, such as *Acacia melanoxylon*, or wind, such as *Hakea*, once a small population has become established it will rapidly increase after a fire until it replaces the natural vegetation. It is hypothesized that moderate fire frequencies will result in dense stands of alien trees and shrubs. Where fire frequency is high there will be an open savanna with a mix of sprouting *fynbos* plants and young alien pines and hakeas. However, if fire is prevented then the fire-adapted invaders will lose their competitive advantage and a mixed stand of indigenous and invading trees is likely.

It has been suggested that alien species are more successful in invading the *fynbos* biome than other Southern African biomes. Several hypotheses have been put forward to test this question of susceptibility to invasion. One proposition is that alien tree species have filled a vacant niche in *fynbos* resulting from an absence of

woody shrubs which take an average of five to six years to reach their seed-producing phase. In communities with fires that occur fairly frequently such woody shrubs would be gradually eliminated. For some members of the genus *Protea*, the median age of seed production is four years; species which take six years or more to flower are usually found in areas relatively protected from fire, such as rock outcrops or moist mountain sites (Rourke, 1980).

Four stages of typical successional response to fire have been identified: youthful, transitional, mature, and senescent. The youthful stage lasts about four or five years and is dominated by restionaceous and graminoid vegetation. During the transitional stage tall shrubs are in the ascendancy until about ten years after the fire. The mature phase can last up to thirty years. During this period trees and shrubs are at their maximum

indigenous trees in the ecosystems. Alien trees such as *Pinus pinaster* may be better able to withstand the summer drought. However, there are a number of arborescent *Protea* which would grow to dominance if fire were excluded (Richardson and van Wilgen, 1992). It is possible that alien trees are able to invade *fynbos* because of a lack of vigorous woody and herb plants immediately after fires. Invaders are thus able to exploit the temporary resources created by gaps opened by fire in *fynbos* (Richardson *et al.*, 1992). Another hypothesis concerns the considerable dispersal ability of invasive species, for example *Acacia cyclops, Pinus pinaster, and Hakea sericea*. Macdonald (1984) questions why their dispersal strategies should be more successful than indigenous species which have had millenia to achieve their optimum distributions; he concludes that it may be that disturbance patterns have changed. For example, a greater frequency of unnatural fires may have given invasive species advantages over indigenous species. The relatively frequent and high-velocity winds common in the region promote dispersal of species with winged seeds. In addition, winds occur at the same time as fires which kill adult pines and hakeas releasing seeds for dispersal (Richardson *et al.*, 1992).

It has also been proposed that many invasive species are successful because their natural predators (in their original localities) are absent. In addition, there is an absence of pathogens in their new environments. The plants therefore become weeds (Macdonald, 1984). There is some indirect support for this proposition in that when such predators are introduced the populations of invaders decline. Macdonald concluded that there is insufficient evidence to establish whether the *fynbos* biome has a disproportionate number of invasive species and is more susceptible to invasion compared with other biomes, especially given the lack of evidence on the state of *fynbos* prior to human occupation of the biome.

Invading plant species may have been introduced to Southern Africa either deliberately or incidentally. Southern Africa has a long history of migration by African pastoralists and agriculturalists (Deacon, 1991). Pastoralism began between 1700 and 2000 BP and the first alien plant species was *Medicago polymorpha* (Eurasian burclover); its association with domestic-sheep remains is dated to about 1200 BP. However, it is difficult to determine the extent to which these people introduced new plant species and what impact they had on the landscape. The major introductions occurred with the European colonization of the Cape from the seventeenth century—a marked contrast to far earlier dates of introduction in the Mediterranean and the Maghreb. Fox (1990) reviews the history of contact between the Cape

and Europe, and other areas with a mediterranean-type climate from which invasive species have originated. Until the opening of the Suez Canal in 1869, trade from Europe to Australia went via the Cape and offered many opportunities for both incidental and deliberate introductions. The most common element of mediterranean-type invaders is from Australia, including *Acacia* species and *Hakea* species. *Acacia saligna* and *A. cyclops* were first introduced to botanical gardens in 1848 and 1857 respectively, and then disseminated for timber and sand stabilization. *Hakea sericea* and *H. gibbosa* were introduced to botanical gardens in 1858 and 1835 respectively and then used for hedging and, in the case of *H. gibbosa*, for sand stabilization (Richardson *et al.*, 1992). Because of separate sea-lanes for trade to Europe, and because South Africa was also colonized by northern Europeans rather than people from the Mediterranean, there was limited contact between South Africa and the mediterranean regions of the Americas: Chile, and California. However, there are examples of invasive plants from California (*Opuntia aurantiaca*) and from Chile (*Bromus unioloides, Xanthium spinosum*). Species from the Mediterranean Basin include *Pinus pinaster, P. halepensis, P. rediata*, all three for timber, with *P. pinaster* introduced as early as the 1680s (Richardson *et al.*, 1992). The flora of South Africa (and Australia) still exhibit summer growth, a feature not found in other mediterranean flora, which have spring growth. It is therefore argued that spring growth gives these invasive plants a competitive advantage in growing immediately after winter rains (Fox, 1990).

Several criteria can be used to measure the invasiveness of plants. Table 17.1 lists the most invasive species according to two criteria. The first column is based on the current area of invasion of the biome as a whole. The second is based on impact (i.e. potential area of invasion), number of species that would be displaced, impact on hydrological cycles, and aesthetic values. For *Hakea* and *Pinus*, there are estimates of 7592 km² as the area invaded by these two genera; *Acacia* and other woody plants which form thickets are estimated to have invaded 8962 km².

Conservation and Management Strategies for the Fynbos Biome

The objectives of conservation and preservation in the *fynbos* biome are based on the same principles given in other parts of the world: i.e. the need to promote biological diversity. However, in the *fynbos* region the impetus has come from recognition of the region's considerable vegetation diversity. The Cape Floristic Kingdom contains the highest concentration of threatened plants

Table 17.1 The ten most important invasive vascular plants of the Fynbos Biome

Based on current area of invasion of the biome as a whole	Based on impact (see text)	
Hakea sericea	Acacia cylcops	
Acacia cyclops	A. saligna	
A. saligna	A. longifolia	Equal impact
Pinus pinaster	A. mearnsii	
A. longifolia	Paraserianthis lophanta	
A. mearnsii	A. melanoxylon	
Mediterranean grasses (e.g. Avena, Briza spp.)	Pinus pinaster	
P. radiata	P. radiata	Equal impact
Leptospermum laevigatum	P. halepensis	
P. halepensis	P. canariensis	
	P. pinea	

Sources: Modified and reproduced with permission from: Wells (1991) 'Introduced Plants of the Fynbios Biome of South Africa', in Groves and di Castri (eds.), Biogeography of Mediterranean Invasions.

in Southern Africa: 68 per cent of the sub-continent's critically rare, threatened and recently extinct taxa (Hall et al., 1984). The main families with a high proportion of threatened taxa are the Restionaceae, Proteaceae, Iridaceae, Penaeaceae, and Bruniaceae. The last two are found only in the Cape Floristic Kingdom and the others are well represented in the region's flora. One of the main threats to the fynbos biome is the invasion of alien vegetation, but other factors also need to be considered, such as land-use changes (afforestation and agriculture), urbanization, water extraction, and dam building (Rebelo, 1992). Figures for loss of, or threatened, land vary with data and methodology but it has been suggested that 56 per cent of the area has been lost to agriculture and that 72 per cent is threatened by invasions. Calculations based on analysis of Landsat imagery indicate that 34 per cent of the natural vegetation of the fynbos biome had been lost to farming between Acocks' 1953 veld maps and 1981 (Moll and Bossi, 1984). Other estimates are that about 61 per cent of the biome had been lost (Hall, 1978). However, these figures are not strictly comparable given the use of different definitions of fynbos.

The main agricultural threats are conversion to wheat, pasture, and vineyards; the area under threat is predominantly in the renoster shrubland. Urbanization is a significant threat in the Greater Cape Town metropolitan area, for which there are estimates of rapid population increase from 2.2 million in 1986, to 3.5 million in 2000, and 6.2 million by 2020 (Rebelo, 1992). Allied

to urban expansion is the increased pressure from the popularity of recreation and tourism in the undisturbed areas.

To counter these threats to fynbos, management strategies are being devised based on empirical research (van Wilgen et al., 1992). The main objectives are for nature and water conservation, fire management, flower harvesting, grazing, and recreation and tourism. A simple hierarchy of management intensity for fynbos communities has been identified. At the lowest level there may be non-management. This is more likely to occur in more remote or arid areas. At the next level, 'natural burning zones' may be implemented, mostly in areas with a relatively low fire-risk from adjacent areas, as in isolated mountain ranges. In these areas the fynbos communities are surrounded by fire-breaks but there is little management, only the removal of alien species. At the next level, there is active management to control invasive plants. Burning takes places at prescribed intervals. This 'block burning' may occur on farms, nature reserves, and in catchment areas, at scales ranging from 2–3 ha for farms and nature reserves, to 500–1500 ha for drainage basins. The final level is intensive management; strategies such as the use of fire, pest management, translocations, seed harvesting, and seed addition will all be used.

Fire is the major management tool and attitudes to fire in the fynbos have changed (Rebelo, 1992). Historically, fire was used extensively to create pasture. Then in the mid-twentieth century the aim was to protect fynbos from fire. However, the occurrence of uncontrollable fires, and subsequent research, has led to the recognition of fire as a conservation and management tool. In using fire as such a tool, knowledge of fire frequency, season, and intensity—i.e. the fire regime—is needed. Experiments carried out at Swartboskloof examined the effects of fire frequency and showed that plant species richness, catchment streamflow, and phytomass are most resilient to fire intervals of twelve to thirty years (Richardson and van Wilgen, 1992). Figure 17.8 shows, hypothetically, that there are increases in biomass and decreases in streamflow with increased post-fire age. Exclusion of fire results in a decrease in diversity as the vegetation becomes senescent and dies. It is possible that new species will invade, but, in general, diversity is likely to be lower compared with more frequently burnt areas of fynbos. Phytomass will also eventually increase, but streamflow will decrease. With a fire frequency of five years, streamflow will increase, but phytomass will decrease as will species diversity as seed-producing shrubs decline. As a result of these experiments, the fire frequency is usually twelve to fifteen years, with burning

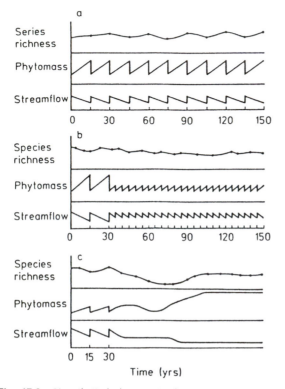

Fig. 17.8. Hypothetical changes in three parameters (plant-species richness, phytomass, and streamflow) at the ecosystem level for three fire frequencies at Swartboskloof.

The fire frequencies depicted are: (a) fire every fifteen years; (b) fire every five years; and (c) fire excluded. Tick marks on the x-axis indicate the occurrence of fire. (Reproduced with permission of Springer-Verlag, from van Wilgen and Richardson (1992), in van Wilgen *et al.*, 1992.)

in late summer or autumn, March and April. Some of the stochastic element is removed with such a prescriptive regime but, in practice, wild fires in summer and escapes of fire from controlled burns in spring will lead to a greater frequency of burn.

Fire is used as a method of controlling the invasion of alien species. For example, young pine seedlings are very vulnerable to fire even though the adults are not. Frequent burning can therefore be used to halt their spread. This can be combined with felling of adult trees a year before burning. Seeds are released and germinate. The seedlings are then killed in the subsequent burn (van Wilgen *et al.*, 1992). With hakeas, seeds are allowed to germinate in order to deplete the seedbank in the soil. Then, between one and two years after germination, the seedlings are cleared and burned; this is the time when they are most sensitive to heat. For both hakeas and pines these controls are effective but costly. With the exception of *Acacia cyclops*, the other main acacia species

all have seeds that are stimulated to germinate by fire. The seeds are relatively long-lived in the soilbank and so seedling populations which grow up after fires need to be cleared in other ways, such as the use of herbicides. This needs to be repeated in order to exhaust the soil seedbank (Kruger and Bigalke, 1984). For *A. cyclops* fire does destroy seeds near the soil surface and the seedbanks are also relatively short-lived (van Wilgen, Bond, and Richardson, 1992).

Other than burning there is some bio-control of invasive species. For example, the pine woolly aphid, *Pinus pini*, was accidentally introduced but does reduce seed production in *Pinus radiata* and *P. pinaster* by causing cone deformation. There are, however, problems with this form of control because of the commercial importance of pine. For acacias there is some control from the bud-galling wasp, *Trichilogaster anaciaelongifoliae*, and the snout beetle, *Melanterius ventralis*. In *Acacia longifolia* these reduce seed production by about 99 per cent. Several species of *Melanterius* may prove effective with other acacias but may not be considered because of the timber value of species like *A. mearnsii*. For *Hakea gibbosa* and *H. suaveolens* there are no available bio-controls, but for *H. sericea*, the larvae of the snout beetle, *Erytenna consputa*, and the larvae of the moth, *Carposina autologa*, destroy fruits, and a fungus, *Colletotrichum gloeosporioides*, reduces vigour and results in death (van Wilgen, Bond, and Richardson, 1992). For effective control of invasive plants there needs to be an integrated management policy using a combination of techniques.

Nature Reserves

Nature reserves in the Cape Floristic Kingdom total 1 797 000 ha, comprising 19 per cent of the total area (Rebelo, 1992). Of these nearly 50 per cent were State Forests transferred in 1987–8 to the Chief Directorate of Nature and Environment Conservation (CDNEC) of the Cape Provincial Administration following government privatization. Much of this area was designated in the 1960s for river-catchment management following research that showed that annual runoff was 50–100 per cent higher in fynbos catchments rather than in forest plantations. A further 38.6 per cent of the preserved area is in the form of Contractual Reserves. These are privately owned and were established from 1970 onwards. Unlike state or provincially owned reserves, in contractual reserves grazing, game farming, flower and plant harvesting on a scale commensurate with ecosystem conservation, are allowed. There are also other, smaller reserves including those owned by the CDNEC prior to transfer of the State Forests.

While the original justification of the State Forests

was water management, the other major aim of preservation lies in the characteristics of the *fynbos* ecosystems themselves. This differs from other areas in South Africa where large mammal species and their habitats have tended to be the focus of conservation.

Some research has looked at the importance of nature reserve size in the Cape Floristic Kingdom. About 20 per cent of the reserves are larger than 10 000 ha and most are in areas of mountain *fynbos*. These may be contiguous and thus some mountain reserves are in fact larger than 100 000 ha. About 43 per cent of the reserves are less than 500 ha and 17 per cent less than 50 ha (Rebelo, 1992). The effect of reserve size on species richness was assessed for 'islands' of limestone *fynbos* surrounded by equally fire-prone *fynbos* on acid sand in the south-western Cape lowlands (Cowling and Bond, 1988). These were compared with species richness in 'mainland' reserves. A similar study was carried out for *fynbos* islands surrounded by Afromontane forest (Bond *et al.*, 1988). The aim was to discover how small a reserve could be before extinction occurred. The results of these studies showed that the fragmentation effect on the limestone *fynbos* islands was not as great as for islands in a sea of forest; probably because the limestone islands were as prone to fire as their surrounding areas, growth forms were little different in the two. In contrast, the Afromontane forests act as barriers to fire in the southern Cape making recurrent colonizations of the islands more difficult. For the islands within the Afromontane forest, it was proposed that nature reserves need to be between 300 and 600 ha to prevent species loss (Bond *et al.*, 1988). For the limestone areas critical reserve sizes of 4–15 ha were calculated (Cowling and Bond, 1991). For the region of the south-western Cape these figures are apparently encouraging as 90 per cent of the reserves are larger than 15 ha. It has also been argued that there is a case for the preservation of areas of *fynbos* ecosystem as small as 4 ha, provided that there is minimal disturbance (Rebelo, 1992).

Conclusions

This review of the biogeography, more specifically the plant biogeography, of the two mediterranean regions of Africa shows that they have broadly similar vegetation-type communities: sclerophyllous, evergreen shrubland communities. Such a similarity raises the issue of convergence and non-convergence of these mediterranean-climate ecosystems (di Castri and Mooney, 1973; Cody and Mooney, 1978). In a comparison of the physical similarities of the mediterranean-climate regions of South Africa, California, Chile, and the Mediterranean

Basin itself, the main feature in common is summer drought, though to varying degrees. Otherwise, the regions each have their own distinctive features. Floristically, the similarities between the four are slight. Endemism is higher in the more isolated Cape Floristic Kingdom at the tip of southern Africa than in the Maghreb, which has had greater potential for contacts with other parts of the Mediterranean Basin and other areas of Africa. A review of the four regions suggests that plant strategy types are in general comparable (Cody and Mooney, 1978).

The two regions have had different histories of environmental change. For the Maghreb, there are marked contrasts between the climates of the late-glacial and the Holocene. In the Cape Province there is evidence which suggests reduced climatic differences between the two periods. In the Maghreb the evidence of later Holocene climatic change recorded in lake sediments is difficult to distinguish from that determined by human activity. Such impact occurred significantly earlier and to a greater degree in the Maghreb than in Southern Africa. However, it can be argued that floristically the implications of human impact have, in one sense, been greater in the Cape with the threats posed by invasive vegetation.

Ecosystems in both regions face challenges from the continued threat of landscape change; the need for conservation strategies is paramount. Such strategies are perhaps better formulated in the Cape Province than in the Maghreb, where more than one government is responsible for such issues.

References

Acocks, J. P. H. (1953), 'Veld Types of South Africa', *Memoirs of the Botanical Survey of South Africa*, 28.

Anon. (1984), 'The Cape's Floral Kingdom', *South African Journal of Science*, 80: 344–5.

Aschmann, H. (1973), 'Distribution and Peculiarity of Mediterranean Systems', in F. di Castri and H. A. Mooney (eds.), *Mediterranean Type Ecosystems* (London), 11–19.

Bews, J. W. (1916), 'An Account of the Chief Types of Vegetation in South Africa, with Notes on the Plant Succession', *Journal of Ecology*, 4: 129–59.

Bond, P. and Goldblatt, P. (1984), 'Plants of the Cape Flora: A Descriptive Catalogue', *Journal of South African Botany*, supp. vol., 13: 1–455.

Bond, W. J., Midgley, J. J., and Vlok, J. (1988), 'When is an Island Not an Island?: Insular Effects and Their Causes in Fynbos Shrublands', *Oecologia*, 77: 515–21.

Booysen, P. de V. and Tainton, N. M. (1984) (eds.), *Ecological Effects of Fire in South African Ecosystems* (Berlin).

Brun, A. (1989), 'Microflores et paléovégétations en Afrique du Nord depuis 30 000 ans', *Bulletin Société Géologique de France*, 8, t.V, no. 1: 25–33.

—— (1991), 'Reflexions sur les pluvieux et arides au Pléistocène supérieur et à l'Holocene en Tunisie', *Palaeoecology of Africa*, 22: 157–70.

—— (1992), 'Pollens dans les series marines du Golfe de Gabès et du Plateau des Kerkennah (Tunisie): Signaux climatiques et anthropiques', *Quaternaire*, 3: 31–9.

Campbell, B. M. (1985), 'A Classification of the Mountain Vegetation of the Fynbos Biome', *Memoirs of the Botanical Survey of South Africa*, 50: 1–115.

—— (1986a), 'Montane Plant Communities of the Fynbos Biome', *Vegetatio*, 66: 3–16.

—— (1986b), 'Vegetation Classification in a Floristically Complex Zone: The Cape Floristic Region', *South African Journal of Botany*, 52: 129–40.

Cheylan, G. (1991), 'Patterns of Pleistocene Turnover, Current Distribution and Speciation Among Mediterranean Mammals', in R. H. Groves and F. di Castri (eds.), *Biogeography of Mediterranean Invasions* (Cambridge), 227–62.

CLIMAP Project members (1976), 'The Surface of the Ice Age Earth', *Science*, 191: 131–7.

Cody, M. L. and Mooney, H. A. (1978), 'Convergence Versus Non-convergence in Mediterranean-climate Ecosystems', *Annual Review of Ecology and Systematics*, 9: 265–321.

Cowling, R. M. (1992) (ed.), *Fynbos: Nutrients, Fire and Diversity* (Cape Town).

—— and Bond, W. J. (1991), 'How Small Can Reserves Be?: An Empirical Approach in Cape Fynbos, South Africa', *Biological Conservation*, 58: 243–56.

—— and Holmes, P. M. (1992), 'Flora and Vegetation', in R. M. Cowling (ed.), *Fynbos*, 23–61.

—— —— and Rebelo, A. G. (1992), 'Plant Diversity and Endemism', in Cowling (ed.), *Fynbos*, 62–112.

Deacon, H. J. (1991), 'Historical Background of Invasions in the Mediterranean Region of Southern Africa', in Groves and di Castri (eds.), *Biogeography of Mediterranean Invasions*, 51–7.

—— Jury, M. R. and Ellis, F. (1992), 'Selective Regime and Time', in Cowling (ed.), *Fynbos*, 6–22.

Deacon, J. and Lancaster, N. (1988), *Late Quaternary Palaeoenvironments of Southern Africa* (Oxford).

di Castri, F. (1990), 'On Invading Species and Invaded Ecosystems: The Interplay of Historical Chance and Biological Necessity', in F. di Castri, A. J. Hansen, and M. Debussche (eds.), *Biological Invasions in Europe and the Mediterranean Basin* (Dordrecht, Neth.), 3–16.

—— and Mooney, H. A. (1973) (eds.), Mediterranean *Type Ecosystems* (London).

—— Hansen, A. J., and Debussche, M. (1990) (eds.), *Biological Invasions in Europe and the Mediterranean Basin* (Dordrecht, Neth.).

Flower, R. J., Stevenson, A. C., Dearing, J. A., Foster, I. D. L., Airey, A., Rippey, B. and Wilson, J. P. (1989), 'Catchment Disturbance Inferred from Paleolimnological Studies of Three Contrasted Environments in Morocco', *Journal of Paleolimnology*, 1, 293–322.

—— and Foster, I. D. L. (1992), 'Climatic Implications of Recent Changes in Lake Level at Lac Agziza (Morocco)', *Bulletin Société Géologique de France*, 163: 91–6.

Fox, M. (1990), 'Mediterranean Weeds: Exchanges of Invasive Plants Between the Five Mediterranean Regions of the World', in di Castri, Hansen, and Debussche, *Biological Invasions*, 179–200.

Goldblatt, P. (1978), 'An Analysis of the Flora of Southern Africa: Its Characteristics, Relationships and Origins', *Annals of the Missouri Botanical Gardens*, 65: 369–436.

Good, R. (1974), *The Geography of Flowering Plants* (London).

Groves, R. H. (1983), 'Nutrient Cycling in Australian Heath and South African Fynbos', in F. J. Kruger, D. T. Mitchell, and J. U. M. Jarvis (eds.), Mediterranean-*type Ecosystems* (Berlin), 179–91.

—— and di Castri, F. (1991) (eds.), *Biogeography of Mediterranean Invasions* (Cambridge).

Hall, A. V. and Boucher, C. (1977), 'The Threat Posed by Alien Weeds to the Cape Flora', *Proceedings of the Second National Weeds Conference of South Africa*, 35–44.

—— (1978), 'Threatened Plants in the Fynbos and Karoo Biomes of South Africa', *Biological Conservation*, 40: 11–28.

—— de Winter, B., Fourie, S. P., and Arnold, T. H. (1984), 'Threatened Plants in Southern Africa', *Biological Conservation*, 28: 5–20.

Hollis, G. E. and Kallel, M. R. (1986), 'Modelling Natural and Man-induced Hydrological Changes on Sebkhet Kelbia, Tunisia', *Transactions of the Institute of British Geographers*, NS 11: 86–104.

Huntley, B. J. (1987), 'The Years of Cooperative Ecological Research in South Africa', *South African Journal of Science*, 83: 72–9.

Kruger, F. J. (1979), 'South African Heathlands', in R. L. Specht (ed.), *Ecosystems of the World* (Amsterdam), 19–80.

—— and Bigalke, R. C. (1984), 'Fire in Fynbos', in P. de V. Booysen and N. M. Tainton (eds.), *Ecological Effects of Fire in South African Ecosystems* (Berlin), 67–114.

—— Mitchell, D. T., Jarvis, J. U. M. (1983), Mediterranean-*type Ecosystems: The Role of Nutrients* (Berlin).

Lamb, H. F., Eicher, U., and Switsur, V. R. (1989), 'An 18 000-year Record of Vegetation, Lake-level and Climatic Change from Tigalmamine, Middle Atlas, Morocco', *Journal of Biogeography*, 16: 65–74.

—— Damblon, F., and Maxted, R. W. (1991), 'Human Impact on the Vegetation of the Middle Atlas, Morocco, During the Last 5000 Years', *Journal of Biogeography*, 18: 519–32.

Le Maitre, D. C. and Midgley, J. J. (1992), 'Plant Reproductive Ecology', in Cowling (ed.), *Fynbos*, 135–74.

Macdonald, I. A. W. (1984), 'Is the Fynbos Biome Especially Susceptible to Invasion by Alien Plants? A Re-analysis of the Available Data', *South African Journal of Science*, 84: 369–77.

Meadows, M. E. and Sugden, J. M. (1991), 'A Vegetation History of the Last 14 000 Years on the Cederberg, South-western Cape Province', *South African Journal of Science*, 87: 34–43.

Moll, E. J. and Bossi, L. (1984), 'Assessment of the Extent of the Natural Vegetation of the Fynbos Biome', *South African Journal of Science*, 84: 355–8.

—— and Jarman, M. L. (1984a), 'Clarification of the Term Fynbos', *South African Journal of Science*, 84: 351–2.

—— (1984b), 'Is Fynbos a Heathland?', *South African Journal of Science*, 84: 352–5.

Mooney, H. A. and Conrad, C. E. (1977) (eds.), *Proceedings of the Symposium on the Environmental Consequences of Fire and Fuel Management in the Mediterranean Ecosystems* (Washington, DC).

Polunin, O. and Huxley, A. (1987), *Flowers of the Mediterranean* (London).

Rebelo, A. G. (1992), 'Preservation of Biotic Diversity', in Cowling (ed.), *Fynbos*, 309–44.

Reille, M. (1976), 'Analyse pollinique de sédiments postglaciaires dans le Moyen Atlas et le Haut Atlas marocains: premiers résultats', *Ecologia Mediterranea*, 2: 153–70.

Richardson, D. M. and van Wilgen, B. W. (1992), 'Ecosystem, Community and Species Response to Fire in Mountain Fynbos: Conclusions from the Swartboskloof Experiment', in B. W. van Wilgen, D. M. Richardson, F. J. Kruger, and H. J. van Hensbergen (eds.), *Fire in the South Africa Mountain Fynbos* (Berlin), 273–84.

—— Macdonald, I. A. W., Holmes, P. M., and Cowling, R. M. (1992), 'Plant and Animal Invasions', in Cowling (ed.), *Fynbos*, 271–308.

Rognon, P. (1987), 'Late Quaternary Climatic Reconstruction for the Maghreb (North Africa)', *Palaeogeography, Palaeoclimatolology, Palaeoecology*, 58: 11–34.

Rourke, J. P. (1980), *The Proteas of Southern Africa* (Cape Town).

Specht, R. L. (1979) (ed.), *Ecosystems of the World. Vol. 9A: Heathlands and Related Shrublands of the World: Descriptive Studies* (Amsterdam).

Stevenson, A. C. and Battarbee, R. W. (1991), 'Palaeoecological and Documentary Records of Recent Environmental Change in Garaet Ichkeul: A Seasonally Saline Lake in NW Tunisia', *Biological Conservation*, 58: 275–95.

Taylor, H. C. (1978), 'Capensis', in M. J. A. Werger (ed.), *Biogeography and Ecology of Southern Africa* (The Hague), 173–229.

Thomas, D., Ayache, F., and Hollis, G. E. (1991), 'Use Values and Non-use Values in the Conservation of the Ichkeul National Park, Tunisia', *Environmental Conservation*, 186: 120–30.

Thrower, N. J. W. and Bradbury, D. E. (1973), 'The Physiography of the Mediterranean Lands with Special Emphasis on California and Chile', in di Castri and Mooney (eds.), Mediterranean-*type Ecosystems*, 37–52.

Till, B. and Guiot, J. (1990), 'Reconstruction of Precipitation in Morocco Since AD 1100 based on *Cedrus atlantica* Tree-ring Widths', *Quaternary Research*, 33: 337–51.

UNESCO–FAO (1963), *Bioclimatic Map of the Mediterranean Zone*, Arid Zone Research Project, 21 (Paris).

—— (1970), *Carte de la végétation de la région mediterranéenne*, Arid Zone Research Project, 30 (Paris).

van der Merwe, P. (1966), 'Die flora van die Swartboschkloof, Stellenbosch, en die herstel van die soorte na 'n brand', *Annale Universiteit van Stellenbosch*, 41 Serie A(14): 691–736.

van Wilgen, B. W., Bond, W. J., and Richardson, D. M. (1992), 'Ecosystem Management', in Cowling (ed.), *Fynbos*, 345–71.

—— B. W., Richardson, D. M., Kruger, F. J., and van Hensbergen, H. J. (1992) (eds.), *Fire in the South African Mountain Fynbos* (Berlin).

Wells, M. J. (1991), 'Introduced Plants of the Fynbos Biome of South Africa', in Groves and di Castri (eds.), *Biogeography of Mediterranean Invasions*, 115–29.

Wengler, L. and Vernet, J.-L. (1992), 'Vegetation, Sedimentary Deposits and Climates during the Late Pleistocene and Holocene in Eastern Morocco', *Palaeogeography, Palaeoclimatology, Palaeoecology*, 94: 141–67.

Werger, M. J. A. (1978) (ed.), *Biogeography and Ecology of Southern Africa* (The Hague).

White, F. (1983), *The Vegetation Map of Africa: A Descriptive Memoir to Accompany the UNESCO Vegetation Map of Africa* (Paris).

18 Soil Erosion

Michael A. Stocking

Introduction

This is *not* a chapter on how terrible erosion is in Africa, of the billions of tonnes of soil sent wastefully to the sea each year, of the millions of hectares of cropland and even more grazing land now irreparably damaged, of forest depletion and the incompetence of humankind. That story is eloquently told elsewhere (e.g. Jacks and Whyte, 1939) and continues to be repeated today (e.g. Lal and Stewart, 1990; Pimentel, 1993). Particularly now as soil conservation budgets are cut in the United States and the major international institutions that deliver aid are under severe financial restraint, the image of an Africa in environmental crisis seems to spur ever more strident demands for action. The 'desertification debate' (described by Warren in Chapter 19) is a notable example of a crisis constructed by the institutions empowered to study it.

This chapter is an alternative view—not so popular; not so dissonant—and a view I believe to be closer to reality than the perception of a continent in terminal decline. This is not to diminish the undoubted human suffering caused apparently by environmental change; but it is to question the stereotyped assumption that we can unambiguously assess soil loss and that causes of soil erosion and land degradation can objectively be set. As the Horn of Africa well testifies, environmental change is often accompanied by conflict, misery, and other indicators of social crisis. It is tempting but misleadingly simple to attach the physical evidence with the social to ascribe cause-and-effect. There is, however, a larger 'political ecology' of the environment which teaches that explanations of physical changes such as soil erosion must be sought at a range of scales, through a variety of perspectives, and from a diversity of sources (see, for example, Blaikie, 1989). Soil erosion is bad, but its badness depends on who you are, how you study it, and what information you care to select. This contemporary phenomenon of soil erosion is, this chapter will argue, as much a social construction as a physical reality.

The 'Scourge of Africa'

Hailey's (1938) African Survey described soil erosion as the 'scourge of Africa'. Supported by other commentators of the day such as Stockdale (1937), an image was painted of smoke-charred tree stumps, sediment-choked rivers, and barren hillsides. Implicit in the message was the view that here we had a rich continent being plundered by the natives; a bounty of vegetation and soil resources being squandered by ignorant peasants; a need for order and control to be imposed by an alliance of modern agriculture and colonial organization. In Tanganyika, for example, a colonial agricultural officer noted that, 'the people are still largely unaware of the necessity for soil conservation and must be constantly supervised to prevent faulty agricultural practices . . . [they] must be taught to realise that with their present wasteful methods of agriculture and stock-raising they are destroying the source of their existence.' (Lunan, 1946: 17 and 21). It was a powerful rationalization of colonialism. It laid the ground for the politicization of environmental degradation, from what most observers would have seen as an objective scientific study to a means of identifying the guilty.

That role for commentaries on soil erosion in Africa continues unabated today. 'Africa's lands are under attack', says FAO (1990: 6). 'We are writing off Africa's future', observes the Worldwatch Institute (Brown and Wolf 1985: 11). 'The enormity of the need for action [to combat land degradation], at local, national and international levels, is such that the scope for Commonwealth action is virtually unlimited', laments the Commonwealth Secretariat (Milner and Douglas, 1989: 81). At a global scale, a triumvirate of international environmental organizations believe we are 'gambling with survival' (IUCN–UNEP–WWF, 1991: 4); the recent Atlas of Planet Management cites the statistic that 75 billion tons of soil are removed from the land by erosion annually (Myers, 1993); and authoritative texts such as that by Lal and Stewart (1990) estimate that 2 billion hectares of arable land have already been destroyed and abandoned because of soil degradation. The intended picture is clear: soil erosion is consuming the resources of Africa and the world; the international community is sitting idly by; corrupt or inept governments are making futile postures; cultivators and pastoralists are enacting Hardin's Tragedy of the Commons; while a small core of dedicated professionals sound the warning but are routinely ignored. The guilty (all but we professionals) are thus clearly exposed. Afropessimism abounds.

The Politicizing of Science

This chapter, then, highlights the theme that soil erosion is as much a political weapon as a real environmental threat. In this it mirrors the approach developed by Warren (Chapter 19) in discussing the related issue of desertification. To deal with soil erosion and land degradation, we have to understand the sociology of science and why the different environmental 'actors' say what they do, just as much as we have to appreciate the process. Scientists themselves are one of the actors; they have needs (promotion of their subject and their careers, research money, publication, prestige); they have prejudices too, just as the colonial agriculturalists did (see Blaikie and Brookfield, 1987, for an account of the politicization of land degradation; and Stocking, 1995, for a view of the various 'actors' in geomorphology who play the soil erosion game).

Take, for example, the analysis of three case-studies of land degradation and soil erosion from sub-Saharan Africa by Biot *et al.* (1992). They claim serious flaws in the scientific process of assessment of erosion: specifically, they say (1) uncertainties in the estimation of erosion have been hidden and 'worst scenarios' are chosen as 'typical'; (2) the impacts of the process are similarly exaggerated in order to support messages of impending doom; (3) unjustified extrapolations have been made from plot-scale measurements to large catchments; and (4) alternative and contradictory models of the interaction

of natural systems with human behaviour are usually ignored. Boserup's (1981) analysis of how societies adapt and introduce new technologies in the face of population pressure and land scarcity is, for example, rarely included in projections of the likely outcome of environmental degradation. For scientists' flawed and exaggerated behaviour, they conclude with possible explanations, ranging from the altruistic to the malign. Some are motivated by genuine concern, where small evidences of degradation spur fears for the rest of the landscape. Some seek attention, where a cry of crisis is far more likely to generate research kudos (and money). Yet others are biased, or prejudiced, or simply do not understand the dynamics of tropical environments and the resilience of rangelands. These challenges themselves are fraught with value judgements, and raise the question as to why Biot *et al.* (1992) themselves make such intolerant claims. However, the point is made: claims of serious soil erosion must be treated cautiously. Where environmental policy is to be affected and development action is to be proposed, erosion assessments should themselves be audited as to who has made them, how, and why.

So, to see *The Physical Geography of Africa* and physical processes such as soil erosion and land degradation, we need an optic that encompasses a healthy scepticism of scientific 'fact' and sees through the hyperbole of a continent in continual environmental crisis. With apologies to William Shakespeare, this chapter will follow the guidance of Hamlet: 'Be not too tame neither, but let your own discretion be your tutor: suit the action to the word, the word to the action; with this special observance, that you o'erstep not the modesty of nature.' (Act III. sc. ii). He concludes that if you overstate your case, it may please the unskilful but the judicious will grieve!

The Extent and Nature of Erosion

Unsurprisingly, the evidence is mixed and contradictory. As IFAD (1992: 17) notes, 'it is difficult to give an indication of the extent of land degradation in sub-Saharan Africa, as there is no simple or commonly agreed measure of degradation. Nor are consistent data available about the state of land in each country.' A salutary lesson in the perils of drawing conclusions from measurements of soil erosion comes from Millington's (1981) research on erosion rates in two small drainage basins in Sierra Leone. Using transects of erosion pins through areas of cultivation, bush regrowth, and forest, by small runoff plots and through monitoring basin sediment yield, he compared results. Not only was there a poor correlation between the erosion pins and the small plots, but estimates of total catchment erosion

Plate 18.1. An erosion 'hot spot'. Typical view of gully from within with fluted, unstable sides. Lesotho, Southern Africa (photo: M. Stocking).

derived from the plots hugely exceeded estimates from both the pins and basin sediment monitoring. Independent techniques of measurement of the same erosion process are giving quite different results. The correspondingly different messages as to the seriousness of the process give us a *smörgåsbord* of choice to suit all tastes. With the warning in place, we have to ask now whether soil erosion in Africa has reached proportions and extents which seriously threaten the economic future of the continent and the livelihoods of its peoples.

The expert consensus is that soil erosion and land degradation are serious throughout Africa but that generalizations of its ubiquity cannot be sustained. For example, in singling out sub-Saharan Africa for particular mention as where erosion is dangerously affecting vegetation productivity, the World Bank (1992: 55) states: 'Estimates of land damaged or lost for agricultural use through soil degradation range from moderate to apocalyptic', noting at the same time that erosion by water, even in arid regions, is the principal process. *The World Map on the Status of Human-Induced Soil Degradation* (Oldeman *et al.*, 1992) supports the picture of a patchy but generally serious condition (see Table 18.1.)

These qualitative assessments, based mainly on expert opinion, give an overview of African degradation as composed of broad swathes of countryside with slight to moderate degradation where water erosion dominates in nearly all zones, wind erosion becomes problematic in the Sahel and Namib regions, and chemical and physical deterioration of soils is also prevalent in the higher

population density areas. Erosion only assumes its most serious proportions in the African mountains and highlands (Messerli and Hurni, 1990) and in the drought-prone Sahel (IUCN, 1986, 1989). Everywhere, however, is characterized by declining yields of the major cereals where agriculture is practised with few inputs, and by a greater difficulty of obtaining livelihoods from the natural resource base. As witnessed by the data of the *1992 World Development Report* (World Bank, 1992), this declining productivity appears to be a peculiarly African problem which we shall examine more fully in the next section.

On a continental-scale comparative basis, the analyses presented by El-Swaify *et al.* (1982) provide a somewhat different view: an Africa with only modest denudation rates. Using data from suspended-sediment yields in major river systems, Africa has an annual mechanical denudation rate of 0.47 tonnes/ha and a chemical denudation rate of 0.25. This compares favourably with Asia (1.66 and 0.42, respectively), South America (0.93 and 0.55), and North and Central America (0.73 and 0.40). Similarly, the mean annual suspended-sediment loads from three major basins (Congo, Niger, and Nile) are two orders of magnitude *less* than for Asian basins such as the Kosi and Ganges. Even the Amazon with its sparse population has an estimated mean field-erosion rate based on sediment yields of 13 tonnes/ha/year, compared with the Congo River at 3 tonnes/ha/year. Part of the explanation for these differences inevitably arises from the differential intensity of land

Table 18.1 Status of soil degradation in African areas selected for relative severity of erosion as derived from the GLASOD Project

Area	Type of degradation	Cause	Seriousness	Extent
Angola, central	Water erosion: topsoil loss	f	3	5
Benin, coastal	Chemical deterioration: nutrients	a	2	5
Botswana, eastern arable zone	Water erosion: topsoil loss	a	2	3
Burkina Faso, Ouagadougou	Water erosion: topsoil loss	g/a	3	5
	mass movement		3	3
Djibouti	Water erosion: topsoil loss	g	3	1
	Wind erosion: topsoil loss		1	3
Ethiopia, central Highlands	Water erosion: topsoil loss	f/g	3	4
	mass movement		3	3
North and Eritrea	Water erosion: topsoil loss	g/a	4	5
	mass movement		3	2
Lesotho	Water erosion: topsoil loss	g	2–3	3–5
Mali, parts of north-east	Wind erosion: topsoil loss	g/e	3	5
Namibia, near Windhoek	Water erosion: topsoil loss	g	2	3
Rwanda	Chemical deterioration: nutrients	a	2	5
	Water erosion: topsoil loss		2	3
South Africa, Karoo	Water erosion: topsoil loss	g	3	5
Southern Transvaal	Physical deterioration: compaction etc.	g	3	3
	Water erosion: mass movement		3	1
W. Transvaal/OFS	Physical deterioration: compaction etc.	g	3	5
Sudan, Darfur	Chemical deterioration: nutrients	g/a	2	5
	Water erosion: topsoil loss		2	2
Tanzania, Arusha (part)	Water erosion: topsoil loss	g	3	4
	mass movement		4	2
Usambara/Pare Mts.	Water erosion: topsoil loss	f	3	4
	mass movement		3	2
Zambia, Central Province Highveld	Physical deterioration: compaction and crusting	a	2	3
Zimbabwe, Middleveld	Water erosion: topsoil loss	a/g	2	4
	mass movement		3	1

Cause: a = agricultural activities; g = overgrazing; f = deforestation; e = overexploitation of vegetation for domestic use.

Seriousness: 1 = light; 2 = moderate; 3 = strong; 4 = extreme.

Extent: 1 = 0–5%; 2 = 5–10%; 3 = 10–25%; 4 = 25–50%; 5 = 50–100% of area affected.

Source: Oldeman *et al.*, 1992.

use between the major continents, the relatively lower and more stable slopes of Africa, and the sediment storage opportunities within drainage basins thereby preventing large net losses.

Specific national studies conducted by scientists and engineers tend to reintroduce the view of a continent in crisis. Most studies are initiated *because* erosion is seen to be a problem. Hence, deductive science has taken the problem, isolated it experimentally on research stations, and predictably has proved the problem exists. In seeking to reduce a complex physical process into a number of component parts, this approach identifies measurable factors of soil, climate, vegetation, and topography that will explain the rate of the process of erosion. Most anglophone and francophone countries of Africa have had their erosion research experiments: for example, at Henderson and Matopos Research Stations in Southern Rhodesia through the 1950s and 1960s; the network of ORSTOM (French Overseas Scientific Research

Agency) stations in Mali, Niger, and Ivory Coast of the 1970s and 1980s. Ethiopia today has seven research stations with plot experiments under the auspices of the Ministry of Agriculture and funded by Swiss aid (see Plate 18.4). The methodologies used—nearly always small-size soil loss and runoff plots—are well known to produce erosion rate measurements that are far higher than actual net loss from a hillside (Stocking, 1987). The measured rate of erosion is an artefact of the bounded experimental plot and the fact that on a small plot virtually all soil that starts to move is collected and measured, whereas on a real slope there will also be deposition. There is, therefore, a tendency to have published in the literature and project reports the extreme and sensational, rather than the typical estimates.

I am myself guilty. In deriving estimates of total nutrient losses by erosion from the soils of Zimbabwe for the UN Food and Agriculture Organization, I took a mean annual erosion rate of 50 tonnes/ha for communal arable

Plate 18.2. Sheet-eroded slope under *Eucalyptus* trees, Ethiopian Highlands. Serious cases such as this can be related to specific site and land-use conditions (photo: M. Stocking).

Plate 18.3. Extreme sheet erosion, Sukumaland, Tanzania. This picture was taken two months *after* the start of the rains.

land based upon an application of the SLEMSA soil-loss model to typical slopes and sandveld soils (Stocking, 1986), but which was subsequently supported by trials carried out by agricultural engineers (Elwell and Norton, 1988). Fifty tonnes was an overestimate of long-term field rates under actual farming conditions, and is definitely excessive as a predictor of net loss (erosion in excess of deposition) of soil from the field. To my occasional protestations, the 50 tonne figure is now widely quoted as the definitive rate of erosion from communal farming systems (e.g. FAO, 1990). From the long series of experiments conducted in the 1930s in Tanganyika

come broadly similar plot data on erosion rates (Christiansson, 1989). Table 18.2 presents a sample of published figures of erosion rates from a number of sources and a variety of measurement techniques. The three factors other than experimental technique that appear most to determine erosion rates are: (1) the status of the vegetation cover; (2) the slope percentage; and (3) the broad climatic belt—the 500 to 800 mm mean annual rainfall zone seems especially hazardous. Beyond these generalizations, is it possible to make more definitive observations about erosion processes and rates?

Lal (1993), using a variety of data sources including

Plate 18.4. Erosion and runoff plot experiments to monitor the impact of different erosion rates on crop yield. Gununo Research Station, Sidamo, Ethiopia (photo: M. Stocking).

Table 18.2 Published erosion rates for African conditions

Location	Measurement details	Mean annual rainfall (mm)	Equivalent field erosion rate (t/ha/yr)
Ethiopia, Central Plateau	Empirical methods; suspended sediment yields	500–800	165
Tanzania, Mpwapwa	Plot experiment; bare, uncultivated	620	146
Ivory Coast, Adiopodoume	Plot experiment; bare, uncultivated, 7% slope	2100	138
Zimbabwe, Henderson Research Station	10-year bare plot experiment, 4.5% slope	750	127
Upper Volta, Gampela	Plot experiment, bare soil, 7% slope	800	126
Nigeria, Ibadan	Plot experiment, bare-fallow, 15% slope	1150	116
South Africa, Ciskei	Reservoir sedimentation over 3 years	550	114
Senegal, Sefa	Plot experiment; cleared open forest, 2% slope	1300	30–55
Dahomey, Cotonou	Plot experiment, cleared dense forest, 4% slope	1300	17–28
Zimbabwe, Matopos Research Station	8-year plot experiment, bare, undisturbed, 4% slope	400	11
Zimbabwe, Umsweswe	220 km² catchment sediment yield	750	10
Zimbabwe, Henderson Research Station	10-year plot experiment; complete grass cover; 4.5% slope	750	0.7
Ivory Coast, Adiopodoume	Plot experiment; natural bush fallow	2100	0.6
Zimbabwe, Umsweswe	1990 km² catchment sediment yield	750	0.25

Sources: Roose, 1975; Lal, 1976; Stocking, 1984; Weaver, 1989.

useful local studies by Lootens and Lumbu (1986) and Kattan *et al.* (1987), gives the following overview for West Africa:

1. *Sahel*:   Large tracts of the West African Sahel region are covered by sand dunes and laterized rock outcrops. The region is prone to wind erosion and the potential rates may range from 10 to 200 t/ha/year. Soil crusting and compaction are severe.

2. *Sub-Sahelian Region*:   This is the most severely affected because of its susceptibility to erosion by both wind and water. Estimates of erosion range from 10 to 50 t/ha/year by wind.

3. *Savanna*:   The sub-humid or savanna region has a rainfall of 800 to 1200 mm which is mostly received in one growing season of four to six months. Water is the principal agent of erosion, although wind erosion also occurs in some regions. In the northern savanna zone, water erosion rates range from 10 to 200 t/ha /year.

4. *Forest*:   Severe erosion in the forest region occurs

Plate 18.5. Overgrazing by cattle is often blamed for serious erosion in dryland areas (photo: M. Stocking).

only when the protective cover of vegetation is removed for intensive crop production. Potential erosion rates are estimated at 10 to 200 t/ha/year. Actual erosion rates can be as high as 100 t/ha/year on steep slopes managed for row-crop farming with plough-till method of seed preparation.

Noting that his estimates do not include deposition, Lal (1993) suggests that the erosion rates correspond to 1, 1.5, 2, and 3 mm for the forest, savanna, sub-Sahelian, and Sahelian regions respectively (note: multiply mm depth of soil by about 13 in order to give a very approximate measure of tonnes/ha). He further states that net erosion is perhaps 50 to 60 per cent of these rates. These assessments based on Lal's research at the International Institute for Tropical Agriculture, Ibadan, Nigeria, deserve serious consideration as they give an overall picture of erosion being an order of magnitude greater than soil formation—i.e. 10 or more tonnes/ha/year as opposed to formation rates of less than 1 tonne. Under such a scenario, the lifespan of the soil must be seriously curtailed (Elwell and Stocking, 1984).

The problem with all such overviews which give definitive rates of erosion is that it must be assumed that the research plot conditions hold for real life slopes. If all measurements were to be conducted under exactly the same experimental conditions, similar plot sizes, and equivalent analytical procedures, we might have a case for comparative assessments of erosion rate. The assumptions, however, do not hold. The best that can be justified

is a broad set of semi-quantitative conclusions for African conditions:

- broad erosion rates are unexceptional and are probably below world averages;
- erosion rates are extremely variable, ranging from field equivalent rates on bounded plots of well less than 1 tonne/ha/year to well over 100 tonnes; specific rates can, therefore, be very high and are mostly concentrated into small geographical areas;
- high erosion rates throughout Africa are associated first and foremost with poor vegetation cover. This factor, whether on cropland, rangeland, or forest, is the single most influential factor in determining the rate of erosion;
- increased erosion through deficient vegetation cover is enhanced by total rainfall amounts and by high intensity rainfall; also by steep slopes and unstable geomorphological situations; and in a few particular locations by some soils which easily lose their structure (e.g. sodic).

Finally, in this review of the extent and nature of erosion, only sheet erosion by water and wind has been considered. There remains to be assessed the extent of rill and gully erosion. By very definition, these are locationally specific: they feature as linear scars in the landscape and, where prevalent, they dominate the view and cause considerable consternation to agriculturalists and conservationists. Rills should perhaps more accurately be seen as closely associated with sheet erosion by water or 'inter-rill' erosion. Gullies, on the other hand, although seen by many authors as 'large rills' (e.g. Poesen and Govers, 1990), feature in two main situations. First, as exemplified by a detailed study in Machakos District, Kenya (Tiffen *et al.*, 1994), many gullies are associated with the wash-out of footpaths, cattle tracks, road drains, and culverts. They are particularly noticeable around watering and dipping points for cattle. Secondly, gullies are the principal feature in 'badlands', a phenomenon of both the dry and sub-humid zones of Africa, where all vegetation has been removed and the slopes consist of bare and rocky rubble (Plates 18.1 and 18.3). Known as *mangalata* in East Africa, badlands can extend over considerable areas. From surveys of south-east Shinyanga Region in Tanzania conducted by this author in 1980, over half of a total area of 25 000 km^2 was estimated to be degraded down to a bare calcrete, punctuated by gullies, where nothing but a few unpalatable herbs would grow.

The problem, however, with gullies is that they have a bad press. They occur along roads where people travel. The eye is naturally drawn to the scars they create, whereas one has to search for the insidious signs of sheet erosion, even though the latter is far more prevalent. Gullies also, many assume, are controllable by structures such as gabions and check dams; and hence they receive far more attention than they perhaps deserve. It is said that the president of one African country is fond of ordering his motorcade to stop whenever he sees a gully. His ministers and attendants, and if the media are there, he too, will throw rocks into the gully to show their support for conservation. He could not do the same for sheet erosion—or else he would fail to reach his destination!

It is extraordinarily difficult to make any assessment of the relative severity of rill and gully erosion compared to sheet erosion by wind and water. Indeed, because the processes are so totally different in their spatial context, it could be argued that it is wrong even to try to compare them. Gullies, especially the badland variety, are *symptomatic* of a degraded catchment; they occur not necessarily because of land use in or immediately around the gully; they are nature's efficient way of removing large quantities of excess runoff and sediment derived from sheet erosion in the catchment. This linkage between sheet and gully erosion was researched in the Umsweswe River catchment in Zimbabwe, data for which appear in Table 18.2. Part of the area is dominated by massive gullies on Karroo sands which have high exchangeable sodium. Yet, it was calculated that only 13 per cent of the total sediment which exits the catchment is derived

from the excavation of gullies; the rest comes from the almost-invisible sheet erosion. Further, the gullies have very particular explanations in the nature of the soil (a very dispersible clay mineral fraction) and the geomorphological position (actively cutting river valleys). Therefore, the comparison is invalid and to be resisted even though tempting in order to score a point for the vital need of soil conservation.

The Impact of Erosion

If the assessment of the seriousness of soil erosion rates is so fraught with methodological and practical uncertainty and if experimental results of erosion rate are so difficult to interpret and subject to exaggeration, it has to be asked whether we need to measure erosion and whether there might be an alternative measure which may assist planning for strategies of soil conservation.

Instead of measuring erosion and quantities of sediment, a more directly useful assessment is the *impact* of erosion and its effect on the utilization of the soil. For example, in commenting on the state of agricultural land in Lesotho, a country famed for its erosion, agricultural economists Nobe and Seckler (1979: 152) noted: 'Under present conditions, the productivity of the agricultural land . . . is so low that the economic conditions of a conventional programme do not adequately exist.' In short, soil erosion affects land use, and a good proxy and practical assessment of erosion would be the extent to which either land quality suffers, or human land-use activities are curtailed. These are known as on-site impacts. Off-site impacts such as downstream pollution, sedimentation, and damage to structures are also major aspects to the seriousness of erosion, but are outside the scope of this chapter.

On-site impacts may be measured in a number of ways, each of which gives a different absolute answer often in financial monetary terms. However, since the measures are more directly related to the use of the land and the impact on economies, it could be argued that they are more relevant. Each major type of measure is considered along with an example of its use in African conditions.

Changing Soil Quality

Some of the most detailed studies on the consequences of erosion on soil fertility come from Africa. In Tanzania, Moberg (1972) compared soil analytical data from two eroded and two non-eroded profiles. He found that erosion had the effect of lowering pH to the extent that aluminium toxicity was commencing; organic carbon and total nitrogen levels were halved in the top 15 cm; zinc and phosphorus were depleted to deficiency levels; and calcium and magnesium were considerably lowered. Fertility appeared to be affected to at least 150 cm in the profile, and Moberg suggests this is brought about by the absence of roots in the eroded profile which prevents recycling. This is one of the few studies that highlight the crucial interrelationship between erosion and loss in potential productivity with explanations for the linkage. The indigenous soil-management strategy of shifting cultivation and bush fallowing (Ruthenberg, 1976) is a rational farmer's response to the known rapid depletion of soil quality under use. Soil erosion and associated washout of organic matter and nutrients is likely the major process in reduction in quality. Young and Wright (1979) have proposed a useful tabulation of the rest period requirements of tropical soils in order to encompass the natural processes of fertility recovery.

Enrichment Ratio

This is a physico-chemical measure of the proportional enrichment of eroded sediments in comparison to the topsoil from which they were derived. An enrichment ratio of 2 based on nitrogen would, for example, indicate that the eroded sediments contain twice the nitrogen than the *in situ* soil. Results from a long-term study in Zimbabwe (Stocking, 1986) where sediments were monitored over five years on four soil types and for up to 3000 storm events gave an overall enrichment ratio of 2.5 to 2.7 based upon nitrogen, phosphorus, and organic carbon. While there was no significant difference between the four soils (all were relatively sandy and potentially lose their fertility easily), there was a large variation between storm events. Some storms in the early part of the growing season had enrichment ratios in excess of 10. In effect, the fine particles of soil were being flushed out. Later in the season as the soil developed a crust, the enrichment ratios declined even though total erosion because of high runoff remained large. The small amount of evidence available does suggest that inherently poor soils have higher enrichment ratios than clay-rich, fertile soils. For example, Tegene (1992), using the FAO-sponsored research design to assess erosion–productivity relationships in Ethiopia, found that on the highly fertile Eutric Nitosols of Gununo (Plate 18.4) that the sediments were very close to an exact representation of the *in situ* soil: ER for nitrogen varied in a narrow band of 0.95 to 1.11.

Nutrient Losses

Because of the selective removal of fine particles by erosion, another measure of impact is to quantify total nutrient losses. Of course, this only represents part of

Table 18.3 Losses of nutrients and organic carbon from Oxisols under small-farm intercropping in Sierra Leone

Erosion phase	Soil loss (t/ha/yr)	Nutrient losses (kg/ha/yr)			
		N	P	K	Org.C
Uneroded	4.8	7.3	0.12	0.19	98.6
Slight	6.7	10.7	0.21	0.45	152.0
Moderate	8.5	15.4	0.41	0.73	176.3
Severe	15.5	23.2	1.02	1.50	233.2

Source: Sessay and Stocking, 1992.

the impact of erosion on the soil but, for commercial agricultural systems where nutrient availability represents the major limitation, it may possibly be the most significant. In a near semi-arid climate in Botswana, Sehuhula and Pain (1991) calculated that in excess of 100 per cent of topsoil content of available nitrogen or over 5 kg/ha in the top 20 cm was lost by erosion. They explain such a significant loss by the extremely high mineralization rates for organic nitrogen. If representative of high temperature conditions in Africa, such losses of the most limiting nutrient are a serious loss to potential productivity, explaining the often-noted crash in crop yields after only one or two years' cultivation. Some typical annual nutrient losses for Oxisols in Sierra Leone are given in Table 18.3 for erosion phases up to 'severe'. The total amounts of lost nutrients are not very remarkable for commercial agriculture where they can easily be compensated for artificially by fertilizer. However, for small-scale agriculture, they are serious and represent a significant proportion of available nutrients for crop growth.

Financial Cost at Farm Level

It is at the farm or household level that the impact of erosion can be seen most clearly. An increasing number of studies are now attempting to calculate the greater difficulty in obtaining a livelihood from soils that are degrading (Bojö, 1991) and the potential benefits of particular soil conservation practices for farm households (Kiome and Stocking, 1993). Table 18.4 outlines some potential financial impacts for different agricultural practices, pointing up the differential value of tied-ridging and mulching which should make these conservation measures attractive to both small and commercial farmers. These figures are generally corroborated in a detailed study of maize and cowpea intercropping in Sierra Leone (Sessay and Stocking, 1992) where the reduction in gross returns because of erosion-related yield losses is $29 per ha without fertilizers and $35 with inputs. With the average farm size of 3 ha in this humid tropical zone,

Table 18.4 Typical annual losses of soil, rainfall and nutrients from Zimbabwe farmlands

Source	Estimated soil erosion (t/ha/yr)	Rainfall loss by runoff (%)	Organic Carbon loss (kg/ha/yr)	Cost of N and P loss ($/ha)
Small-scale farming on well-drained sands:				
current rates	50	30	535	94
tie-ridging	2	5	21	4
Commercial farming on other soils:				
continuous cotton	75	35	1155	245
tie-ridged cotton	5	7	77	16
continuous maize	15	20	231	50
trashed maize	1	5	15	3

Sources: Based on data in Stocking, 1986 and experiments at the Institute of Agricultural Engineering, Harare.

the farm household loss of income of about $100 annually is a major impact on the average household income from agricultural produce of $270. In a largely subsistence economy, there is little wonder that food crises continue and that a part of the reason can be ascribed to erosion.

National Economic Cost

The depletion of the quality of natural resources jeopardizes the potential for future production and can, therefore, be seen as a drain on national biophysical resources. Only a few studies in Africa to date have attempted to place a notional cost on erosion processes. In the Zimbabwe study (see Table 18.4), it was estimated that erosion is costing approximately 3 per cent of the Gross Domestic Product annually. Similarly, a study carried out for the World Bank in Malawi puts the impact on GDP at between 0.5 and 3.1 per cent (Bishop, 1990). Many assumptions have to be brought into play in such calculations: for example, introducing the cost of replacement of lost nutrients into the assessment would not be contentious (Stoorvogel and Smaling, 1990) but the cost for increased irrigation because of a claimed decrease in plant-available water or for greater mechanized power to overcome soil compaction consequent upon erosion might be seen as highly speculative. Economists also have a reputation for picking the statistics that make their case. Some authors (e.g. de Graaff, 1993) argue that land degradation in Africa is primarily related to

soil exhaustion and desertification: for example, Zambia and Zaire have low population densities and a low nutrient-balance deficit per hectare. Isolating the primary physical and chemical causes of reduced plant production consequent upon erosion is difficult. There is a further danger of double-counting: for example, it would not be valid to count both the fertilizer-replacement cost of lost nutrients and the value of crops forgone because of erosion—these are merely two ways of measuring the same process. To give an idea of the quite incredible monetary sums potentially involved on a national basis for the cost of soil erosion, an analysis of the national cost to Zimbabwe (Stocking, 1986) gave the following:

- Zimbabwe's arable lands are losing annually $85 million worth of nitrogen and $65 million of phosphorus;
- the loss of nutrients from arable lands is approximately three times greater than the actual application of nutrients through fertilizers;
- all land, including range, is losing $900 million of N and $700 million of P a year;
- on a per hectare basis by major farming system, commercial arable lands are losing $20 annually, and communal arable $50, while degraded communal rangeland suffers over $80 worth of lost nutrients.

These figures are based upon relatively meagre data and extrapolations that require verification and should therefore be treated cautiously. Yet, even if exaggerated, such estimates of the economic cost of erosion must explain a significant part of the reason why African agricultural output has failed to match population growth and why social disruption can be traced to the effects of human-induced environmental change.

Social Impact

The ultimate impact of erosion is its effect on society. In its most extreme form the consequence of erosion is the collapse of agriculture and major migrations of people. Evidence is usually anecdotal. An increasing number of studies now document the enforced abandonment of land (e.g. Misana, 1992, for the Iringa Region, Tanzania). In urban environments, soil erosion has also been disruptive as witnessed by a FINNIDA-supported project in Mozambique (Ojanperä, 1994) which has shown the 'winners and losers' both in degradation and in the efforts to combat it. Other good African examples are cited by Blaikie and Brookfield (1987) and in the Proceedings of the International Soil Conservation Conference held in Kenya and Ethiopia (Tato and Hurni, 1992). However, it is extraordinarily difficult to isolate

Plate **18.6.** Sorghum being grown commercially in Lesotho. The financial cost of lost productivity determines land use and crop types (photo: M. Stocking).

Plate **18.7.** Small coffee farm, Kenya Highlands, with *fanya juu* terraces (photo: M. Stocking).

cause-and-effect, and simplistic assumptions that demographic increase contributes to environmental degradation, which further induces population change, have to be questioned in the face of evidence of the adaptability of society to changing circumstances (e.g. Tiffen *et al.*, 1994).

Approaches to Combating Erosion

Responses to soil erosion have varied over time according to the perception of the problem and political forces.

Up to about 1945 the main arbiters of conservation policy were colonial administrators and field officers. Many of these had an intimate knowledge of local farming and their attempts at introducing more formalized techniques of conservation often included traditional practices and names. One of the best examples is 'The Soil Conservation Orders' in Sukumaland, Tanganyika, brought in about 1941 which legitimized the practice of setting aside dry season grazing reserves and used local names such as *Ikarusi* and *Ibushi* for soils. Sukumuland's plough rules also provided for ridging practices well known to local people, as well as measures which would have been less popular such as the insistence on contour grass strips. However, as Wood (1992) notes for Zambia, African farmers regarded conservation as an extra imposition of colonial rule rather than an essential part of their farming systems. It was true that the amount of effort expended in conservation by the colonial authorities seemed to be in proportion to population density and the contribution of local agriculture to export crops. Enforcement of conservation rules would be attached to tours of inspection to collect Hut Taxes or levy funds for the 'Native Maize Pool' and similar unpopular extractions from the local communities. Resentment inevitably built up. Colonial conservation efforts which centred around tree planting, regulations against cultivating steep slopes and certain soils, destocking, and rotational grazing were widely flouted and often induced overt resistance. For example, Tiffen *et al.* (1994) quote how the Akamba, when faced with compulsory culling of livestock, marched in protest to Nairobi and camped around Government House for six weeks. In Sukumaland, the levying of fines for non-compliance of conservation orders led by 1956 to severe civil disturbances and ultimately to communal riots in which contour banks, terraces, and hedges were systematically destroyed. Even in 1980 a resources survey in the same area conducted by this author revealed almost no evidence of the former measures other than a few *minyara* hedges (*Euphorbia tirucali*) which doubled as both a fence against wind erosion and as a source of easily propagated firewood. To greater or lesser extent, the same story can be repeated throughout colonial Africa.

It was not surprising that African liberation movements capitalized on the resentment engendered by coercive conservation. The Tanganyika African National Union which brought Julius Nyerere to power campaigned to denounce soil conservation as the repressive face of colonialism. Elsewhere, conservation measures were neglected and structures fell into disrepair. However, under the influence of environmental agencies, most notably the UN Environment Programme based in Nairobi and the UN Sudano-Sahelian Office, established in 1973 and now administered by the UN Development Programme from New York, post-independence African governments have re-embraced soil and water conservation. As in colonial times, the emphasis remains on control of erosion rather than prevention with the use of structures and barriers and specific campaigns of tree-planting and destocking. The messages are disturbingly familiar. Local reactions are predictable. For example, in one part of Sukumaland this author found people planting tree seedlings upside down—the regulations insisted that two trees per family would be planted

annually, and so to register protest but not break the law, this incongruous outcome resulted.

One of the most remarkable rediscoveries to come to the aid of approaches to combating erosion is the role of indigenous technical knowledge and societal adaptation by local people to their changing environment. The important anthropological study of Kondoa Region, Tanzania by Östberg (1986) shows how ecological change has been accommodated in the views of the Irangi about their own environment and the practices they adopt. A study conducted for the International Fund for Agricultural Development (IFAD, 1985) documents a wealth of indigenously developed land-managment practices which have arisen through enforced change in the environment and the necessity to modify agricultural practices and crops. Many of these practices such as the Tanzanian *matengo* pits compensate for a decline in soil quality through erosion on hillsides by increasing plant-available water and trapping sediment from upslope. On the slopes of Mt Kilimanjaro, Maro (1990) relates how people have migrated towards the drier lowlands which have poorer soils and have acquired drought-tolerant crops such as the *mkojosi* variety of banana.

There is not the space or purpose in this chapter to review the many different soil and water conservation technologies that have been promoted at various times in Africa: these are described comprehensively elsewhere (e.g. Unger, 1984; Hudson, 1987; Sheng, 1989). A sample of three measures which have been (and continue to be) strongly promoted might include:

Grass strips. Up to about two m wide, aligned along the contour, these act as infiltration and sediment deposition strips. They have been widely accepted in some African countries (e.g. Swaziland) where the strips double as a source of dry season grazing which helps to counteract the most important farming constraint in the perception of local people, the lack of fodder just before the commencement of the rains. So, far from being seen as a soil conservation measure, the strips have an important production function for the local economy. Elsewhere in Africa, grass strips are seen as taking up valuable land and are resented. In eastern Uganda, strips are being ploughed out and planted to maize because they constitute one of the last reserves of reasonably fertile soil.

Fanya juu terracing. Named after the Swahili term to indicate the action of throwing soil uphill, these terraces are constructed from an initial contour ditch. The land is reformed eventually into a set of bench terraces on which crops are planted, water infiltration is maximized and land preparation should be easier. Technically, it is argued that no land is wasted with these terraces

Plate 18.8. Soil conservation strip-cropping with grass strips and woody shrubs. ICRAF field station, Machakos, Kenya (photo: M. Stocking).

because the risers can be used for fodder grasses and the ditches for tree planting. There is some evidence that *fanya juu* can increase evaporation and may not always be best for plant growth and economic yields (see Kiome and Stocking, 1993). More importantly, however, the terraces are extremely labour-demanding. Evidence from Kenya suggests that in the high potential areas (Plate 18.7), labour investment in such terracing is worthwhile (Table 18.5), whereas in the medium and low potential-rainfall zones any additional crop yield fails to give sufficiently attractive returns to pay for the efforts of constructing and maintaining the measures.

Agroforestry is an age-old set of techniques used by African farmers to (1) maximize production per unit area; (2) provide a variety of subsistence outputs; (3) minimize risk of wholesale failure of a monocrop. Through the work of the Africa-based International Centre for Research in Agroforestry and its Africa Network (AFRE-NA), agroforestry has been strongly promoted as a means of soil conservation (e.g. Young, 1989). It has to be said that there are potential technical problems with agroforestry technologies, not least the competition between species for limited nutrients and plant-available water. But there are also major management problems with some agroforestry systems such as hedgerow intercropping. The timing of pruning and the use of leaves and branches as a mulch can cause severe short-term stresses in the crops, meaning that for the technique to be successful, the farmer must possess the labour resources to respond at exactly the right time—an impossibility for most resource-poor households.

Approaches are now changing gradually, partly in recognition of the widespread failure of purely technical strategies of soil and water conservation (Hudson, 1991), and partly in response to new development thinking as embodied in 'farmer-first' approaches (Chambers *et al.*, 1989) and their application to soil conservation (Lundgren *et al.*, 1993). 'Land husbandry' is now the vogue term for soil conservation: its composite meaning takes in the care, management, and improvement of land resources in a context of primary support to farmers' livelihoods. The development of the land-husbandry approach can be charted through a succession of four 'phases of participation' (Lundgren *et al.*, 1993) of one of the few long-term soil conservation programmes in Africa that is acknowledged to be successful, Kenya's National Soil Conservation Project:

(1)   From 1974, an alternative model of soil conservation was agreed between the aid donor (SIDA) and the government of Kenya in which training and extension, labour-intensive but low-cost methods, and concentration on high potential areas were to be emphasized. Subsidies were employed at the outset to induce farmer participation in the construction of cut-off drains, artificial waterways, and gully control. Job creation was stressed for communal areas, but on private land, terracing was to be done by the farmer without payment on the grounds that, if successful, the measure should more than adequately repay the private investment of effort. Experience from this first phase showed that payments for any soil-conservation measures were counter-productive,

Table 18.5 Economic value of soil conservation for farmers in Kitui District, Kenya (figures in Kenyan shillings per farm) and labour requirements (person-days/farm)

Economic factor	Model 1. Farm without soil conservation	Model 2. Average soil conservation practices	Model 3. 'Superior' soil conservation practices
Total revenue	7550	12 187	14 732
Variable costs			
food crops	853	853	853
soil conservation crops	—	170	230
TOTAL VARIABLE COSTS	853	1023	1083
Gross margin	6697	11 164	13 649
Annualized costs			
investment in soil conservation	—	165	165
Gross margin minus annualized investment (net return)	6697	10 999	13 484
Labour requirement (person-days)			
cultivation	585	636	692
soil conservation work	—	45	45
TOTAL LABOUR REQUIREMENT	585	681	737
Net return per person-day of labour:	11.44	16.15*	18.29*

* The net return per person-day to the *additional* labour required under Model 2 is 44.8 Kshs/farm and under Model 3 it is 44.65 Kshs/farm.

Source: Holmberg, 1990.

leading to lack of maintenance of the works and the construction of unnecessary structures as a way of extracting further wage labour.

(2)   In the second phase through the early 1980s, subsidies were abandoned. A modified Training-and-Visit system of extension was employed to sustain farmer participation. Most importantly, soil conservation was integrated with other agricultural messages. But progress was considered to be too slow.

(3)   From 1987 a new extension approach was adopted where socio-economic as well as ecological differences were incorporated, leading to a more flexible and responsive design of soil-conservation measures according to local circumstances. But still the procedures were evaluated as too 'top-down' and over-technical.

(4)   Hence, from 1990 a full Participatory Rural Appraisal (PRA) approach was adopted as a tool to increase the understanding between extension staff and farmers. PRA is based upon interactive learning, shared knowledge, and flexible yet structured analysis (Pretty and Gujit, 1992). In the field, multidisciplinary teams attempt to generate new information quickly on resource constraints and opportunities. The process of gaining the information is as much a part of prioritizing soil and water conservation as the results of the analysis. Techniques adopted by the Kenyan Soil and Water Conservation Branch now include: semi-structured interviews; transect walks; participatory mapping; seasonal calendars; matrix ranking; farm sketches; and Venn

diagrams (see RRA Notes, published by the International Institute for Environment and Development (London), for empirical testing of the various techniques). It is reckoned that this open approach to local-level planning for soil and water conservation has increased the credibility of extension staff, obviated the need for subsidies, and changed local attitudes. As Lundgren *et al.*, 1993: 35) have noted, 'the central justification for using PRA has been the mobilization of farmers to do it themselves. The adoption of soil and water conservation in catchments planned using PRA has been very significant, with increased yields, and regeneration of local resources and economies.'

In the Kenyan case, matters have progressed a long way. Where at first soil conservation advice would simply be a set of recommended techniques, quick-fix solutions, and measures to stop soil already entrained in runoff, now the emphasis is on biological techniques which combine prevention of erosion with (hopefully) increased production. Instead of soil conservation being taken as an end in itself, it now comprises part of an overall strategy of sustainable farming systems, sustainability being seen as encompassing not only ecological protection but also socio-economic and institutional viability. Today, the private economic benefit of soil conservation is often considered as perhaps the single best measure for local acceptance or rejection of measures. In a detailed study from Kitui District of Kenya, Holmberg

(1990) demonstrates that soil conservation involves higher labour costs but that the return on labour is also increased, resulting in significantly enhanced annual revenues from the same piece of land (Table 18.5). Where once it was simply seen as self-evident that soil erosion should be prevented, research and implementation projects in Africa now accept that any introduced technology (including soil conservation) must be to the private benefit of the land user and must not increase risk or foreclose other farm or non-farm income-generation opportunities.

Conclusion

Soil erosion is serious in Africa—for some people, some places, and some national economies. In any physical geographic analysis of contemporary processes such as soil erosion, explanations for particular occurrences of serious degradation must incorporate a human dimension. That dimension is, this chapter has argued, complicated by the biases of the professionals who have a stake in exaggerating the seriousness of erosion in Africa. The whole picture is further clouded by the various responses to stories of erosion, and the use of soil degradation as a vehicle for other things such as justification of colonialism. It is fascinating that some post-independence politicians have rediscovered the political usefulness of soil erosion; it is equally interesting to observe the reaction of the international community in demanding more resources to fight the menace of erosion and desertification (see next chapter) in Africa.

In the final analysis, soil erosion and conservation are two sides of the same coin. We should be concerned not that soils are eroding but that the potential for future production is being mortgaged at the expense of current production. Similarly, soil conservation as a stand-alone technical exercise has never been appreciated in Africa; instead, the more realistic view is that livelihoods depend on plant production, and plant production depends in part on soil quality. Hence, soil conservation is really about the maintenance and management of soil quality under the various forms of land use that people want. How do we minimize the threat to land resources while allowing people the choice and the ownership of decisions that they themselves have made? These are the big questions for the use and management of Africa's soil resources.

References

Biot, Y., Lambert, R. and Perkin, S. (1992), 'What's the Problem? An Essay on Land Degradation, Science and Development in Sub-Saharan Africa', Development Studies Discussion Paper 222 (Norwich).

Bishop, J. (1990), *The Cost of Soil Erosion in Malawi*, Report to the World Bank (Washington, DC).

Blaikie, P. (1989), 'Explanation and Policy in Land Degradation and Rehabilitation for Developing Countries', *Land Degradation and Rehabilitation*, 1: 23–37.

—— and Brookfield, H. (1987) (eds.), *Land Degradation and Society* (London).

Boardman, J., Foster, I. D. L., and Dearing, J. A. (1990) (eds.), *Soil Erosion on Agricultural Land* (Chichester).

Bojö, J. (1991), 'Economics and Land Degradation', *Ambio* 20: 75–9.

Boserup, E. (1981), *Population and Technology* (Oxford).

Brown, L. R. and Wolf, E. C. (1985), 'Reversing Africa's Decline', Worldwatch Paper 65 (Washington, DC).

Chambers, R., Pacey, A., and Thrupp, L. A. (1989) (eds.), *Farmer First: Farmer Innovation and Agricultural Research* (London).

Christiansson, C. (1989), 'Rates of Erosion in East African Savanna Environment', in D. B. Thomas, E. K. Biamah, A. M. Kilewe, and L. Lundgren (eds.), *Soil and Water Conservation in Kenya* (Nairobi), 99–115.

De Graaff, J. (1993), *Soil Conservation and Sustainable Land Use: An Economic Approach* (Amsterdam).

Dixon, J. A., James, D. E., and Sherman, P. B. (1990) (eds.), *Dryland Management: Economic Case Studies* (London).

El-Swaify, S. A., Dangler, E. W., and Armstrong, C. L. (1982), 'Soil Erosion by Water in the Tropics', Research Extension Series 024 (Hawaii).

Elwell, H. A. and Norton, A. J. (1988), *No-Till Tied Ridging: A Recommended Sustained Crop Production System* (Harare, Zimbabwe).

—— and Stocking, M. A. (1984), 'Estimating Soil Lifespan for Conservation Planning', *Tropical Agriculture*, 61: 148–50.

FAO (1990), *The Conservation and Rehabilitation of African Lands: An International Scheme* (Rome).

Hailey, Lord (1938), *An African Survey* (Oxford).

Haskins, P. G. and Murphy, B. M. (1992) (eds.), *People Protecting Their Land*, Proceedings of the 7th ISCO Conference (Sydney).

Holmberg, G. (1990), 'An Economic Evaluation of Soil Conservation in Kitui District, Kenya', in J. A. Dixon, D. E. James, and P. B. Sherman (eds.), *Dryland Management* (London), 56–71.

Hudson, N. W. (1987), 'Soil and Water Conservation in Semi-arid Areas', FAO Soils Bulletin 57 (Rome).

—— (1991), 'A Study of the Reasons for Success or Failure of Soil Conservation Projects', FAO Soils Bulletin 64 (Rome).

IAHS (1981), *Erosion and Sediment Transport Measurement*, Publication No. 133 (Wallingford).

—— (1984), *Challenges in African Hydrology*, Proceedings of the Harare Symposium, Publication No. 144 (Wallingford).

IFAD (1985), 'Soil and Water Conservation in Sub-Saharan Africa: Issues and Options', Report prepared by the Centre for Development Cooperation Services, Free University, Amsterdam (Rome).

—— (1992), 'Soil and Water Conservation in Sub-Saharan Africa: Towards Sustainable Production by the Rural Poor', Report prepared by the Centre for Development Cooperation Services, Free University, Amsterdam (Rome).

IUCN (1986), *The IUCN Sahel Report: A Long-term Strategy for Environmental Rehabilitation* (Gland, Switz.).

—— (1989), *The IUCN Sahel Studies* (Nairobi).

IUCN–UNEP–WWF (1991), *Caring for the Earth: A Strategy for Sustainable Living* (Gland, Switz.).

Jacks, G. V. and Whyte, R. O. (1939), *The Rape of the Earth: A World Survey of Soil Erosion* (London).

Kattan, Z., Jry Gac, and Probst, J. L. (1987), 'Suspended Sediment

Load and Mechanical Erosion in the Senegal Basin: Estimation of the Surface Runoff Concentration and Relative Contributions of Channel and Slope Erosion', *Journal of Hydrology*, 92: 59–76.

Kiome, R. M. and Stocking, M. A. (1993), 'Soil and Water Conservation in Semi-arid Kenya', Bulletin 61 (Chatham).

Lal, R. (1976), 'Soil Erosion on Alfisols in Western Nigeria', *Geoderma*, 16: 363–431.

—— (1993), 'Soil Erosion and Conservation in West Africa', in D. Pimentel (ed.), *World Soil Erosion and Conservation* (Cambridge), 7–25.

—— and Stewart, B. A. (1990), *Soil Degradation* (New York).

Lootens, M. and Lumbu, S. (1986), 'Suspended Sediment Production in a Suburban Tropical Basin', *Hydrological Sciences Journal*, 31: 39–49.

Lunan, M. (1946), 'Soil Erosion', in N. V. Rounce (ed.), *The Agriculture of the Cultivation Steppe of the Lake, Western and Central Province* (Cape Town), 17–21.

Lundgren, L., Taylor, G., and Ingevall, A. (1993), *From Soil Conservation to Land Husbandry: Guidelines Based Upon SIDA's Experience* (Stockholm).

Maro, P. S. (1990), 'Agricultural Land Management under Population Pressure: The Kilimanjaro Experience, Tanzania', in B. Messerli and H. Hurni (eds.), *African Mountains and Highlands* (Berne), 311–26.

Messerli, B. and Hurni, H. (1990) (eds.), *African Mountains and Highlands: Problems and Perspectives* (Berne).

Millington, A. C. (1981), 'Relationship between Three Scales of Erosion Measurement on Two Small Basins in Sierra Leone', in IAHS, *Erosion and Sediment Transport Measurement* (Wallingford), 485–92.

Milner, C. and Douglas, M. G. (1989), *Problems of Land Degradation in Commonwealth Africa* (London).

Misana, S. B. (1992), 'A Report on Soil Erosion and Conservation in Ismani, Iringa Region, Tanzania', in K. Tato and H. Hurni (eds.), *Soil Conservation for Survival* (Ankeny, Ia.), 391–8.

Moberg, J. P. (1972), 'Some Soil Fertility Problems in the West Lake Region of Tanzania, including the Effects of Different Forms of Cultivation on the Fertility of Some Ferralsols', *East African Agricultural and Forestry Journal*, 38: 35–46.

Myers, N. (1993), *Gaia: An Atlas of Planet Management* (Garden City, NY).

Nobe, G. K. and Seckler, D. (1979), 'An Economic and Policy Analysis of Soil and Water Problems in the Kingdom of Lesotho', LASA Research Report No. 3 (Maseru, Lesotho).

Ojanperä, S. (1994), *When People Have to Move Away: Resettlement as Part of Erosion Control in Nacala, Mozambique*, Nacala Integrated Urban Development Project (Helsinki).

Oldeman, L. R., Hakkeling, R. T. A., and Sombroek, W. G. (1990), *World Map of the Status of Human-Induced Soil Degradation* (Wageningen, Neth. and Nairobi).

Östberg, W. (1986), *The Kondoa Transformation: Coming to Grips with Soil Erosion in Central Tanzania*, Research Report No. 76 (Uppsala, Sweden).

Pimentel, D. (1993) (ed.), *World Soil Erosion and Conservation*, Cambridge Studies in Applied Ecology and Resource Management (Cambridge).

Poesen, J. and Govers, G. (1990), 'Gully Erosion on the Loam Belt of Belgium: Typology and Control Measures', in J. Boardman, I. D. L. Foster, J. A. Dearing (eds.), *Soil Erosion on Agricultural Land* (Chichester), 513–30.

Pretty, J. and Gujit, I. (1992), 'Primary Environmental Care: An Alternative Paradigm for Development Assistance', *Environment and Urbanization*, 4: 22–36.

Roose, E. J. (1975), *Erosion et ruissellement en Afrique de l'Ouest*, ORSTOM, Centre d'Adiopodoumé (Abidjan, Ivory Coast).

Rounce, N. V. (1946) (ed.), *The Agriculture of the Cultivation Steppe of the Lake, Western and Central Province* (Cape Town).

Ruthenberg, H. (1976), *Farming Systems in the Tropics* (Oxford).

Sehuhula, A. and Pain, A. (1991), 'Study of the Effects of Soil Erosion on Soil Productivity: Preliminary Results', Proceedings of the First Annual Scientific Conference of the SADCC Land and Water Management Research Programme (Gaborone, Botswana).

Sessay, M. F. and Stocking, M. A. (1992), 'Financial Cost of Soil Erosion on a Small Farm Intercropping System in Sierra Leone', in P. G. Haskins and B. M. Murphy (eds.), *People Protecting Their Land* (Sydney), 562–7.

Sheng, T. C. (1989), 'Soil Conservation for Small Farmers in the Humid Tropics', FAO Soils Bulletin 60 (Rome).

Stockdale, Sir F. (1937), Soil Erosion in the Colonial Empire. *Empire Journal of Experimental Agriculture*, 5: 1–18.

Stocking, M. A. (1984), 'Rates of Erosion and Sediment Yield in the African Environment', in IAHS, *Challenges in African Hydrology and Water Resources* (Wallingford), 285–95.

—— (1986), 'The Cost of Soil Erosion in Zimbabwe in Terms of the Loss of Three Major Nutrients', Consultants' Working Paper No. 3 (Rome).

—— (1987), 'Measuring Land Degradation', in P. Blaikie and H. Brookfield (eds.), *Land Degradation and Society* (London), 49–63.

—— (1995), 'Soil Erosion in Developing Countries: Where Geomorphology Fears to Tread', *Catena*, 25: 253–67.

Stoorvogel, J. J. and Smaling, E. (1990), 'Assessment of Soil Nutrient Depletion in Sub-Saharan Africa, 1983–2000', Report prepared by the Winand Staring Centre (Wageningen, Neth. and Rome).

Tato, K. and Hurni, H. (1992) (eds.), *Soil Conservation for Survival* (Ankeny, Ia.).

Tegene, B. (1992), 'Erosion: Its Effects on Properties and Productivity of Eutric Nitosols in Gununo Area, Southern Ethiopia, and Some Techniques of its Control', African Studies Series A9 (Berne).

Thomas, D. B., Biamah, E. K., Kilewe, A. M., and Lundgren, L. (1989) (eds.), *Soil and Water Conservation in Kenya*, Proceedings of the Third National Workshop, Kabete, Nairobi, 1986 (Nairobi).

Tiffen, M., Mortimore, M. and Gichuki, F. (1994), *More People, Less Erosion: Environmental Recovery in Kenya* (Chichester).

Unger, P. W. (1984), 'Tillage Systems for Soil and Water Conservation', FAO Soils Bulletin 54 (Rome).

Weaver, A. V. B. (1989), 'Soil Erosion Rates in the Roxeni Basin, Ciskei', *South African Geographical Journal*, 71: 32–7.

Wood, A. P. (1992), 'Zambia's Soil Conservation Heritage: A Review of Policies and Attitudes towards Soil Conservation from Colonial Times to the Present', in K. Tato and H. Hurni (eds.), *Soil Conservation for Survival* (Ankeny, Ia.), 156–71.

World Bank (1992), *World Development Report: Development and the Environment* (New York).

Young, A. (1989), *Agroforestry for Soil Conservation* (Wallingford and Nairobi).

Young, A. and Wright, A. C. S. (1979), 'Rest Period Requirements of Tropical and Subtropical Soils Under Annual Crops', Land Resources for Populations of the Future Project, Consultants' Working Paper No. 6 (Rome).

19 Desertification

Andrew Warren

Introduction

Three themes recur in this chapter. The first is the depth of the crisis in dry Africa. The continent endured its worst ever environmental crisis after the intense drought of 1984 (Grove, 1986). The lowest rainfalls on record triggered the movement of 10 million environmental refugees (UNOEOA, 1985), most of them from the semi-arid country stretching between Senegal to Ethiopia. Another intense crisis hit Southern Africa in 1991–2. These emergencies, though precipitated by drought, were in both cases the culmination of long runs of dry years and slow declines in per capita agricultural production. It is this brutal mix of long and short drought and the declining productivity of the land that adds up to 'desertification'.

These were only two, albeit the most serious events in a sequence of similar crises in the last few decades, each displacing many thousands if not millions of people, let alone the very many more who remained at home and yet suffered severe distress during and after each event. The accumulated damage is driving already impoverished African countries into becoming the poorest on Earth (Figure 19.1). The problem is the most significant of Africa's, let alone Earth's environmental crises, for no other has yet produced such terrible statistics (even in combination).

The second recurring theme is the increasing severity of the crises. Though desertification is not new in Africa, its effects are progressively more damaging. The problem occurs at two very different scales. At the large scale, it can be said to have begun when a massive and as yet unreversed episode of desertification produced the modern Sahara some 8000 years ago, though there were comparable or even more severe events earlier in prehistory in Northern, Southern, and Eastern Africa. At the other end of the time-scale, harsh drought, creating temporary, though real enough deserts, strikes somewhere in Africa almost every year, as it has done throughout human history. These phenomena, the long and the short, are covered in detail in other parts of this book, but they must be a constant background to a discussion of problems at the medium scale, that of households and communities, which is the focus here. At whatever scale, droughts are wreaking more havoc because of increasing demands (for better standards of living, for more people, for more cash crops) and closing options (among them international borders that are closed to trade and migration, and less access to land).

The third theme is the neglect of the crisis on the global stage. For all its significance, desertification is often downplayed. It commands only a very small proportion of the research funding that is applied to global warming, the ozone hole, or even biodiversity (all, unlike desertification, potential rather than actual catastrophes); it has been able to call on little international aid; and it is constantly sidelined in forums like the United Nations Conference on Environment and Development (UNCED) at Rio de Janeiro in 1992 (although the African nations did win a Desertification Convention out of that debate). These snubs are possible, in large measure, because, of the very poverty of dry Africa, and of its increasing political marginality. A vicious downward spiral of deprivation is the result. The indifference, however, has two other causes whose discussion must also pervade the discussion of desertification in this chapter.

The first cause is the complexity of the predicaments of dry rural economies. This makes them more difficult to define and deal with. Most other global environmental

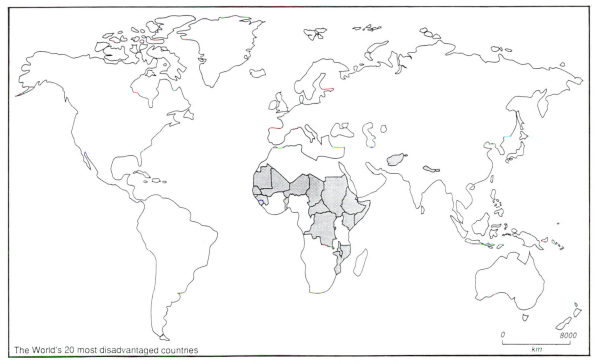

The World's 20 most disadvantaged countries

Fig. 19.1. The World's twenty most-disadvantaged countries.
(*Source*: UNDP, 1990.)

concerns are linked to relatively few and unambiguous scientific measurements (atmospheric O_3 or CO_2 levels, pH of rainfall and fresh waters and so on). With desertification (and with soil erosion), the problem is revealed through all sorts of distress among millions of people, trying to make a living from environments that are extremely variable from year-to-year and village-to-village. Desertification is a 'bottom–up' environmental issue, unlike these other 'top–down' ones (Warren, 1993). Second is the relativity of response, which also makes answers difficult to find. Each village or encampment, by virtue partly of different cultures, has its own yardsticks by which it measures distress, and its own mix of factors that determines its resilience in the face of environmental challenges.

The Issues

Desertification has often been claimed as a distinct environmental problem in its own right (UNEP, 1992*a*), but this is too facile. The environmental problems of dry lands must be seen as a closely interacting set of processes, as the events of 1984 showed. As problems to be solved, the most appropriate way to subdivide them is according to the kind of intervention they demand. Three kinds of issue—drought, desiccation, and dry-land degradation—suggest three distinct kinds of help (Warren and Khogali, 1992).

Drought

Drought is a dry period that brings trauma, but one with which dry-land ecosystems and economic systems can cope. They may be damaged, but survive more or less intact. Drought is a dry period that causes ecosystems and economies to reel, but not to tumble. Indeed, the systems that survive are those that have evolved to cope.

Studies of African ecosystems weathering droughts are few, but give vivid accounts. The ecosystem of the Ferlo in central Senegal was very severely damaged by the droughts of the late 1960s and early 1970s, but slowly pulled itself back to almost its former self when the rains returned (Bille and Poupon, 1974). The vegetation of the Gourma in Mali recovered from the more severe drought of 1984 in an essentially similar way (Hiernaux *et al.*, 1990). In each case, there was a retreat

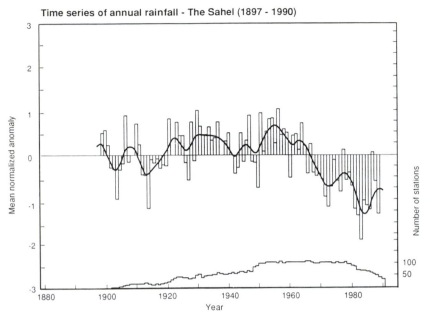

Fig. 19.2. The recent history of drought in the Sahel. (*Source*: UNEP, 1992*b*.)

of vegetation from the poorer, drier sites, but a survival of seed parents in damper hollows, and of seed banks in the soil. In the Ferlo, many wild and domestic animals moved away to better places as the drought intensified, but returned when it ended.

A similar pattern occurs in the social sphere. A village community in north-eastern Nigeria copes with a severe drought in many ways (Mortimore, 1989). Some of the men leave on pilgrimage, in search of work elsewhere, or on commercial adventures in the wetter parts of the country. Cultivators withdraw from poorer, more distant fields to concentrate their efforts and the diminishing supply of manure on fields near the village. They turn to other sources of income, such as basket-making. The number of domestic animals is reduced and rationalized. This and most other Sahelian communities have inherited their responses to drought, for they have had to resort to them before, and will again, although circumstances are never exactly the same in each drought, or in each community.

This way of defining drought cannot be based on a prescribed length of the period of rainfall deficiency or its severity. An ecosystem accustomed to extreme desert can survive a drought of great severity and over many years: one in a rain forest may collapse after a few months of dry weather. A village that has excellent storage facilities, or uncommonly good access to other resources might survive a three-year absence of rain with equanimity: one that is impoverished and isolated

might collapse after a year. These are facets of the cultural relativity of desertification, and of the interaction between its various components. None the less there are few rural communities in Africa that can face a drought of more than two years without radical readjustment, so that two years of very dry conditions can be taken as the threshold between a drought and desiccation, in most cases.

Drought requires a distinct kind of public policy response. It is the coping strategies of the land users themselves that need to be strengthened: insurance, communications, diversification, storage, and so on. In almost every case, it is existing systems that need to be strengthened, rather than new ones introduced.

Desiccation

In this system of nomenclature, 'desiccation' is a dry period that is extended to the point where it destroys natural or cultural communities. The word (with nearly the same sense) is borrowed from Hare (1987) who used the term to describe what has happened in the Sahel over the last two decades. Rainfall, almost everywhere from Cape Verde in the west to the Sudan in the east, began to decline in the late 1960s and has never fully recovered, although the some years in the early 1990s have in general been kinder (Figures 19.2 and 19.3). This desiccation has destroyed many natural communities to the extent that they will take many

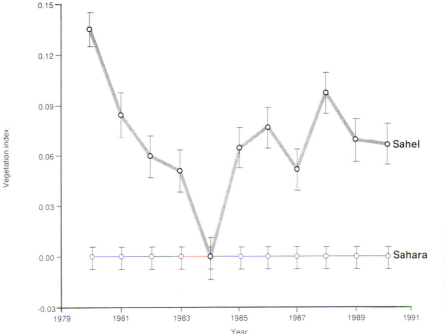

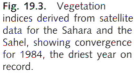

Fig. 19.3. Vegetation indices derived from satellite data for the Sahara and the Sahel, showing convergence for 1984, the driest year on record.

(*Source:* Tucker *et al.*, 1991.)

years to recover (if they ever do). In ecosystems, desiccation eliminates seed parents entirely, so that plant communities take many years to re-establish when the good rains return. It is the trees in many of the 'tiger-bush' communities of Mali and Niger that have been destroyed during the latest period of desiccation and they can only reseed and again reach maturity after many years (Hiernaux *et al.*, 1990).

The Neolithic pastoralist communities of the central Sahara suffered very substantial desiccation, being effectively eliminated, and sequences of less severe, though still very marked periods of desiccation have punctuated history since then (Grove, 1978). Desiccation has brought shifts in population and changes in life-style, as African historians are now showing (Johnson and Anderson, 1988). The latest episode, in the last two decades, has destroyed economic systems on a massive scale. In the Sudan, pastoral communities that inhabited the edges of the desert have moved south into what was once dense bush, or they have resorted to the outskirts of the towns (Khogali and Johnson, 1982). In Mali, the Tuareg camel nomads of the north have had to move south among the Wodaabe cattle nomads, and the Wodaabe, in turn, south among cultivators, into Burkina Faso and beyond. Hausa who cultivated parts of central Niger have been forced to abandon their villages and move south, some into Nigeria (Mortimore, 1989).

Helping those who suffer from desiccation is much more problematical than intervention to alleviate drought. There are two closely related issues: recognition and response. The period of desiccation of the 1970s and 1980s has only belatedly been recognized by the international community, and, in the absence of a much better forecasting system than exists, the same fate awaits the next period of desiccation. Response to desiccation would be difficult even if it were realized quickly, but, if recognition is belated, response is *post hoc*, and even more problematical. The response, belated or not, will always have to be on a large scale, and involve many communities (unlike the response to drought, which can be on a community-by-community basis). New settlements must be built or planned for; there must be long-lasting relief schemes until new communities are established; inter-ethnic tensions will have to be managed; and so on.

Dry-land Degradation

Dry-land degradation (which includes soil erosion, discussed in Chapter 18) can be defined most generally as a 'reduction in the potential of land to produce crops' (UNEP, 1992a), but this is not very precise or useful

Plate 19.1. Sheet erosion exposing tree roots in the Baringo District of Kenya (photo: A. S. Goudie).

formulation. Land contributes many factors in agricultural production: soil-nutrient-holding capacity; soil-nutrient stores (both characteristics differing for the main nutrients, nitrogen, potassium, and phosphorus); soil-water-holding capacity; soil-water-release and water-retention characteristics; rates of renewal and removal of soil by wind and water; content of salts and toxins; and so on. If we include vegetation in 'land', as do some commentators, then we must add the species-mix in the vegetation, its nutrient value, and trace-element content and productivity. Each factor, in soil or vegetation, fluctuates at many scales from the daily (as with the soil-nitrogen store), through the annual, to the longer term. Each is very much at the mercy of the cultivator or grazier, who can add or remove nutrients, water, or toxins, or accelerate or decelerate erosion. With all this complexity it is little wonder that so little is known about land degradation (as the next section will make plain). Here, the definition will merely be refined by calling dry-land degradation a process that reduces land productivity to the extent that natural recovery can only happen over many decades, or where artificially accelerated recovery is beyond the capital and technical resources of existing communities.

Dry-land degradation differs from land degradation in more humid climates by being brought about by processes that are very much more variable in space and time, and by having a distinctive mix in which wind erosion and salinization are much more prominent. It is its combination with drought and desiccation that create special problems. Though water erosion is a major

component of the degradation process in semi-arid lands, as it is elsewhere, it too is given a special character by this variability. Some authorities believe that rates of water erosion are higher in semi-arid lands than elsewhere, mainly because of the sparseness of the vegetative cover (Plates 19.1 and 19.2) (see discussion by Stocking, Chapter 18).

A discussion of the practical consequences of dry-land degradation brings us up against some central problems in the understanding of desertification, for, not only is it consummately difficult to assess (as described below), but, for all the theoretical desirability of distinguishing degradation as a process, there are great practical difficulties in disentangling its effects from those of drought, desiccation, and a host of social issues.

The entanglement of causes and effects can be illustrated by describing the degradation of the traditional mode of cultivation on the mostly sandy and almost always nutrient-poor soils of dry Africa. The management of infertility in most of these soils has been the swidden system, where fields are left to recover fertility over anything up to ten years or more. In central Sudan this was known as the *hashab* cycle (Jackson and Shawki, 1950). The *Acacia senegal*, or hashab tree regrew from stumps in fields that had been abandoned. As the coppiced trees recovered after the fields had been abandoned, their roots began again to fix nitrogen in the soil. They also provided a crop of gum arabic.

The last two decades in central Sudan have seen a shortening of the fallow and a decline in gum production, but the reasons are not simple to discover (Khogali,

Plate 19.2. Badland gully erosion developed in colluvial sediments developed on Karroo Formation in the Umsweswe River catchment in Zimbabwe (photo: A. S. Goudie).

1991). Partly it is because of desiccation (the long drought), which may have killed some trees, but more certainly has brought the need for a greater cultivated acreage to provide the same harvest at lower yield per acre; partly it is because of rising population, requiring more food; partly it is because of pests, parasites, and diseases, whose proliferation may be encouraged by the erratic nature of the climate (which may eliminate predators but not pests); and partly it is because of a catastrophic decline in the market for gum arabic, itself due to a host of reasons including corruption, poor quality-control, poor market management, and so on (Pearce, 1988). Diagnosis of the undoubted decline in productivity, and especially of the role of soil degradation in this decline is therefore very hazardous. The persistence of the decline is even harder to predict.

These forms of degradation, and the complexity of their causality are not confined to the Sudan. Population has been growing across the dry zone, at rates that are as high, if not higher than in wetter areas. Available agricultural land is being used up. The swidden cycle is shortening everywhere. Some authorities believe that the consumption of land by cultivation is fuelling an impending crisis. It will leave little grazing for the domestic stock, which provide manure, traction, and protein and this may well induce further degradation (Breman *et al.*, 1990). But here too the extent and persistence of decline are very hard to estimate. There are some who point, on the contrary, to the positive effects of increased labour availability that growing populations provide, for this allows more intensive and careful use of the land, as in the heavily populated periphery of Kano in Nigeria (Mortimore, 1989), in the Machakos District of Kenya (Tiffen *et al.*, 1994), and in some other parts of semi-arid and sub-humid Sahel (Pieri, 1989).

Intervention to mitigate land degradation, unlike that to mitigate drought, calls for policies centred on the land itself, but the complexity of the processes of agricultural production calls for complex forms of intervention. Simply to advocate techniques, such as terracing, windbreaks, or the use of manure, is far too simplistic. The labour relations and the social structures that determine responses to the land and its problems also need to be implicated, and this raises very sensitive issues. This complexity, and the absence of its recognition, explains why so little intervention has been successful, as will be explained below. The unsuccessful introduction of narrow-based terraces in the Machakos District of Kenya is a case in point (Tiffen *et al.*, 1994).

Interaction

It is worth re-emphasizing that these troubles (drought, desiccation and, dry-land degradation) never come singly, and that they amplify each other, this composite problem being 'desertification'. A village that has already lost the fertility of its soils, succumbs more quickly to a drought than one whose fertility is intact. A pastoral group whose herds have been quartered by desiccation, will keel over in a drought that might be barely noticed by a group untouched by the desiccation. Interaction creates major problems for the assessor and the interventionist.

Assessment

General Problems

A fundamental problem in assessing the extent of desertification has been constant redefinition by the authorities. Assessment is almost impossible if the standards are constantly changing (Verstraete, 1986; Warren and Agnew, 1988). Until the late 1980s, desiccation (as defined above) was excluded from the definition. The definition now accepted by most official bodies, and produced after many long hours of negotiation at UNCED, is less dogmatic, though it still does not yield measurable criteria:

> Desertification is land degradation in arid, semi-arid and dry sub-humid areas resulting from various factors, including climatic variations and human activities.

But the problems of assessment go well beyond those of definition. The principal difficulty has been in pinpointing problems that can respond to intervention. In addition to the recurrent problems of complexity and relativity, there are two further reasons: prejudice and ignorance.

This pernicious mixture has a long pedigree, for fear of the desert is very ancient and has begotten deep-seated preconceptions. The juxtaposition of the complete barrenness of the Eastern and Western Deserts of Egypt with the ancient and widely influential culture of the Nile Valley, has produced a very long succession of well-publicized, threatening accounts. The Bible, for one, blamed the major droughts in Egypt on human transgression, and Roman records of decline on the desert frontier put blame on barbarian raiders (Vita-Finzi, 1969). These accounts, and the rediscovery of an apparently crumbling East (which in many Western minds included Africa) by an awakening West, fuelled the notion that Africa and Asia suffered terminal, self-inflicted decay (Said, 1978). These seductive pictures, propagated in the biblical and classical bias of Western education in the nineteenth century, gave the colonial officials who briefly held some power in the dry parts of Africa a host of preconceptions. They added them to others about the people who inhabited the continent, particularly about those in the dry areas, who were usually the last to be brought to heel, and were the least understood. Add to this a poor comprehension of these environments and the result is a considerable clouding of colonial judgments.

These legacies are still paying dividends. Very few of the countries of dry Africa are yet fully democratic and many of their leaders are inclined to listen more to the messages of their Western tutors in matters environmental than to those of their own people. Most dry African countries are very poor, and in constant receipt of environmental advice from foreigners, whose acceptance has financial inducement, and which comes mostly from strangers to Africa. There has been very little investment in discovering the nature and location of environmental problems, or more seriously, in seeking the opinions of those who have to deal with them at the sharp end. All this reinforces the momentum of prejudice. Prejudice feeds and in turn feeds off widespread ignorance of dry lands. Those who are convinced by education and culture of the profligacy of the African dry-land inhabitants will not subscribe to research or new ideas that might challenge their beliefs. There are, however, other reasons that the dry lands are poorly understood.

First, they are sparsely populated and of low productivity per acre, and although they often show high productivity per unit of labour input, which can yield healthy profit to pastoralists in good years, this is seldom appreciated. These misapprehensions have meant that there has been much less investment in scientific research than that in other environments. The data-bases and concept-bases on dry Africa are not well stocked.

Second, the desert edges are environments of extreme variability. Chance (or some cryptic cycle) may bring a decade or more of good years in which all seems well, to be followed by a short or a long drought that is all the crueller for having followed plenty. Opinions tend to form round one or other parts of these cycles, and either outlast the conditions themselves or take part in cycles of their own (which are seldom in phase with the environmental ones). If a dry or a wet period coincides with a critical period in political history, it leaves a long shadow of opinion down the decades. The 1960s were good rainfall years, as well as being the period when post-war colonial administrations, and then newly independent African countries, were looking to expansion; the Sahel then looked ripe for investment. The terrible 1970s and 1980s coincided with the political disintegration of many African nations. The Sahel now looked a bleak prospect for any kind of investment. In prejudiced minds, the blame was to be placed more at the doors of the indigenous cultivators and pastoralists, or even their fractious governments, than at that of the climate (let alone their own misjudgements).

Drought

Drought, scarcely surprisingly, is the best understood of the three elements of desertification listed above. Patterns of rainfall across Africa in recent decades are reviewed elsewhere in this volume by Hulme (Chapter 5), and the occurrence of droughts in recent centuries by Nicholson (Chapter 4). However, even here there are problems. There can be no doubt that droughts occur and that they cause problems, but where exactly, and with what exact effect is less certain. For a start, information is not always perfect, and is getting worse, as the fall in the number of stations recording rainfall in Figure 19.2 shows. Moreover, it is becoming more and more apparent that rainfall in the dry zone is very patchy (Sicot, 1991), so that rainfall stations, scattered ever more widely as they are, pick up less and less of the fine detail. The village with the remaining rain-gauge may not experience the drought (or rains) that another village does. This deficiency is being overcome, to an extent, by the monitoring of the rainfall through surrogate measures using indices (such as the NDVI) derived from satellite data (Konaté and Traoré, 1987). NOAA-AVHRR satellite images give data on a 1 km × 1 km grid, and danger spots can be picked with some accuracy, but it does not yield quantitative data on rainfall. More quantitative ways of measuring rainfall from satellites may soon be available, but they have yet to be fully deployed.

But what do these data mean, even when perfect? Total rainfall is only one element in crop production. The first complication is timing. If there is a false start to the season, farmers may plant only to see the crop die in a subsequent dry spell, and then have no seed for the real rains when they come. The issue, however, is seldom timing on its own. The complex ways in which the timing of the rains interacts with other factors to produce famine or plenty can be seen in a study by Newhouse (1987). In 1987 there was a false start to the rains in the north of the Sahel and then a break to the end of July, so that a fresh sowing was necessary. Because rainfall thereafter was enough for good growth, the result was not so much lower yields, as much lower acreages than had been sown in 1986, especially north of 13 °N, partly because farmers had run out of seed. But the reduction in acreage was also a consequence of non-environmental processes, such as the low prices following the record crop of 1986 (even though that year had produced plentiful supplies of seed). Moreover, because there was less flooding (for complex reasons), there was less high-yielding flood-cropping. In other words, a deficiency of food is only partly a function of any one particular drought; many other interactions are at work.

There are also differential effects of drought depending on soil and crop. A given rainfall on a sandy soil produces a different crop response than if it falls on a clay. It brings one yield on a field of millet, another on a field of sorghum (Agnew, 1989). A field that yields well in a dry year may produce nothing in a wet year if there is waterlogging. The same rainfall means different things for farmers with different techniques of cultivation. Agnew argued that some of the droughts of the 1970s were only destructive in the drier parts of the Sahel in Niger, for rainfall in the south did not fall below the threshold where it seriously affected production. The distress in the south, in other words, had other causes. None of these issues is insurmountably difficult to measure or evaluate, and the farmer, to survive, has to have made, at least, a partly successful attempt to evaluate and surmount them. But they have yet fully to be built into drought strategies by central authorities.

Desiccation

Desiccation can be detected in much the same way as drought, as Tucker's studies have shown (Figure 19.3; Tucker *et al.*, 1991), but, as with drought, the problem is to recognize and interpret the signals. Desiccation can probably only be recognized by some clever combination of climatic and socio-economic observations.

A problem with desiccation is to understand the mechanism that produces it, for comprehension might improve policy. Other chapters in this book (3, 4, and 5) explore the possible causes of desiccation, be they regular processions in Earth's orbital parameters; oscillations such as the ENSO (El Niño-Southern Oscillation) in the Pacific Ocean; or global warming. If desiccation were to be unavoidable, or if it could be forecast, then policies might be formulated accordingly, though with what effectiveness is debatable (Glantz, 1977).

One alarming proposal is that desiccation may be linked to devegetation in the dry lands themselves, by way of the so-called Charney effect (Warren *et al.*, 1993). This is a positive, self-reinforcing process in which it is postulated that devegetation increases the reflectiveness of the ground surface, which in turn cools the air above it and so discourages rainfall. There is little field evidence to support the hypothesis and some well-controlled field experiments strongly challenge it, finding lower ground temperatures where there is more vegetation, probably as a consequence of greater evapotranspiration (Balling, 1992). The idea lingers by virtue of numerical experiments in large Global Climatic Models. Newer notions attribute lower rainfall to the reduction in surface roughness as trees are removed; or to a reduction in soil moisture and so of feedback of moisture to the atmosphere, both ideas that are also supported more by modelling than by empirical observation. These speculations are no less important for being largely theoretical, for if change at the surface does trigger lower rainfall, then the preservation of vegetative cover may be an important policy objective. But the processes are not yet well enough understood to be so influential.

Pasture Degradation

The story of the assessment of African dry-land pastoralism (a form of land use that occupies a far greater area of the dry lands than any other) is a good example of the ways in which ignorance and prejudice have functioned. Indigenous pastoral people were not a cause of great concern before the 1930s. They were colourful, truculent folk, nervously applauded for their resistance to the colonialists, but even in the rare instances where they challenged colonial designs, as in the parts of Kenya and Algeria, where their land was coveted by European settlers, they could be easily marginalized.

In the middle part of the present century, however, these attitudes were transformed by three new, provocative concepts (Warren, 1995). First, early anthropologists characterized the pastoralists as people whose land-use strategy was driven by ritual and magic: they had an 'irrational' approach to pasture and herd management (Widenstrand, 1975). Second, Hardin (1968) introduced the phrase 'tragedy of the commons' to describe the ways in which the common resources of pastoralists might be misused by ruthless individuals to the detriment of the communal good. Third, ecologists developed a model of grazing in which there was a stable 'ecological balance' (another contagious concept) between the grazing animal and its pasture, and believed that they could establish the number of animals per hectare at which this occurred (pasture-by-pasture) (Behnke *et al.*, 1993).

Driven by these three slogans (none of which had any good empirical support), by deep biblical, classical, and even racist prejudices about the inhabitants and environments of the dry lands, and by the anxiety about the possibility of another Dust Bowl after the traumas of the Midwest of the United States (Anderson, 1984), colonial administrators rounded on pastoralists. The characterization of African pastures as 'overgrazed' came to infest almost all writing about them, in official reports, scientific papers, anthropological accounts, school and university textbooks, travel writing, and journalism. More perniciously, it informed policies towards the pastoralists, in which attempts were made to reduce the numbers of their stock; to privatize their lands; and even to replace them their methods with systems based on European and North American models (Halderman, 1985; Hogg, 1988).

Mercifully, few pastoralists heeded the demands. Some, close to the centres of power, like the Maasai in Kenya,

acceded to privatization, but very few reduced their stocking rates to anything like those called for by the ecologists. In the 1970s and with growing force in the 1980s, the pastoralists' disdain for ecological and economic models began to be heeded by the scientific community.

Furthermore, range scientists began to notice that their models were being flatly contradicted by much of the evidence. First, stock numbers kept on growing, despite the warnings. The 'scientifically' deduced capacities were exceeded decade-by-decade in Zimbabwe, for example, and though there were many reasons, like the opening up of new pastures, there was little doubt that more cattle were being kept alive than the model of ecological balance allowed (Livingstone, 1991). Second, stock numbers appeared to recover fairly quickly after severe droughts and soon exceeded the pre-drought numbers, a pattern that would have been impossible had the pastures been overgrazed, as claimed (Penning de Vries and Djeteyé, 1982; Figure 19.4). Third, many accounts were published in which it was affirmed that the indigenous sector in pastoralism was more efficient than the introduced methods such as ranching (although the criteria for judging success were often different) (Western, 1982; Breman and de Wit, 1983; Behnke *et al.*, 1993).

These empirical observations are now supported by theory. In what is probably now the dominant paradigm among ecologists, dry ecosystems are no longer seen as capable of balance (see also the discussion of savanna ecology by M. Adams in Chapter 12 of this volume). They are said to lurch from state to state as they are buffeted by recurrent drought and other shocks like fires and pests (Westoby *et al.*, 1989). Stock numbers may well grow after they have been ravaged in a drought, but their numbers will usually again be devastated by the next drought, well before they can cause permanent damage to the pastures (Ellis and Swift, 1988; Figure 19.5). Dry season pastures, moreover, are seldom capable of maintaining enough stock permanently to damage wet season pastures. In these ways most dry, seasonal ecosystems are actually protected by drought. The peripheries of watering points, the delight of photographers intent on demonstrating desertification, are now seen as permissible 'sacrifice zones', or even as places whose pasture quality may be enriched by the nutrients imported in cattle dung. It is probable that it is only where there is consistently enough rain to allow the 'density-dependent' relationship of the old ecological model, or where feed is imported, that there can be serious and long-lasting 'overgrazing'. Even here, intense grazing may encourage scrub at the expense of grass, and so produce another kind of self-regulation.

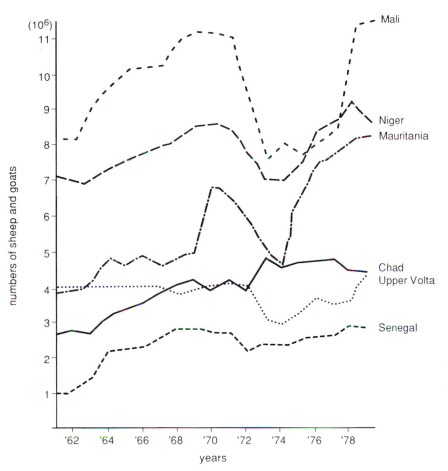

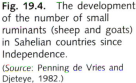

Fig. 19.4. The development of the number of small ruminants (sheep and goats) in Sahelian countries since Independence.
(*Source*: Penning de Vries and Djeteye, 1982.)

The old, accepted wisdom about rationality and communal management has now also been challenged. Many examples have been described in which pastoralists seem rather more rational than range scientists (Sandford, 1983) and in which they manage their communal resources rather well (Runge, 1986). The outcome is that traditional, indigenous modes of pasture management are being endorsed, tragically, in many cases, too late for their survival.

The story of the treatment of African dry-land pastoralists is therefore one of arrogance, or ignorance, or both. Ironically, there is almost a superfluity of data, but all of the wrong sort. The new paradigm in pastoral management, as it is termed, may itself be displaced, but it may have established a tradition of seriously listening to what those on the ground have to say. Yet, despite this realignment of thought about pastoralism in dry Africa, indigenous pastoralism is still characterized as damaging in many influential publications (for

example FAO, 1986). The recently published *Atlas of Desertification* (UNEP, 1992b) maps large areas of pasture degradation, but this assessment must now be regarded with great suspicion.

Degradation in Dry-land Agriculture

The *Atlas of Desertification* also contains maps of soil degradation in various categories (accelerated erosion by wind and water, soil salinization, and so on). The maps may become very influential, for they are virtually the only comprehensive attempts at assessment of the problem, yet they are based almost entirely on opinion, not measurement. Large areas of the Sahel, for example, are shown in the Atlas to be suffering severe wind erosion, but there is not one reliable measurement of the rate of wind erosion in the whole zone (let alone in Africa). It is true that dust production seems to have increased several fold in the last few years (Middleton, 1985), and this must mean increasingly severe wind

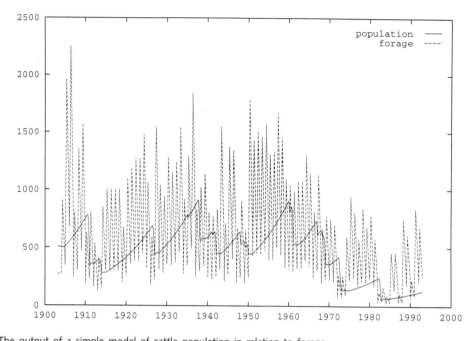

Fig. 19.5. The output of a simple model of cattle population in relation to forage.

The forage is driven by an annual cycle, and real rainfall data for Mali. The cattle population follow the forage according to rules derived from their individual intake, individual growth-rates, age to reach reproductive capacity, fatness (which in turn controls their reproductive capacity and their death rates). There are no assumptions about the effect of grazing on forage, and many other critical assumptions could be questioned (and their numerical values varied). None the less, it is seen that cattle numbers only rarely surpass the capacity of the forage to support them, and then only for very short periods. In other words, seasonal and longer-term droughts apparently have the capacity to protect pastures for much of the time. (I thank Philip Lewis at UCL for the model.)

erosion, but the suspicion is that this is very localized, probably round the towns where visibility recordings are made.

The Atlas shows great swathes of southern Mali as suffering from intense water erosion, yet Bishop and Allen (1989) could find no good data for water erosion in Mali, and were forced to use very dubious models in their evaluation of the costs of land degradation. Thus assessments of problems and projects for their alleviation are being based on extremely inadequate information. As Stocking suggests in Chapter 18, this problem is far from unusual. Although Bishop and Allen, albeit on this shaky basis, reached the conclusion that degradation could be costing Mali an important part of its agricultural production, and that it was probably cost-effective for the government to introduce programmes to combat erosion, other findings on soil erosion have been much more equivocal. In Botswana, for example, Biot (1990) estimated that soil erosion on pastoral land was a very minor problem (despite the findings of the Atlas). In Ethiopia, one of the very few careful experimental assessments concluded that soil was being replaced as fast as it was being eroded, in other words that present systems of cultivation were quite sustainable (Mulugeta Tesfaye, 1988). It is hard to escape the suspicion that dry-land cultivators may be being made the victims of the same kinds of preconceptions as the pastoralists, based again on inadequate data and models.

Interventions

Assessment and intervention are intimately linked: if assessment is misled, intervention is likely to fail. In view of the discussion of assessment in the last section, it is not surprising that intervention has also not been very effective. There are, of course, other elements in failure, among them shortages or maldistribution of resources.

A suite of international institutions intent on intervention materialized after the great drought of the late 1960s and early 1970s. First in the field was UNSO (the United Nations Sudano–Sahelian Office) whose task was initially to combat drought, but which later took on an additional anti-desertification role. Its headquarters are in the UNDP in New York, where it sees its chief task to be the mobilization of funds to tackle problems in the dry parts of Africa from Cape Verde to Tanzania. It has regional offices in Ouagadougou and Nairobi. The United Nations Conference on Desertification, held in Nairobi in 1977, passed on the responsibility for monitoring and managing desertification on a global scale to what became the Desertification Control Programme Activity Centre of the United Nations Environment Programme (UNEP), based also in Nairobi. This office also took responsibility for the Conference's 'Plan of Action' which included the provision that each country should produce its own plan (most of these had appeared by about 1992). In some countries, as in Mali, there are special desertification bureaux, but in most the plan has been produced as a co-operative effort between ministries.

Regional organizations aiming to co-ordinate research and action to combat drought and desertification have now appeared, encouraged by UNSO and UNEP. In West Africa, the Comité permanente Inter-états pour la Lutte contre la Sécheresse au Sahel (CILSS) is based in Ouagadougou, and co-ordinates specialist institutions such as AGRYMET (mainly monitoring drought) in Niamey, and INSAH (researching issues like population and crops, in Bamako). More recently the Inter-Governmental Authority on Drought and Development (IGGAD) has been formed by the Eastern Africa dry countries, while in the south the older Southern African Development Co-ordinating Conference (SADCC), has also assumed drought co-ordination policies. In addition to these multilateral institutions, many bilateral bodies became involved, notably USAID, CIDA from Canada, the GTZ from Germany, ORSTOM from France, the ODA from Britain, SIDA from Sweden, Noraid from Norway, and equivalent bodies from the Netherlands and Japan. Each has its own publications, its own assessment machinery, and its own project management organizations.

This impressive panoply of institutions is, however, facing somewhat of a crisis of confidence. The very fact that so many million environmental refugees were created by the 1984 drought, when many of the institutions had been in place for a few years, is the worst testimony, but there are many more signs of failure or malfunction. The countries that these organizations seek to help, as noted above, are still among the poorest in the world, and few are self-sufficient in food. Investment has only been a fraction of that requested (Tolba, 1986), and there are signs among donors both of disaster fatigue and impatience with the unreliability and poor returns on investment in dry environments. Debt has grown to the point where in some years Africa exports funds to the developed world, and the dry countries are some of the biggest debtors. Civil and international wars have made

the task no easier. Even the few schemes on which so much hope was placed in the 1970s have not succeeded.

The fashion in intervention has changed very radically in the last quarter century. In the 1960s there was still a belief in large schemes, particularly for irrigation, planned to bring Africa to 'take-off point'. There had been some successes, notably in the Gezira in the Sudan (for all its faults), but many more failures. It is surprising, therefore that this sort of thinking survives, as in the recent Manantali Dam on the Senegal River, the planned Hadeija scheme in north-eastern Nigeria, and the Tana Scheme in Kenya (Adams and Hughes, 1989). All have already displaced or may displace thousands of low-tech, but successful, indigenous farmers, fisherman, and pastoralists with almost certainly wasteful, inefficient, and environmentally hazardous high-tech schemes of irrigation. Mercifully, the lack of funds (if nothing else) has now shifted the emphasis to more modest, but still bold conceptions in which trees are planted, fields terraced, pastoralists' herds controlled and corralled, and wells dug. But even these activities have not brought the hoped-for relief from poverty.

There is widespread dismay. Pastoralists have shown deep disinterest in schemes which propose to decimate their stock, or settle them on small parts of their range (Halderman, 1985; Homewood and Rodgers, 1987). Few of the many schemes to halt soil erosion have been successful in the Sahel (Reij, 1988). Many, like those in which bulldozers were used to create bunds across many slopes in Burkina Faso, actually increased the rate of erosion if the bunds were not exactly aligned, and now sit forlornly in fields unheeded by the indigenous cultivators. Fuelwood plots have been plundered or neglected.

The last decade has seen a series of enquiries into failure. The most commonly acknowledged cause is neglect to involve those who are supposed to benefit (Gorse and Steeds, 1985; Catterson, 1988; Kane and Ouedraogo, 1989). The new orthodoxy is that the land users themselves need to be carefully consulted (Chambers, 1983). This growing consensus may be paying off, but only in the last few years (Harrison, 1987). Where simple techniques that fill a real need and produce rapid results are introduced, as when lines of stones were placed along contour lines in the Yatenga area of northern Burkina Faso by an Oxfam project, then the response is massive (Wright, 1986; Rochette, 1989). Incidentally, the technique has been popular not because of soil, but because of water conservation; in other words, the local farmers do not perceive soil erosion to be the main problem they face.

But all too many project appraisals speak of rigidity in conception and time-horizons, driven by the donor's need for accountability, and above all their obdurate attachment to their prejudged assessments of the problems (Skinner, 1988). Aid projects, successful and unsuccessful, have in any case barely scratched the surface of the vast acreages of Africa that are suffering. The only real hope is for there to be grass-roots movements for change, and although these are springing up in some places (Richards, 1985), poverty and growing populations will make their task very difficult.

Desertification is still Africa's main environmental challenge. The African experience of desertification will probably remain the world's most destructive environmental process. It is one of the most complex, and it is by no means a problem under control.

References

Adams, W. M. and Hughes, F. M. R. (1990), 'Irrigation Development in Desert Environments', in A. S. Goudie (ed.), *Techniques for Desert Reclamation* (Chichester), 135–59.

Agnew, C. T. (1989), 'Sahel Drought: Meteorological or Agricultural?', *Journal of Climatology*, 9: 371–82.

Anderson, D. M. (1984), 'Depression, Dust Bowl, Demography and Drought: The Colonial State and Soil Conservation in East Africa', *African Affairs*, 83: 321–43.

Balling, R. C., Jr. (1992), *The Heated Debate: Greenhouse Prediction vs Climatic Reality* (San Francisco).

Behnke, R. H., Jr., Scoones, I., and Kerven, C. (1993) (eds.), *Range Ecology at Disequilibrium: New Models of Natural Variability and Pastoral Adaptation in African Savannas* (London).

Bille, J. C., and Poupon, H. (1974), 'Recherches écologiques sur une savane sahélienne du Ferlo septentrionale, Sénégal: La régénération de la strate herbacée', *Terre et Vie (Revue de L'Histoire naturelle appliquée)*, 28: 21–48.

Biot, Y. (1990), 'Can Livestock Production be Sustained in the Hardveld of Botswana?', *Pedologie*, 40: 243–55.

Bishop, J., and Allen, J. (1989), *On-Site Costs of Soil Erosion in Mali*, World Bank Environmental Department, Working Paper, 21 (Washington, DC).

Breman, H., and de Wit, C. (1983), 'Rangeland Productivity and Exploitation in the Sahel', *Science*, 221: 1341–7.

—— Ketellaars, S. G., and Traoré, N. (1990), 'Un remède contre le manque de terre?', *Sécheresse*, 1: 109–17.

Catterson, T. (1988), 'Mechanisms to Enhance Effective Population Participation', in *Desertification Control and Renewable Resource Management in the Sahelian and Sudanian Zones of West Africa*, World Bank Technical Paper, 70: 28–41.

Chambers, R. (1983), *Rural Development: Putting the Last First* (London).

Ellis, J. E., and Swift, D. M. (1988), 'Stability of African Pastoral Eco-systems: Alternate Paradigms and Implications for Development', *Journal of Range Management*, 41: 450–9.

FAO (1986), *African Agriculture: The Next 25 Years* (Rome).

Glantz, M. H. (1977), 'The Value of a Long-range Weather Forecast for the Sahel', *Bulletin of the American Meteorological Society*, 58: 150–8.

Gorse, J., and Steeds, D. R. (1985), *Desertification in the Sahelian and Sudanian Zones of West Africa* (Washington, DC).

Grove, A. T. (1978), 'Geographical Introduction to the Sahel', *Geographical Journal*, 144: 407–15.

—— (1986), 'The State of Africa in the 1980s', *Geographical Journal*, 152: 193–203.

Halderman, J. M. (1985), 'Problems of Pastoral Development in East Africa', *Agricultural Administration*, 18: 199–216.

Hardin, G. (1968), 'The Tragedy of the Commons', *Science*, 162: 1243–8.

Hare, F. K. (1987), 'Drought and Desiccation: Twin Hazards of a Variable Climate', in D. Wilhite and W. Easterling (eds.), *Planning for Drought* (London), 3–10.

Harrison, P. (1987), *The Greening of Africa* (London).

Hiernaux, P., Diarra, L., and Maiga, A. (1990), *Dynamique de la végétation sahélienne après sécheresse: Un bilan du suivi des sites pastoreaux du Gourma en 1989*, Document de travail, ILCA (Bamako, Mali).

Hogg, R. (1988), 'Changing Perceptions of Pastoral Development: A Case Study for Turkana District', in D. Brokensha, and P. P. Little (eds.), *Anthropology of Development and Change in East Africa* (Boulder, Colo.), 183–99.

Homewood, K., and Rodgers, W. A. (1987), 'Pastoralism, Conservation and the Overgrazing Controversy', in D. Anderson, and R. Grove (eds.), *Conservation in Africa* (Cambridge), 111–28.

Jackson, J. K., and Shawki, M. K. (1950), *Shifting Cultivation in the Sudan*, Memoirs of the Forestry Division, 2, Ministry of Agriculture, Sudan Government (Khartoum).

Johnson, D. H., and Anderson, D. M. (1988) (eds.), *The Ecology of Survival: Case Studies from North-east African History* (London).

Kane, A., and Ouedraogo, S. B. (1989), *Approche des politiques des etats sahéliens en matière de développement rural integré et lutte contre la désertification* (CILSS) (Ouagadougou, Burkina Faso).

Konaté, M., and Traoré, K. (1987), 'Agroclimatic Monitoring During the Growing Season in Semi-arid Regions of Africa', in Wilhite and Easterling (eds.), *Planning for Drought*, 13–129.

Khogali, M. M. (1991), 'Famine, Desertification and the 1988 Rainfall: The Case of Umm Ruwaba District in the Northern Kordofan region, Sudan, *Geojournal*, 25: 91–98.

—— and Johnson, D. L. (1982), 'Pastoralism and Desertification', in J. A. Mabbutt and S. M. Berkowicz (eds.), *The Threatened Drylands* (Sydney), 133–7.

Livingstone, I. (1991), 'Livestock Management', *Ambio*, 20: 80–5.

Middleton, N. J. (1985), 'Effect of Drought on Dust Production in the Sahel', *Nature*, 316: 431–4.

Mortimore, M. J. (1989), *Adapting to Drought: Farmers, Famines and Desertification in West Africa* (Cambridge).

Mulugeta Tesfaye, M. (1988), *Soil Conservation Experiments on Cultivated Land in the Maybar Area, Welo Region*, Soil Conservation Research Report, 16 (Addis Ababa, Berne, and Tokyo).

Pearce, D. W. (1988), 'Natural Resource Management and Anti-Desertification Policy in the Sahel-Sudan Zone: A Case Study of Gum Arabic, *Geojournal*, 17: 349–55.

Newhouse, P. M. (1987), *Les conditions de croissance dans les pays sahéliens et les pays de l'ouest côtiers de l'Afrique occidentale* (Rome).

Penning de Vries, F. W. T., and Djeteyé, M. A. (1982) (eds.), *La productivité des pâturages Sahéliens: Une etude de sols, des végétations et de l'exploitation de cette resource naturelle* (Wageningen, Neth.).

Pieri, C. (1989), *Fertilité des terres de savannes: Bilan de trente ans de recherche et de développement agricoles au Sud du Sahara* (CIRAD—IRAT) (Montpellier, France).

Reij, C. (1988), *L'état actuel de la conservation des eaux et du sol au Sahel* (CILSS)–OECD–Club du Sahel, D(89)329 (Paris and Ouagadougou, Burkina Faso).

Richards, P. (1985), *Indigenous Agricultural Revolution* (London).

Rochette, R. M. (1989) (ed.), *Le Sahel en lutte contre la désertification, leçons d'experiences* (CILSS) / (PAC) (Paris and Ouagadougou, Burkina Faso).

Runge, C. F. (1986), 'Common Property and Collective Action in Economic Development', *World Development*, 14: 623–35.

Said, E. W. (1987), *Orientalism* (London).

Sandford, S. (1983), *Management of Pastoral Development in the Third World* (Chichester).

Sicot, A. M. (1991), 'Example de dispersion spatiale des pluies au Sahel: La pluviosité du bassin de la mare d'Oursi au Burkina Faso', in M. V. K. Sivakumar, J. S. Wallace, C. Renard, and C. Giroux (eds.), *Soil Water Balance in the Sudano-Sahelian Zone, International Association for Scientific Hydrology (IAHS) Publication*, 199: 75–83.

Skinner, J. R. (1988), 'Towards Better Woodland Management in the Sahelian Mali', *Pastoral Development Network Paper*, 25d (ODI) (London).

Tiffen, M., Mortimore, M., and Gichuki, F. (1993), *More People, Less Erosion: Environmental Recovery in Kenya* (Chichester).

Tolba, M. K. (1986), 'Desertification in Africa', *Land Use Policy*, 3: 260–8.

Tucker, C. J., Dregne, H. E., and Newcomb, W. W. (1991), 'Expansion and Contraction of the Sahara Desert From 1980 to 1990', *Science*, 253: 299–301.

UNEP (1992a), *Status of Desertification and Implementation of the United Nations Plan of Action to Combat Desertification*, Report of the Executive Director, UNEP, UNEP GCSS.III.3 (Nairobi).

—— (1992b), *World Atlas of Desertification* (London).

UNOEOA (1985), *Status Report on the Emergency Situation in Africa as of 1st September, 1985*, 3rd monthly report (New York).

Verstraete, M. M. (1986), 'Defining Desertification: A Review', *Climatic Change*, 9: 5–18.

Vita-Finzi, C. (1969), *The Mediterranean Valleys: Geological Changes in Historical Time* (Cambridge).

Warren, A. (1993), 'Desertification as a Global Environmental Issue', *GeoJournal*, 31: 11–14.

—— (1995), 'Changing Understandings of African Pastoralism and the Nature of Environmental Paradigms', *Transactions of the Institute of British Geographers*, 20: 193–203.

—— and Agnew, C. T. (1988), 'An Assessment of Desertification and Land Degradation in Arid and Semi-Arid Areas', Report for Greenpeace International, Shortened Version, in *International Institute for Environment and Development Paper*, 2 (London).

—— and Khogali, M. (1992), *Desertification and Drought in the Sudano-Sahelian Region 1985–1991* (New York).

—— Sud, Y. C., and Rozanov, B. (forthcoming), 'The Future of Deserts', *Journal of Arid Environments*.

Western, D. (1982), 'The Environment and Ecology of Pastoralists in Arid Savannas', *Development and Change*, 13: 183–211.

Westoby, M., Walker, B., and Noy-Meir, I. (1989), 'Opportunistic Management of Rangelands Not at Equilibrium', *Journal of Range Management*, 42: 266–74.

Widenstrand, C. G. (1975), 'The Rationality of Nomad Economy', *Ambio*, 4: 146–53.

Wright, P. (1986), 'Water and Soil Conservation by Farmers', in H. Ohm, and J. Nagy (eds.), *Appropriate Technologies for Farmers in Semi-Arid West Africa* (West Lafayette, NY), 54–6.

20 Biodiversity and Biodepletion

Norman Myers

Introduction

Sub-Saharan Africa is well known for its biodiversity. Not nearly so well endowed biotically is North Africa, so it is not considered here. Sub-Saharan Africa's reputation applies especially to its large mammal life, which amounts to the last sizeable remnant of its sort from the Pleistocene megafauna (Bigalke, 1978). Unlike most other major biogeographic regions, sub-Saharan Africa did not lose much of its large vertebrate communities during the environmental vicissitudes—whether climate- or human-derived—of the late Pleistocene (Kingdon, 1990; Martin and Klein, 1984). So what we see in the region today is an insight into the most recent apogee of large vertebrates.

The region harbours the great majority of the world's antelopes, 54 species (East, 1988), plus a good number of other Bovidae, and 35 species of mammal carnivore, including 10 of the world's 35 Felidae species (Dorst and Dandelot, 1970). This contrasts with far smaller numbers in tropical South America and Asia. The region also features an abundance of other species assemblies, certain of which feature a noted tendency toward gigantism (MacDonald, 1984). The greatest concentrations occur in East Africa. Kenya, no more than two and a half times the size of the United Kingdom, contains at least 7500 plant species, whereas the United Kingdom contains only 1620 species; 314 mammals against 50; 1067 birds against 219; 191 reptiles against 11; and 88 amphibians against 7 (World Resources Institute, 1992).

In some respects, however, sub-Saharan Africa is relatively depauperate. Of 175 000 plant species found in the tropics, fewer than one-quarter occur in this region (almost half in Latin America). If we look at countries worldwide with the greatest diversity of mammals, birds, reptiles, and amphibians, we find the top ten include African countries only in the case of mammals—Zaire fifth, and Kenya tenth. Conversely, Indonesia, four-fifths the size of Zaire, ranks in the top five on all counts, and Mexico, with much the same expanse as Indonesia, ranks among the top four in three categories.

The Biodiversity Array

1. Savannas and Their Mammals

The savanna wildlife is so well known that there is little need to reiterate its abundance and variety. Suffice it to say that certain localities support exceptional communities of large mammals. The Serengeti ecosystem with its 25 000 km^2, for instance, contains 4 million large herbivores: mostly wildebeest, zebras, and gazelles, but also sizeable numbers of buffalo, giraffe, impala, and the like (plus several thousand large carnivores). The United States has no more than 20 million large herbivores in its 9.4 million km^2, a 376-times larger expanse.

2. Birdlife

Equally remarkable is the region's avifauna. It is also known so well that there is no point in further describing it here—except to note that Kenya features 1067 bird species in 583 000 km^2, or 11 per cent of Earth's avifauna in 0.4 per cent of Earth's land surface. Two of Kenya's Rift Valley lakes, Lakes Nakuru and Naivasha, the first saline and the other freshwater, each support more than 400 bird species, or almost twice as many as in the whole of the United Kingdom. An additional spectacle at Nakuru is the flamingos, generally half a million and sometimes three times as many. Certain of Kenya's smaller parks, e.g. Nairobi and Amboseli, likewise feature 400 species, albeit with much overlap. One can readily check 150 species in a weekend, and several dozen before breakfast.

3. Lakes and Swamps

However remarkable the region for its large-mammal and birdlife communities, it is notable too for the fish diversity of its three East African lakes. Yet this form of biodiversity receives little relative attention, so it merits a more detailed account than for items 1 and 2 above. (There is a general account of African lakes in Chapter 6, and of wetlands in Chapter 15.)

By virtue of their isolation, lakes often form 'ecological islands'. This situation can lead to a high degree of speciation and endemism, producing exceptional biodiversity of fish faunas, especially in the tropics. They can also feature sizeable invertebrate faunas alongside the fish communities, presumably with parallel levels of endemism—though all too little is known about this aspect of their species richness.

Outstanding from the standpoint of the region's fish faunas are Lakes Victoria, Tanganyika, and Malawi (Barel *et al.*, 1985; Barlow and Lisle, 1987; Coulter *et al.*, 1986; Hughes, 1986; Lowe-McConnell, 1987; Payne, 1986). The evolution of these Rift Valley lakes has produced a phenomenon of explosive speciation, generating exceedingly rich 'fish swarms' with far greater differentiation than in any other tropical lakes. Indeed the chain of lakes, encompassing almost 700 endemic cichlid fish species (plus almost 100 other endemic fish species), must be regarded as more significant for the study of evolution than even the Galapagos Islands (Echelle and Kornfield, 1984; Greenwood, 1981). Yet the basic biology of the leading lake in the chain, Lake Malawi, has yet to be elucidated, and there is next to no scientific programme for long-term research of a substantive and systematized sort.

Lake Malawi's expanse, 28 231 km², contains around 500 cichlid species, 495 of them (99 per cent) endemic; plus roughly 40 other fish species, approximately 20 of them (50 per cent) endemic. Each year a good number of new species is discovered, which means the true totals could be rather higher than the accepted figures (the same applies to the other two lakes). Yet Lake Malawi is only one-eighth the size of North America's Great Lakes, which feature only 173 species, fewer than 10 of them endemic. Lake Malawi is suffering through pollution from industrial installations, and could be markedly impoverished through the proposed introduction of alien species.

In Lake Victoria's 69 485 square kilometres there are —or rather, there used to be—250-plus cichlid species (more than in the whole of Europe), around 130 (52 per cent) of them endemic; plus almost 40 other species, approximately 20 of them (50 per cent) endemic. On top of the remarkable speciocity of Lake Victoria's fish flock, the evolutionary youth of the biota, and its ecological diversity make the fish community an unrivalled subject for comparative biology and evolutionary study of vertebrates. The lake is only 750 000 years old, and the cichlids are thought to have evolved within the last 200 000 years (Avise, 1990). Yet introduced predators among other problems are likely to reduce the flock of endemics by 80–90 per cent within another decade at most. A good number of cichlid species, conceivably as many as 150, have already been eliminated or are on the verge of becoming extinct (Kauffman, 1992).

In Lake Tanganyika's 28 399 square kilometres there are around 140 cichlid species, more than 70 (50 per cent) of them endemic; plus 110 other fish species, approximately 55 of them (50 per cent) endemic. The least rich lake in the chain, Lake Tanganyika, none the less features almost one-quarter as many more fish species as Europe's total of 192 species. (When one considers invertebrates as well, the lake may turn out to be the most species-rich of the three (Coulter, 1991).) Fortunately Lake Tanganyika is subject to little pervasive degradation as yet, though there are plans for various forms of disruptive development in the lake and its environs, notably in the form of oil extraction.

Indeed both Lake Malawi and Lake Tanganyika could rapidly follow the depauperizing experience of Lake Victoria, unless precautionary conservation measures are put in place ahead of time. To this extent, and on grounds of their spectacular cichlid faunas alone, there is much scope for 'creative' (as opposed to salvage-type) conservation. It is likely too that conservation of these lake ecosystems would safeguard large numbers of invertebrate species as well, though the scale of this spin-off benefit is impossible to determine at this stage.

In sum, these three lakes feature 1080 fish species (possibly a good many more), or 16 per cent of the world's 6650 freshwater fish species in 0.09 per cent of the Earth's land surface. Some 790 species (73 per cent) are endemic.

Other lakes in the region are not nearly so rich biotically. Nor are other freshwater ecosystems, though the Sudd, Okavango, and Bangweulu Swamps are outstanding as wildlife spectacles. The Sudd makes up an expanse of 16 500 square kilometres of permanently flooded lands and an additional 15 000 square kilometres of seasonally flooded lands. The swamp contains at least 500 plant species (only 100 or so in the permanently flooded central sector), though only one is known to be endemic. Over 500 bird species have been recorded, and almost 100 fish species. The area also supports 400 000 people and 800 000 cattle.

Much of the Sudd's biota is threatened by the Jonglei Canal, a 360-km conduit that would cause much of the Nile's water to bypass the Sudd, reducing the permanent swamps by 21–5 per cent and the floodplains by 15–17 per cent. A fully operational canal could eventually cause the Nile to decline to its early 1950s level, whereupon there would be an 80 per cent reduction in the swamps's expanse and a 58 per cent loss of floodplains, hence entraining all manner of adverse repercussions for the biota. As it happens, the Canal's construction has been suspended since 1983, due largely to the civil war that persists in southern Sudan (Carp, 1988; Driver and Marchand, 1985; Howell *et al.*, 1988).

4. Tropical Forests

The region's forests are the main locus for biodiversity. The classification of forests, their exploitation and conservation and discussed in detail by Grainger (Chapter 11). To gain an idea of the species richness of African forests, and of the extinction threats imminent, two prominent areas are considered here.

(a) South-western Ivory Coast

The south-western sector of Ivory Coast features part of a centre of biodiversity that straddles the border with Liberia, and coincides with the so-called Upper Guinea Pleistocene Refuge. Its forest area supports 2770 of the country's 4700 plant species, perhaps 200 of them being endemic (Guillaumet, 1984; Hamilton, 1976).

Of the country's original forest cover totalling 160 000 km² or 50 per cent of national territory, 118 000 km² remained as recently as 1956. But today there are only 16 000 km² at most, or 10 per cent of the original expanse; and of this, only 4000 km² can be classified as primary forest (Maitre, 1987; Myers, 1989a; Spears, 1985). The remnants are being cut at a rate of up to 2000 km² per year; and if present trends continue, there will surely be next to nothing left outside protected areas by the late 1990s.

This makes the Tai National Park all the more important, as the only large tract of original forest left, and with 870 plant species, around 150 (17 per cent) of them endemic (Assi, 1988; Bousquet, 1978; Dosso *et al.*, 1981; Roth, 1984). Nominally the park covers 3300 km²; but it is being steadily reduced by logging (both legal and illegal), by gold prospecting, and most of all by encroaching cultivation. Less than 2000 km², or 61 per cent, are reported to be in primary-forest form, comprising half of the remaining primary forests in the country.

The main cause of deforestation lies with slash-and-burn farmers, who in turn reflect recent immigration patterns. Following the Sahel droughts, Ivory Coast has received 1.4 million foreigners from countries to the north, and by the mid-1980s every fifth person in the country was a foreigner (Ahonzo, 1984). Given the continuing drought situation in the Sahel, we can surely anticipate a persistent influx of environmental refugees, maintaining the country's annual population growth-rate of more than 4 per cent.

(b) The Eastern Arc Forests of Tanzania

Ranging along central-eastern Tanzania is an arc of montane forests. They include, from north to south, the Pare, Usambara, Nguru, Uluguru, Ukaguru, Rubeho, Uzungwa, Mahenge, and Matengo forests. Their expanse amounts to 15 000 km², out of an original expanse thought to have totalled some 31 000 km². Moreover, less than half, and probably only about 40 per cent, i.e. 6000 km², is considered to be primary forest. Being located on ancient hills and mountains, these forests have mostly enjoyed lengthy periods of isolation, leading to much speciation and high levels of endemism: in the East Usambara mountains, close to 80 per cent of millipedes, gastropods, and amphibians are endemic. Some biologists consider the region is of paramount importance at global level for its potential insights into biogeography and evolution. (This review is based on Hawthorne, 1984; Lovett, 1989 and 1990; Polhill, 1989; Rodgers *et al.*, 1986.)

Altogether these forests contain an estimated total of 1600 plant species, of which 535 species (33 per cent) are endemics. They thus contain almost 15 per cent of Tanzania's 11 000 plant species. Further, they contain a greater wealth of endemic species than the forest expanse of West Africa covering 150 000 square kilometres; and they feature 1600 of the 30 000 plant species in all of tropical Africa with its 20 million square kilometres, i.e. over 5 per cent of species in 0.075 per cent of land area.

Population pressures on the forests are high and rising rapidly, expressed in the form of agricultural encroachment, fuelwood gathering, and commercial logging (Lovett, 1989; Rodgers and Hall, 1986). The formerly continuous forest of the East Usambaras is believed to have lost 70 per cent of its original expanse in just the last thirty years, and has been fragmented into a patchwork of some thirty remnants of mixed primary and secondary forest. The Usambara mountains feature over 300 people per km², one of the highest densities in the world for agriculture-based economies. Much the same can be said of most of the forest tracts listed.

This is all the more regrettable in that certain of the

plant species offer marked potential for economic development (Lovett, 1988). For instance, the African violet, an ornamental flower popular in affluent nations, enjoyed a retail trade of $30 million in 1983; of twenty known species of the violet, eighteen are confined to Tanzania's montane forests. Only three species have so far been exploited for cultivation, but others could lend themselves to development, including a hardy variety that grows at 2000 m altitude on the Uluguru mountains where it has become adapted to occasional frosts. Tanzania's forests also contain sixteen species of wild coffee, ten of them endemic; only three have been exploited as commercial crops. The worldwide trade in coffee was worth $12 billion as far back as 1981.

By the late 1980s the protected-areas system in Tanzania was in gross disarray as a result of the country's declining economy and extreme poverty (average GNP per head today, $100, one of the lowest in the world).

At the same time there is a 'good news' area in the form of southern Cameroon and western Gabon, extending also into parts of Congo, and covering roughly 300 000 km² (Gentry, 1992; Reitsma, 1988). Southern Cameroon supports probably the richest flora in continental tropical Africa, and Gabon, while little explored botanically, is reputed to harbour around 8000 plant species, of which 1750 (22 per cent) are endemic. Congo contains an estimated 4000 plant species, 880 (22 per cent) of them endemic; and the western portion of Congo's northern forests is considered to be one of the most species-rich areas in the entire Zaire Basin. With human populations in Gabon and Congo totalling only 3.5 million in 610 000 square kilometres, and no more than 5 million in 150 000 km² in southern Cameroon, population pressures on the forests are slight, though in Gabon there is some threat from logging interests (Myers, 1989a).

5. Cape Floristic Province

The Cape flora is so remarkable that it has been accorded a floristic kingdom of its own, one of six in the world. In this 89 000–km² area with almost 78 000 km² of *fynbos* heathlands, there occur 8600 plant species, over 6300 (73 per cent) of them endemic; and in the south-western tip of the Cape, only 18 000 km² in extent, there are about 6000 plant species, at least 4200 (70 per cent) of them endemic. This review is based on Bond and Goldblatt (1984), Cowling *et al.* (1989), Gibbs-Russell (1987), Hall (1987), and Jarman (1986). *Fynbos* vegetation in discussed in detail by Allen (Chapter 17).

Those species known to be rare or threatened (plus 26 extinct) total around 1500 (over 17 per cent), or nearly as many as the entire flora (whether endemic or not) of the British Isles. The true total could well be more

like 2000 species, or as many threatened plants as in all the United States. One-third of the original *fynbos* has already been lost to agriculture among other forms of development, or to invasion of exotic plants; and much of the remaining two-thirds suffers accelerating attrition. Much wild-land habitat is broken up into a patchwork of relics dispersed among farmlands and urban areas. Most species occur in areas of less than 1 km², and many feature fewer than the 500 individuals often regarded as a minimum for genetic viability.

If one anticipates that original habitats will shortly be reduced to 10 per cent of their former extent, and if one applies the findings of island biogeography to the total of 6300 endemic plant species, then one must expect the early extinction of 3150 species in this area. There would be numerous extinctions too of associated animal species, though whether one can here apply the ratio of 1 plant : 20 animal species is debatable and unascertainable. All one can assert is that animal extinctions would surely total tens of thousands of species.

In sum, this area qualifies as one of the hottest of all hot spots for biodiversity in the world (Myers, 1990), both in terms of its species richness and its extreme threats of habitat destruction. Fortunately the area is exceptional in one further respect: there is a large scientific presence, there are numerous conservation bodies, and the government may be starting to show more readiness to tackle the challenges at issue. Conservation and management strategies for *fynbos* are discussed by Allen (Chapter 17).

6. Madagascar

This ancient island is famous for its unique biota. Indeed it can be regarded as a kind of global epicentre for biodiversity. Only twice the size of the United Kingdom, Madagascar harbours at least 8500 plant species, more than 80 per cent of them endemic, the latter figure amounting to 2.7 per cent of the Earth's species in 0.4 per cent of Earth's land surface (also endemic to the island are 2.1 per cent of the world's mammals, 3.7 per cent of reptiles, 3.4 per cent of amphibians, and 1.9 per cent of swallowtail butterflies). Even more remarkable, Madagascar features seven species of baobab, whereas the whole of continental Africa musters only one; it possesses two-thirds of the world's chameleons; and it is the sole home of the lemurs, new species of which are still being found.

Regrettably only 5 per cent of original vegetation remains (Jenkins, 1987; Jolly *et al.*, 1984; Lowry, 1986). Fortunately the situation is somewhat better in a strip of moist forest along the island's eastern side, where more than 6000 plant species occur, of which 4900 (82 per cent) are endemic. Virtually all of the narrow coastal plains and the lower slopes have been deforested, though on the higher slopes some tree-cover persists, amounting today to rather less than 20 000 km², or 32 per cent of the original 62 000 km² (Sussman *et al.*, 1988). Around half of these relict patches have been degraded through intensive slash-and-burn cultivation, which is now eliminating 1000–1500 km² of forest per year. The forest remnants that feature greatest species diversity and endemism, notably in the northern part of the strip, are precisely those forest tracts under most pressure from agricultural settlers.

It is a realistic prognosis that little original forest will survive into the early part of next century. The country's population growth-rate remains high at 3.3 per cent per year, a rate that if maintained means the total would double every twenty-one years. Rural poverty is pervasive; over three-quarters of the populace lives in the countryside, and per capita GNP is a mere $210. Existing farmlands are among the most eroded anywhere, and because they are generally unable to support the rural population, growing throngs of peasants are seeking new lands to cultivate, notably in the forests.

Ironically, Madagascar has fewer than ten herbarium specimens for each of its 8500 plant species, whereas the United Kingdom with 1620 plant species, 16 of them endemic, has an average of 1000 or more herbarium specimens per species.

Extinction and Biodepletion

1. Patterns and Trends

Many governments of sub-Saharan Africa have made exemplary efforts to establish networks of protected areas. In Tanzania, for instance, the total expanse of parks, reserves, and conservation units amounts to well over one-quarter of national territory. Much the same applies in Zambia and several other countries. This remarkable record notwithstanding, the region's biodiversity has long been experiencing progressive depletion both inside and outside protected areas (Myers, 1972; Western, 1989). How many species have already been eliminated is difficult to say. But in the case of Madagascar, it is estimated that if only two-thirds of forest cover has been destroyed thus far (a very conservative reckoning), at least 500 of the country's 1000-plus orchids must have been eliminated or be on the verge of extinction (Koopowitz, 1992).

Moreover, many species that survive with sizeable numbers are suffering progressive depletion of populations. The lion, for instance, probably totalled hundreds of thousands of individuals as recently as 1950, but by the late 1970s it had experienced such widespread loss of habitat that its numbers had surely fallen to 50 000 at most (Myers, 1982b). Perhaps the largest population remaining is that of the Serengeti ecosystem, no more than 2000 (Sinclair and Norton-Griffiths, 1979).

Even more significant from a long-term perspective is that the decline of population numbers means we are effectively closing off the scope for future evolution in a region that has featured much radiation of large vertebrates. So far as we can discern from the experience of Madagascar, the area required for speciation among large mammals and birds is at least half a million square kilometres, yet the region's largest parks such as Tsavo and Kruger are only 20 000 km². By fortunate contrast, however, we can note that four species of fish appear to have speciated in just the past few thousand years in a small cut-off sector of Lake Victoria totalling only a few thousand square kilometres.

The conservation outlook for the region's biodiversity is dealt with in detail in Chapter 21, so it is not examined further here. The rest of this chapter will consider the principal threat, that of fast-growing human communities and their impact on remaining habitats.

2. Human Population Pressures

By far the greatest threat to biodiversity comes from human population pressures. The issue is worth exploring at some length.

Already sub-Saharan Africa is grappling with multiple problems of population, together with related problems of environment and development. The present population of 550 million is projected (not predicted, still less forecast) to grow to 1.3 billion by 2025, and to quadruple or even quintuple by the time it attains stationary growth roughly 150 years hence (United Nations, 1992; World Bank, 1991). Yet the region shows many signs of difficulty in supporting its present human numbers (Myers, 1989b, 1993; Schreiber, 1992; World Bank, 1988, 1989). Almost 200 million people receive less than 90 per cent of the minimum of 2200 calories a day needed to maintain an active working life; that is, they are chronically undernourished. Another 150 million are subject to acute food deficits, and 30 to 50 million are actually starving (Chen *et al.*, 1990; Mellor *et al.*, 1987; Schreiber, 1992). Today's average per-capita income of roughly $250 is only 95 per cent of real income in 1960. Of the thirty-six poorest countries in the world, twenty-nine are in the region, and at least 62 per cent of the populace endure absolute poverty (Chidzero, 1988; DeCuellar, 1988). This is the only region of the developing world, however, where fertility rates show little sign of decline. Indeed sub-Saharan

Africa is projected to feature almost half of the increase in humanity's numbers during the foreseeable future (Acsadi and Acsadi, 1990; Caldwell and Caldwell, 1990; Ohadike, 1990; Sinding, 1993). As a result, there are already severe land-hunger pressures on protected areas and other biodiversity habitats—pressures that appear set to become much more pronounced throughout the foreseeable future.

Food output per head, which has declined by 20 per cent since 1970, is anticipated to decline by a further 30 per cent during the next twenty-five years (United Nations World Food Council, 1988; World Hunger Program, 1989). Because of unpromising environmental conditions generally, and particularly as concerns harsh climate and widespread water shortages (Falkenmark, 1989), together with pervasive land degradation and poor agricultural policies, the persistence of only moderately adverse weather conditions can quickly trigger broad-scale famine—leading to the desperation that drives multitudes of enfamished destitutes into remaining wildlife areas. As much as 80 per cent of croplands and 90 per cent of stock-raising lands are affected by land degradation of various forms (World Bank, 1989). If adverse weather conditions continue indefinitely (they could well be aggravated by the climatic vicissitudes soon to be entrained by global warming (Farmer and Wigley, 1985; Glantz, 1987), the enfamished throngs that total 30 to 50 million today could increase to well over 100 million by the year 2000 (Goliber, 1989; McNamara, 1990). This means that the proportion of starving people would expand from less than 7 per cent of the region's population in 1985 to 15 or even 18 per cent by the year 2000.

The current food deficit of 12 million tonnes per year is expected to rise to 50 million tonnes (one third of anticipated consumption) by the year 2000 and to 250 million tonnes by 2020 (Braun and Paulino, 1990; Pinstrup-Andersen, 1992; see also Kendall and Pimentel, 1993). (To put these figures in perspective, note that North America's corn crop totals around 200 million tonnes; and that food aid worldwide today is no more than 12 million tonnes.) Worse, the region's hopes of purchasing food from outside are meagre in light of its trade relations. Commodities comprise 90 per cent of exports, and while the volume of these exports increased by 25 per cent during the 1980s, the revenues declined by 30 per cent due to deteriorating prices on world markets (Harrison, 1992). In addition, the region pays out more than $1 billion every month (twice as much as it receives in aid) to service a debt burden proportionally three to four times heavier than Latin America's (Grant, 1993; World Bank, 1992).

Were the region enabled, however, to double its annual increase in food production and to cut its population growth-rate in half by 2020, it would then become food self-sufficient (Pinstrup-Andersen, 1992). In other words, although the region's problems are dire, solutions are available. The resources in shortest supply are human innovation, appropriate technology, external funding among other supports, policy responses backed by political will, and above all, time to mobilize these crucial resources.

To consider the prospect in a little further detail, let us briefly look at the case of one of the prime wildlife countries, Kenya, and with particular respect to the concept of carrying capacity. Kenya's economic performance since independence in 1963 has surpassed that of virtually all other sub-Saharan countries. But the benefits of a strongly growing GNP have been markedly reduced by population growth. During the late 1980s the growth-rate was 4 per cent per year or even more, the highest ever recorded for natural increase in a single country. By 1993 it has declined to 3.7 per cent, but this still means the current total of 28 million people will double in less than nineteen years. Because of the 'youthful profile' of its population pyramid (49 per cent of people are under the age of 15, by contrast with an average of 22 per cent in developed countries), there is much demographic momentum built into Kenya's population future. Even were the family size to come down to two children forthwith, the population would still keep on growing for another two generations at least and double in size before attaining zero growth. As it is, the present population of almost 28 million is projected to reach 62 million by 2025, before expanding to an ultimate total of around 120 million early in the twenty-second century (Frank and McNicoll, 1987; McNamara, 1990; Ominde, 1987).

Yet the Food and Agriculture's study on Land Carrying Capacity (1984) postulates that even with high-level farming inputs, Kenya could not feed more than 51 million from its land resources. Meanwhile agricultural production averaged only a 3.5 per cent annual growth-rate during the 1980s, well below the population growth-rate of 4.0 per cent or higher. As a result, per-capita agricultural output has fallen: with an index of 100 in 1976–8, it slumped to 87 in 1985 (Juma and Munro, 1989; US Department of Agriculture, 1986), and it has been declining even more since then (World Bank, 1992; World Resources Institute, 1992).

A main reason why agricultural production has been declining is that the environmental support base has been overloaded by fast-rising human numbers in the rural areas where two-thirds of the population lives.

Despite a strong soil-conservation effort, most farmland areas have long featured soil erosion to some degree; in several densely populated areas, potential food output could eventually decline by as much as 50 per cent if soil loss cannot be reversed forthwith. Severe desertification has already overtaken 19 per cent of the country and is spreading fast. Forests have been reduced to 3 per cent of national territory, precipitating gross disruption of upland water flows to valley-land farming areas (Odhiambo and Odada, 1988).

This environmental rundown affects sectors other than agriculture. Deforestation is triggering a crisis in energy. Fuelwood and charcoal account for around three-quarters of energy supplies nationwide, and at least 95 per cent of all wood cut is for this purpose. But as forests diminish, the prospect is that demand for fuelwood and charcoal in the year 2000 (from an additional 7 million people on top of the present 28 million who already impose unsustainable pressures on forests) will be twice as large as can be supplied through incremental wood growth (Brokensha *et al.*, 1985; Hyman, 1987; O'Keefe and Raskin, 1985)—with all that implies for Kenya's forests and their exceptional concentrations of biodiversity.

Overall there must be a doubling of all manner of facilities and services—school places, teachers, health-centres, doctors, administrators and the like—every nineteen years merely to keep abreast of an annual population growth-rate of 3.7 per cent. That the nation has hardly managed to keep up with the challenge (even with a smaller population) is demonstrated by the fact that its per-capita GNP has shown scarcely any advance for the great majority of people since 1970, and many people have less to eat today than at the time of independence in 1963. The mutually reinforcing problems of population growth and environmental decline are further compounded by the lack of socio-economic infrastructure: too few trained leaders, professional staffers and general managers, and too few of all these in almost every sector and at almost every level (Kiriro and Juma, 1989). Despite some recent admirable efforts to foster family planning, leading to a sudden decline in the fertility rate from over 8.0 to 6.5 (Sinding, 1993), demographic inertia from exceptionally and even uniquely high population growth-rates in the past mean that Kenya will face still greater difficulties if it is merely to hold the line on living standards. The present total of 28 million people surely exceeds the capacity of the country's environmental-resource base and its planning capabilities alike. What when there will be another 34 million people, as is projected within just another thirty-two years?

Clearly Kenya will have to depend on steadily increasing amounts of food from outside to support itself. But in large part because of the country's high population growth-rate, its per-capita economic growth in recent years has been less than 2 per cent. Worse again, Kenya's terms of trade have been declining throughout the past ten years until they are barely positive today, meaning the country faces the prospect of diminishing financial reserves to purchase food abroad. Its export economy will have to flourish permanently in a manner far better than it has ever achieved to date if the country is to buy enough food to meet its fast-growing needs. Worst of all, the country will have to undertake this massive challenge with a natural-resource base from which forests have almost disappeared, watershed flows for irrigation agriculture are badly depleted, and much topsoil is gone with the wind.

Kenya shows many signs, then, of already being in an 'overshoot' situation as concerns its carrying capacity. The best time to tackle the situation was during the far-back period when its population was only starting to grow rapidly—and when all seemed well in terms of its capacity to feed itself for a while. The source of its population/food dilemma was becoming entrenched. Other countries with currently satisfactory capacity to ensure their food supplies might ponder Kenya's experience. The main opportunity remaining for Kenya to relieve its situation lies with an immediate and vigorous effort to slow its population growth. Were the two-child family to be achieved in 2010 instead of the projected 2035, Kenya's ultimate population could be held to 72 million, 53 million fewer than expected.

Much the same applies to another biodiversity-rich country, Ethiopia. Ethiopia epitomizes those countries with high population growth-rates that must work exceptionally hard to satisfy their basic food needs today, let alone in the future. When a nation is impoverished to start off with, it may well find the task all but beyond its means. So the need grows faster still, until eventually it threatens the very structure and stability of the nation. In Ethiopia the essentials of everyday life, in terms of food supplies alone, are increasingly maintained by outside agencies rather than by the government. The country now imports two-thirds of its food.

As a consequence of the severe imbalance between population and food supplies, plus associated political upheavals, there are now at least 5 million Ethiopians threatened by starvation, and a total of almost 15 million people, or one-quarter of the population, chronically undernourished, i.e. they do not receive enough food to support an active working life. As a further result, there are now at least 3 million displaced people within Ethiopia, and another half a million in refugee camps in Sudan. Again, we must anticipate a prospect of acute and fast-growing pressures from land-hungry peasants on biodiversity's remaining habitats.

Let us briefly review the past record that has led to this situation. By the early 1970s, as much as 470 000 km^2 of Ethiopia's traditional farming areas in the high-land zone—home to 88 per cent of the population—were seriously or severely eroded, to the extent that they were losing an estimated 1 billion tonnes of topsoil per year (a more recent and refined estimate (Hurni, 1990) puts the loss at 1.5 to 3.5 billion tonnes per year). This massive soil erosion was due partly to rudimentary agri-cultural practices, partly to inequitable land-tenure sys-tems, and partly to pressures generated by a population that increased from 18 million in 1950 to 25 million in 1970. The results included a marked fall-off in agri-cultural production accompanied by food shortages in cities, with ensuing disorders that precipitated the over-throw of Emperor Hailie Selassie in 1974 (Farer, 1979; Shepherd, 1975; Tareke, 1977).

Environmental breakdown and food shortages have now become endemic in Ethiopia (Clarke, 1986; Clay and Holcomb, 1986; Eshetu and Teshome, 1984; Gamachu, 1988; Hancock, 1985; Jansson *et al.*, 1987). The mid-1980s drought was no more than a triggering factor, precipitating a crisis that had been building up through the pressure of population growth and agri-cultural mismanagement (Kumar, 1987; Tato and Hurni, 1991). To relieve the longer-term crises, there is now need for agricultural-development inputs to be doubled, soil-conservation and other environmental programmes to be increased fourfold, livestock-husbandry inputs to be expanded sixfold, and population-planning activities to be increased sufficiently to limit the ultimate popu-lation to considerably less than the projected 200-plus million (Gaschen, 1990; Hurni, 1990).

While we cannot conclude that population growth has been the prime or direct cause of Ethiopia's desti-tute state today, it has certainly worsened the situation markedly. Successive governments have been unable to resolve the problem of growing human numbers seek-ing to survive off a deteriorating natural-resource base. Even though Ethiopia has had one of the lowest popu-lation growth-rates in sub-Saharan Africa—no more than 2 per cent per year for most of the time between 1950 and 1980, rising to 2.9 per cent in 1990 (where-as the average for much of sub-Saharan Africa has long been 3 per cent or more)—this has still caused Ethiopia's population to increase from 18 million in 1950 to 57 million today.

Consider what could have been achieved on the population front. If Ethiopia had gone ahead with a strong family-planning programme, proposed for 1970 onwards, and if the programme had achieved only mod-erate success, it might well have prevented 1.7 million births by 1985. The programme would likely have cost around $170 million. Emergency food for Ethiopia dur-ing the 1985 famine crisis amounted to 300 000 tonnes and supplied enough food to sustain 1.7 million people —at a cost of about $170 million (Ehrlich, 1985).

So much for the illustrative examples of Kenya and Ethiopia. True, certain other countries of the region appear to feature far less in the way of population pres-sures. Botswana has only 1.4 million people in 550 000 km^2, an area more than twice the size of the United Kingdom. But because of its arid climate and because the government chooses to base much of the economy on cattle husbandry, the country suffers increasing deser-tification. Zaire today contains 41 million people in 2.3 million km^2. But the impoverished population (per-capita GDP, $220) is projected to reach 172 million before zero growth, and most of the country's farmlands are on soils that are poor or worse. Zimbabwe with 10.7 million people in 387 000 km^2 has achieved some suc-cess in feeding itself (it has recently become a net food exporter), and it has reduced its population growth-rate from 3.4 per cent in 1984 to 3.0 per cent today. But the population is projected to increase to 28 million before it attains zero growth: one wonders how the coun-try will support so many people at an acceptable level of living.

Of course one could propose that Botswana shift its economy away from the emphasis on cattle raising, that Zaire require President Mobutu to return the funds he has sequestered abroad (he is reputed to be richer than Zaire itself), and that Zimbabwe take all manner of measures to make its economy more efficient and pro-ductive. But the three countries, like all other countries in the region, could most practically resolve their devel-opment dilemmas by taking whatever initiatives are available to reduce their population growth-rates.

Fortunately there is much scope, and apparently much political commitment (at long splendid last!), to tackle the population problem. In Kenya, only 17 per cent of women in the late 1970s wanted to have no more children or to delay further pregnancies. But the figure almost tripled within just ten years to 49 per cent. During the same period, the contraception prevalence rate jumped from 7 per cent to 27 per cent; it is probably 32 per cent today, and could well rise as high as 40 per cent by 1995. This remarkable breakthrough has largely been due to a massive effort to get girls into school. The desired family size has now declined to 4.4

children; if contraceptive use indeed soars to 40 per cent by 1995, fertility could well slump from 6.7 children in 1990 to 4.0 children just four years later. Moreover, were all Kenya's unmet demand for family planning to be satisfied, contraceptive use would climb to 57 per cent (Sinding, 1993).

Similar family-planning breakthroughs could be imminent in countries as disparate as Botswana, Namibia, Sudan, Senegal, Togo, and southern Nigeria.

Conclusion

Sub-Saharan Africa contains biodiversity assemblages of exceptional scope and scale. For sheer spectacle—no mean attribute, especially when conservation depends crucially upon public support—the savanna throngs of large herbivores and their attendant carnivores cannot be matched for immediate visual impact. Here we can have a last glimpse of the heyday of large mammal life that characterized the megafauna of the late Pleistocene. At the same time, we can enjoy remarkable concentrations of birdlife, also of freshwater fish, invertebrates, and plants. While the region's biodiversity may not be so rich in species numbers as that of South America or of Asia, it has its own distinctive features that make it uniquely valuable whether in senses biological, ecological, evolutionary, or aesthetic.

Fortunately the biodiversity also possesses marked economic value that in many areas enables conservation to serve as a competitive form of land use. In tourism terms, an elephant herd in Kenya's Amboseli Park is worth $610 000 of renewable income per year (McNeely, 1988). Let us remember too that the two alkaloids from Madagascar's rosy periwinkle generate annual commercial sales from their two anti-cancer drugs worth $200 million in the United States alone, with economic benefits to American society worth twice as much—though let us remember too that Madagascar has not received a single cent from drug sales during the past thirty years (Rasoanaivo, 1990).

Regrettably this biodiversity faces mounting threats, mainly in the form of pressures from fast growing and increasingly impoverished human populations. The established conservation measures, principally centred on set-aside areas protected from human influence, has long served us well—but today it falls far short of needs. A conservation strategy of 'the same as before only more so and better so' is no longer pertinent as an overall response, even while it still has a prime part to play. What is required is an approach that meshes biodiversity's needs with those of human communities—an option that is amply available provided the correct

guiding spirit is there. There is still time to turn profound problems into fine opportunities.

References

Acsadi, G. T. and Acsadi, G. J. (1990) (eds.), *Population Growth and Reproduction in Sub-Saharan Africa* (Washington, DC).

Ahonzo, J. (1984), *La Population de la Côte d'Ivoire* (Abidjan, Ivory Coast).

Assi, A. L. (1988), 'Espèces rares et en voie d'extinction de la flore de la Côte d'Ivoire', in P. Goldblatt and P. P. Lowry (eds.), *Modern Systematic Studies in African Botany* (St. Louis, Mo.).

Avise, J. C. (1990), 'Flocks of African Fishes', *Nature* 347: 512–13.

Barel, C. D. N. *et al.* (1985), 'Destruction of Fisheries in Africa's Lakes', *Nature*, 315: 19–20.

Barlow, C. G. and Lisle, A. (1987), 'The Biology of the Nile Perch', *Biological Conservation*, 39: 269–89.

Baskin, Y. (1992), 'Africa's Troubled Waters', *BioScience*, 42: 476–81.

Bigalke, R. C. (1978), 'Evolution of Mammals of the Southern Continents: The Contemporary Mammal Fauna of Africa', *Quarterly Review of Biology*, 43: 265–300.

Bond, P. and Goldblatt, P. (1984), 'Plants of the Cape Flora', *Journal of South African Botany*, Supplement, 13: 1–455.

Bousquet, B. (1978), 'Un parc de forêt dense en Afrique: La Parc National de Tai (Côte d'Ivoire)', *Bois Forêts Tropicales*, 179: 27–46.

Braun, J. von and Paulino, L. (1990), 'Food in Sub-Saharan Africa: Trends and Policy Challenges for the 1990s', *Food Policy*, 15: 505–17.

Brokensha, D. W., Riley, B. W., and Castro, A. P. (1985), *Fuelwood Use in Rural Kenya: Impacts on Deforestation* (Binghampton, NY).

Caldwell, J. C. and Caldwell, P. (1990), 'High Fertility in Sub-Saharan Africa', *Scientific American*, 262: 82–9.

Carp, E. (1988), *Directory of Afro-Tropical Wetlands of International Importance* (Cambridge).

Chen, R. S., Bender, W. H., Kates, R. W., Messer, E., and Millman, S. R. (1990), *The Hunger Report: 1990* (Providence, RI).

Chidzero, B. T. (1988), 'Africa and the World Economy', *World Futures*, 25: 157–62.

Clarke, J. (1986), *Resettlement and Rehabilitation: Ethiopia's Campaign Against Famine* (London).

Clay, J. W. and Holcomb, B. K. (1986), *Politics and the Ethiopian Famine 1984–85* (Engelwood Cliffs, NJ).

Coulter, G. W. *et al.* (1986), 'Special Features of the African Great Lakes and Their Susceptibility to the Effects of Human Activities', *Bulletin of the J. C. B. Smith Institute of Ichthyology* (Grahamstown, South Africa).

Coulter, G. W. (1991) (ed.), *Lake Tanganyika and Its Life* (Oxford).

Cowling, R. M. (1991) (ed.), *The Ecology of Fynbos: Nutrients, Fire and Diversity* (Oxford).

—— Gibbs-Russell, G. E., Hoffman, M. T., and Hilton-Taylor, C. (1989), 'Patterns of Plant Species Diversity in Southern Africa', in B. J. Huntley (ed.), *Biotic Diversity in Southern Africa* (Oxford), 19–50.

DeCuellar, P. (1988), *Africa's Economic Situation* (New York).

Dorst, J. and Dandelot, P. (1970), *A Field Guide to the Larger Mammals of Africa* (Boston).

Dosso, H., Guillaumet, J. L., and Hadley, M. (1981), 'The Tai Project: Land Use Problems in a Tropical Rain Forest', *Ambio*, 10: 120–5.

Driver, C. A. and Marchand, M. (1985), *Taming the Floods: Environmental Aspects of Floodplain Development in Africa* (Leiden, Neth.).

East, R. (1988) (compiler), *Antelopes: Global Survey and Regional Action Plans* (Gland, Switz.).

Echelle, A. A. and Kornfield, I. (1984), *Evolution of Fish Species Flocks* (Orono, Me.).

Ehrlich, A. H. (1985), 'Critical Masses', *The Humanist*, 45: 18–22, 36.

Eshetu, C. and Teshome, M. (1984), *Land Settlement in Ethiopia: A Review of Developments* (Addis Ababa, Ethiopia).

Falkenmark, M. (1989), 'The Massive Water Scarcity Now Threatening Africa: Why Isn't it Being Addressed?', *Ambio*, 18: 112–18.

FAO (1984), *Potential Population Supporting Capacities of Lands in the Developing World* (Rome).

Farer, T. J. (1979), *War Clouds on the Horn of Africa* (Washington, DC).

Farmer, G. and Wigley, T. M. L. (1985), *Climatic Trends for Tropical Africa* (Norwich).

Frank, O. and McNicoll, G. (1987), 'An Interpretation of Fertility and Population Policy in Kenya', *Population and Development Review*, 13: 209–40.

Gamachu, D. (1988), *Environment and Development in Ethiopia* (Geneva, Switz.).

Gaschen, D. (1990), *Ressourcen-Modell der Regionen Aethiopiens für die Nächsten 50 Jahre* (Berne, Switz.).

Gentry, A. H. (1992), 'Diversity and Floristic Composition of Lowland Tropical Forests in Africa and South America', in P. Goldblatt (ed.), *Biological Relationships between Africa and South America* (New Haven).

Gibbs-Russell, G. E. (1987), 'Preliminary Floristic Analysis of the Major Biomes in Southern Africa', *Bothalia*, 17: 213–27.

Glantz, M. H. (1987) (ed.), *Drought and Hunger in Africa: Denying Famine a Future* (Cambridge).

Goliber, T. J. (1989), *Africa's Expanding Population: Old Problems, New Policies* (Washington, DC).

Grant, J. P. (1993), *Children and Women: The Trojan Horse Against Mass Poverty* (New York).

Greenwood, P. H. (1981), *The Haplochromine Fishes of the East African Lakes* (Munich, Germany).

Guillaumet, J. L. (1984), 'The Vegetation: An Extraordinary Diversity', in A. Jolly, *et al.* (eds.), *Key Environments: Madagascar* (Oxford), 27–54.

Hall, A. V. (1987), 'Threatened Plants in the Fynbos and Karoo Biomes, South Africa', *Biological Conservation*, 40: 29–52.

Hamilton, A. C. (1976), 'The Significance of Patterns of Distribution Shown by Forest Plants and Animals in Tropical Africa for the Reconstruction of Upper Pleistocene and Palaeoenvironments: A Review', in E. M. Van Zinderen Bakker (ed.), *Palaeoecology of Africa, the Surrounding Islands, and Antarctica* (Cape Town), 63–97.

Hancock, G. (1985), *Ethiopia: The Challenge of Hunger* (London).

Harrison, P. (1992), *The Third Revolution: Environment, Population and a Sustainable World* (London).

Hawthorne, W. D. (1984), 'Ecological and Biogeographical Patterns in the Coastal Forests of East Africa', Ph.D. thesis, Oxford.

Howell, P., Lock, M., and Cobb, S. (1988) (eds.), *The Jonglei Canal: Impact and Opportunity* (Cambridge).

Hughes, N. F. (1986), 'Changes in the Feeding Biology of the Nile Perch, *Lates niloticus* (L) (Pisces, Centropomidae), in Lake Victoria, East Africa, Since Its Introduction in 1960, and Its Impact on the Native Fish Community of the Nyanza Gulf', *Journal of Fish Biology*, 29: 541–8.

Huntley, B. J. (1989) (ed.), *Biotic Diversity in Southern Africa: Concepts and Conservation* (Oxford).

Hurni, H. (1990), *Towards Sustainable Development in Ethiopia* (Berne, Switz.).

Hyman, E. L. (1987), 'The Strategy of Production and Distribution of Improved Charcoal Stoves in Kenya', *World Development*, 15: 375–86.

Jansson, K., Harris, M. and Penrose, A. (1987), *The Ethiopian Famine* (Atlantic Highlands, NJ).

Jarman, M. L. (1986), 'Conservation Priorities in Lowland Regions of the Fynbos Biome, South Africa', *National Scientific Programmes Report*, 87 (Pretoria).

Jenkins, M. D. (1987), *Madagascar: An Environmental Profile* (Cambridge).

Jolly, A., Oberle, P. and Albignac, E. R. (1984) (eds.), *Key Environments: Madagascar* (Oxford).

Juma, C. and Munro, R. (1989), *Environmental Profile: Kenya* (Nairobi).

Kauffman, E. (1992), 'Catastrophic Change in Species-Rich Freshwater Ecosystems: The Lessons of Lake Victoria', *BioScience*, 42/11: 846–58.

Kayanja F. I. B. (1979) (ed.), 'Proceedings of 4th East African Wildlife Symposium: Ecological Islands and Their Conservation in the Tropics', *African Journal of Ecology*, 7/4.

Kendall, H. and Pimentel, D. (1993), 'Constraints on the Expansion of the Global Food Supply', *Ambio*, 23: 198–205.

Kingdon, J. (1990), *Island Africa: The Evolution of Africa's Rare Animals and Plants* (London).

Kiriro, A. and Juma, C. (1989) (eds.), *Gaining Ground: Institutional Innovations in Land-Use Management in Kenya* (Nairobi).

Koopowitz, H. (1992), 'A Stochastic Model for the Extinction of Tropical Orchids', *Selbyana*, 13: 115–22.

Kumar, G. (1987), *Ethiopian Famines 1973–1985: A Case Study* (Helsinki).

Lovett, J. C. (1988), 'Practical Aspects of Moist Forest Conservation in Tanzania', in P. Goldblatt and P. P. Lowry (eds.), *Modern Systematic Studies in African Botany, Monographs in Systematic Botany*, 25: 491–6.

——(1989), 'Tanzania', in D. G. Campbell and H. D. Hammond (eds.), *Floristic Inventory of Tropical Countries* (Bronx, NY), 232–5.

——(1990), 'Classification and Status of the Moist Forests of Tanzania', Mitteilungen aus dem Institut fuer Allgemeine Botanik, Hamburg, 23A: 287–300.

Lowe-McConnell, R. H. (1987), *Ecological Studies in Tropical Fish Communities* (Cambridge).

Lowry, P. P. (1986), 'A Systematic Study of Three Genera of Araliaceae Endemic to or Centered in New Caledonia', Ph.D. diss. Washington University, St. Louis, Mo.

MacDonald, D. (1984) (ed.), *The Encyclopaedia of Mammals* (London).

McNamara, R. S. (1990), 'Population and Africa's Development Crisis', *Populi*, 17/4: 35–43.

McNeely, J. A. (1988), *Economics and Biological Diversity: Developing and Using Economic Incentives to Conserve Biological Diversity* (Gland, Switz.).

——Miller, K. R., Reid, W. V., Mittermeier, R., and Werner, T. B. (1989), *Conserving the World's Biological Diversity* (Washington, DC and Gland, Switz.).

Maitre, H. (1987), 'Natural Forest Management in Côte d'Ivoire', *Unasylva*, 39: 53–60.

Martin, P. S. and Klein, R. G. (1984) (eds.), *Quaternary Extinctions: A Prehistoric Revolution* (Tucson, Ariz.).

Mellor, J. W., Delgado, C. L., and Blackie, M. J. (1987) (eds.), *Accelerating Food Production in Sub-Saharan Africa* (Baltimore).

Moreau, R. E. (1966), *The Bird Faunas of Africa and Its Islands* (London).

Myers, N. (1972), 'National Parks in Savannah Africa', *Science*, 178: 1255–63.

——(1982*a*), 'Forest Refuges and Conservation in Africa', in G. T. Prance (ed.), *Biological Diversification in the Tropics* (New York), 658–72.

——(1982*b*), 'Conservation of Africa's Cats: Problems and Opportunities', in S. D. Miller and D. D. Everett (eds.), *Cats of the World: Biology, Conservation and Management* (Washington, DC), 437–46.

—— (1989a), *Deforestation Rates in Tropical Forests and Their Climatic Implications* (London).

—— (1989b), 'Population Growth, Environmental Decline and Security Issues in Sub-Saharan Africa', in A. Hjort af Ornas and M. A. Mohamed Salih (eds.), *Ecology and Politics: Environmental Stress and Security in Africa* (Uppsala, Sweden), 211–31.

—— (1990), 'The Biodiversity Challenge: Expanded Hot Spots Analysis', *The Environmentalist*, 10/4: 243–56.

—— (1993), 'Population, Environment and Development', *Environmental Conservation*, 20/3: 205–10.

Odhiambo, L. O. and Odada, J. E. O. (1988) (eds.), *Kenya's Industrial and Agricultural Strategies towards the Year 2000*.

Ohadike, P. O. (1990) (ed.), *The State of African Demography* (Liege, Belg.).

O'Keefe, P. and Raskin, P. (1985), 'Crisis and Opportunity: Fuelwood in Kenya', *Ambio*, 14: 220–4.

Ominde, S. H. (1987) (ed.), *Issues in Population Growth and Resources in Kenya* (London).

Payne, A. I. (1986), *The Ecology of Tropical Lakes and Rivers* (New York).

Pinstrup-Anderson, P. (1992), *Global Perspectives for Food Production and Consumption* (Washington, DC).

Polhill, R. M. (1989), 'East Africa (Kenya, Tanzania and Uganda)', in Campbell and Hammond (eds.), *Floristic Inventory of Tropical Countries*, 217–31.

Rasoanaivo, P. (1990), 'Rain Forests of Madagascar: Sources of Industrial and Medicinal Plants', *Ambio*, 19: 421–4.

Rebelo, A. G. and Siegfried, W. R. (1990), 'Protection of Fynbos Vegetation, Ideal and Real-World Options', *Biological Conservation*, 54: 15–31.

Reitsma, J. M. (1988), 'Forest Vegetation of Gabon', *Tropenbos Technical Series*, 1: 5–142.

Rodgers, W. A., Mziray, W., and Shishira, W. (1986), *The Extent of Forest Cover in Tanzania Using Satellite Imagery* (Dar es Salaam, Tanz.).

—— and Hall, J. B. (1986), 'Pole-Cutting Pressure in Tanzanian Forests', *Forestry Ecology and Management*, 14: 133–40.

Roth, H. H. (1984), 'We All Want the Trees: Resource Conflict in the Tai National Park, Ivory Coast', in J. A. McNeely and K. R. Miller (eds.), *National Parks, Conservation and Development: The Role of Protected Areas in Sustaining Society* (Washington, DC), 127–9.

Schreiber, G. A. (1992), *The Population, Agricultural and Environment Nexus in Sub-Saharan Africa* (Washington, DC).

Shepherd, J. (1975), *The Politics of Starvation* (Washington, DC).

Sinclair, A. R. E. and Norton-Griffiths, M. (1979) (eds.), *Serengeti: Dynamics of an Ecosystem* (Chicago).

Sinding, S. W. (1993), *The Demographic Transition in Kenya: A Portent for Africa?* (New York).

Spears, J. (1985), *Malaysia Agricultural Sector Assessment Mission: Forestry Subsector Discussion Paper* (Washington, DC).

Stuart, S. N. and Adams, R. J. (1990), *Biodiversity in Sub-Saharan Africa and Its Islands* (Gland, Switz.).

Sussman, R., Green, G., and Gentry, A. H. (1988), 'A Study of Deforestation in Eastern Madagascar Using Satellite Imagery', *Science*.

Tareke, G. (1977), 'Rural Protest in Ethiopia 1961–1970: A Study of Rebellion', unpub. doctoral diss. Syracuse University, New York.

Tato, K. and H. Hurni (1991) (eds.), *Soil Conservation for Survival* (Berne, Switz.).

United Nations (1992), *Long-Range World Population Projections: Two Centuries of Population Growth, 1950–2150* (New York).

United Nations World Food Council (1988), *The Global State of Hunger and Malnutrition: 1988 Report* (New York).

United States Department of Agriculture (1986), *World Indices of Agricultural and Food Production 1976–85* (Washington, DC).

Western, D. (1989), 'Population, Resources and Environment in the Twenty-First Century, and Conservation Without Parks: Wildlife in the Rural Landscape', in D. Western and M. C. Pearl (eds.), *Conservation for the Twenty-First Century* (Oxford), 11–25, 158–65.

World Bank (1988), *The Challenge of Hunger in Africa: A Call to Action*. (Washington, DC).

—— (1989), *Sub-Saharan Africa: From Crisis to Sustainable Growth*. (Washington, DC).

—— (1991), *World Development Report 1991: The Challenge of Development* (New York).

—— (1992), *World Development Report 1992: Development and the Environment* (New York).

World Hunger Program (1989), *Beyond Hunger: An African Vision of the 21st Century* (Providence, RI).

World Resources Institute (1992), *World Resources Report 1992–93* (Washington, DC).

21 Conservation and Development

William M. Adams

Introduction

Development and conservation are closely interrelated. The urge to transform and improve peoples' lives and living conditions, and also to preserve and maintain particular attributes of the environment, have been alternative aspects of the same approach on the part of outsiders to Africa, its people, and environments. Ideas about 'development' (in the sense of systematic economic, cultural, and social change) and 'conservation' (in the sense of the considered regard for non-human species, communities of species, or physical or biological processes) are mostly discussed outside rural Africa, in international meetings and agencies, or in the urban offices of government. Both sets of ideas, however, seek and claim roots in nature and rural society, and are part of an attempt to plan and bring about new and better ways to manage and transform African economies, societies, and environments.

It is an illusion that either development or conservation began in Africa only in the present century. Compared to the diversity of knowledge about the rise of conservation and development in colonial Africa and since independence, we know relatively little about the ways people related to other species, and indeed little about how they organized economic life and relations with each other, in former periods. However, archaeology and history have already begun to reveal something of the history of former states, and specifically about poverty, pastoralism, and agriculture (e.g. Iliffe, 1987; Connah, 1988; Sutton, 1989; Robertshaw, 1990). These studies are illuminating periods that remain lost for many scientists and others in alternative notions of pre-colonial Africa as either a 'Dark Continent' (Jarosz, 1992) or a place of idealized harmony between people and nature, the 'Merrie Africa' that historians are now at pains to avoid invoking (Hopkins, 1973).

Indigenous Environmental Management

It is clear that notions of an Eden-like pre-colonial past in Africa, with people and nature in some kind of harmony, are misplaced. Nature was frequently exploited, controlled, and contested in ways driven by both utilitarian and religious imperatives, although there are few formal studies of the history, geography, or ecology of these practices (Grove, 1990a). There is now a substantial literature on certain aspects of 'indigenous knowledge', particularly with reference to agriculture. This has emerged in recent decades as fashion has swung to counter earlier denigration by colonial and post-colonial technicians of the skills and enterprise of African peasant farmers. Mixed cropping and intercropping, mixing of crop varieties, planting across soil catenas, or across the boundaries between different ecosystems (particularly between wetland and dryland), are now seen to be important and effective adaptations to drought, uncertainty, labour shortage, and other realities of rural life (e.g. Richards, 1985, 1986), and earlier voices describing indigenous knowledge are now recognized (e.g. Jones, 1935; Stamp, 1938). The adaptability of farmers in the face of uncertain and unproductive conditions has been shown to be important, and the failure of development initiatives to recognize and build upon that adaptability has been a major cause of their depressingly low success rate (Mortimore, 1989; Adams, 1992a).

Similar changes have taken place in understanding of the ways pastoralists relate to and manage the environment. The once near-universal view was that pastoralists inevitably overstocked rangelands, and that savanna ecosystems are fragile ecosystems very easily and permanently degraded (derived from an uncritical adoption of ideas about the 'tragedy of the commons' (Hardin, 1968)), has been replaced by a less dismissive and more scientifically informed understanding of pastoral ecology (Horowitz and Little, 1987). Belatedly, the difference between open access and common-property resources have been recognized, and new ideas on the dynamics of savanna ecosystems stress their resilience (Behnke and Scoones, 1991; Mace, 1991; Behnke *et al.*, 1993). There is now a much greater understanding of the skill with which pastoralists extract subsistence from semi-arid ecosystems, and deal with variability of rainfall (e.g. Western, 1982*a*; Coughenour *et al.*, 1986). Once again, development projects are being seen as a threat to the flexibility and adaptability of pastoralists and pastoral systems.

There is less information on indigenous use of other elements within the environment than there is on agriculture and pastoralism. However, it is clear that similar principles of indigenous knowledge are important in the context of both artisanal fisheries (e.g. Welcomme, 1979), and fuel and fibre use (e.g. Hughes, 1987). Work in particular on wetlands within Africa has highlighted the economic importance of hydrological, fibre, fish, herbaceous, and land resources, and the continuity between indigenous management in the historical period and at the present day (e.g. Dugan, 1990, 1993; Adams, 1993, see also Hughes this volume).

Most research on indigenous environmental management has focused on direct productive use of ecosystems and physical resources. Some of these have direct impacts on particular species or the structure of the ecosystem, but there have been few research studies of 'indigenous conservation' *per se*, even though this has become an important issue to those concerned with the relations between conservation policy in protected areas and local people today (e.g. Kemf, 1993). It is likely that further research on religious and cosmological aspects of environmental management will reveal a far richer inventory of non-utilitarian relations with nature than exists at present (Grove, 1990*a*). Like more directly consumptive uses of nature in Africa, even where they persist these relations are now everywhere going to be in a process of transformation by modernization, a money economy, and the impacts of development initiatives.

The Development of Conservation

Imperial conquest and colonial rule transformed the human geography of tropical Africa, and imposed new and highly structured approaches both to society and nature. There is debate about the impact on the environment of Africa of the imposition (often by force) of imperial rule. Kjekshus (1977) argued that imperial assault disrupted the control over ecology that had previously been maintained, ushering in livestock and human disease and economic disruption. While this view probably paints far too idealistic (and selective) a picture of pre-colonial society, it is clear that in East Africa smallpox, rinderpest, and sleeping sickness had a vast impact on economy, society, and ecology at the end of the nineteenth century. Rinderpest struck the Ethiopian Highlands in June 1888, and swept south, killing 90 per cent of cattle. Crop failure through drought and pest attack, wars of imperial conquest, and human disease (typhus, smallpox, cholera, and influenza) brought widespread famine (Pankhurst and Johnson, 1988). In Maasailand, epidemics of bovine pleuropneumonia and rinderpest first weakened and then killed livestock in the 1880s and 1890s, and an epidemic of smallpox brought famine and total disruption of social and economic organization (Waller, 1988). In subsequent decades the aftermath of these events, compounded by colonial alienation of land in Kenya, promoted the advance of tsetse fly and of sleeping sickness (Lamphrey and Waller, 1990; Waller, 1990). Causation and impacts were undoubtedly complex, but certainly previous ecologies and political economies were transformed and severely disrupted by the advent of colonial rule (Johnson and Anderson, 1988). The exploration by Watts (1983, 1984) of the impacts of the unholy alliance between colonial state and capitalism in Nigeria on the rural economy, and the moral economy that underpinned it, demonstrate how completely colonial rule transformed both peoples' lives and the rural environment.

Part of the impact of colonialism lay in the importation of ideas about conservation. From the mid-eighteenth century, ideas about the right management of the environment began to emerge in the colonial periphery as part of the fruit of a clash between the extractive thrust of imperial trade, a rising sensitivity to Romanticism and the growth of science (Grove, 1990*b*, 1992). These ideas were developed first in India, the oceanic islands of the Indian and Atlantic Oceans (Mauritius and St. Helena), and the West Indies. They involved in particular a concern for the loss of forests and the perceived links between forests and rainfall, and they led to policies of control of forests and the creation of forest reserves. The notion that removal of forests caused 'desiccation' was very widely held in the emerging ranks of colonial scientists. These ideas reached Africa in the Cape Colony in the nineteenth century, fostered in particular by alarm at the droughts of 1847 and 1862 (Grove, 1987). A Chief Botanist to the Cape Colony was appointed in 1858. Forest reserves had been established on Tobago and St. Vincent in the West Indies in the eighteenth century, and in India in the nineteenth century. The same model was adopted in the Cape. The first legislation to reserve land had been passed in 1846, to prevent soil erosion on open areas close to Cape Town. Forest reserves were established at Knynsa and Tsitsikamme in 1858 (Grove, 1987), the first protected areas in Africa.

The institutional organization of conservation in the imperial powers of Western Europe and in North America also began in the nineteenth century (Allen, 1976; Sheail, 1976; Nash, 1983; Worster, 1985), and the now-familiar pattern of National Parks began to appear in the 1880s and 1890s in Canada, South Australia, and New Zealand (Fitter, 1978). In Africa, conservation activities became increasingly concentrated on the alienation of land for game reserves. The first such area in Africa was created in 1892 when the Sabie Game Reserve was established in the Transvaal. In 1899 a game reserve was established in Kenya enclosing the present Amboseli National Park (Lindsay, 1987). The Etosha National Park and the Namib-Nakluft National Park in Namibia both date back to the establishment of game reserves in 1907. In 1925 King Albert created the first African National Park, the Parc National Albert (now the Virunga National Park, Fitter, 1978; Boardman, 1981), and in 1926 the Sabie Game Reserve became the Kruger National Park.

The creation of game reserves stemmed directly from concern about the extinction of the remarkable 'big game' of Africa. Hunting was seen as the main threat, initally that by Europeans and latterly that by Africans. Commercial hunting for ivory and skins in the nineteenth century grew into hunting to feed railway workers and missions, and to finance imperial enterprises of many kinds (Mackenzie, 1987). In time, sport hunting came to replace commercial hunting, and with it came the ideological baggage and ritual of hunting, with its trophies and mythology of sportsmanship and other ideals of British boys' education (e.g. Thesiger, 1987). The closure of the frontier in the USA brought wealthy Europeans and North Americans to Africa to hunt, and they killed game in astonishing quantities (Graham, 1973; Nash, 1983; Mackenzie, 1987, 1988). By the end of the nineteenth century white hunters had depleted large areas of Africa of large animals, and sent the Quagga into extinction.

Reservation of game for shooting grew progressively into the establishment of reserves for conservation

(Graham, 1973, Mackenzie, 1987, 1988; Steinhart, 1989). Converted big-game hunters (the so-called 'penitent butchers') established the Society for the Preservation of the Wild Fauna of the Empire, with a powerful list of members, including President Theodore Roosevelt and the British Secretary of State for the Colonies (Fitter, 1978). Concern about the depredations of white hunters led, somewhat paradoxically perhaps, towards measures to prevent Africans from hunting. The Cape Act for the Preservation of Game was passed in 1886 and extended to the British South African Territories in 1891 (Mackenzie, 1987). In 1900 the Kenyan Game Ordinance was passed, effectively banning all hunting except by licence (Graham, 1973). Continued hunting by Africans became by definition 'poaching', and dominated the work of newly established Game Departments.

International concern for conservation in Africa grew in the 1930s. The African colonial powers (Britain, Germany, France, Portugal, Spain, Italy, and the Belgian Congo) had signed a Convention for the Preservation of Animals, Birds, and Fish in Africa in 1900, although it was never implemented (Fitter, 1978). The International Office for the Protection of Nature (IOPN) was established in 1934, and in the 1930s it published a series of reports on African colonial territories. As IUCN (International Union for Conservation of Nature) it put a biologist into the field in Africa to report on threatened mammals. Other work was developed by the Frankfurt Zoological Society through the work of Bernard Grzimek in Kenya and Tanzania, and the Conservation Foundation.

Several National Parks were declared in the late 1940s, for example the Nairobi National Park in Kenya in 1946 and Tsavo two years later, Serengeti in Tanganyika in 1951, and Murchison Falls in Uganda in the same year (Fitter, 1978). In 1953 a special conference on African conservation problems convened at Bukavu in the Belgian Congo. By 1960 Africa had become 'the central problem overshadowing all else' for IUCN (Boardman, 1981: 148): Africa was becoming independent, and political control was shifting as poachers turned gamekeepers. IUCN and FAO launched the African Special Project in 1961. The 'Arusha Declaration on Conservation' was adopted in 1961, stressing both a commitment to wildlife conservation and wider concerns about resource development (Plate 21.1). The African Special Project was followed up by IUCN missions to seventeen African countries (Fitter, 1978), and eventually (after considerable political haggling between the IUCN and FAO) a new African Convention on the Conservation of Nature and Natural Resources, was adopted by the Organization of African Unity in Addis Ababa in 1968 (Boardman, 1981).

Plate 21.1. Bird-viewing hide, Dagonna Ox-Bow Bird Sanctuary, north-east Nigeria. The Hadejia-Nguru Wetlands are a major area of seasonally and permanently flooded land mixed with dry farmland, formed where the Hadejia and Jama'are rivers drain north-eastwards through north-east Nigeria. They eventually reach Lake Chad as the Komadugu Yobe. The wetlands are of recognized international importance for wintering Palaearctic migrant birds, within a system of sub-Saharan wetlands, and for their local bird populations. The Borno State Government declared a bird sanctuary in a deeply-flooded ox-bow lake at the village of Dagonna in 1989. A bird hide was constructed, and opened by the President of the Worldwide Fund for Nature. By 1991 the limited infrastructure (road, bird-watching hide, patrol huts, signboards, and commemorative plaque) were in disrepair and there was no conservation management on the ground. Such conventional approaches to conservation have now been replaced by a more people-orientated approach (led by the Hadejia-Nguru Wetland Conservation Project), centred on the question of the sustainable management of the water resources of the area and attempts to integrate upstream development and downstream human and wildlife needs for floodwater (photo: W. M. Adams).

The Development of Development

Imperial dreams about Africa, and particularly perhaps the development of its water resources, were both bold and fertile (e.g. Heffernan, 1990). They were also almost all unrealized. Prior to the Second World War the impact of economic investment or infrastructure on the environments of Africa (certainly those large parts of tropical Africa without extensive European settlement) were slight, often amounting to little more than the loose exercise of political control, promotion of cash crops, and taxation. This changed in the 1940s. The Colonial Development and Welfare Acts were passed in the British Parliament in 1940 and 1945, and explicit development began to increase in pace and scope in British African territories (Low and Lonsdale, 1976). This development was complementary to concerns about conservation, in that it focused on the accelerated exploitation and better management of resources. It was also driven by

notions of progress and continuous improvement, and on the modernization of African life, particularly through the application of science and technology (Anderson, 1992). The new developmentalism was also prosecuted in a very ordered way. By the end of the 1940s all British colonies in Africa were producing development plans, and for the first time interventions were spearheaded by government staff with technical training, particularly in agriculture. Late-colonial agricultural schemes, for example attempts to grow groundnuts in Nigeria or Tanzania (Baldwin, 1957; Wood, 1950) or to irrigate large areas in the Niger Inland Delta, were often unsuccessful (Adams, 1992a).

Development programmes reflected the importance of science to the development of Africa. This was first emphasized before the Second World War in *Science in Africa*, and restated in the 1950s (Worthington 1938, 1958). Several scientific research organizations were established in Africa in the 1940s, for example the British

Colonial Research Council, the French Office de Récherche Scientifique et Technique d'Outre Mer (ORSTOM), and the Belgian Institut pour la Récherche Scientifique en Afrique Central (IRSAC) (Worthington, 1983).

By the 1960s, as many African countries neared independence, it was clear that science held few quick or easy answers. In discussing the need for wildlife research in East Africa, Boyd (1965) commented that 'most of the management problems in East Africa are not simply scientific, they usually have political implications' (p. 4). Biological research of an intensity that is in retrospect remarkable was carried out on the many reservoirs built in the 1960s on major African rivers, but one commentator declared 'research effort has been on too small a scale and has started too late to provide either useful predictions, or firm conclusions about the course of biological events consequent upon impounding water in the tropics' (White, 1969).

By the 1960s, as the colonial powers backed rapidly out of Africa, critics were pointing to the failure of development, notably René Dumont in *L'Afrique noire est mal partie* (Dumont, 1962). Two decades later, in the 1980s Africa's problems were seen to be legion, and its progress towards the chimera of 'development' slight at best (O'Connor, 1991). Economic growth had been uneven, both over time (with significant industrial growth in the 1960s, for example, but decline and even de-industrialization in the 1980s), and between countries (World Bank, 1990). Both oil-importing countries like Kenya and oil-exporting countries like Nigeria were plagued by debt, and debt-servicing rates were poor. Several African countries endured debt burdens more than 30 per cent of the value of exports. Africa's share of world trade fell from 2.4 per cent in 1970 to 1.7 per cent in 1985, despite the volume of oil exported by some countries. Primary commodities (including oil) still represented 88 per cent of exports by value in 1988. The value of agricultural exports had declined, although the World Bank continued to stress the importance of agricultural commodities like tea, cocoa, coffee, edible oils, and cotton to African economies, and most countries remain dependent on such products.

The World Bank's analysis of the state of economic development across Africa describes declining industrial output, poor export performance and weak agricultural growth, growing debt, deteriorating socio-economic and environmental conditions, and deteriorating physical infrastructure. Institutions, from universities to river-basin authorities, are starved of funds and ineffective. Agricultural output has grown by 1.5–2 per cent per year, but growth in food production has been less than this, and food production per capita fell in the 1970s.

Despite recovery in many places it remains depressed and many countries in Africa are in serious food deficit, particularly in the Sahel. Food imports have been rising by 7 per cent per year. Drought has been a repeated problem, and malnutrition is widespread. Warfare is causing destruction and famine in many countries, and high levels of military spending saps weak national economies.

Rates of population growth are rapid in sub-Saharan Africa (see Myers, this volume). Life expectancy at birth has improved, and crude death rates, levels of infant and child mortality, and maternal death in childbirth, have all fallen, but birth rates have barely fallen at all. The population of the nine countries of the Sahel (Somalia, Ethiopia, Sudan, Chad, Niger, Mali, Burkina Faso, Mauritania, and Senegal) doubled between 1950 and 1980, and may do so again by the year 2020 (World Bank, 1990), reaching 263.3 million. The current rate of growth is 2.8 per cent, but this is rising, and will do so for at least the next decade. Over half the population of the Sahel is under 15. In countries such as Kenya, the rate of population growth has been above 4 per cent per year. There are now 450 million people in Africa, twice the number at independence. By 2010, there could be 1 billion. The provision of basic services (shelter, food, health, education) in Africa in the future will be difficult (O'Connor, 1991). The AIDS pandemic will undoubtedly have an impact on rates of population growth, although this is likely to be complex and to vary regionally, locally, and socially (Barnett and Blaikie, 1992; Smallman-Raynor *et al.*, 1992).

Conservation and Protected Areas in Africa

Arguments for conservation embrace both utilitarian and non-utilitarian ideas (IUCN, 1980, 1991). Caro (1986) describes six approaches to conservation in Africa, one of which is a response to a directly aesthetic argument about the attractiveness of animals, and five others which involve the realization in a more or less direct way of economic values of wildlife, either through non-consumptive use (game-viewing and tourism), as a feed-stock for the creation of products using genetic diversity, or through direct consumption in the form of licence fees for big-game hunting, cropping of wildlife for meat, or the farming of wildlife animals. He argues that no one strategy is likely to be best in all circumstances. Utilitarian arguments for conservation have been given increasing emphasis in recent years, reflecting the growing perception of the need to link conservation and

Plate 21.2. Elephant, Yankari Game Reserve, Nigeria. The Yankari Game Reserve holds one of the largest remaining large populations of elephants in Nigeria. Nigeria probably has about 1300 elephants, but the number is declining. In total about 38 000 km² is in designated protected areas in Nigeria, 4.1 per cent of the country. Until recently there was only one National Park in Nigeria, at Lake Kainji, but in 1991 the number was increased to five (photo: W. M. Adams).

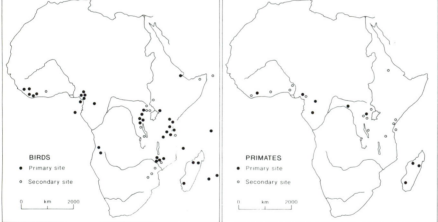

BIRDS
• Primary site
○ Secondary site

0   km   2000

PRIMATES
• Primary site
○ Secondary site

0   km   2000

Fig. 21.1. Sites of primary and secondary importance for bird and primate conservation in Africa
(*Source*: MacKinnon and MacKinnon 1986.)

development (Adams, 1990; IUCN, 1991), and a growing vigour in environmental economics (e.g. Barbier *et al.*, 1990). The debate has revolved in particular around the question of international trade in ivory (Plate 21.2), and the decision to place the African elephant on Appendix 1 of CITES (the Convention on International Trade in Endangered Species). One view (broadly an East African position and pursued strongly by Kenya) argues than only by preventing legal trade can poaching and illegal trade be stopped, and that unless this is done

elephants will become extinct in most areas. The opposing view (held by the CITES Secretariat itself, and widely in central/southern Africa, particularly in Zimbabwe) is that it is only by realizing the economic value of elephants by allowing local people and hunters to kill them and trade their body products that elephant survival can be sustained: without trade, poaching will recommence and there will be no money to pay either local people or guards to prevent it (Gavron, 1993; Bonner, 1993).

Table 21.1   IUCN Protected-Area categories

Category I	*Scientific Reserve/Strict Nature Reserve* (to protect nature [communities and species] and maintain natural processes in an undisturbed state in order to have ecologically representative examples of the natural environment available for scientific study, environmental monitoring, education, and for the maintenance of genetic resources in a dynamic and evolutionary state)
Category II	*National Park* (to protect natural and scenic areas of national or international significance for scientific, educational, or recreational use)
Category III	*Natural Monument/Natural Landmark* (to protect and preserve nationally significant features because of their special interest or unique characteristics and, to the extent consistent with this, provide opportunities for interpretation, education, research, and public appreciation)
Category IV	*Nature Conservation Reserve/Managed Nature Reserve/Wildlife Sanctuary* (to assure the natural conditions necessary to protect nationally significant species, groups of species, biotic communities, or physical features of the environment where these may require special manipulation for their perpetuation)
Category V	*Protected Landscape or Seascape* (to maintain nationally significant natural landscapes which are characteristic of the harmonious interaction of people and land while providing opportunities for public enjoyment through recreation and tourism within the normal life style and economic activity of the area)
Category VI	*Resource Reserve* (to restrict use of these areas until adequate studies have been completed on how to best utilize these remaining resources; to protect the natural resources of the area for future use and prevent or contain development activities that could affect the resource pending the establishment of objectives that are based upon appropriate knowledge and planning)
Category VII	*Natural Biotic Area/Managed Resource Area* (to allow the way of life of indigenous societies living in harmony with the environment to continue undisturbed by modern technology)
Category VIII	*Multiple Use Management Area/Managed Resource Area* (to provide for the sustained production of water, timber, wildlife (including fish), pasture, or marine products, and outdoor recreation)

Source: MacKinnon and MacKinnon (1986).

The dominant strategies for achieving conservation in Africa have to date involved non-consumptive approaches to nature, and measures to protect individual species and particular areas (Nelson, 1987). As the history of conservation discussed above shows, concern for species conservation has long been most strongly focused on the survival of the large mammals of Africa. The establishment of protected areas grew out of concerns about hunting, and modern popular classics of African conservation (e.g. Guggisberg, 1966; Grzimek, 1960) emphasize the links between the importance of individual conspicuous species and the places where they are found and conserved. The challenge of biodiversity conservation is discussed by Myers (this volume). Knowledge of the status of different taxa, and concern about them, varies. There is perhaps particular concern about primates (Oates, 1985), and birds (Collar and Stuart, 1985), although awareness and knowledge of other taxa (particularly perhaps plants) are increasing through the work of the IUCN/Species Survival Commission Specialist Groups. Priority lists of places where conservation action is particularly urgent have been drawn up for several taxa (e.g. Figure 21.1), and a Red Data Book of plant sites in need of protection in Africa is in preparation (MacKinnon and MacKinnon, 1986).

There is now an extensive protected-area system in all African countries. The IUCN Commission on National Parks and Protected Areas has long recognized various different categories of protected area, ranging from 'Strict Nature Reserve' (Category I) through 'National Park' and 'Protected Landscape' to 'multiple-use management area'. These embrace a wide range of objectives (Table 21.1; IUCN Commission, 1984). From 1994, IUCN reduced the number of main categories that it recognized to six, although the definitions of Category I–IV have remained broadly the same. In total, protected areas cover some 2.4 M ha in sub-Saharan Africa (about 10 per cent of the land area), of which about half are in Categories I–IV (Olindo and Mankoto, 1993; Sulayem et al., 1993). In addition to this a further 3 per cent of sub-Saharan Africa is covered in forest reserves. Several countries have a total of more than 100 000 km^2 designated in all categories (Algeria, Botswana, Chad, Ethiopia, Namibia, Sudan, and Zaire) and two have more (Tanzania with 365 000 km^2 and Zambia with 295 802 km^2). Twenty countries have designated more than 10 per cent of their land area, and five have designated more than 20 per cent (Benin, Seychelles, Tanzania, Uganda, and Zambia). With the exception of the Seychelles (with large areas of marine reserves), Tanzania and Zambia stand out with vast areas of designated land. Just under 14 per cent of Tanzania's land surface is in Category I–IV protected areas, and a further 25 per cent in category VI–VIII and other areas (Olindo and Mankoto, 1993; Fig. 21.2). Apart from one or two island states without any IUCN category reserves, low levels of protected area designation are found in a number of countries. Fifteen mainland states have less than 5 per cent of their land area in recognized protected areas of some kind (including Congo, Gambia, Guinea, Madagascar,

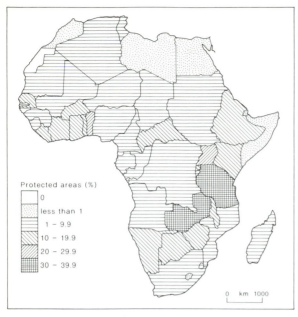

Fig. 21.2. Percentages of each African and North African state in Protected Areas of all kinds
(IUCN Categories I–VIII and others.)

Mali, Mauritania, Nigeria, Sierra Leone, and Sudan). All North African states have small protected area systems, only Algeria (5.3 per cent) having more than 2 per cent of its land area designated.

Of the fifty countries in sub-Saharan Africa, thirty-three have signed the World Heritage Convention, although only thirteen have designated sites. Twenty countries participate in the UNESCO Man and the Biosphere (MAB) Programme (with thirty-six sites covering 0.13 m km², and seventeen countries have signed the Ramsar Convention on wetlands (Olindo and Mankoto, 1993; Fig. 21.3). North Africa is influenced by a number of additional programmes associated with the Mediterranean and the Middle East. All the countries of the Maghreb have signed the World Heritage Convention, and all except Libya have signed the Ramsar Convention (Sulayem et al., 1993; Fig. 21.3).

Although impressive, the protected-area system within Africa contains gaps. Terrestrial habitats believed to be under-represented include most forest environments (montane, lowland, and coastal), mountain environments, Sahelian savanna and desert environments, grasslands and miombo woodlands, and swamps and freshwater lakes (Olindo and Mankoto, 1993). Many coastal and marine environments are also under-represented and vulnerable (see Orme this volume). Nationally, the coverage of habitat types varies, with

some countries (like Malawi, Kenya, or Zimbabwe) with good coverage, and others (like Sengal, Zaire, or Cameroon) with important habitats excluded. Furthermore, it is recognized that protected areas are not well adapted to maintaining populations of migratory species (for example African-Palaearctic migratory birds or migratory herbivores such as wildebeest), and that climatic change presents a particular threat to any habitat 'islands', whether protected or not (MacKinnon and MacKinnon, 1986; Spellerberg, 1991). In addition to this there are institutional weaknesses in protected-area systems, and the government wild-life administrations which operate them (Olindo and Mankoto, 1993). Legislation is often weak or confusing, and conservation or protected-area policies are frequently divorced from policy development elsewhere in government (e.g. in agriculture, forestry, river-basin planning, or tourism). Conservation staff are often undertrained and badly paid, and government bureaucracies are frequently immobilized for lack of operating expenses or means of transport. Institutional weakness can be tackled by training and targeted external support, and institutional overlap and inertia through innovative approaches to planning. One example of this approach is the national wetland seminars and working groups supported in countries like Uganda and Kenya by the IUCN Wetlands Programme.

People, Protected Areas, and Development in Africa

The biggest threats to the effectiveness of protected areas in achieving their conservation objectives relate to the conflict that such land designation creates with the economic interests and rights of local people (Plate 21.3). As Adams and McShane (1992) comment 'conservation will either contribute to solving the problems of the rural poor who live day-to-day with wild animals, or those animals will disappear' (p. xix). The institutional emergence of conservation departments in Africa from a concern with preventing hunting and poaching, and the influence of the North American model of national parks without consumptive use of nature, have both led to a history of exclusion of people from African protected areas. This can have severe impacts on those excluded. Turton (1987) discusses the impact of the Omo National Park which was established in 1966 on the Mursi of the Omo Valley in Ethiopia. This area is perceived by the Wildlife Conservation Department as 'wilderness', but its ecology is in fact created and maintained by the Mursi economy, based on cattle-herding, dry-season

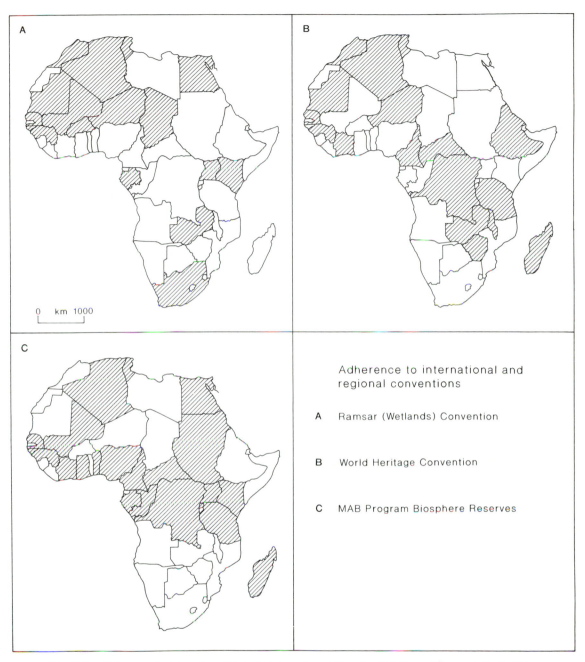

Fig. 21.3.   Adherence to International Protected Area Conventions in sub-Saharan Africa and North Africa
(Ramsar Convention, World Heritage Convention, MAB Program.)

Plate 21.3. Giraffe, Nairobi National park, Kenya. The Nairobi National Park is very close to the edge of Nairobi, and is increasingly isolated from other open land by agriculture and urban development. It is extensively visited both by foreign tourists and Kenyan nationals. It is the most accessible of the Kenyan National Parks, and hence forms an important part of the wildlife-viewing tourist industry (photo: W. M. Adams).

cultivation, and flood-retreat farming along the Omo River, as well as (traditionally) hunting for ivory and skins for trade. In the 1970s the Mursi were driven south into the Mago area by drought, and some uses intensified. In practice the area is remote, and conservation management is minimal. Even so, Turton suggests that the park promoted the marginalization of the Mursi in a time of drought, and he believes that if the Mursi were effectively excluded from the park, their economy would be destroyed. A second park to the east, the Mago National Park, was proposed in 1978, and the planning report saw the Mursi as a threat to conservation, and proposed exclusion and resettlement.

Gordon (1985) describes similar impacts of conservation areas on the Bushmen of Namibia following establishment of game reserves early this century. In some reserves (notably what became the Etosha National Park) Bushmen were tolerated, and used as trackers and piece labourers. Elsewhere, they were moved out. The Gemsbok National Park in South Africa was declared, in 1931, on land occupied by Bushmen. The National Parks Board banned hunting, and prosecuted 'poachers'. In 1941 land was set aside for Bushmen to hunt adjacent to the park, but there and elsewhere cultural contact and economic integration brought cultural change and attempts to move Bushmen who were no longer 'true' Bushmen living in a traditional manner, out of the parks and reserves.

Ideas about the place of people in national parks has been changing rapidly in the 1980s, as debates at the Third and Fourth World Congresses on National Parks and Protected Areas in Bali in 1982, and Caracas in 1992, show (McNeely and Miller, 1984; McNeely, 1993). In 1984, Blower wrote about national parks in developing countries 'having gained the support of government planners and decision-makers, the most important task is to win the understanding and co-operation of the local people in the vicinity of the proposed parks' (p. 725). At the 1992 Congress the adverse impacts of parks on local people were widely recognized and discussed, not least by delegates from indigenous groups who came and spoke. A much higher priority was accorded to the need both to integrate parks into economic development planning frameworks and to prioritize the interests (and recognize the rights) of local people (McNeely, 1993; Kemf, 1993).

In practice, those interested in exploring ways to 'integrate conservation and development' have a relatively small number of case studies to draw on (see Hannah, 1992; Wells and Brandon, 1992), notably perhaps the Amboseli National Park in Kenya, the Luangwa Valley in Zambia, and the East Usumbara Mountains in Tanzania. The Amboseli experience is perhaps particularly interesting in that it reveals both the intractable nature of the problems of integrating wildlife conservation and the interests of people, and the way in which the combination of a lack of incisive and in-depth research, and an almost blind commitment on the part of conservationists to the notion of integrating conservation and development, makes it difficult to derive hard and fast lessons from experience.

Amboseli lies in land grazed by the Maasai in the dry season, with cattle comprising up to 60 per cent of large animal biomass at this season (Lindsay, 1987). Amboseli is within a very extensive area designated as a game reserve by the colonial government in southern Kenya in 1899, a few years after the Maasai had been hit by famine following the rinderpest pandemic (Waller, 1988, 1990). Amboseli National Reserve was created in 1952 as one of a series of smaller reserves within this area. Hunting was banned, but Maasai grazing continued. In 1961 the District Council assumed control of the area, but local Maasai continued to fear loss of further grazing rights, and demanded formal ownership of the area. Although the Council received some entrance-fee revenue and hunting-licence fees, conflict developed and large game animals began to be killed. Because of this 'poaching', Amboseli was declared a National Park in 1974, under a complex agreement whereby Maasai gave up the right to graze within the park in return for joint ownership of surrounding bushlands in group ranches. These ranches received water supplies, compensation for lost production through wildlife grazing, the development of lodges and wildlife-viewing circuits, and developments such as a school and a dispensary. Developments were funded by the World Bank (Hannah, 1992). The Kajiado District Council retained control of lodges on 160 ha in the heart of the Park.

Initial reviews suggested that the Amboseli programme was a success (Western, 1982b), and it is often cited as a model example of the ways in which conservation and economic development can coexist through the generation of revenues from non-consumptive use of wildlife. However, more recent reviews have been more critical (Lindsay, 1987; Hannah, 1992; Wells and Brandon, 1992). There have undoubtedly been revenues generated by the park from campsite fees and the sale of wood and gravel. However, although a school and cattle dip have been built, the borehole and pipeline system constructed to supply water outside the Park has broken down and the financial benefit to the Maasai has been small (Lindsay, 1987), particularly as tourist numbers levelled off in Kenya in the 1970s, and in the 1980s when park-entrance fees were retained by the Treasury. Killings of rhinoceros and elephant began again in the early 1980s. Lindsay (1987) argues that the Amboseli Park Plan did not offer an acceptable and sustainable alternative for Maasai excluded from the Park. Income

to the Maasai was too little and too unpredictable, and continuing cultural, social, and economic change among the Maasai undermined static assumptions about the long-term acceptability and sustainability of the group-ranching system.

There are three closely linked concepts commonly proposed as strategies to build harmonious relations between protected areas and surrounding people. The first is the notion of separating a core area where human use is prohibited, from outside areas where human use is unregulated by an intermediate or buffer zone. The second idea is to derive financial flows from wildlife within the protected area (by either consumptive or non-consumptive use) and use these to compensate or reward those living near the reserve. The third strategy is to do away with the notion of exclusion altogether and explore the potential of local people *as* conservationists. To a considerable extent these three approaches have now become merged into a single broad and rather generalized concern to forge new relations between 'people and parks' (Wells and Brandon, 1992).

Buffer zones are discussed in the literature of conservation biology as a means of extending management favourable to wildlife beyond the strict reserve boundary, but in Africa they have also been seen as a place to focus 'appropriate' development activities, often labelled 'integrated conservation and development' projects (Brandon and Wells, 1992) or 'conservation-with-development' projects (Stocking and Perkin, 1992). Such activities might be used to deflect the economic interests of people living close to a protected area from economic opportunities within it that are now illegal (e.g. hunting), or a means of compensating people forced to leave the reserve and locate outside.

The idea of buffer zones was an integral part of the Biosphere Reserve concept introduced by the Man and the Biosphere (MAB) Programme in the 1970s. In biosphere reserves, exclusion is restricted to a core area within a unit of sufficient size 'to accommodate different uses without conflict' (Batisse, 1982). Productive activities are prohibited in the core, but existing activities may continue under supervision in the surrounding buffer zone. Of course, this zoning approach does not in any sense automatically solve problems of existing rights of human use. Schoepf (1984) argues that the Lufira Valley Biosphere Reserve in Zaire was imposed against the advice of local chiefs, and that the core zone contains several thousand people. The Zaire MAB Project was established in 1979, and placed in 1982 under the Department of Agriculture and Rural Development. Schoepf argues that MAB has become caught up in efforts by the state to increase agricultural production through obligatory cultivation of cassava and maize in the experimental zone, failing either to appreciate or build upon the indigenous knowledge and skills of Lemba farmers. He predicts that the reserve will increase inequality in the Lufira Valley and promote rural emigration.

Experience with the first generation of integrated conservation and development projects (ICDPs) are mixed. Many of the problems experienced mirror those of the integrated rural development projects developed in Africa in the 1970s and 1980s, particularly by the World Bank (Brandon and Wells, 1992; Stocking and Perkin, 1992; Wells and Brandon, 1992). Conservationists may have been naïve in assuming that a commitment on paper to sustainability and participation or 'bottom-up' planning would yield successful projects where more conventional development projects have a poor record. Wells and Brandon note that 'linking conservation and development objectives is in fact extremely difficult, even at a conceptual level' (p. 567).

Stocking and Perkin (1992) describe the East Usambaras Agricultural Development and Environmental Conservation project in Tanzania. The East Usambaras reach an altitude of 1500 m and support submontane forests with a very high level of endemism (see Myers, this volume). The IUCN project began in 1987 with three aims, to improve the living standards of the people, to protect the functions of the forest (particularly its role as a catchment for downstream water supply), and to preserve biological diversity. The traditional conservation objectives were deliberately downgraded to stress revenue generation and development. After four years, achievements were modest. A vast range of activities had been undertaken, from agricultural extension to attempted control of illegal pit-sawing. Lack of funds was a major problem, leading to a lack of breadth in technical expertise, dissipation of capital and energy in too many activities, and lack of proper feasibility study. Stocking and Perkin conclude that ICDPs are hard to transfer from paper to reality, that they are inherently highly complex and demand high levels of skill on the part of project staff and substantial funds, and that their overall success is vulnerable to the failure of particular components because of its impact on local perceptions of the project. They conclude that ICDPs need clear and precise objectives, demand careful evaluation of the costs and benefits of project components at the level of the individual household (since this will make it possible to predict who may be able to participate), that there must be a long-term commitment to funding and strong local participatory linkages. It is obvious that projects of this sort will not be cheap to implement, and will not yield results quickly. Practical experience with development

'micro projects' in support of conservation and sustainability objectives bears out this conclusion (e.g. Adams and Thomas, 1993).

An alternative way to compensate local communities or subsidize their economy is to seek to capture the economic value of wildlife. This was attempted in the Amboseli case discussed above, through wildlife tourism. There may be risks attached to the linkage of remote rural economies to the global tourist market (for example through the influence of AIDS or economic recession), but this non-consumptive use is seen to have an important role in many Third World protected areas, and is obviously very well developed in a number of African countries, notably Kenya where tourism is of central importance to the economy. However, not all African protected areas share the attractiveness of the grassland savannas of East Africa or the gorillas of Zaire, and tourism is therefore unlikely to provide a way forward everywhere.

The other obvious source of revenue is through controlled hunting. The idea of controlling hunting in buffer zones around protected areas (discussed by Worthington, 1960) has been attempted in the Luangwa Valley in Zimbabwe in the Luangwa Integrated Resource Development Project, begun in 1987. Revenue from safari hunting in the Lupande Game Reserve is used for development projects in the local area, as well as to finance the cost of game guards in the South Luangwa National Park. Game guards are locally recruited, and both safari hunting and community game harvesting (particularly of hippopotamus) take place legally (Wells and Brandon, 1992). This approach was obviously judged a success, since some of its principles were extended in 1987 to ten other game-management areas in Zambia under the ADMADE programme. Revenue from safari and other hunting fees are used to meet wildlife management costs (40 per cent to these activities within the game-management area itself, plus 15 per cent to the National Parks system, and 10 per cent to the Zambian Tourist Bureau), and to generate revenue for local community projects (35 per cent).

Wildlife utilization has been developed further in Zimbabwe. Wildlife ranching has become important economically, and also in conservation terms since populations of species such as black rhinoceros and cheetah within Zimbabwe are mostly on private land. There has also been development of communal wildlife utilization projects under the CAMPFIRE programme (Barbier, 1992). This is similar to Zambia's ADMADE programme, but places greater emphasis on communal initiation and control of the programme and hunting activity. In Zimbabwe people, either in communities or large agribusiness enterprises, are starting to generate wildlife conservation as a by-product of economic activity. CAMPFIRE lies at the opposite end of the continuum that began with the idea of generating benefits from a pre-existing protected area to help offset local opposition. The relative success of CAMPFIRE and other community-based consumptive-use projects in Africa is reviewed by Barbier (1992). This may be a way forward for conservation, at least for those areas of Africa with extensive savannas, good tourist infrastructure, and a ready supply of wealthy safari hunters.

Clearly in some circumstances, not only can economic development and conservation go hand-in-hand to the potential benefit of local people, but those people can be the conservationists themselves, without great intervention from national or international conservation bureaucracies. This casts an interesting light on the growing debate about the extent to which this can be reversed, and people allowed to enter national parks (the 'core' areas of zoned protected areas) to pursue economic activities, and the parallel question of the extent to which local communities or local governments can be left to organize conservation locally. This debate is sharpest in the Serengeti–Mara ecosystem that lies across the Kenya/Tanzania border. Parkipuny (1991) argues that the Maasai-Mara National Reserve functions perfectly well in conservation terms under the ownership of the Narok County Council, and generates sizeable economic benefits for local people. He then turns to the Ngorogoro Conservation Area in Tanzania and argues that Maasai graziers both have a right to be there and do no harm. The latter point is taken up and substantially corroborated by Homewood and Rogers (1991). However, not all observers would agree: Prins (1992) argues vehemently that pastoralism and conservation are incompatible: 'Nature reserves have little to do with "wise use of natural resources": we are closing our eyes if we think that allowing people to invade protected areas can result in a harmonious relation between them, their livestock and wildlife' (p. 121).

Conclusion

Whatever the outcome of the growing debate about people and parks, three things stand out. The first is that protected areas will not survive in Africa without substantially increased participation of local people, and furthermore that this participation will have to be far more thorough and extensive than has conventionally been the case in the past (Zanen and de Groot, 1991; Wells and Brandon, 1992; McNeely, 1993). This raises interesting questions about the extent to which the

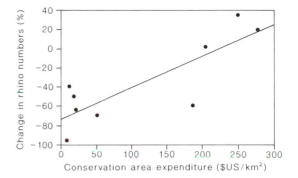

Fig. 21.4. Expenditure on rhino conservation and changing numbers, and poaching losses and catch effort in the Luangwa Valley.

(*Source*: Leader-Williams and Albon, 1988.)

conventional international (and hence 'Western' or 'Northern') ideas about conservation (e.g. notions about rarity and about the value of particular species) are going to find acceptance with local communities. Empowering local communities to achieve their own conservation is a popular idea in international conservation circles, but it might not bring the exact conservation results that have been pursued in the past.

Second, it is clear that only those conservation areas and species will survive that can find a place within the legitimate development plans of African communities and states. Wildlife that can generate a revenue, dead or alive, may pay its way, but the development imperative in Africa leaves little room for luxuries that are not explicitly financed. Effective conservation costs money, and in some cases at least there is a clear relationship between expenditure and success, for example in the conservation of black rhinos (Leader-Williams and Albon, 1988; Figure 21.4). Debt-for-nature swaps, direct support for communities whose livelihoods are impaired by protected areas, wildlife tourism or consumptive use of wildlife, all offer different ways to generate revenue for development. Without this, very little wildlife will survive, even inside protected areas.

Third, wildlife conservation needs to be thoroughly integrated into development programmes (McNeely, 1993), and moreover those programmes need to be far more sensitive to the dynamics of the environment, and the ways that the needs of the rural and urban poor are linked to environmental change and degradation. Hackel (1993) analyses the poor performance of conservation in north-east Swaziland in the context of rapid social and economic change, and concludes that conservationists must address rural-development needs if they are to have sufficient influence to maintain significant wildlife populations into the future. Wildlife conservation therefore must become an extension of wider debates about environment and development, or 'sustainable development'. The World Conservation Strategy offers one approach to this synthesis (IUCN, 1980, 1991), and the various products of the 1992 United Nations Conference on Environment and Development another (Holmberg *et al.*, 1993).

Both the lack of development and the growth of economic activity bring environmental problems (World Bank, 1992). In Africa the former problems in the past have outweighed the latter, and it is the conflict between the economic needs of the rural poor and the conventional aims and strategies of wildlife conservation that has been the chief focus for debates about conservation and development. Those debates are now broadening to embrace wider issues of environmental utilization and degradation, of drought, deforestation, desertification, and the sustainability of water resource use, and also the political economic structures and institutional frameworks within which decisions are made about people, development, and environment.

References

Adams, J. S. and McShane, T. O. (1992), *The Myth of Wild Africa: Conservation without Illusion* (New York).

Adams, W. M. (1990), *Green Development: Environment and Sustainability in the Third World* (London).

—— (1992*a*), *Wasting the Rain: Rivers, People and Planning in Africa* (London).

—— (1992*b*) 'Indigenous Use of Wetlands and Sustainable Development in West Africa', *Geographical Journal*, 159: 209–18.

—— and Thomas, D. H. L. (1993), 'Mainstream Sustainable Development: The Challenge of Putting Theory into Practice', *Journal of International Development*, 5: 591–604.

Allen, D. E. (1976), *The Naturalist in Britain* (Harmondsworth).

Anderson, D. M. (1992), 'Environment and Development in Rural Africa: Towards a History of Ideas', unpub. MS, April 1992.

Barbier, E. B., Burgess, J. C., Swanson, T. M., and Pearce, D. W. (1990), *Elephants, Economics and Ivory* (London).

Barbier, E. B. (1992), 'Community-based Development in Africa', in T. M. Swanson and E. B. Barbier (eds.), *Economics for the Wilds: Wildlife, Wildlands, Diversity and Development* (London), 103–35.

Baldwin, K. D. S. (1957), *The Niger Agricultural Project: An Experiment in African Development* (London).

Barnett, T. and Blaikie, P. (1992), *Aids in Africa: Its Present and Future Impact* (London).

Batisse, M. (1982), 'The Biosphere Reserve: A Tool for Environmental Conservation and Management', *Environmental Conservation*, 9: 101–11.

Behnke, R. H. and Scoones, I. (1991), *Rethinking Range Ecology: Implications for Range Management in Africa*, ODI/IIED Pastoral Network Paper, Technical Meeting on 'Savanna Development and Pasture Production', 19–21 Nov. 1990, Woburn, UK (London).

—— and Kerven, C. (1993) (eds.), *Range Ecology at Disequilibrium: New Models of Natural Variability and Pastoral Adaptation in African Savannas* (London).

Blower, J. (1984), 'National Parks for Developing Countries', in A.

McNeely and K. R. Miller (eds.), *National Parks, Conservation and Development: The Role of Protected Areas in Sustaining Society* (Washington, DC), 722–7.

Boardman, R. (1981), *International Organisations and the Conservation of Nature* (Bloomington, Ind.).

Bonner, R. (1993), *At the Hand of Man: Peril and Hope for Africa's Wildlife* (New York).

Boyd, J. M. (1965), 'Research and Management in East African Wildlife', *Nature*, 208/5013: 828–30.

Brandon, K. E. and Wells, M. (1992), 'Planning for People and Parks: Design Dilemmas', *World Development*, 20: 557–70.

Caro, T. (1986), 'The Many Paths to Wildlife Conservation in Africa', *Oryx*, 20/4: 221–9.

Collar, N. J. and Stuart, S. N. (1985), *Threatened Birds of Africa and Related Islands* (Cambridge).

Connah, G. (1988), *African Civilisations: Precolonial Cities and States in Tropical Africa: An Archaeological Perspective* (Cambridge).

Coughenour, M. B., Ellis, J. E., Swift, D. M., Coppock, D. L., Galvin, K., McCabe, J. T., and Hart, T. C. (1986), 'Livestock Feeding Ecology and Resource Utilisation in a Nomadic Pastoral Ecosystem', *Journal of Applied Ecology*, 23: 573–83.

Dugan, P. J. (1990), *Wetland Conservation: A Review of Current Issues and Required Action* (Gland, Switz.).

—— (1993) (ed.), *Wetlands in Danger* (London).

Dumont, R. (1962), *L'Afrique noire est mal partie* (Paris) pub. in English (1966) as *False Start in Africa* and repr. (1988) (London).

Fitter, R. S. R. (1978), *The Penitent Butchers* (London).

Gavron, J. (1993), *The Last Elephant: An African Quest* (London).

Gordon, R. J. (1985), 'Conserving Bushmen to Extinction in Southern Africa', in Survival International, *An End to Laughter? Tribal Peoples and Economic Development*, Review No. 44 (London), 28–42.

Graham, A. (1973), *The Gardeners of Eden* (Hemel Hempstead).

Grove, R. H. (1987), 'Early Themes in African Conservation: The Cape in the Nineteenth Century', in D. M. Anderson and R. H. Grove (eds.), *Conservation in Africa: People, Policies and Practice* (Cambridge).

—— (1990*a*), 'Colonial Conservation, Ecological Hegemony and Popular Resistance: Towards a Global Synthesis', in J. M. MacKenzie (ed.), *Imperialism and the Natural World* (Manchester), 15–50.

—— (1990*b*), 'The Origins of Environmentalism', *Nature*, 345/6270: 11–14.

—— (1992), 'Origins of Western Environmentalism', *Scientific American*, 267: 42–7.

Grzimek, B. (1960), *Serengeti Shall Not Die* (London), pub. in German (1959) (Berlin).

Guggisberg, G. A. W. (1966), *S. O. S. Rhino* (London).

Hackel, J. D. (1993), 'Rural Change and Nature Conservation in Africa: A Case Study From Swaziland', *Human Ecology*, 21: 295–312.

Hannah, L. (1992), *African People, African Parks: An Evaluation of Development Initiatives as a Means of Improving Protected Area Conservation in Africa* (Washington, DC).

Hardin, G. (1968), 'The Tragedy of the Commons', *Science*, 162: 1243–48.

Heffernan, M. (1990), 'Bringing the Desert to Bloom: French Ambitions in the Sahara Desert During the Late Nineteenth Century. The Strange Case of 'la mare intérieure', in D. Cosgrove and G. Petts (eds.), *Water, Engineering and Landscape: Water Control and Landscape Transformation in the Modern Period*, (London), 94–114.

Holmberg, J., Thomson, K., and Timberlake, L. (1993), *Facing the Future: Beyond the Earth Summit* (London).

Homewood, K. and Rodgers, W. A. (1991), *Maasailand Ecology* (Cambridge).

Hopkins, A. G. (1973), *An Economic History of West Africa* (London).

Horowitz, M. M. and Little, P. D. (1987), 'African Pastoralism and Poverty: Some Implications for Drought and Famine', in M.

the Lower Tana River Basin of Kenya', in Anderson and Grove (eds.), *Conservation in Africa*, 211–28.

Iliffe, J. (1987), *The African Poor: A History* (Cambridge).

IUCN (1980), *The World Conservation Strategy* (Geneva).

—— Commission on National Parks and Protected Areas (1984), 'Categories, Objectives and Criteria for Protected Areas', in McNeely and Miller (eds.), *National Parks, Conservation and Development*, 47–53.

—— (1991), *Caring for the Earth: A Strategy for Sustainable Living* (Gland, Switz.).

Jarosz, L. (1992), 'Constructing the Dark Continent: Metaphor as Geographic Representation of Africa', *Geografiska Annaler*, 74 B: 105–15.

Johnson, D. and Anderson, D. M. (1988) (eds.), *The Ecology of Survival: Case Studies from North East African History* (London).

Jones, G. H. (1936), *The Earth Goddess: A Study of Native Farming in the West African Context* (London).

Kemf, E. (1993) (ed.), *The Law of the Mother: Protecting Indigenous Peoples in Protected Areas* (San Francisco).

Kjekshus, H. (1977), *Ecology Control and Economic Development in East African History* (Berkeley).

Lamphrey, R. and Waller, R. (1990), 'The Loita–Mara Region in Historical Times: Patterns of Subsistence, Settlement and Ecological Change', in P. Robertshaw (ed.), *Early Pastoralists of South-western Kenya*, British Institute in Eastern Africa, Memoir No. 11 (Nairobi).

Leader-Williams, N. and Albon, S. D. (1988), 'Allocation of Resources for Conservation', *Nature*, 336/6199: 553–5.

Lindsay, W. K. (1987), 'Integrating Parks and Pastoralists: Some Lessons from Amboseli', in Anderson and Grove (eds.), *Conservation in Africa*, 149–68.

Low, D. A. and Lonsdale, J. M. (1976), 'Towards a New Order 1945–1963', in D. A. Low and J. M. Lonsdale (eds.), *History of East Africa* (Oxford), iii. 1–63.

Mace, R. (1991), 'Overgrazing Overstated', *Science*, 339 (24 Jan.), 280–1.

Mackenzie, J. M. (1987), 'Chivalry, Social Darwinism and Ritualised Killing: The Hunting Ethos in Central Africa up to 1914', in D. M. Anderson and R. H. Grove (eds.), *Conservation in Africa*, 41–62.

—— (1988), *The Empire of Nature: Hunting, Conservation and British Imperialism* (Manchester).

MacKinnon, J. and MacKinnon, K. (1986), *Review of the Protected Areas System in the Afrotropical Realm* (Gland, Switz.).

McNeely, J. A. (1992) (ed.), *Parks for Life*, Report of the 4th World Congress on National Parks and Protected Areas, Caracas, 1992 (Gland, Switz.).

—— and Miller, K. R. (1984) (eds.), *National Parks, Conservation and Development: The Role of Protected Areas in Sustaining Society* (Washington, DC).

Mortimore, M. (1989), *Adapting to Drought* (Cambridge).

Nash, R. (1983), *Wilderness and the American Mind* (New Haven).

Nelson, J. G. (1987), 'National Parks and Protected Areas, National Conservation Strategies and Sustainable Development', *Geoforum*, 18: 291–320.

Oates, J. F. (1986), *Action Plan for African Primate Conservation 1986–90* (Gland, Switz.).

O'Connor, A. (1991), *Poverty in Africa: A Geographer's Perspective* (London).

Olindo, P. and Mankoto, M. M. (1993), *Regional Review of Protected Areas in Sub-Saharan Africa* (Cambridge).

Pankhurst, R. and Johnson, D. H. (1988), 'The Great Drought and Famine of 1888–1892 in North-East Africa', in D. Johnson and D. M. Anderson (eds.), *The Ecology of Survival: Case Studies from North-East African History* (London).

Parkipuny, M. S. (1991), *Pastoralism, Conservation an Development in the Greater Serengeti Region*, IIED Dryland Network Issues Paper 26 (London).

Prins, H. H. T. (1992), 'The Pastoral Road to Extinction: Competition Between Wildlife and Traditional Pastoralism in East Africa', *Environmental Conservation*, 19/2: 117–23.

Richards, P. (1985), *Indigenous Agricultural Revolution: Ecology and Food Production in West Africa* (London).

—— (1986), *Coping with Hunger: Hazard and Experiment in an African Rice-farming System* (London).

Robertshaw, P. (1990) (ed.), *Early Pastoralists of South-western Kenya*, British Institute in Eastern Africa, Memoir No. 11 (Nairobi).

Schoepf, B. G. (1984), 'Man and the Biosphere in Zaire', in J. Barker (ed.), *The Politics of Agriculture in Tropical Africa* (Beverley Hills, Calif.), 269–90.

Sheail, J. (1976), *Nature in Trust: The History of Nature Conservation in Great Britain* (Glasgow).

Smallman-Raynor, M. R., Cliff, A. D., and Haggett, P. (1992), *Atlas of Aids* (Oxford).

Spellerberg, I. F. (1991), 'Biogeographical Basis of Conservation', in I. F. Spellerberg, F. B. Goldsmith, and M. G. Morris (eds.), *The Scientific Management of Temperate Communities for Conservation* (Oxford), 293–322.

Stamp, L. D. (1938), 'Land Utilisation and Soil Erosion in Nigeria', *Geographical Review*, 28: 32–45.

Steinhart, E. (1989), 'Hunters, Poachers and Gamekeepers: Towards a Social History of Hunting in Colonial Kenya', *Journal of African History*, 30: 247–64.

Stocking, M. and Perkin, S. (1992), 'Conservation-with-development: An Application of the Concept in the Usambara Mountains, Tanzania', *Transactions of the Institute of British Geographers*, NS 17: 337–49.

Sulayem, M., Saleh, M., Dean, F., and Drucker, G. (1993), *Regional Review of Protected Areas in North Africa and the Middle East* (Cambridge).

Sutton, J. E. G. (1989) (ed.), 'History of African Technology and Field Systems', *Azania*, spec. vol. 24.

Thesiger, W. (1987), *The Life of My Choice* (London).

Turton, D. (1987), 'The Mursi and National Park Development in the Lower Omo Valley', in Anderson and Grove (eds.), *Conservation in Africa*, 169–86.

Waller, R. D. (1988), 'Emutai: Crisis and Response in Maasailand 1884–1904', in Johnson and Anderson (eds.), *The Ecology of Survival*, 73–112.

Waller, R. (1990), 'Tsetse Fly in Western Narok, Kenya', *Journal of African History*, 31: 81–101.

Watts, M. J. (1983), *Silent Violence: Food, Famine and Peasantry in Northern Nigeria* (Berkeley).

—— (1984), 'The Demise of the Moral Economy: Food and Famine in the Sudano-Sahelian Region in Historical Perspective', in E. P. Scott (ed.), *Life before the Drought* (London), 128–48.

Welcomme, R. (1979), *Fisheries Ecology of Floodplain Rivers* (London).

Wells, M. and Brandon, K. (1992), *People and Parks: Linking Protected Area Management with Local Communities* (Washington, DC).

Western, D. (1982a), 'The Environment and Ecology of Pastoralists in Arid Savannas', *Development and Change*, 13: 183–211.

—— (1982b) 'Amboseli National Park: Enlisting Landowners to Conserve Migratory Wildlife', *Ambio*, 11: 302–8.

White, E. (1969), 'The Place of Biological Research in the Development of Resources of Man-made Lakes', in L. E. Obeng (ed.), *Man-Made Lakes: The Accra Symposium* (Accra, Ghana), 37–49.

Wood, A. (1950), *The Groundnut Affair* (London).

World Bank (1981), *Accelerated Development in Sub-Saharan Africa* (Washington, DC).

World Bank (1981), *Accelerated Development in Sub-Saharan Africa* (Washington, DC).

—— (1990), *Sub-Saharan Africa: From Crisis to Sustainable Growth: A Long-term Perspective Study* (Washington, DC).

—— (1992), *Development and the Environment: World Development Report 1992* (New York).

Worster, D. (1985), *Nature's Economy: A History of Ecological Ideas* (Cambridge).

Worthington, E. B. (1938), *Science in Africa: A Review of Scientific Research Relating to Tropical and Southern Africa* (London).

—— (1958), *Science in the Development of Africa: A Review of the Contribution of Physical and Biological Knowledge South of the Sahara* (London).

—— (1960), 'Dynamic Conservation in Africa', *Oryx*, 5/6: 341–5.

—— (1983), *The Ecological Century: A Personal Appraisal* (Cambridge).

Zanen, S. M. and de Groot, W. T. (1991), 'Enhancing Participation of Local People: Some Basic Principles and an Example from Burkina Faso', *Landscape and Urban Planning*, 20: 151–8.

Index